1998

1998

ENCYCLOPEDIA OF GARBAGE

Steve Coffel

Introduction by William L. Rathje

Facts On File, Inc.

Encyclopedia of Garbage

Facts On File, Inc.
11 Penn Plaza
New York NY 10001

Library of Congress Cataloging-in-Publication Data

Coffel, Steve.
    Encyclopedia of garbage / Steve Coffel.
        p.   cm.
    Includes index.
    ISBN 0-8160-3135-5
    1. Refuse and refuse disposal—Encyclopedias.   2. Pollution—Encyclopedias.
3. Organic wastes—Encyclopedias.   I. Title.
TD785.5.C64   1996
62B.4'4—dc20   96-4345

Jacket design by Robert Yaffe

INNO VB 10 9 8 7 6 5 4 3 2 1

TABLE OF CONTENTS

INTRODUCTION

William L. Rathje

When I think back, I'm always surprised that I learned one of my most important environmental lessons by watching tourists at a restaurant/bar during the early 1970s. At the time, I was in charge of an archaeological survey and excavation program. The restaurant/bar's open verandah (also the town's sidewalk) was where the students working with me headed after they returned form the field. There, amongst the constant buzz of tenacious flies, the tired archaeologists would sip beers and sodas and gaze out at picturesque vistas.

When I joined them one day, I noticed that they drank directly from bottles and not from the glasses provided by waiters. I was about to ask why, when one of the students directed my attention to the table of tourists at the farthest end of the sidewalk. A man and a woman, clad in the latest tourist chic and bedecked in cameras, bulging shopping bags and floppy straw hats stitched with hot-pink yarn, were stalwartly defending their table against a twelve-year-old employee wielding a huge canister with which he intended to spritz their table.

"No, no, please!" the tourists begged. "Don't spray this table. No DDT here. Spray the next one!"

After a few moments, the boy conceded the first table's opposition and prepared to assault the next table in the line. The response of the tourists there, however, was the same. Arms and cries of "No! No!" were both flying about with equal vigor. The movements of the boy and his DDT could be tracked by a slow-motion "wave" movement (like that in the stands at sports events) as each table's occupants (including ours) stood up in turn to protect their territory from being defiled.

Finally, the stoic youth retreated from view. As the confrontations' survivors gave out audible sighs of relief, one of my students arose, and motioning to me, walked through the doorway that led to the kitchen.

When I followed him, I was struck by a stifling mist. I remember the unmistakable sound of a hand-operated spray can, and soon I could make out the twelve-year-old enthusiastically dousing pots, pans, cooking surfaces, glasses, plates, bowls, silverware, raw food, cooking food, cooked food . . . everything. That too was his job.

Now I knew why my students drank from the lips of their bottles (lips that had been covered by a cap), and why the tourists' gallant stand against DDT spray at their tables was so often a source of jokes and laughter among my students. Furthermore, I became convinced that what you don't know *can* hurt you!

Having reached this conclusion, I can find no more satisfying antidote to ignorance in the environmental arena than this book. I personally evaluated it by looking up items with which I had some acquaintance, such as "disposable diapers" and "grocery bags." The entries I found were knowledgeable, up-to-date, and—most important—even handed; they presented the good, the bad and the ugly realities of the environmental choices available to us all. There, in a few concise statements, was the information which I had personally gone to great lengths to acquire so that I could make informed decisions that were environmentally friendly.

In the process of looking up what I knew, I often stopped to read about what I didn't know. Take "deep-well disposal"—"a method of disposing of industrial liquid waste by pumping it into aquifers that may be anywhere from 1,000 to 20,000 feet below the surface." I had no idea that "more than half of the liquid hazardous waste produced in the United States is now disposed of in approximately 600 operating injection wells."

As recycling has become accepted as "the right thing to do," I have begun to spend more time promoting "source reduction." Increasing source reduction, or just plain "using less stuff," would mean that less stuff would be manufactured and less waste would be created. How could I come to feel better about source reduction than I already did? Deepwell disposal!

Soon I began to focus on my own immediate environment. My home, Tuczon, Arizona, is having a few problems with groundwater contamination. A series of environmental villains have been named—THMs, TCE, PCE, benzene, toluene and nitrates—but I hadn't known much about them. A quick read of this reference work provided some valuable enlightenment.

For THMs (trihalomethanes), I found that Tucson is certainly not alone. Eighty of eighty sites tested in a 1975 EPA study had them as well. The reason was straightforward enough. THMs form when city water is chlorinated; they have also been linked through epidemiological studies to a few varieties of cancer, especially colon. So life appears to be a trade-off, in this case between a slow-acting carcinogen and fast-acting intestinal and other bacteria.

TCE (trichloroethylene) and PCE (perchloroethylene) are both sweat-odor solvents dumped in the desert near Tucson by an Air Force base, an aircraft factory and a large dry cleaning plant. Toluene is a similar solvent found in gasoline, a large quantity of which leaked from a city storage tank. All three are "non-persistent" in surface water. Benzene, another constituent of the leaking gas, tends to float on top of surface water. The problem is that Tucson is the largest city in the United States whose drinking and other water is drawn almost solely from groundwater aquifers, and the people who figured that the passage through hundreds of feet of aggregate would cleanse the potential contaminants figured wrong. The litany of potential health problems associated with these chemicals is almost as long as they are hard to eradi-

cate. At least I know what are considered safe levels in drinking water. If the water gets me, I'll know what did it.

I found more reason for optimism in what I read about the problem of nitrates from fertilizer in agricultural runoff. I was happy to learn that while nitrates are difficult for conventional sewage treatment plants to process, they can be removed effectively by normal biological processes in artificial marshes. A plan for constructing one in the vicinity of Tucson would get my support on a referendum.

And, of course, there are interesting surprises lurking in practically every entry. I already knew that in combination with the chemicals in some foods nitrates are dangerous and capable of forming nitrosamines, which have been linked to stomach and other cancers. How many of us, on the other hand, are aware that nitrates occur naturally in rather high concentrations in such common vegetables as celery, lettuce and spinach. I can just imagine the conversation between a well-read child and a parent that begins with the child saying, "Mom, I'd like to talk to you about diet and cancer risk. Shall we start with spinach?" I hope the parent will be equally well read!

This whole encyclopedia is highly fortified food for thought. With it and a little common sense you may really be able to avoid what can hurt you and your environment—for example dichlorodiphenyltrichlorethane (otherwise known as DDT)!

William L. Rathje is professor of anthropology and director of the Garbage Project at the University of Arizona. He is the coauthor, with Cullen Murphy, of Rubbish! The Archaeology of Garbage.

A

Absorption a process by which a liquid or gas is held within a solid without changing the chemical properties of either substance. Molecules are absorbed when they penetrate the spaces between the molecules in the absorbing surface. Absorption plays a key role in metabolism, and in the pollution of air and water (see AIR POLLUTION, POLLUTION, WATER POLLUTION). The filtering and absorptive abilities of soils purifies water as it seeps down to the water table, and the ability of the materials of which the aquifer is composed to absorb pollutants is one measure of the ease with which the water it contains can become contaminated (see LE GRAND RATING SYSTEM). Scrubbers use absorption to remove gases and PARTICULATES from contaminated air (see INDUSTRIAL AIR POLLUTION TREATMENT, SCRUBBERS), and gas absorption is used to remove hydrogen sulfide and mercaptans from natural gas. See also DISSOLVED SOLIDS. Compare ADSORPTION.

Acetaldehyde a colorless liquid or gas with a pungent, fruity odor. Acetaldehyde is used in the production of other chemicals including paraldehyde, acetic acid and butanol, and in the production of perfumes, flavors, aniline dyes, PLASTICS and synthetic rubber. It is also used in the silvering of mirrors and in the hardening of gelatin fibers. Acetaldehyde gets into air and water from manufacturing EFFLUENTS and spills. It is highly water soluble and highly toxic to aquatic organisms. Tobacco smoke (see ENVIRONMENTAL TOBACCO SMOKE) and auto exhaust (see AUTOMOTIVE AIR POLLUTION, MOBILE SOURCES) both contain acetaldehyde.

The ACUTE EFFECTS of exposure to acetaldehyde can include rashes or a burning feeling on contact. It can cause severe eye burns leading to permanent eye damage. Inhaling acetaldehyde vapors can irritate the lungs causing coughing and/or shortness of breath. Breathing higher concentrations of the vapor can cause a build-up of fluid in the lungs, and can cause sleepiness, dizziness, unconsciousness and death. Acetaldehyde is a proven MUTAGEN and a possible TERATOGEN and CARCINOGEN. It is highly flammable either as a liquid or a gas (see FLAMMABLE LIQUID, FLAMMABLE GAS), and is moderately reactive (see REACTIVITY).

The OCCUPATIONAL SAFETY AND HEALTH ADMINISTRATION has established a permissible exposure limit for airborne acetaldehyde of 100 parts per million (ppm), averaged over an eight-hour workshift, with concentrations not to exceed 150 ppm during any 15-minute work period. It is also cited in AMERICAN CONFERENCE OF GOVERNMENT INDUSTRIAL HYGIENISTS, Department of Transportation and ENVIRONMENTAL PROTECTION AGENCY regulations. Acetaldehyde is on the Hazardous Substances List, the Special Health Hazard Substance List and the HAZARDOUS AIR POLLUTANTS LIST. Businesses handling significant quantities of acetaldehyde must disclose use and releases of the chemical under the provisions of the EMERGENCY PLANNING AND COMMUNITY RIGHT TO KNOW ACT.

Acid any substance that gives up protons when combining with other substances. A narrower definition of an acid—any compound that dissociates upon contact with water to form *hydronium ions* (H_3O^+)—has been used in the past. Acids have a pH of less than 7 and a sour taste. They occur in liquid, solid or gaseous form at room temperature, and are moderately to strongly corrosive. Acidic solutions are electrolytes—they conduct electricity. The hydronium ION is responsible for these characteristics. The more strongly a substance dissociates, the more hydronium ions are formed and the more acidic is the resulting compound. A *fatty acid* contains the *carboxyl group*, the *carbohydrates* found in plant and animal fats (see ESTERS, GREASE, OILS). See ACID RAIN, ACIDIC WASTE.

Acid Rain rainwater that has become more acidic as the result of AIR POLLUTION. *Acid precipitation* includes rain, snow, sleet and hail with unnaturally high ACID levels (see PRECIPITATION). Acid rain is the best-known form of the AERIAL DEPOSITION of POLLUTANTS, and occurs to some degree in every part of the world. Scandinavia, central and eastern Europe, eastern Canada and the northeastern United States have particular problems with acid precipitation. Rainwater is naturally moderately acidic, with a pH of about 5.6 (7 is neutral). This is caused by atmospheric CARBON DIOXIDE (CO_2) being absorbed by a raindrop as it falls through the air to form a weak carbonic acid. The sulfur dioxide released into the ATMOSPHERE by volcanoes and hot springs is another natural source of atmospheric acid. Rain from a storm in Wheeling, West Virginia in 1980 had a pH (see ACIDIC WASTE, PH) of 1.5—the lowest yet recorded. That is approximately 10,000 times the acidity of normal rainfall and about five times the acidity of a lemon.

The combustion of FOSSIL FUELS is responsible for most acid rain, and areas with significant environmental damage from acid precipitation are generally downwind of industrial areas where lots of fossil fuels are used. When oil, gas or coal is burned, some of the NITROGEN and oxygen molecules in the high-temperature exhaust gases combine to form NITROGEN OXIDES (NO_X). The airborne NO_X molecules, in turn, are converted to nitric acid after sufficient exposure to other atmospheric gases. Impurities in fuels combine with oxygen to produce a variety of other acids. Burning coal that contains sulfur will form sulfur oxides (see SULFUR DIOXIDE). Some sulfur oxides will break down further in the atmosphere to form *sulfuric acid*.

Some climatologists believe that reducing acid rain by cutting atmospheric emissions of sulfur dioxide could accelerate GLOBAL WARMING. They say that because sulfur dioxide makes clouds more reflective—bouncing more sunlight back into space and thereby reducing temperatures at the earth's surface—therefore, its presence in the atmosphere is helping to keep temperatures at the earth's surface lower. The reduction of atmospheric sulfur dioxide could cause global warming to come in what one scientist called "a tremendous heat pulse" rather than through a more gradual warming. But because failing to control acid rain will definitely cause further harm to lakes, streams and forests—and would no more than delay the so-far theoretical greenhouse warming—the effort to control all acid-rain precursors continues.

The Effects of Acid Rain

The effects of acid precipitation depend on when and where it comes to earth. Rainfall of a certain acidity may do no harm in a region with thick soils and bedrock of a base mineral such as limestone, while devastating nearby hills with thin, acidic soil and granitic bedrock. As precipitation's acidity increases, fish, insects and water creatures may suffer reproductive failure; plants—including trees and commercial crops—may grow more slowly; soil erosion may increase; the rate at which minerals and HEAVY METALS are leached from soil, rocks and drinking-water aqueducts may increase. Even the acidity of normal rainwater is sufficient to slowly etch the surface of rocks. Acid rain accelerates the process to the extent that statues and buildings made of rock deteriorate much more quickly. Mountains located downwind from urban areas and other major sources of sulfur dioxide frequently are blanketed with acidic snowfall. The acidic snow has little environmental effect while it is still frozen in the winter, but with the spring thaw, a torrent

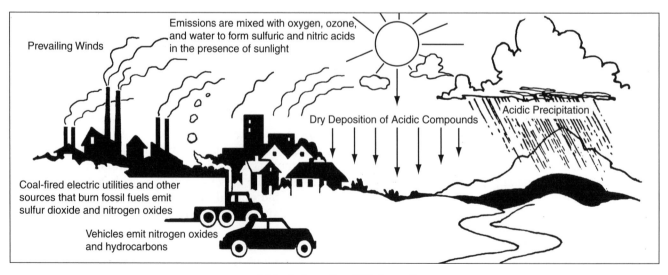

Acid rain formation (*EPA Journal*)

of acidic water is released by the melting snow at a critical time for the lake ecosystem. During the spring of 1993, melting snow in Massachusetts caused the worst acid shock to lakes and streams in more than a decade.

History of Acid Rain
Although acid precipitation has been in existence since the advent of the industrial revolution in Europe during the 19th century, the first scientific studies of its effects started in Scandinavian countries in 1952. Researchers in Scandinavia and North America documented that the average acidity of rain increased by 200 times between 1960 and 1980. Monitoring programs were undertaken in the late 1960s by Cornell University in the Adirondack Mountains of New York and by the University of Toronto in the La Cloche Moutons of northern Ontario. These studies had revealed by 1981 that 180 lakes in the Adirondacks and 140 in the La Cloche Mountains were devoid of life, apparently the result of low pH levels caused by acid precipitation. Half of the acid rain that falls in Canada is spawned in the United States, largely from coal-fired power plants in the midwestern and central Atlantic states. Quebec sued the ENVIRONMENTAL PROTECTION AGENCY (EPA) in 1988 in an attempt to force action on the problem. The government claimed in the suit that acid rain was endangering 14,000 of the province's lakes.

Solutions for Acid Rain
The United States Geological Survey announced in July 1993 that average concentrations of sulfates and nitrates had fallen in the preceding decade, the result primarily of improved air pollution control at coal-burning power plants. The CLEAN AIR ACT Amendments of 1990 called for a 50% reduction in sulfur-dioxide emissions by the year 2000, and an eventual reduction of 10 million tons a year below 1980 SO_2 levels. It also called for a reduction of nitrous-oxide emissions of two-million tons a year. The EPA was charged with compiling a *National Acid Lakes Registry* that would list all U.S. lakes that have become more acidic as the result of acid precipitation. Calcium carbonate, the same ingredient that soothes an acid stomach, has been used to reduce the acidity of some streams and lakes and thereby revive their biological viability. British and American researchers reported in 1994 that acidity of precipitation was not declining as quickly as predicted because the emission of alkaline particles that help to neutralize airborne sulfates is on the decline along with the emission of sulfates.

See AIR POLLUTION TREATMENT.

Acidic Waste   waste that has a PH of less than 7. Industrial effluents, runoff from mines and the tailings from mines and mills and drilling brine are all forms of acidic waste. STEEL production, metal plating, leather tanning and the production of pickles also produce acidic by-products. Coal mines and coal-fired power plants both create acidic waste. Bacteria that metabolize IRON and SULFUR in the presence of oxygen break down the PYRITE found in the TAILINGS from mines, and contribute to acid runoff from mines and tailings piles. The largest part of the hazardous waste produced annually in the United States is the corrosive materials (see CORROSIVITY), spent acids and alkaline wastes from the metal-finishing, chemical-production and petroleum-refining industries. The toxic brew inside many of the nation's most seriously polluted superfund sites is highly acidic. Highly acidic waste kills aquatic life rendering receiving waters unfit for use as drinking water or for recreation. If the pH of the soup inside an anaerobic digester gets too low or too high, or if the temperature of the whole mix strays outside a relatively narrow range, the digester will produce a malodorous, highly acidic sludge.

Removing Acidity from Waste
An important step in the pretreatment of industrial effluent and hazardous waste is neutralization—mixing acids and bases into the waste as needed to neutralize ACIDITY or ALKALINITY. *Acid-gas scrubbers* spray a mixture of lime and water into acidic flue gas to form calcium salts and thereby reduce the acidity of exhaust gases. The metabolism of microbes can reduce the acidity of wastewater through their metabolism (see MICROORGANISMS). See also SPOIL.

Acidity   the capacity to donate protons. The acidity of wastewater is an important factor in the treatment of sewage and industrial waste (see ACIDIC WASTE, INDUSTRIAL LIQUID WASTE, SEWAGE TREATMENT). Acidity increases water's solvent effects (see SOLUBILITY, SOLVENT) and can contribute to CORROSION. AGGRESSIVE WATER is soft water (see DRINKING WATER) that is also acidic, and is very corrosive. Mineral acids and hydrolizing salts are the principal cause of acidity in wastewater. Acidic compounds in the air cause chemical weathering and can contaminate soil and water (see ACID RAIN). PH is a measurement of acidity/alkalinity. See ACIDIC WASTE.

Action on Smoking and Health   a non-profit charitable trust that brings legal action to restrict smoking in public places and the workplace. Action on Smoking and Health (ASH), formed in 1967, has been involved in every legal decision restricting smoking and cigarette advertising in the United States, and has convinced many businesses to voluntarily restrict or eliminate smoking on their premises. ASH publishes papers on the health threats of tobacco smoke and on the non-smoker's legal rights, and offers assistance to the non-smoking public interested in limiting smoking. The organization is supported entirely by tax-deductible donations and does not accept government grants. Address: 2013 H Street, NW, Washington, DC 20006. Phone: 202-659-4310.

See ENVIRONMENTAL TOBACCO SMOKE, INDOOR AIR POLLUTION.

Activated Carbon   carbon from which most of the impurities have been removed by heating and exposure to steam. Activated-carbon air and water filters are especially effective at capturing very small particles such as most HYDROCARBONS and can improve the odor and taste of drinking water. The carbon used in the activation process is derived primarily from coconut husks, apricot and peach pits and other woody by-products of food production. The carbon source is charred, and is then treated with heat and steam to increase its affinity for toxic compounds.

Chemicals are removed from water or air that is passed through an activated-carbon filter through ADSORPTION. Activated carbon is extremely porous—one pound has a surface area of approximately one acre. This honeycomb of micropores chemically attracts and latches onto minute impurities that come near them. The particles are said to be "adsorbed": the molecules chemically or physically removed from the air or water passing through the filter, and are attached to the surface of the carbon. Activated carbon is manufactured as a powder, in granules and in block form. Most water filters (see WATER TREATMENT) use granular carbon or are molded in a solid block, sometimes along with other filter media materials. Powdered carbon was once used in water filters, but problems with carbon granules escaping into finished water has largely eliminated the practice. Municipal treatment plants often use powdered carbon to remove organic chemicals from drinking water.

The Egyptians and other ancient cultures used charcoal, which is mostly carbon, to purge objectionable tastes and odors from their drinking water, so knowledge of the purifying powers of carbon is far from new. Activated carbon is, however, a product of modern technology. First manufactured during World War I for use in gas masks, it wasn't until the 1960s that a handful of municipal water treatment plants began to use activated carbon, for much the same reasons that the Egyptians had used charcoal. As organic chemicals in community drinking-water supplies became of increasing concern during the 1980s, an increasing number of water utilities installed activated-carbon treatment facilities to remove toxic chemicals from drinking water.

Effectiveness

The effectiveness of activated carbon as a filter depends on several factors: the amount of time water is exposed to the carbon (the longer the better); the pattern in which water flows through the filter medium (the more diffuse the path, the better); the number and range in size of the micropores in the carbon material (the greater the range of micropore size, the greater the range of contaminants that will be removed); and the amount of activated carbon in the filter (a filter with more carbon will last longer than another, smaller filter of the same configuration). A carbon filter will lose its efficiency with use, and must be replaced occasionally. The concentration of contaminants in the air or water being filtered and the volume of gas or liquid being filtered determine the filter's life expectancy. As the filter is used, more of its micropores pick up chemical hitchhikers and are thereby no longer available to adsorb contaminants. Carbon water-filter manufacturers specify that the life expectancy of their product can typically be expected to process, and can provide chemical test kits to help determine when a filter is due for replacement. If a filter is used too long, some of the contaminants it has initially trapped may be bumped out of their niche by other compounds with a stronger chemical affinity for the micropore. For instance, molecules of the insecticide aldrin, which has a very strong bond, could displace molecules of BENZENE, which has a much lower affinity for activated carbon. If the benzene isn't adsorbed by another micropore on its way out of the filter, it is then free to enter finished air or water. Activated carbon can be reconditioned for reuse by heating or through the use of a chemical solvent.

Activated Sludge   a microbial mass formed after wastewater is exposed to oxygen and microorganisms for several hours in an aeration tank at a SEWAGE TREATMENT plant. The activated sludge—also called *floc* or *zoogloeal* ("living glue")—is the frothy mass of biota created primarily by the growth of the aerobic bacteria *Zoogloea ramigera*, its sticky, multi-branched, interlocking fingers laced throughout the microbial soup. Floc is removed from wastewater in a settling tank, with a portion of the sludge going back into the aeration tank to activate further microbial growth. Activated sludge is supercharged with bacteria, and more than 60 species have been identified in a single sample. See FLOCCULATION, SECONDARY TREATMENT.

Acute Effects   the short-term health effects of exposure to a toxic substance. Most acute effects occur immediately or within a few hours of exposure, although certain toxins' acute effects may occur days after exposure. Effects observable within four days of exposure are technically termed acute. Acute effects vary from relatively minor irritation of the skin, eyes, digestive tract or lungs to convulsions, coma or death, depending on the severity of the poisoning. Compare CHRONIC EFFECTS. See LD$_{50}$.

Adsorption a process in which a liquid or gas adheres to the surface of a solid without penetrating it. The film that adheres to the outer surface of the solid is generally only one molecule thick. Adsorption is one of the methods by which a gas can be removed from an airstream. The contaminants intercepted by an adsorption system can be removed by heating or by lowering air pressure, after which the sorbent is said to be regenerated. ACTIVATED CARBON is the most commonly used sorbent, although clays, chars, gels, oxides and silicates are also used. Since the removal of contaminants through adsorption occurs only at the surface of the sorbent, a large surface area is essential for efficiency. Compare ABSORPTION.

Aeration the process of passing air through a liquid, thereby increasing its oxygen content. In secondary SEWAGE TREATMENT (see SECONDARY TREATMENT), aeration kills ANAEROBIC BACTERIA by exposure to oxygen, and encourages the growth of AEROBIC BACTERIA, which typically remove 95% or more of the effluent's biological oxygen demand by consuming the organic material in the wastewater. Exposure to oxygen neutralizes many toxic compounds, converting them to a less-toxic oxide (see OXIDATION, OXIDIZING AGENT). Small particles are also removed from the wastewater when they are caught in the surface tension on air bubbles as they rise to the top to be skimmed off the wastewater.

Aeration is also used in the treatment of industrial liquid waste. Wastewater is allowed to stand in large, shallow *oxidation ponds* where natural aeration and photosynthesis break down suspended organic material. Aeration is accomplished in three basic ways:

• Compressed air is pumped into the bottom of the wastewater vessel and through a diffuser, which disperses the bubbles
• Effluent is stirred vigorously, literally to a froth
• Effluent is sprayed out of a nozzle through the air

Volatile organic compounds (see VOLATILE ORGANIC COMPOUNDS, VOLATILITY) evaporate and pass into the air as a gas when exposed to the air. When water containing volatile chemicals is sprayed out of a faucet or shower head or agitated in a washing machine, part of the volatiles will evaporate and pass into indoor air. *Packed tower aeration* takes advantage of this phenomenon to remove volatile toxins from water by cascading

it through a tower designed to maximize EVAPORATION of volatile compounds. Aeration occurs naturally during a rain storm and in water falls and rapids. AIR POLLUTION at Niagara Falls, where a very polluted river receives a very thorough aeration, has prompted concern about the health of tourists and of people who work around the falls. See also ACTIVATED SLUDGE, FLOCCULATION, SEWAGE TREATMENT.

Aerial Deposition the deposition of airborne contaminants on the earth's surface. Industrial smokestacks, gravel roads, plowed croplands, sand-and-gravel plants and the DUST along roads are all potential sources of airborne PARTICULATES that can contaminate land, lakes or streams. Chemical air pollutants such as PESTICIDES and SULFUR DIOXIDE can also contaminate land and water via polluted precipitation. The CLEAN AIR ACT requires the ENVIRONMENTAL PROTECTION AGENCY to assess the extent to which the Great Lakes, Lake Champlain and coastal waters are being contaminated by aerial deposition. Pesticides used only in Asia have been found in the Great Lakes (see GLOBAL POLLUTION).

Some of the heavier SUSPENDED SOLIDS will settle onto land or water fairly near their point of origin (see DUSTFALL), while lighter particles may settle many miles from the point of emission. Many airborne pollutants come to earth only after becoming part of a raindrop or a crystal of ice or hail. Chemical contaminants such as the precursors of ACID RAIN combine with atmospheric moisture, which falls as acid precipitation, often hundreds of miles from the point of emission. Other lightweight contaminants such as pesticides may travel around the globe before becoming part of precipitation and falling onto a body of water or onto land. State, federal and local regulation of AIR POLLUTION over the last two decades has greatly reduced aerial deposition in the U.S. from its peak, which was around the turn of the century. Aerial deposition of toxic particulates on heavily populated areas remains a serious problem in eastern Europe and in developing countries, which frequently rely on low-efficiency coal combustion for much of their energy needs.

Mill Creek

Uncontrolled aerial deposition of contaminants has, in the past, caused serious long-term damage. Between 1986 and 1988, the federal government relocated 37 families that lived in the Mill Creek drainage near Anaconda, Montana, because the soil around their homes was found to be tainted to a depth of three feet or more with ARSENIC and other HEAVY METALS. Aerial deposition was responsible for the tainted soil. A 585-foot smokestack looms just upwind of the neighborhood, and some of the heavier constituents of the smoke that belched out of the stack in the eight decades before the smelter was closed in 1980 fell to earth in the Mill Creek neighborhood. The Clean Air Act amendments of 1990 call for a program to assess the extent of aerial deposition of pollutants into the Great Lakes and coastal waters. See also CLARK FORK RIVER.

Aeroallergens airborne POLLENS, mold spores, ALGAE, animal dander and other biological substances that can cause an allergic reaction in sensitive individuals. The feces of the tiny dust mites found on furniture and bedding can be swept into the air by people's movements and cause a reaction when inhaled by an allergic individual. Bursting bubbles at the surface of water can propel microorganisms such as the green algae, Chlorella, into the air, where it can cause respiratory allergy and skin irritation for susceptible people. The mold and pollen count, which tells what kind of day for allergy sufferers to expect, is a regular feature of the weather report in warmer parts of the U.S. See INDOOR AIR POLLUTION, MICROORGANISMS.

Aerobic Bacteria any species of BACTERIA that requires oxygen to live. Aerobic bacteria cause the decomposition of organic material in a compost pile, which must be turned regularly to expose the bacteria and the material they are consuming to air. They are used to break down SEWAGE in SECONDARY TREATMENT, and are used in HAZARDOUS WASTE TREATMENT. See also AEROBIC STABILIZER, MICROORGANISMS. Compare ANAEROBIC BACTERIA.

Aerobic Stabilizer any method of using AEROBIC BACTERIA to remove BIOCHEMICAL OXYGEN DEMAND (BOD) from wastewater. In SEWAGE TREATMENT plants, the trickling filter and the activated sludge tank are the most commonly employed aerobic stabilizers. Both treatment methods are capable of removing 85% to 95% of the BOD from a waste stream in as little as two hours. Aerobic stabilizers are a method of SECONDARY TREATMENT. See SEWAGE, WATER POLLUTION.

Aerosol a suspension of minute drops of liquid or solid particles in a GAS. Although defined as liquid or solid particles, the term *aerosols* is more often used to denote liquid particles, while *particulates* is used in reference to solid particles. Aerosols comprise a dynamic system that is impacted by changes in temperature, sunlight and humidity. Even relatively clean ambient air typically contains thousands of suspended particles per cubic centimeter. The matter and energy tied up in aerosols is in constant flux, with solid and liquid particles colliding, bouncing off one another, combining,

breaking apart and recombining in yet new forms. Droplets of liquid combine to form bigger drops, or evaporate to become gaseous water vapor. Solids cling to solids to form larger particulates. Gases precipitate to become liquid particles, or dissolve to become part of an airborne drop of liquid. Chemical reactions occur on the surfaces of particles and inside liquid particles. Fog, SMOKE, PARTICULATES and SMOG are aerosols. Volcanoes can gush enormous clouds of aerosols into the upper ATMOSPHERE when they erupt, which can alter planetary weather for years. Liquid pollutants in aerosol form are particularly dangerous because they can be inhaled deep into the lungs where they can cause direct tissue damage, or can be absorbed into the blood stream. Aerosols are treated as a form of particulate in air pollution control (see AIR POLLUTION, AIR POLLUTION TREATMENT). The aerosol formed when water vapor condenses on airborne sulfur dioxide to form dilute sulfuric acid is the precursor of ACID RAIN.

Aerosol Spray a spray composed of microscopic drops of a liquid that are light enough to remain suspended in air. An *aerosol spray can* is the pressurized can from which a liquid product such as deodorant or PAINT is dispensed. Such spray cans employ pressurized gas as a propellant to drive the liquid product out of the spray nozzle at the top where it is aerosolized. The propellant is usually introduced into the can in solution with the product to be dispensed. As the liquid is dispensed out of the bottom of the container, more of the gas evaporates to the top of the chamber. CHLOROFLUOROCARBONS (CFCs) were the most commonly employed gas used as a propellant in aerosol spray cans until the 1970s, when concern about their harmful effects on the stratospheric ozone layer led to a ban on their use in the United States in 1978. Manufacturers then switched to other HYDROCARBONS such as propane, butane and iso-butane for use as spray-can propellants. Bottles in which air is pumped in to force the product out, and products that are pumped directly out of the bottle, have seen increasing use since that time.

In 1989, the CALIFORNIA AIR RESOURCES BOARD ordered a 20% cut in the smog-producing hydrocarbons used as propellants in antiperspirants and deodorants by 1992 and another 60% reduction by 1995. The board estimated that 19 million cans containing such products are sold each year in the state, and that five tons of hydrocarbon propellants are released into the atmosphere when the products are used. The board estimates that the use of paints in aerosol cans releases 29 tons of volatile organic compounds into California's air each day. See AIR POLLUTION, GREENHOUSE EFFECT, INDOOR AIR POLLUTION, OZONE LAYER.

Agency for Toxic Substances and Disease Registry see PUBLIC HEALTH SERVICE

Aggressive Water water with a high acidity and a low concentration of dissolved minerals. Aggressive water can quickly corrode metal pipes and vessels, and DRINKING WATER must be treated to reduce its acidity and therefore its aggressiveness so that it will not pick up lead or copper in the water-distribution system. See COPPER, CORROSION, CORROSIVE MATERIAL, LEAD, SAFE DRINKING WATER ACT.

Agricultural Wastes wastes produced in farming, ranching and forestry. Agricultural wastes are a major source of pollution of both groundwater and surface water, and agricultural operations are a primary source of DIFFUSE POLLUTION. PESTICIDES, SALTS and NUTRIENTS from FERTILIZER and ANIMAL WASTES often pollute irrigation return water (see IRRIGATION WATER), which in turn contaminates the stream it enters (see BIOCHEMICAL OXYGEN DEMAND [BOD], EUTROPHICATION). A variety of hazardous waste are produced on farms including empty pesticide containers, solvents, paints and wood preservatives. Erosion can be caused by exposing soil for planting on farms, by overgrazing on ranches and public rangeland and by the construction of roads and skid trails in logging. This leads to a high sediment load in RUNOFF (see TOTAL SUSPENDED SOLIDS), which can in turn lead to SEDIMENTATION. Air pollution can be caused by the atmospheric application of pesticides on farms and forests, and by the erosion of exposed surfaces by the wind. The burning of the stacks of limbs, tree tops, dead trees and brush that accumulate during a logging operation is a source of air pollution and countless wildfires.

Air see ATMOSPHERE

Air Changes Per Hour the percentage of indoor air that is replaced with fresh outdoor air in an hour. As much as two air changes per hour may be required to prevent the concentration of some contaminants from exceeding acceptable levels, but ventilation rates can be much lower in those indoor spaces with no major sources of contamination (such as ENVIRONMENTAL TOBACCO SMOKE).
See INDOOR AIR POLLUTION, VENTILATION.

Air Current a persistent wind caused by environmental factors unrelated to a storm. Air currents are driven by differences in temperature and pressure. Where storm winds can be violent and changeable, air currents are more steady and predictable from day to day and season to season. See also AIR MASS, ATMOSPHERIC INVERSION.

Air Filters a device that removes impurities from air by capturing airborne particles on a filter medium (see FILTRATION). Air filters are used to purify indoor air (see INDOOR AIR POLLUTION), to remove samples of PARTICULATES from industrial smokestacks for testing, to purify combustion air being drawn into an engine and to protect delicate equipment from dust. Some air filters fit into heating, ventilating and air-conditioning system ductwork; others are cylindrical, with contaminated air being drawn in through the center and filtered air exiting from the outside of the tube. Filters using a variety of synthetic and natural fibers in every conceivable weave and molded form are on the market. Common air filters include:

- *dust-stop filters*—Air is forced through tiny openings in a fibrous material or metal mesh, and particles too large to pass through the openings are left behind, a process called *straining*. Such inexpensive filters intercept only the largest DUST particles—those 10 microns and larger in diameter, or about 10% of the contaminants in a typical sample of air. Such filters tend to operate with about the same efficiency until they become heavily clogged and must be replaced (sometimes monthly). About 90% of the filters installed in furnaces and air conditioners are dust-stop filters.
- *media filters*—Air is forced through a finer filter material, often a PAPER fabric that is pleated to maximize surface area. Media filters remove contaminants primarily through straining and *impaction*, a process in which particles are driven by turbulence in the airstream into filter fibers, to which they stick. A media filter can actually become more efficient with use because, as it gets dirty, it can become harder for particles to find a way through. The filter still must be replaced when it picks up enough contaminants to impede airflow through the system in which it is installed. Dust-stop and media filters are both *mechanical filters*.
- *moving curtain filters*—Fresh filter fabric is constantly fed into the air filter, and the used fabric is stored on a take-up reel. When the entire roll has been fed through the filter, it is cleaned or discarded. Another moving-curtain design utilizes a constantly-moving metal mesh fabric that goes through a cycle of immersion in a solution that leaves fresh adhesive on the media fabric, which is then exposed to airborne contaminants in the air filter. When the dirty filter media again enters the solution tank, the media releases its load of dirt and receives a fresh coat of adhesive.

- *self-charging mechanical filter*—This filter uses a blend of mechanical and *electrostatic* forces to remove contaminants from air. The fibers in a self-charging filter pick up their electrical charge from the friction generated when air blows through its fibers. The efficiency of such filters is lower with humid air because it is more difficult to maintain the necessary electrical charge.
- *charged-media filter*—This filter uses mechanical forces and electrical attraction to trap contaminants. High dc voltage is passed through the media fibers to create an electrostatic field capable of polarizing and capturing airborne material.
- *high-efficiency particulate-arresting (HEPA) filter*—This type of filter is made of densely packed fibers that are pleated to increase their surface area. The filters, developed during World War II to remove radioactive dust from the exhaust of nuclear plants, are used in laboratory clean rooms and hospital operating rooms. Vacuum cleaners with HEPA filters are used to clean up buildings contaminated with lead and other fine toxic dust.
- *automotive air filters*—The engine air cleaner is mounted on top the motor, and most automotive ventilation and air-conditioning systems include at least a DUST filter. An increasing number of new cars come equipped with high-capacity filters capable of removing the carbon monoxide and petrochemical molecules typically encountered by commuters. To ensure safety, air filters in a vehicle ventilation system must be replaced regularly to ensure adequate airflow to defog windows. An ELECTROSTATIC PRECIPITATOR—which zaps foreign material in the air passing through it with a much stronger electrostatic field than a charged-media filter—is frequently called an air filter, and although air filters may be used in an electronic air cleaner system, the precipitator itself is not an air filter.

Air-Cleaner Tests

There are five standard tests of air-cleaner efficiency:

1) In the *atmospheric dust-spot* test, a filter is installed in a clean ductwork and outdoor air is blown through it at a constant rate. A high-efficiency paper filter is exposed to the airstream before it passes through the filter being tested, and another filter paper mounted downwind catches most of the contaminants that pass through the test filter. The upstream filter paper is then compared with its downstream counterpart using an optical-scanner to measure what percentage of airborne matter the filter removed.

2) The *average atmospheric dust-spot* test uses the same techniques, but averages the results of several test runs conducted over a period of time.

3) In the *initial synthetic dust weight arrestance* test, a measured quantity of synthetic dust is blown through a clean test filter. Before the test, a high-efficiency filter paper of known weight is installed downstream from the test filter. After the test, the paper is removed and weighed to determine the percentage of the dust removed by the test filter.

4) The *average synthetic dust weight arrestance* test uses the same techniques applied over a series of tests to determine the average performance of the filter being tested.

5) The *fractional efficiency* test measures the test filter's ability to remove particles of a uniform size from an airstream. Particles of a given size represent only a fraction of those normally present in air, hence the name. The efficiency rating of a given filter for removing a certain fraction of the range of contaminants typically present in air may be quite high even though the filter in question has a low overall efficiency rating.

Easy cleaning or replacement of filter media is essential in any air filter, especially in industrial applications where the load of suspended particles in the wastestream being processed can be heavy (see INDUSTRIAL AIR POLLUTION TREATMENT). See also AIR POLLUTION TREATMENT, PARTICULATES, VENTILATION. Compare NEGATIVE ION GENERATOR, SCRUBBER.

Air Freshener a product intended to reduce the effect of objectionable household odors by covering them with stronger, more agreeable odors. Most commercial air fresheners come in AEROSOL SPRAY cans, but incense, scented candles and potpourri are also commonly used for the purpose.

Air Mass a body of air that tends to act as a coherent, separate unit. Air circulation within an air mass is largely independent of other air masses. A given air mass may differ from neighboring air masses in temperature, barometric pressure, pollutant content, relative humidity and total water content. The interplay of and boundaries between air masses is one of the building blocks of weather and a primary ingredient in AIR POLLUTION patterns. See also AIR CURRENT, ATMOSPHERIC INVERSION.

Air Pollution the contamination of air with natural or man-made pollutants. In order to be considered air pollution, the airborne contaminants must be harmful to humans, animals or plants. The damage may be either direct, as when an airborne toxin is ingested and causes a disease; or indirect, as when SULFUR DIOXIDE and NITROGEN OXIDES are transformed in the ATMOSPHERE to ACID RAIN. Air pollution in American cities has accounted for an estimated 60,000 deaths a year, according to an ENVIRONMENTAL PROTECTION AGENCY (EPA) survey released in 1991.

A *primary pollutant* is emitted from its source in the same form in which it exists in the atmosphere, while a *secondary pollutant* is created by the reaction of a primary pollutant with atmospheric gases and other pollutants. OXIDATION creates most secondary pollutants. Water vapor in the atmosphere increases its ability to carry pollutants, and atmospheric moisture can react with airborne contaminants to form new pollutants (see PHOTOCHEMICAL SMOG). Airborne contaminants may be a liquid (commonly called an AEROSOL), a GAS or a solid (see PARTICULATE). Gases account for almost 95% by weight of all reported air pollution in the United States. CARBON MONOXIDE, NITROGEN OXIDES, SULFUR DIOXIDE and VOLATILE ORGANIC COMPOUNDS are the gaseous air pollutants that are of greatest concern. SUSPENDED SOLIDS represent another 5% of reported air pollution, and aerosols and radiation account for the remaining fraction of a percent. SMOG, the brown soup of smoke, dust, airborne contaminants and water vapor that sometimes gathers in cities, is largely the result of a photochemical process in which nitrous oxides, volatile organic compounds (VOCs) and other airborne contaminants combine in the presence of sunlight to produce new toxins such as ozone.

Air Pollution Sources

Natural sources of air pollution include salt from oceans, microorganisms, animals, particulates and gases from volcanoes. Emissions resulting from incomplete COMBUSTION of hydrocarbons (see HYDROCARBON) in automobiles is the principal source of air pollution in urban areas. The EVAPORATION of automotive FUELS adds to the problem. Automobiles are a MOBILE SOURCE of air pollution. Power plants, garbage incinerators, factories and similar large centralized facilities are POINT SOURCES of air contaminants. Wood smoke from stoves and slash burning, and dust from roads and farmlands are diffuse air-pollution (see DIFFUSE POLLUTION) sources. Pesticides used to control unwanted insects, plants and microorganisms; PAINTS that give off VOCs to the atmosphere as they cure; and evaporation from toxic waste sites and accident scenes are also typical sources of air pollution. Stratospheric emissions from aircraft are of special concern because they occur in the stratosphere, where little mixing occurs.

Weather and Air Pollution

Weather patterns dictate the concentration of airborne contaminants in a given AIRSHED. *Dispersion* is the spread and dilution of pollutants. Wind and the vertical mixing of air in the troposphere helps disperse and dilute pollutants, and wind can carry pollutants for hundreds of miles, even around the world. When wind hits an obstacle such as a tree, areas of turbulence are formed, increasing the air's ability to disperse pollutants. Wind blowing over a jagged urban skyline typically has four times the dispersive ability of wind blowing over a flat surface.

Rain can rinse contaminants out of the atmosphere, often with negative environmental effects (see ACID RAIN). ATMOSPHERIC INVERSIONS, which occur when warmer air aloft traps a layer of cooler air next to the ground, can allow dangerous levels of air pollution to build up by blocking the passage of cleansing winds through the affected area, and even without an inversion, weather with minimal air movement can lead to extremely contaminated air in regions with abundant sources of pollution. Air in the troposphere can be quite turbulent when the sun heats surfaces at different rates during the day, which causes vigorous vertical mixing of air and the dispersal of pollutants, but turbulence drops off above 300 feet. Weather systems such as thunder storms can carry contaminants directly into the stratosphere. Hot, sunny weather facilitates the creation of OZONE, the principal contaminant found in urban smog. The mixing of pollutants into the stratosphere can lead to ozone depletion (see STRATOSPHERIC OZONE).

Health Damage from Air Pollution

Disasters such as those in the Meuse Valley in Belgium in 1930, in Donora, Pennsylvania in 1948 and in London in 1952 occasionally occur, but most of the damage caused by air pollution is much less dramatic. The very old, the very young and those already weakened by existing disorders are the most vulnerable to the effects of polluted air. Heavily polluted air can cause severe distress for individuals with chronic lung disorders such as emphysema or asthma. Toxins in GASES, AEROSOLS and RESPIRABLE suspended particulates can be inhaled and then either become lodged in lung tissue or be passed directly into the blood stream. Vehicle exhaust is thought to cause up to 8% of the lung cancer in nonsmoking urban residents. Air pollution can upset atmospheric dynamics by changing the way in which long-wave and short-wave radiation pass through it (see GLOBAL WARMING). Visibility in many U.S. cities was less than 0.6 mile (1 km) before pollution controls were instituted. This is no longer the case.

Crop and Tree Damage

Trees in forests downwind from major pollution sources can be weakened or killed by airborne pollutants, especially ozone, which destroys plant tissue that it comes in contact with. Acid precipitation can sterilize lakes and streams and reduce the soil's capacity to support trees or crops. Research has demonstrated that both pine trees and alfalfa may be damaged by sulfur dioxide. In California's San Joaquin Valley, ozone pollution has reduced cotton yields an estimated 5 to 29% and has stunted the growth of sweet corn and lemons by as much as 50%. The growth of alfalfa, beans, grapes, peppers, potatoes, radishes, soybeans, spinach, tomatoes and wheat has been similarly affected. A July 1993 report by the American Lung Association estimated that two-thirds of Americans live in areas that fail to meet current air-quality standards. See also AERIAL DEPOSITION, CLEAN AIR ACT, INDOOR AIR POLLUTION, OPEN BURNING, POLLUTION.

Air Pollution Control   actions that reduce air pollution. Any measure or device that reduces the emission of solid, liquid or gaseous pollutants into outdoor air is a form of air pollution control. These actions may be direct, as with the treatment of atmospheric emissions from a POINT SOURCE (see AIR POLLUTION TREATMENT, INDUSTRIAL AIR POLLUTION TREATMENT), or they may be indirect, as with the building of a rapid-transit system in an urban area to reduce commuter mileage (see AUTOMOTIVE AIR POLLUTION, MOBILE SOURCES) or switching to a cleaner burning fuel or more-efficient combustion equipment in a manufacturing operation. A CATALYTIC CONVERTER, cleaner-burning engines, car pooling or the application of a dust-control agent on an unpaved road are all ways of limiting air pollution from automobiles. Air pollution can be controlled by switching to cleaner FUELS. Electric power plants and other coal-burning industries can reduce SULFUR DIOXIDE emissions, for instance, by switching to coal with a lower sulfur content. This can be achieved by using coal with a naturally low sulfur content or by cleaning high-sulfur coal. Similarly, an automobile will emit less CARBON MONOXIDE and other contaminants when switched to gasoline or diesel that has been reformulated to minimize emissions (see GASOLINE, METHANOL, METHYL-TERTIARYBUTYLETHER).

Air Pollution Measurement   the measurement of contaminants found in air. Contaminants are commonly measured in outdoor and indoor air, and in the effluent coming from wood stove smokestacks and industrial installations (see INDOOR AIR POLLUTION, POLLUTION MEASUREMENT, INDUSTRIAL AIR POLLUTION MEASUREMENT). The monitoring of air pollution sources is required under the CLEAN AIR ACT, and air monitoring is also required at toxic waste cleanup sites (see COMPREHENSIVE ENVIRONMENTAL RESPONSE COMPENSATION AND LIABILITY ACT). The monitoring of AMBIENT AIR and of weather conditions is part of the measurement of atmospheric pollution plumes. Ambient air quality and weather data are frequently recorded as a part of a pollution monitoring effort. Far more instruments and methods than can be described here are used in the sampling and analysis of air contaminants, but some of the principal methods are described. See also CARBON MONOXIDE, RADON.

Air Pollution Treatment   the control of the POLLUTANTS found in outdoor air (see INDOOR AIR POLLUTION for details on the purification of indoor air, and POLLUTION TREATMENT for general information on the removal of contaminants from air or water). Gases and vapors can be removed from air through ABSORPTION, ADSORPTION, INCINERATION and catalytic combustion (see CATALYTIC CONVERTER), and particulates can be removed through the use of filters (see BAGHOUSE), through absorption (see SCRUBBER), through ionization in an ELECTROSTATIC PRECIPITATOR and in a centrifugal cleaner (see PARTICULATE REMOVAL).

Pollution-control devices that remove particulates and soluble gases from industrial smokestacks from air through *absorption* in a liquid are called scrubbers. *Adsorption* is used primarily to remove very small molecules such as gases and vapors that cause odors or that are potentially toxic, flammable, explosive or corrosive, to reclaim SOLVENTS from the industrial gaseous waste (see INDUSTRIAL AIR POLLUTION) and to remove volatile organic compounds (VOCS) from indoor air. Small hydrocarbon molecules are removed through adsorption. In *catalytic combustion*, exhaust gases are passed through a bed of catalytic material (see CATALYST), which can reduce the temperature at which substances in the waste stream burn by as much as 500°F below what would be required for incineration, thereby considerably reducing the cost of disposal while increasing combustion efficiency to as high as 98%. Platinum alloys are widely used as a catalyst because they produce the lowest available catalytic ignition point. Catalytic combustion is used to dispose of wastes from oil refineries, chemical and plastics production facilities and plants that produce fish oil and animal fat. Particulates such as FLY ASH, inorganic solids and vaporized metals can lead to fast deterioration of the catalyst and excessive maintenance and replacement costs. See also AIR FILTERS, AUTOMOTIVE AIR POLLUTION, CLEAN AIR ACT, HEPA, ION EXCHANGE, POLLUTION.

Air Quality Classes   an air-quality rating system established in the 1970 amendments to the CLEAN AIR ACT. Under the system, the United States is divided into regions according to AIRSHED. Each airshed is assigned to one of three classes. In Class 1 airsheds, which include national parks, wilderness areas and other pristine regions, no deterioration of air quality below December, 1974, levels is allowed. In Class 2 airsheds, which are regulated by the states, some deterioration is allowed. In Class 3 airsheds, which are primarily industrial regions and large urban areas, AMBIENT AIR quality standards are ignored as long as individual pollutants do not exceed standards set by Congress or the ENVIRONMENTAL PROTECTION AGENCY. See also AIR POLLUTION.

Air Quality Control Region   an AIRSHED containing two or more cities, counties or other government units. An Air Quality Control Region is designated by the federal secretary of Health and Human Services and is defined in the CLEAN AIR ACT. Local governments in the airshed are required to adopt a consistent air-pollution control strategy that must be enforced by the state or states that are at least partially in the region. If state enforcement is deemed inadequate, the ENVIRONMENTAL PROTECTION AGENCY is authorized to set and enforce standards for the airshed. More than 200 Air Quality Control Regions have been established. See also AIR POLLUTION, AIR POLLUTION TREATMENT.

Air Stripping   a process used to remove ammonia from wastewater at SEWAGE TREATMENT plants. Most of the NITROGEN found in domestic wastewater comes either in the form of AMMONIA (NH_3) or in another organic form that gets converted to ammonia during the treatment process. The overall nitrogen content of the water can be reduced either before or after SECONDARY TREATMENT by passing process water through an air-stripping tower (see WASTEWATER TREATMENT). Before passing through the stripping tower, the wastewater must be low in carbon and have a healthy population of nitrifying bacteria. Between the pH of 6 and 8, most of the ammonia found in sewage comes in the form of positive nitrate ions. All the nitrogen can be converted into a form of ammonia that is removable through agitation by raising the pH of process water above 10. The vigorous agitation necessary to drive the ammonia out of solution can be provided by running wastewater through a forced-draft, counter-current stripping tower. See also AIR POLLUTION TREATMENT, DRINKING WATER.

Air-to-Air Heat Exchanger   see HEAT-RECOVERY VENTILATOR.

Airshed   an area that, due to prevailing winds and topography, generally shares the same air and air currents. Contaminants emitted in one part of an airshed will typically pollute other parts of the airshed, and will be carried into other airsheds only sporadically. See also AIR POLLUTION, AIR POLLUTION CONTROL, AMBIENT QUALITY STANDARDS.

Alachlor   an herbicide used to kill annual grasses and broad-leaved weeds. Alachlor is the mostly widely used herbicide in the United States, with approximately 85 million pounds applied annually, primarily to corn and soybean crops. It is produced by Monsanto under the tradename *Lasso*. It is a cream-colored or white solid that is slightly soluble in water. High concentrations of alachlor have been found in GROUNDWATER in nine states, primarily in the corn belt. The herbicide is especially likely to leach out of porous soils (see PERCOLATION) and soils with a few microorganisms. In healthy soils, alachlor is broken down by microorganisms within about two months of application. Alachlor is a proven animal CARCINOGEN and a probable human carcinogen. Alachlor was reclassified by the ENVIRONMENTAL PROTECTION AGENCY in 1987 as a restricted-use pesticide. Aerial application of alachlor is prohibited, and it can not be used east of the Mississippi except on potatoes. Alachlor is banned in Kern County, California, and in Massachusetts. It was banned in Canada after it was found in Ontario's drinking water. PUBLIC WATER SYSTEMS are required under the SAFE DRINKING WATER ACT to monitor concentrations of alachlor (MAXIMUM CONTAMINANT LEVEL = 0.002 milligram per liter). The BEST AVAILABLE TECHNOLOGY for the removal of alachlor is granular ACTIVATED CARBON.

Alcohol   an ORGANIC COMPOUND in which one or more hydrogen atoms have been replaced by a hydroxyl group (OH-), with a hydrogen atom or a HYDROCARBON radical at the other end of the molecule. Most alcohols are clear, flammable liquids (see FLAMMABLE LIQUID) with a distinctive odor and a strong taste. Alcohols are used as a SOLVENT for many substances. Production of ethyl alcohol (see ETHANOL) and methyl alcohol (see METHANOL) as motor fuels has increased exponentially in recent years. Their use as fuel additives is mandated for cities that fail federal standards for airborne CARBON MONOXIDE mandated by the CLEAN AIR ACT. To one degree or another, all alcohols are poisonous. Ethyl alcohol is more poisonous than equivalent doses of methyl alcohol. However, the body is capable of metabolizing small amounts of ethyl alcohol into CARBON DIOXIDE and water, while the metabolism of methyl alcohol produces toxic formic acid. ETHYLENE GLYCOL—also called *dihydric alcohol*—has two hydroxyl groups replacing hydrogen atoms. *Glycerin* is trihydric: It has three hydroxyls per molecule. Branched-chain alcohols have hydrocarbon radicals at each side of the carbon chain instead of at one side, as is the case with ethylene glycol or glycerin. *Isopropyl alcohol*, or rubbing alcohol, is of the branched-chain type. Its molecular formula, C_3H_7OH, is the same as that of normal propyl alcohol.

Aldehydes a class of ORGANIC COMPOUNDS formed when alcohols are partially oxidized (see OXIDATION). The most familiar aldehyde is FORMALDEHYDE, which is obtained from methyl alcohol. See also ACETALDE-HYDE.

Aldicarb a carbamate pesticide (see CARBAMATE, PESTICIDE) that is used on potatoes, peanuts, sugar beets, citrus crops and especially cotton. Aldicarb is manufactured by Union Carbide under the tradename *Temik*. It is applied below the soil's surface in granular form. Compared to other carbamates it is relatively water soluble, and gets into water from its use on farms and from spills and industrial discharges. It has been found in GROUNDWATER in most parts of the United States. Groundwater contamination is most acute in areas with sandy, acidic soils and in moist, warm climates.

Aldicarb is one of the most toxic pesticides now in use. It damages human and animal nervous systems. The ACUTE EFFECTS of human exposure to aldicarb include dizziness, weakness, stomach cramps, diarrhea, sweating, nausea and vomiting. A U.S. Centers for Disease Control study revealed immune-system changes in 23 women suffering aldicarb poisoning. It is highly poisonous to honey bees, certain freshwater fish, invertebrates and birds.

Aldicarb is banned in Wisconsin, Rhode Island, New York and Florida, but is not regulated nationally. The ENVIRONMENTAL PROTECTION AGENCY recommended the pesticide not be used on potatoes or imported bananas after concluding that it poses an unreasonable risk to children's health. PUBLIC WATER SYSTEMS are required under the SAFE DRINKING WATER ACT to monitor concentrations of aldicarb (proposed MAXIMUM CONTAMINANT LEVEL [MCL] = 0.007 milligram per liter). Aldicarb sulfoxide (proposed MCL = 0.007 mg/L) and aldicarb sulfone (MCL and MCLG = 0.007 mg/L) must also be monitored in public water systems. The total allowable concentration of two or more of the three regulated forms of aldicarb is 0.009 mg/L. The BEST AVAILABLE TECHNOLOGY for the removal of aldicarb, aldicarb sulfone and aldicarb sulfoxide is granular ACTIVATED CARBON.

Aldrin an organochlorine INSECTICIDE (see ORGANO-CHLORIDE) used primarily to control termites. Aldrin gets into air and water from improper application, industrial discharges, urban RUNOFF and spills. It is slightly soluble and highly persistent in water, with a HALF-LIFE of more than 200 days. It has high acute and chronic toxicity (see ACUTE EFFECTS, CHRONIC EFFECTS, TOXICITY) to aquatic life and high acute toxicity to birds. Aldrin has a tendency to bioaccumulate in aquatic organisms. The concentration of aldrin found in fish tissues is usually much higher than the average concentration of aldrin in the water from which the fish was taken. Aldrin is oxidized (see OXIDATION, OXIDIZING AGENT) to the highly persistent insecticide dieldrin when sufficient oxygen is present.

Aldrin can affect health when inhaled or by passing through the skin. The acute health effects of exposure to aldrin can include headaches, dizziness, nausea and vomiting, muscle jerks, severe seizures and, in extreme cases, death. Aldrin has been shown to cause liver cancer in animals and is considered a possible human CARCINOGEN. Experiments with animals have proven aldrin to be a TERATOGEN, and it is rated as a possible human teratogen. Aldrin may also decrease human fertility. Prolonged exposure may cause liver damage.

The ENVIRONMENTAL PROTECTION AGENCY (EPA) restricted the use of aldrin and the related insecticide dieldrin in 1974. The OCCUPATIONAL SAFETY AND HEALTH ADMINISTRATION (OSHA) has established an airborne permissible exposure limit for aldrin of 0.25 milligrams per cubic meter, averaged over an eight-hour workshift. It is also cited in AMERICAN CONFERENCE OF GOVERNMENT INDUSTRIAL HYGIENISTS, NATIONAL INSTITUTE OF OCCUPATIONAL SAFETY AND HEALTH and Department of Transportation and EPA regulations. Aldrin is on the Hazardous Substances List and the Special Health Hazard Substance List. Businesses handling significant quantities of aldrin must disclose use and releases of the chemical under the provisions of the EMERGENCY PLANNING AND COMMUNITY RIGHT TO KNOW ACT.

Algae a group of organisms that contain chlorophyll and carry out photosynthesis. Algae were once classified as plants, but the approximately 25,000 known varieties are now grouped in eight different divisions of two biological kingdoms, *monera* and *protista*. Most algae consist of a single cell, although multicellular kelp or seaweed, which are forms of algae, may grow to several hundred feet in length. Algae are aquatic, except for those species that combine with fungi to form lichens, and are found in fresh and salt water. Algae, primarily ocean phytoplankton, create more than 70% of the oxygen in the earth's atmosphere. They thrive everywhere from high mountain snow fields to hot springs with 160°F (70°C) water. Algae are used in the BIOLOGICAL TREATMENT of SEWAGE (see

SEWAGE TREATMENT) and industrial wastewater (see INDUSTRIAL LIQUID WASTE, WATER TREATMENT).

BLUE-GREEN ALGAE (Cyanophyta) is the simplest form of algae, and the one primarily responsible for ALGAL BLOOMS in polluted water. Other types of algae include green algae (Chlorophytes); brown algae (Phaeophytes), which include seaweed such as kelp; red algae (Rodophytes—see RED TIDE); golden algae (Chrysophytes), which include diatoms; yellow algae (Xanthophytes); dinoflagellates (Pyrrophytes); euglenoids (Euglenophyta). See also EUTROPHICATION, MICROORGANISMS, THERMAL POLLUTION, TRICKLING FILTER.

Aliphatic Hydrocarbon   a class of ORGANIC COMPOUNDS characterized by carbon-to-carbon links that form an open chain structure. Aliphatic hydrocarbons have no BENZENE RING or similar aromatic ring in molecular structure. Paraffin, olefin and acetylene HYDROCARBONS and the derivatives are the principal aliphatic hydrocarbons. See also DISTILLATION, PETROLEUM.

Alkalinity   the ability of water or other liquids to accept protons—also called basicity. The balance of alkalinity and ACIDITY in wastewater (see PH) is a critical factor in the treatment of sewage and industrial waste (see SEWAGE TREATMENT). About a half a pound of alkalinity is removed for each pound of BIOCHEMICAL OXYGEN DEMAND removed from wastewater. Alkalinity in wastewater is primarily the result of hydroxide ions (OH–), hydrogen carbonate ions (HCO_3–), and carbonate ions (CO_3––).

Allergen   any substance that causes an allergic reaction in an organism. Common allergens include POLLENS, mold spores, algae, animal dander, foods and drugs. An allergic reaction is triggered when the body's immune system mistakes an allergen for a PATHOGEN and produces antibodies to fight off the invader. The antibodies, in turn, react with the allergen to form *histamines*, which cause the localized swelling of tissue associated with allergies. Rashes and constricted airways result from the swelling. Allergens can also interfere with the absorption of food and water, leading to diarrhea. Asthma and hay fever are allergic reactions of the respiratory system. Humidifier fever and hypersensitivity pneumonitis are influenza-like allergic reactions. An individual with multiple chemical sensitivities can have severe reactions to allergens and environmental toxins that would have no effect on a healthy individual. Sensitivity to specific allergens varies widely among individuals. Most allergens are proteins. While most people are allergic to poison oak, for instance, relatively few are allergic to dust-mite feces. See also AIR POLLUTION, INDOOR AIR POLLUTION, MICROORGANISMS.

Alpha Particles   particles consisting of two protons and two neutrons that are emitted by radioactive decay (see RADIOACTIVITY, RADIONUCLIDE). Alpha particles—also called *alpha rays*—are essentially the nucleus of a helium atom that has been stripped of its electrons resulting in a double positive charge, although they are not necessarily derived from helium. Alpha particles move at about $\frac{1}{10}$ the speed of sound and have a low penetrating ability because of their relatively slow speed and high mass—they can be stopped by a sheet of paper. Alpha particles can cause flesh burns and can ionize air molecules (see ATMOSPHERE, MOLECULE). Alpha radiation tends to accumulate in the bones and exposure can lead to bone cancer. PUBLIC WATER SYSTEMS are required under the SAFE DRINKING WATER ACT to monitor concentrations of alpha particles (MAXIMUM CONTAMINANT LEVEL = 15 picocuries per liter (pCi/L). See also ION, IONIZING RADIATION. Compare BETA PARTICLES, GAMMA RAYS, NUCLEAR REACTOR, RADIOACTIVE WASTE.

Alternative Motor Fuels Act   a law passed by Congress in 1988 that directed the Department of Energy to establish a panel to devise a national strategy for the increased use of alternatives to gasoline and diesel as fuel. Methanol, ethanol, natural gas, liquefied petroleum gas and electricity were among the fuels to be considered. The law was passed primarily because of the role of petroleum imports in the U.S. trade deficit. Alternative fuels requirements in the 1990 amendments to the CLEAN AIR ACT superseded the act.

Aluminum   a silver-colored element that is the most abundant metal in the crust of the earth; the only elements more abundant than aluminum are oxygen and silicon. The mining and smelting of aluminum are sources of AIR POLLUTION and WATER POLLUTION. Almost 6.8 million metric tons of aluminum were produced by American businesses in 1992, with a little more than 4 million metric tons produced from virgin ore, and the other almost 2.8 million tons coming from recycled materials. The U.S. Can Manufacturer's Institute reported that a little more than 63% of aluminum beverage cans were recycled in 1993.

Aluminum is a very reactive material, and it never occurs as a free metal. When aluminum is exposed to air, a tough, clear oxide that resists further corrosive action quickly forms. Tarnish and rust can not penetrate the tough oxide coating, which is why the metal does not turn dark when exposed to the elements.

Aluminum is valued because of its strength, durability and light weight, its resistance to weathering and its conductivity of both electricity and heat. Gemstones such as the ruby and the sapphire consist primarily of crystalline aluminum oxide. Alum is a colorless aluminum salt, which is also called *potassium alum*. Alum and aluminum sulfate are similar and are used in similar ways. Alum is used as a FLOCCULATION agent in the purification of wastewater, and in the manufacture of textiles, baking powders and medicines. Alum is a powerful astringent, and the styptic pencils used to stop bleeding from small cuts are often pure alum. Aluminum compounds have been shown to cause epilepsy and other nerve disorders in animal tests. There is limited evidence of a link between Alzheimer's disease and the use of aluminum. Aluminum compounds are associated with the onset of fatal dementia in dialysis patients. Aluminum silicate is the most common compound of aluminum, but it is chemically difficult and therefore expensive to extract pure aluminum from the compound, and bauxite, an impure hydrogenated aluminum oxide, is used instead as the principal aluminum ore. Aluminum is obtained from bauxite by dissolving the aluminum oxide in the ore in fused cryolite, and then using electrolytic action to extract the aluminum. The process produces 99.5% pure aluminum, and further processing can produce aluminum that is 99.99% pure.

Properties and Uses

Aluminum has several unique properties that make it valuable:

- A given volume of aluminum weighs only one-third as much as STEEL. Only lithium, beryllium and magnesium are lighter.
- It is highly reflective of both light and heat.
- It can produce strong, lightweight alloys that are strong enough to be used as armor on tanks.

Because of their strength and light weight, aluminum alloys are used extensively in aircraft and to a somewhat lesser degree in automobiles and trains. The energy invested in aluminum parts such as aluminum pistons is repaid many times over the life of the vehicle in the form of fuel savings. Aluminum is widely used in boat hulls because of its resistance to saltwater corrosion and decay. It is also used in electrical wire. An aluminum wire conducts only 63% as much electricity as does one made of copper, but it weighs only half as much. As a result, an aluminum wire will be thicker but lighter than a copper wire that carries the same amount of current, an important factor in the design of large-scale electrical transmission lines. Aluminum wire was once widely used in the United States to wire

mobile homes, but numerous fires resulting from poor connections between electrical wires put an end to the practice.

Aluminum foil is used extensively in PACKAGING and for wrapping food and other items. Household aluminum foil is typically about 0.007 inches thick. Aluminum is also used extensively in containers and packaging. Composite aluminum/plastic packaging (see COMPOSITE PACKAGING) can be next to impossible to recycle, but aluminum cans are one of the most recycled items (see METAL RECYCLING, RECYCLING). The amount of aluminum used to produce a beverage container has declined markedly over the years due to a process called LIGHTWEIGHTING.

A shiny aluminum surface reflects virtually the entire electromagnetic spectrum including both visible light and infrared radiation. The foil backing on batts of building wall and floor insulation reflects infrared radiation, helping to reduce the flow of heat. Aluminum parts in heat-producing appliances similarly help to control the flow of infrared radiation. Because of its excellent conductivity of heat, aluminum is used in the production of pots and pans, heat exchangers and solar heat collectors. Aluminum products are widely used architecturally and ornamentally because of its bright color and its resistance to weathering. As aluminum is cooled it actually becomes stronger; nor does it get brittle as do many metals. It is therefore widely employed in cryogenics. It is also used in the production of cosmetics and deodorants. Anhydrous aluminum chloride is a salt that is used in the production of petroleum products and synthetic chemicals.

History

The process used to produce aluminum was independently and almost simultaneously discovered in 1886 by Charles Martin Hall in the United States and Paul L. T. Héroult in France, and is called the Hall-Héroult process. A large amount of electricity is needed to produce aluminum, the reason for its relatively high cost. Aluminum refineries are located in areas with low electricity prices, such as the western United States and Canada with their cheap hydropower, and ores are often shipped from many thousands of miles away for processing. Most of the bauxite processed in the Pacific Northwest, for instance, is mined in South America. Aluminum plants and hydroelectric dams were built along the Columbia River by the government during the Second World War to drastically increase America's production of the metal, the primary ingredient of the aircraft that were seen as essential to the war effort. Many analysts say that the output of the northwest's aluminum smelters may have made a critical difference, especially in the Pacific, where aircraft were essential. The plants were sold to private industry at Army-Navy store prices after the war. Rising electricity prices and fluctuating world aluminum prices have strained the industry in the Northwest and other parts of the world in recent years.

Aluminum was removed from the list of PRIORITY POLLUTANTS monitored under the SAFE DRINKING WATER ACT in 1988. A secondary standard of 0.05 milligram per liter has been proposed.

Ambient Air   the air that is present throughout a vicinity, as opposed to that at any single location. See AMBIENT QUALITY STANDARDS.

Ambient Quality Standards   the air quality standards in effect for an entire locality. Ambient air standards specify the level of AIR POLLUTION to be tolerated area-wide, not the concentrations found at particular spots. Ambient air quality is measured by taking spot samples from numerous locations, then averaging the results. See also AIR QUALITY CLASSES, CLEAN AIR ACT.

American Conference of Governmental Industrial Hygienists (ACGIH)   a private, not-for-profit association that focuses on the administrative and technical aspects of worker health and indoor air quality. The ACGIH publishes *Threshold Limit Values of Airborne Contaminants*, a regularly updated listing of acceptable limits for contaminants commonly found in the workplace. See THRESHOLD LIMIT VALUE.

Address: 6500 Glenway Ave., Bldg D-7, Cincinnati, Ohio 44311. Phone: 513-661-7881.

American Industrial Hygiene Association   an international association of professionals in the occupational safety and health field with more than 11,000 members and 78 local chapters. The American Industrial Hygiene Association (AIHA), founded in 1939, publishes pamphlets, technical papers and books on occupational health, environmental toxins and indoor air quality. More than 30 technical committees set up by the AIHA monitor scientific research relating to the workplace environment. American Board of Industrial Hygiene was established by the AIHA in 1961 to set industrial health and safety standards and to certify professional hygienists. The AIHA also certifies laboratories and sponsors meetings.

Address: 2700 Prosperity Dr., Suite 250, Fairfax, VA 22031. Phone: 703-849-8888.

American Ref-Fuel   a general partnership between Air Products and Browning Ferris Industries formed in 1983 to design, build, own and operate municipal garbage incinerators.

Address: 777 N Eldridge, Houston, TX 77079. Phone: 800-727-3835.

American Society of Heating, Refrigerating and Air-Conditioning Engineers (ASHRAE)   an association of more than 50,000 engineers and other professionals involved with heating, refrigeration, air conditioning, indoor air quality and ventilation. ASHRAE sponsors research, organizes meetings and publishes technical data relating to healthful, comfortable indoor environments. The society also develops standards relating to heating, cooling, human comfort and indoor air quality. ASHRAE Standard 62-1989 significantly increased recommended VENTILATION rates in all types of buildings.

Address: ASHRAE, 1791 Tullie Circle, NE, Atlanta, GA 30329. Phone 404-636-8400.

American Water Works Association   a trade association of public water supply professionals formed in 1881. Members of the American Water Works Association (AWWA) include engineers, chemists, bacteriologists, water utilities and water system management, manufacturers and distributors of water treatment equipment and water system components, government officials, students and individuals interested in DRINKING WATER. The AWWA sponsors conferences, and publishes books, technical papers and a variety of periodicals. Address: 666 W Quincy Ave, Denver, CO 80235. Phone: 303-794-7711.

Ammonia   an alkaline compound of HYDROGEN and NITROGEN that occurs naturally as a clear, lighter-than-air gas with a sharp, penetrating odor. About 80% of the 15 million tons of ammonia (NH_3) produced annually in the U.S. becomes fertilizer. Ammonia is also used as a refrigerant, a SOLVENT and in the production of household cleansers, PLASTICS, resins, PAPER, animal feed, rubber and explosives. Ammonium hydroxide, ammonium bicarbonate, ammonium carbonate and ammonium phosphate are widely used compounds containing ammonia. It is produced commercially from natural gas or coke, but natural processes produce most ammonia. The decomposition of organic material produces ammonia, and it is a byproduct of metabolism in most animals. Ammonia is excreted as urea or uric acid, either of which decompose immediately upon release into the environment, and release ammonia gas. Levels of manufactured ammonia exceed natural background levels only in isolated instances such as near a feed lot or chemical factory or as the result of a spill. At high concentrations, ammonia gas is explosive. It liquefies under pressure and is usually transported in liquid form. Pure liquid ammonia is called *anhydrous ammonia.*

Ammonia is very soluble in water, and is usually found in landfill LEACHATE. It is often found in GROUNDWATER, with highest concentrations found in shallow AQUIFERS underlying fertilized farmland and feedlots. Ammonia is removed from water by AIR STRIPPING. The metabolism of chemolithotrophic microorganisms oxidizes ammonia to produce NITRATE.

Ammonia was called "spirit of hartshorn" during the Middle Ages. Ammonia gas was obtained at the time by heating the horns and hooves of oxen. Ammonia gas was employed as a chemical warfare agent during World War I.

Ammonia is sometimes mixed with bleach to produce a more powerful cleaning solution, but the practice is dangerous because it produces chloramines, which are more dangerous than either ammonia or bleach fumes alone. Mixing ammonia cleansers with lye or sodium hydroxide can release toxic quantities of ammonia gas. Ammonia can get into indoor air from cleansers and from blueprint machines.

Exposure to ammonia can cause a burning sensation to the skin, eyes, mouth, throat and respiratory system. Pain in the lungs, headache, nausea, tearing and an increased respiration rate are symptoms of exposure. Pulmonary endema, pneumonia and permanent structural changes in the lungs can follow severe exposure. Water containing dissolved ammonia is highly alkaline and can cause caustic burns. Ingestion can cause corrosive effects to the mouth, esophagus and stomach. Ammonia can break down in water, releasing nitrogen in the process. Excess nitrogen can lead to EUTROPHICATION. Although ammonia is a base, it makes soils more acidic because it is quickly converted to nitrate (NO_3), which release hydrogen ions just like other acids. The ENVIRONMENTAL PROTECTION AGENCY has set a limit of 35 milligrams per liter for aquatic ammonia for the protection of human health and 0.02 mg/L for the protection of aquatic life. The OCCUPATIONAL SAFETY AND HEALTH ADMINISTRATION has established a limit for airborne ammonia in the workplace of 25 parts per million. The Food and Drug Administration sets limits for ammonia compounds found in food. Businesses handling significant quantities of ammonia, ammonium nitrate or ammonium sulfate must disclose use and releases of the chemical under the provisions of the EMERGENCY PLANNING AND COMMUNITY RIGHT TO KNOW ACT. Ammonia is on the Hazardous Substances List.

Anaerobic Bacteria   bacteria that do not require oxygen for growth. Anaerobic bacteria can use SULFUR or NITROGEN instead of oxygen to initiate the chemical conversions essential to METABOLISM. HYDROGEN SULFIDE (H_2S), METHANE and AMMONIA are the principal metabolic byproducts rather than the WATER and CARBON DIOXIDE produced by oxygen-based metabolism. Many anaerobic bacteria are pathogens that thrive in puncture wounds and in the bladder, kidneys and lower intestinal tract.

Peat is a metabolic byproduct of anaerobic bacteria. Anaerobic bacteria are used in the treatment of SEWAGE, SEWAGE SLUDGE, LEACHATE and HAZARDOUS WASTE and the production of methane (see ANAEROBIC DIGESTER). *Aerotolerant anaerobes* can metabolize in the presence of oxygen, although they generally do better without it. *Facultative anaerobes* can use oxygen for metabolism, and can also metabolize without it, but they tend to function best when oxygen is present. *Obligative anaerobes,* also called *strict anaerobes,* die when exposed to oxygen. Anaerobic bacteria thrive on the oxygen-starved beds of eutrophic lakes and streams where they break down organic wastes to produce methane, also called *marsh gas* or *swamp gas.* (Compare DECAY.) See also AERATION, PUTREFACTION, SEPTIC SYSTEM.

Anaerobic Digester   a sealed vessel in which anaerobic bacteria break down organic material to produce methane gas, also called *methane digester.* SEWAGE SLUDGE is treated in anaerobic digesters to remove odors and to condition the sludge for disposal. Bacterial action on the organics in sludge first produces acids, which are, in turn, consumed by bacteria that produce METHANE (*methanobacteria*). The dark, crumbly sludge that is produced by anaerobic action is full of organic material, is sweet-smelling, and contains lots of nitrogen—ideal for use as a fertilizer. Maintaining the balance between the acid-producing bacteria, which can tolerate a wide range of conditions, and the methane-producing bacteria, which cannot, is a delicate operation (much like avoiding indigestion on Thanksgiving). If the PH of the soup inside the digester gets too low or too high, or if the temperature of the whole mix strays outside a relatively narrow range, the methanobacteria will die off, and the digester will produce a malodorous, highly acidic sludge. The bacteria are happiest at about human body temperature (98°F/35°C), so digesters must be heated for best results. Methane already produced by the digester is usually burned under the tank to keep it at the desired temperature. As the digestion process proceeds, the sludge separates into a liquid constituent, called *supernatant,* and solid sludge. The supernatant is normally drawn off and run through the digester again.

Methane digesters are used on countless farms to produce biogas from animal feces, solving what can be a severe waste disposal problem at a feedlot or other operation where many animals are confined in a small space. Such digesters typically consist of a round steel or concrete tank with a cover that floats on top of the sludge to exclude air. Methane gas is extracted through a gas dome in the center of the lid. The larger, more elaborate digesters used in sewage treatment plants often employ mechanical paddles to stir the sludge and thereby mix bacteria more thoroughly and speed up the decomposition process. Mixing is also sometimes achieved by bubbling gases up through the tank.

Anaerobic Stream a stream with a dissolved oxygen content of zero. Anaerobic streams result from the introduction of too much BIOCHEMICAL OXYGEN DEMAND, which in turn leads to the oxygen-starved condition. Anaerobic bacteria thrive under such conditions, and the AMMONIA and HYDROGEN SULFIDE emitted by their metabolism can permeate the air in the area of the stream. The water in an anaerobic stream may appear to be black as a result of floating and suspended sludge made up primarily of dead anaerobic bacteria. The only way of clearing up these conditions is the return of dissolved oxygen to the stream's water. See also NUTRIENT, SEDIMENTATION, THERMAL POLLUTION, WATER POLLUTION.

Animal Wastes wastes produced by animals, including their bodies after death. The volume of animal waste was enormous in urban areas before the coming of the automobile. In New York City, for instance, 15,000 dead horses were removed from city streets every year before the turn of the century. Bones, bone fragments and shells of animals and shellfish are among the principal pieces of evidence examined by archaeologists studying ancient cultures. In the wild, animal wastes maintain the soil's fertility and are a source of nutrition for life forms from microorganisms to scavengers. The waste of domesticated animals can foster the spread of pathogenic microorganisms, and are a source of WATER POLLUTION (see EUTROPHICATION, NUTRIENTS, RANGE MANAGEMENT). The largest volume of waste from domestic livestock comes in the form of FECES and urine, often mixed with bedding and straw. When many animals are kept in a small space such as a feedlot or a stable, the volume of excrement to be disposed of can be quite large. Such barnyard waste is sometimes spread on fields as FERTILIZER; COMPOSTING and processing in an ANAEROBIC DIGESTER are also used to reduce the volume—and to produce a higher-quality fertilizer in the process. Surface runoff from a feedlot or other area with a high concentration of feces and urine can be highly contaminated with

nutrients, and is a major source of diffuse water pollution (see DIFFUSE POLLUTION). The bodies of animals that die on the farm are generally either buried or taken to a rendering plant. Internal organs, blood, bones and other unused body parts are slaughterhouse waste products. The meat processing industry is a major source of infectious waste. See also FOOD SCRAPS, ZOODOO.

Appliances household equipment such as stoves, refrigerators, dish washers, clothes washers and dryers and microwave ovens that are designed to perform routine household tasks—collectively called *white goods* in the scrap metal industry. The fate of used appliances offers an interesting portrait of reuse in action. Researchers at the GARBAGE PROJECT had noticed that very few appliances and pieces of furniture were being found in their excavations at Tucson's landfill, although demographic data suggested that an average of 65 major appliances would be discarded every day in a city of Tucson's size. Two researchers spending a week at the dump and recording every major appliance and large piece of furniture that arrived. As it turned out, very few of the items they were looking for did, in fact, materialize, and every large appliance or stick of furniture that was dumped was carted off within hours by other dump patrons.

An archaeologist at the University of Arizona decided to follow up on this surprising data by working with a group of anthropology students in a project they dubbed the Reuse Project to try to find out what was happening to the used furniture and appliances if they weren't being dumped. They selected a random sample of Tucson's households and conducted a survey of 184 homes, asking residents whether they had recently disposed of any of the following items: a washer or dryer, a refrigerator, a stove, a stereo, a television, a couch or an armchair, a kitchen or dining room table or accompanying chairs, a dresser, a bookcase or a bed. The researchers found that 743 of the mentioned items had been replaced in the homes in their sample, and that of these: 30% had been kept in the house, 29% had been sold or given to friends or relatives, 34% had been given or sold to stores or people they didn't know and only 46 of the 743 items replaced—or a little more than 6%—had been thrown away. Further research revealed that much of what was thrown away didn't make it to the dump because enterprising individuals cruised the alleys in pickup trucks before the GARBAGE TRUCK was scheduled to come and hauled them away, presumably for reuse or resale.

Aquifer a geologic formation under the ground's surface that is capable of holding water (see GROUNDWA-

TER). Sand and gravel of glacial or alluvial origin are the most common materials found in an aquifer, although fractured rock also has the capacity to hold groundwater. Many aquifers are continually recharged by the PERCOLATION of water through overlying soil, although the rate of the withdrawal of water through wells for agricultural, industrial and domestic use frequently exceeds the rate of natural recharge. Others, such as the OGALLALA AQUIFER, which underlies a large part of the Great Plains, have been largely or entirely shut off from recharge by geologic action, in which case the groundwater they contain is termed *fossil water*. Most groundwater use is from relatively shallow unconfined aquifers, which are not shielded from the penetration of surface water by an impenetrable layer of clay above the water-bearing formation. Such an aquifer is like a sponge. Precipitation soaks into the ground quickly to recharge the aquifer, but contaminants in the water can also easily be carried along.

Groundwater overdraft can lead to saline intrusion, which results when a water table in an aquifer near an ocean or other source of salt water falls to a level at which salt water can move in to replace it. As the water table falls, the cost of pumping water out of an aquifer increases, and it usually becomes impractical to pump it from much more than a few hundred feet, especially for irrigation. When groundwater is removed from aquifers composed of relatively fine materials, the materials compact, leading to SUBSIDENCE—the fall of the overlying ground's surface. Once the materials in the aquifer have settled together, they are unaffected by the reintroduction of groundwater, and the aquifer's capacity is permanently reduced. Between Tucson and Phoenix in Arizona the water table has fallen 15 feet as the result of subsidence, and cracks caused by the uneven settling of the ground's surface run for miles across the desert. If current rates of subsidence continue in Houston, the top of a 45-story building will be under water by the beginning of the 23rd century. The vulnerability of aquifers to pollution depends on patterns of groundwater flow and the makeup of overlying soils. Aquifers are increasingly being used to store water for later use in PUBLIC WATER SYSTEMS. Water that has been recycled from sewage effluent (see EFFLUENT, SEWAGE, WATER RECYCLING) has been stored in aquifers, primarily in California, and has also been used to stop saline intrusion. See also AQUIFER SENSITIVITY, DRINKING WATER, GROUNDWATER POLLUTION, IRRIGATION WATER, PERCHED AQUIFER, PROCESS WATER.

Aquifer Sensitivity   the vulnerability of an aquifer to contamination. The nature of the material between the top of the WATER TABLE and the ground's surface is the most important factor in aquifer sensitivity. Impermeable clay layers will protect an aquifer from pollutants released on the surface, but will also block the aquifer recharge. A soil composed of fine materials that allow the slow passage of water down to the water table will filter out many contaminants before they become part of the groundwater. Coarse sand and gravel will allow the quick penetration of poorly filtered water from the surface into an aquifer, greatly increasing the likelihood of contamination (see PERCOLATION, PERMEABILITY, POROSITY). The presence of major sources of pollution in an aquifer's recharge area also increases its vulnerability to pollution. The amount of water contained in an aquifer and the rate at which it flows also affect its sensitivity to pollution. Aquifer sensitivity is expressed as a number between 1 and 9, with aquifers that are not sensitive to pollution rated from 1 to 3; moderately sensitive aquifers, from 4 to 6; and highly sensitive aquifers from 7 to 9.

Archaeology   (also spelled *archeology*); the scientific study of the ARTIFACTS left behind by ancient cultures. Archaeologists attempt to get an idea of religions, economies and lifestyles of ancient cultures in large part by sorting through the garbage they left behind. The classification and analysis of materials found in MIDDENS (ancient garbage piles), is an important window into the human past.

The emphasis of archaeological study has shifted since the mid-20th century from finding out how cultures change to trying to understand why they change. *Critical archaeology* assumes that modern man's attempt to understand ancient cultures by examining their artifacts is flawed because everything is seen through the lens of modern attitudes, modern research and education techniques and modern tools and materials, and therefore a complete understanding of the cultures of ancient and prehistoric people is not possible. GARBOLOGY is the use of archaeological techniques to analyze modern garbage, as pioneered by the GARBAGE PROJECT at the University of Arizona. See also BONES, FOOD SCRAPS, FORMATION PROCESS, HORIZON MARKER, STRATIGRAPHIC LAYER.

Aromatic Hydrocarbon   a HYDROCARBON with an attached BENZENE RING or similar cyclical structure. BENZENE, TOLUENE, XYLENE, NAPTHALENE and PHENOL are some of the better known aromatic hydrocarbons, which got their name because of their fragrant odor. Aromatic hydrocarbons are added to GASOLINE to increase its octane rating. When they are burned, benzene is released. See also POLYAROMATIC HYDROCARBONS.

Arsenic a naturally occurring ELEMENT that is a silver-gray, brittle, crystalline solid or a black or yellow amorphous material. It is used in fungicides, wood preservatives, growth stimulants for plants and animals, in medicines, in the making of alloys from HEAVY METALS, and in special solders, GLASS, cloth and electrical semiconductors. It gets into air and water primarily from emissions from coal-fired power plants and from its use as a PESTICIDE. Where soil is highly contaminated with arsenic (usually near smelters and farmlands where arsenic-based pesticides have been used), it can accumulate in plants to toxic levels, and plant growth and crop yields will be significantly decreased. Although arsenic has a low solubility in water, it is highly persistent once dissolved, with a HALF-LIFE of more than 200 days. If arsenic is exposed to ACIDS or acid mists, deadly arsine gas can be released.

Arsenic is a confirmed CARCINOGEN, associated with skin and lung cancer, and a suspected TERATOGEN. Some arsenic-based compounds are confirmed human teratogens. Contact with exposed skin can cause a burning, itching sensation, rashes and thickening and color changes in the skin. Arsenic can enter the bloodstream by passing through the skin or via inhalation. Brief exposure to high concentrations of arsenic or repeated exposure to lower concentrations can damage nerves (see NEUROTOXIN), with "pins and needles," numbness, and weakness of arms and legs. Other symptoms include poor appetite, nausea, stomach cramps, nose ulcers, hoarseness, or damage to the liver and blood vessels. Arsenic reduces bone marrow's ability to produce red blood cells.

The OCCUPATIONAL SAFETY AND HEALTH ADMINISTRATION has established an airborne permissible exposure limit for arsenic of 0.01 milligrams per cubic meter averaged over an eight-hour workshift. PUBLIC WATER SYSTEMS are required under the SAFE DRINKING WATER ACT to monitor concentrations of arsenic (MAXIMUM CONTAMINANT LEVEL = 0.05 milligrams per liter. A lower MCL of between 0.002 and 0.020 mg/L has been proposed, but uncertainties about the severity of arsenic's health effects have postponed promulgation of a proposed MCL). It is also cited in AMERICAN CONFERENCE OF GOVERNMENT INDUSTRIAL HYGIENISTS, NATIONAL INSTITUTE OF OCCUPATIONAL SAFETY AND HEALTH, International Atomic Regulatory Commission and Department of Transportation regulations. Businesses handling significant quantities of arsenic or arsenic compounds must disclose use and releases of the chemical under the provisions of the EMERGENCY PLANNING AND COMMUNITY RIGHT TO KNOW ACT. Arsenic is on the Hazardous Substances List and the Special Health Hazard Substance List. Inorganic arsenic compounds are on the HAZARDOUS AIR POLLU-TANTS LIST. Arsenic disulfide, arsenic pentoxide, arsenic trichloride, arsenic trioxide and arsenic trisulphide are on the Hazardous Substances List.

Artifact a man-made article from an ancient culture. Artifacts found during the excavation of ancient structures and MIDDENS are the principal clues available to archaeologists attempting to decipher prehistoric cultures. Buildings, tools, weapons and ornamental and spiritual objects are all artifacts. A practitioner of the science of GARBOLOGY would contend that an artifact need not be old to be of archaeological interest. See also ARCHAEOLOGY, RELIC.

Asbestos a naturally occurring family of fibrous minerals that is waterproof, sound absorbent and resistant to heat, friction and CORROSION. Asbestos fibers may be white, green, brown, or blue. Most commercial-grade asbestos is chrysolite or "white" asbestos, a fibrous form of the metamorphic rock *serpentine*. About three-quarters of the known world supply is found in the province of Quebec, Canada. Chrysolite is used in 95 percent of the 3,000 commercial products that contain asbestos. Worldwide asbestos production increased from 200,000 tons in 1920 to 5 million tons in the mid-1980s. A pound of asbestos can yield six miles of flexible fiber with the tensile strength of STEEL and the ability to withstand temperatures of almost 400°C. It is these properties that have made the mineral so popular. Flameproof cloth has been woven from asbestos fiber for more than 2,000 years. Thermal insulation, brake linings, vinyl floor tiles and sheets, plaster, gypsum wallboard, drywall compound, textured walls and ceilings, electrical insulation, appliances, roofing and siding are just a few of the products that have used asbestos. About 14,000 metric tons of asbestos were produced by U.S. minerals companies in 1993.

An outbreak of cancer during the 1960s and 1970s among workers exposed to very high levels of airborne asbestos fiber in ship yards and factories during World War II sparked concern about its toxicity. A report issued by the secretary of Health Education and Welfare in the late 1970s predicted that asbestos would cause two million cancer deaths over the next three decades. Based on this information, Congress initiated a massive removal of asbestos from schools in the ASBESTOS HAZARD EMERGENCY RESPONSE ACT of 1984. Subsequent research established that asbestos' toxicity been greatly overrated. This is because chrysolite, the most commonly encountered asbestos, has a much lower toxicity than other, less commonly used versions, and is dangerous only when inhaled in extremely large doses. It is possible that the widespread removal of asbestos—and the consequent dispersal of asbestos fibers that would never otherwise have gotten into indoor air—may have caused more health problems than would have resulted if the materials had simply been left alone. Although old, friable asbestos that is releasing fibers into indoor air is dangerous, sealing it in place is often a better solution than removal. Crocidolite and amosite asbestos have been linked to mesothelioma, a cancer of the lining of the lung or abdomen, which, once contracted, is always fatal. Unlike other cancers, which may take three or four decades to cause a malignancy, mesothelioma can be quick-acting. Children have died of the disease within three years of exposure. But these forms of asbestos are rarely found inside schools and homes. U.S. Gypsum Co., a subsidiary of Chicago-based USG Corporation, and the company that manufactured much of the asbestos now in place in buildings, faces an estimated $600 million in property damage suits from school districts and others trying to recover the costs of removing asbestos.

It is well established that the regular inhalation of high concentrations of asbestos can lead to asbestosis, a scarring of the lungs that causes shortness of breath and can lead to disability and death. Asbestos is also a confirmed carcinogen in humans. It has been shown to cause lung cancers (including mesothelioma) as well as stomach, colon, rectal, vocal cord and kidney cancers. It has no known acute health effects (see ACUTE EFFECTS).

PUBLIC WATER SYSTEMS are required under the SAFE DRINKING WATER ACT to monitor concentrations of asbestos (MAXIMUM CONTAMINANT LEVEL = 7 million fibers per liter). The BEST AVAILABLE TECHNOLOGY for the removal of asbestos from DRINKING WATER is coagulation-filtration, direct and diatomite filtration and corrosion control (see WATER TREATMENT). Emissions of asbestos are regulated under the CLEAN AIR ACT, and asbestos identification and removal of asbestos construction materials in schools is regulated under the TOXIC SUBSTANCES CONTROL ACT. The ENVIRONMENTAL PROTECTION AGENCY (EPA) recommends that asbestos from household construction materials be double-bagged, labeled as asbestos and disposed of in a landfill, since INCINERATION or chemical treatment are not practical. The EPA initiated a ban on the production of most products containing asbestos in 1989. A federal appeals court overturned the ban in 1992, ruling that the agency had overlooked other, less burdensome, means of controlling the atmospheric release of asbestos particles. Asbestos is on the Workplace Substance List, the Hazardous Substances List, the Special Health Hazard Substance List and the HAZARDOUS AIR POLLUTANTS LIST. The OCCUPATIONAL SAFETY AND HEALTH ADMINISTRATION has established a permissible exposure limit for asbestos fibers longer than 5 micrometers of 0.2 fibers/cc averaged over an eight-hour workshift. Asbestos is also cited in NATIONAL INSTITUTE OF OCCUPATIONAL SAFETY AND HEALTH, AMERICAN CONFERENCE OF GOVERNMENT INDUSTRIAL HYGIENISTS and National Toxicology Program regulations. Businesses handling significant quantities of friable asbestos must disclose use and releases under the provisions of the EMERGENCY PLANNING AND COMMUNITY RIGHT TO KNOW ACT. See also ASBESTOS HAZARD EMERGENCY RESPONSE ACT, MINERAL FIBERS, PARTICULATE.

Asbestos Hazard Emergency Response Act   a law passed in 1986 requiring the inspection of ASBESTOS-containing products in all public and private schools. If asbestos was found, the Act required that parents be notified of its presence and that the school develop a management plan for its removal or control. More than $10 billion has been spent on asbestos removal from schools since that time. Experts project that a comparable sum would be required to remove asbestos from other public buildings. In 1989, the ENVIRONMENTAL PROTECTION AGENCY initiated a ban that called for a 10-year phase out of all uses of asbestos.

Ash   the inorganic remains left when organic materials such as wood, coal, plant fiber or animal bones are burned. The residue of nonvolatile oxides and SALTS of METALS such as sodium, calcium, magnesium and iron; nonmetallic atoms such as silica; and pure metals such as platinum that are left behind when combustible substances such as plants and foods are thoroughly oxidized (see OXIDATION) by a wet OXIDIZING AGENT such as nitric acid are also called ash. Ash from coal power plants and garbage incinerators sometimes contains high enough levels of toxins such as DIOXIN, MERCURY, LEAD, CADMIUM and zinc that must be disposed of as toxic waste. The fine mineral particles that are blown out of vents during a volcanic eruption are *volcanic ash*. FLY ASH consists of fine particles of ash from 1 to 200 microns in diameter that are released by the COMBUSTION of FUELS such as COAL. Fly ash is carried into the atmosphere by FLUE GAS. BOTTOM ASH, which is generally less toxic than fly ash, consists of particles that are too heavy to be carried away by the turbulence in the combustion chamber and that drop into the ash pit where they periodically collected. More than 8 million tons of ash is generated each year by GARBAGE INCINERATOR.

In May 1994 the Supreme Court upheld a 1993 appeals court ruling against the city of Chicago, finding that ash from its garbage incinerator must be handled as hazardous waste because of the high, leachable levels of lead and cadmium that it contains.

The city had held that ash from its Northwest MUNICIPAL SOLID WASTE (MSW) incinerator—which burns about 350,000 tons of garbage and produces about 130,000 tons of ash each year—was exempt from regulation as hazardous waste, which is managed under the dictates of the federal RESOURCE CONSERVATION AND RECOVERY ACT (RCRA). Plaintiffs in the suit, the Environmental Defense Fund and Citizens for a Better Environment, held that the city's practice of hauling the ash to a municipal landfill was illegal and that the ash should be treated as hazardous waste because it had routinely failed lead and cadmium tests. The decision meant that cities across the country would no longer be able to dump ash from municipal garbage incinerators in landfills designed for municipal trash if the ash is above standards for hazardous waste. Cities have opposed the classification since it is much more expensive to dispose of hazardous wastes. The ruling noted that toxic incinerator ash was still being disposed of in ordinary, and in some cases unlined, sanitary landfills, and that it was being transported in open or leaking trucks or being used as landfill cover and in other applications where it could get into air and water.

Operators of waste incinerators can minimize the lead and cadmium content of ash by ensuring that materials that contain the toxic metals are not sent to incinerators. Segregating more-toxic fly ash removed from flue gases with pollution-control devices (see AIR POLLUTION TREATMENT) from less-toxic bottom ash—rather than combining the two as had usually been the case in the past—can further reduce the toxicity of ash that must be handled as toxic waste. Ash is sometimes recycled to produce concrete and masonry products. See also HAZARDOUS MATERIAL, HAZARDOUS WASTE DISPOSAL, POLLUTION.

Ashfill   a landfill built for the disposal of FLY ASH and BOTTOM ASH from an incinerator. The ENVIRONMENTAL PROTECTION AGENCY estimates that a little more than about 36% of the ash from U.S. incinerators is dumped in an ashfill. See also INCINERATION.

Asphalt   a black, sticky material that may be a solid or semi-solid at room temperature; it is the principal component of "blacktop" pavement. Asphalt occurs naturally as a residue that collects in places where crude PETROLEUM has seeped through cracks in the earth. The La Brea Tar Pits near Los Angeles are actually made up of asphalt. The ancient Babylonians used asphalt in building blocks, and its use as caulking is mentioned in the Old Testament of the Bible. About 90% of the paved roads in the United States have an asphalt surface. It is also used to make roofing shingles and waterproofing products, in the manufacture of paint and electrical insulation. Asphalt from natural deposits was used for surfacing roads and producing building materials until the refinement of petroleum became common at the beginning of the 20th century.

Asphalt from paving and shingles is often recycled and used for the production of new pavement and in the stabilization and dust-proofing of gravel roads, the first step in the *cold planing* process used to resurface asphalt pavement, a milling machine removes an existing asphalt surface to a specified depth, leaving a smooth textured surface that may be resurfaced or used as is. The *recycled asphalt pavement* removed by the carbide-tipped blades of the rotary-drum cold planing machine can be combined with new aggregate and an asphalt cement or recycling agent to produce new hot-mix asphalt. Asphalt shingles can also be used.

Atmosphere   the gaseous envelope of air and water vapor that surrounds the Earth. The atmosphere is composed primarily of NITROGEN (78%) and oxygen (21%) gas. Other gases present include argon (0.9%), CARBON DIOXIDE (0.03%), and traces of helium, krypton, neon and xenon. Half of the atmosphere's mass is within four miles of the earth. Atmospheric pressure, the weight of the gases over a given surface, is about 14.7 pounds per square inch at sea level. The HYDROLOGIC CYCLE is the primary driver of the atmosphere, and atmospheric moisture in the form of water vapor, hail, snow, clouds, sleet, rain, fog, mist and dew are the most observable parts of atmospheric weather. See AIR CURRENT, AIR MASS, ATMOSPHERIC INVERSION, GLOBAL WARMING.

Atmospheric Inversion   an atmospheric condition in which warmer air is trapped near the Earth's surface—usually in a valley—by cooler air aloft. Because cool air settles and warm air rises, there is little or no mixing of air across the boundary between the two bodies of air. Winds that would otherwise dilute and disperse AIR POLLUTION near the ground are thereby blocked, resulting in the pocket of trapped air becoming increasingly polluted as the inversion persists. Inversions are most severe in valleys that are surrounded by mountains. An inversion may start on a calm night when air near the ground cools more quickly than air aloft. In the absence of air currents to restore thermal equilibrium, a layer of cooler air will form near the ground by morning. An inversion can also be formed when a warmer AIR MASS overrides a cooler air mass. Inversions are broken up when the sun warms the ground and the cool air mass sufficiently to cause mixing, or when winds strong enough to break through the inverted layer arrive. The worst OZONE and CARBON MONOXIDE contamination occurs in areas that regularly experience atmospheric inversions. See also SMOG, PLUME.

Atomic Energy Commission   a federal agency, created by the Atomic Energy Act of 1946, that was responsible for research, promotion, development and regulation of nuclear energy. Because there was no commercial nuclear power industry for the first two decades of the Atomic Energy Commission (AEC), the agency focused on oversight of the nuclear programs of the Department of Defense and on research and development and promotion of atomic energy. The AEC spent the better part of two decades attempting to get U.S. utilities interested in building nuclear power plants, but—given the very low price of coal and oil, and the uncertainties and risks involved with the new technology—had little success. Utilities finally started to order nuclear plants in the late 1960s and early 1970s because the cost of fossil fuels and the air pollution equipment required to burn them increased significantly, and because the Arab oil embargo made supply of the fuels uncertain. Passage of the PRICE-ANDERSON ACT of 1957, which limited the damage settlements that could be awarded in the event of an accident at a nuclear facility, also contributed to the growth of the nuclear industry. In 1974 the regulatory side of the AEC was renamed the NUCLEAR REGULATORY COMMISSION, and its research and promotion functions were taken over by the newly created Energy Research and Development Administration.

Automobile   a passenger vehicle used for the land transit of people and materials powered by an internal combustion engine. A typical passenger car consists of about 70% steel and iron by weight, and more than 6% plastic. GLASS, rubber, glues and fabrics account for most of the remaining mass. See also AUTOMOTIVE AIR POLLUTION, AUTOMOTIVE PRODUCTS, BATTERY, JUNKYARD, TIRES.

Automotive Air Pollution   AIR POLLUTION that comes from vehicles' tailpipes and evaporated fuels. Emissions from automobiles are the primary source of air pollution in most urban areas. The majority of the CARBON MONOXIDE, OZONE and VOLATILE ORGANIC COMPOUNDS and much of the PARTICULATES in polluted urban air comes from vehicle exhaust. Uncombusted HYDROCARBONS and NITROGEN OXIDES produced by INTERNAL COMBUSTION ENGINES and other urban sources enter into a complex chemical interaction when exposed to sunlight, one result being elevated ozone levels (see PHOTOCHEMICAL SMOG, SMOG). Automotive fuels that evaporate as tanks are refilled and during spills and accidents are another significant

source of air pollution. California's South Coast Air Quality Management District (see CALIFORNIA AIR RESOURCES BOARD, CALIFORNIA CLEAN AIR ACT) estimates that the average gasoline-fueled automobile will, when driven 100,000 miles, emit 2,623 pounds of organic gases, 2,574 pounds of carbon monoxide and 172 pounds of nitrogen oxides. Fuels released by leaking pipelines and storage tanks (see LEAKING UNDERGROUND STORAGE TANKS) and by vehicle accidents are a major source of GROUNDWATER POLLUTION, and SURFACE WATER POLLUTION). Pollution from motor vehicles, according to ENVIRONMENTAL PROTECTION AGENCY (EPA) estimates, is responsible for about half of the cancers caused by air pollution. The average automobile built since the early 1980s emits 90% less carbon monoxide and hydrocarbons, due primarily to tighter emissions standards in the CLEAN AIR ACT, but exhaust fumes still account for 36% of the volatile organic compounds in urban air and about 45% of nitrogen oxides (NO_x), the two primary pollutants that form ozone. New automobiles are required to meet emission standards for carbon monoxide, hydrocarbons and nitrogen oxides (see CATALYTIC CONVERTER).

A standard GASOLINE engine operates at about 18% efficiency when idling and about 28% under full load. Most of the remaining 70 to 80% of gasoline's energy is lost to friction and heat, which is removed primarily through the radiator and exhaust pipe. Regular maintenance and tune-ups ensure that an engine will run as efficient as possible, and emissions testing programs require drivers to have an up-to-date emissions sticker and to keep their engine in good running order. Gasoline engines that get better mileage can reduce the amount of pollution produced per mile, even if engine efficiency stays the same. A catalytic converter reduces air pollution by converting uncombusted hydrocarbons and other automotive emissions to less toxic compounds.

Improved FUELS are another way of reducing emissions from internal combustion engines. Areas that exceed federal limits for carbon monoxide, ozone or volatile organic compounds are required under the CLEAN AIR ACT to use fuels that contain more oxygen than standard gasoline. During 1992, 58 U.S. cities began using such oxygenated fuels, gasoline to which ETHANOL, METHANOL or METHYL-TERTIARYBUTYLETHER (MTBE) is added. Compressed NATURAL GAS, which produces approximately one-third of the pollutants per vehicle mile as gasoline, already fuels an estimated 700,000 vehicles in the United States, and its use is increasing.

In 1990, the California Air Resources Board established Low-Emission Vehicle and Clean Fuels regulations that set tight emission standards for four classes of light- and medium-duty vehicles: transitional-low-emission vehicles (TLEVs), low-emission vehicles (LEVs), ultra-low-emission vehicles (ULEVs), and zero-emission vehicles (ZEVs). Beginning in 1994, auto manufacturers that sold their cars in the state were allowed to produce any combination of TLEVs, LEVs, ULEVs and ZEVs they choose, as long as the fleet average requirement is met. Beginning in 1998, 2% of each manufacturer's light-duty fleet must be ZEVs.

Electric vehicles are the only automobiles now available that emit no pollutants, but limited range, high price and lack of availability has held such cars to an experimental status to date. The overall efficiency of an electric vehicle depends on how and where the power to recharge its batteries was generated. Hydrogen powered fuel cells hold promise as another power source for zero-emission vehicles, although considerable research remains to be done before they will be available.

The EPA announced plans in April 1993 to limit the nitrogen oxide emissions of vehicles that do not travel the highways. Included were DIESEL-powered equipment such as tractors, fork lifts and heavy equipment used in road-construction. See also ALTERNATIVE MOTOR FUELS ACT, GASOHOL, MOBILE SOURCE, POLLUTION.

Automotive Products automobiles and their individual parts, accessories, maintenance products and replacement parts. Automobiles contain FIBERS, FABRICS, METAL, WOOD, GLASS, PLASTICS, PAPER and CARDBOARD (not to mention the FOOD SCRAPS, FAST-FOOD PACKAGING, JUNK MAIL and assorted TRASH that collects inside many cars). Car-care products include waxes, PAINTS, GREASE and SOLVENTS. The improper disposal of automotive BATTERIES and ENGINE OIL can lead to water pollution. Used TIRES can cause problems if they are disposed of whole in a municipal landfill, and are a significant fire hazard when stockpiled (see WINCHESTER TIRE FIRE). Automotive fuel may burn or cause an explosion in an accident, and FUELS from ruptured gas tanks can pollute SURFACE WATER and GROUNDWATER. A Garbage Project study investigation found that more automotive products are disposed of from low-income households than from middle- or upper-income homes. Wrecked and otherwise retired automobiles are stripped when they go to a JUNKYARD, valuable parts are resold and STEEL, ALUMINUM and other metals are recycled (see METAL RECYCLING). Scientists at General Motors Corp. have used PYROLYSIS to convert "car fluff," the mixture of plastics, glass, rubber, rust and fluids into gaseous and oil fuels and other waste products left over from the manufacture of automobiles into a black, powdery

material that can be used to make car parts, cement, roofing shingles and other products. See also CATALYTIC CONVERTER.

Automotive Recycler's Association an association of JUNKYARD owners founded in 1943. The Automotive Recycler's Association has 2,000 direct members and another 3,500 member companies that are part of one of 54 affiliate chapters located worldwide. The association coordinates lobbying and testimony on pending legislation that would affect its members, and serves as a clearinghouse of information about automotive recycling to the public. The ARA has been working with the auto industry in recent years on the development of vehicles that can be more easily and more efficiently recycled. See also METAL RECYCLING.

Automotive Recycling see JUNKYARD

Avoided Cost the economic, social and environmental costs that are avoided by choosing a particular product or course of action (see ENVIRONMENTAL COST, RESOURCE DEPLETION COST, SOCIAL COST). Avoided cost is applicable to the production of any product, although data on the comparative avoided costs associated with many industrial processes is scant.

Garbage and Avoided Cost

Any action that reduces the volume of garbage:

- avoids the dollar cost of collecting and disposing of waste—called *base costs*
- extends the life of the landfill, thereby putting off the expense of building a new one—called the *value of diversion*
- recovers valuable materials for reuse or recycling, thereby reducing the amount of energy and the volume of virgin materials that must be used in the production of new products
- reduces the pollution associated with the collection and disposal of trash, the cost of cleaning up environmental contamination, and of closing and maintaining a former dump site
- reduces fines and the cost of compliance with environmental regulations
- reduces the *opportunity cost* that limits future uses of land that has been used as a landfill

Social costs include the number and kind of jobs created, injuries and illness associated with trash collection, both among workers and the general public, although a few communities have used the concept of avoided cost in setting up their solid-waste disposal programs, none have included all the avoided costs associated with source reduction. A UCLA study concluded that the avoided cost of reducing the volume of garbage is at least $72.70 per ton.

Electricity and Avoided Cost

Avoided cost is used by public utility commissions to evaluate the cost electricity generated by various techniques under the terms of the PUBLIC UTILITIES REGULATORY POLICY ACT (PURPA), where it is defined as the cost of electricity produced at the most cost-effective conventional power plant available to a utility. The law requires utilities to buy power from qualifying facilities—small-scale power producers that meet requirements laid out in the act—at a rate equivalent to that of electricity that would have been generated at a new power plant. Avoided cost can reflect environmental costs through the use of adders or subtractors that can assess factors such as the environmental and health effects of air or water pollution, the effects of resource depletion or the danger of overreliance on a single source of energy (see EXTERNALITIES). Avoided cost has gone down dramatically in recent years due to the availability of cheap NATURAL GAS. Gas turbines can be built quickly, and in small increments that minimize the utility's long-term debt. Many GARBAGE INCINERATORS and REFUSE DERIVED FUEL plants have been financially stressed by the reduction in avoided cost, resulting in increased tipping fees and decreased profits. Municipalities that guaranteed operators of waste-to-energy facilities a certain rate for the electricity they produce have been faced with extra payments. Although gas turbines emit NITROGEN OXIDES (NO_X) and other atmospheric contaminants, their emissions pale compared to those from a typical coal-fired power plant. The resource planning and acquisition process in most states does not recognize the full value of investments in renewable resources.

California and Avoided Cost

California is in a class by itself when it comes to the use of avoided cost and the promotion of renewable energy sources as a way of building a more-reliable power system. The state has the strictest air quality standards in the country, one reason for its official interest in nonpolluting energy sources. California utilities must consider the environmental costs of running old, dirty generating facilities when deciding whether new resources are needed. A new energy is defined as "needed" if it will reduce the net present value of the system's future costs. California leads the nation in the generation of electricity from every renewable source except hydroelectricity. Non-fossil power generation has gone from 5 megawatts in 1979 to almost 10,000 megawatts today, with a bit more than half that total coming from cogeneration, the rest from renewables. Each of California's three IOUs have at least one approved renewable resource in their plan, and half of that capacity is set aside for renewables. Plant and animal derived fuels (biomass) provide 3% of California's electricity—the highest percentage in the nation. In 1991, the legislature passed a "green set aside" requiring that a portion of the next round of electric-generating resource acquisitions come from renewables. Concern that the CPUC was not giving adequate consideration to environmental externalities other than air pollution or to the risk-reduction value of fuel diversity moved the legislation. Backers said that renewables would fare poorly in competitive bidding—even with emissions values being considered—given the low price of natural gas. The law requires the commission to continue with the set-aside program until all environmental externalities have been quantified and rolled into avoided cost. The commission has quantified air pollution, but not the contamination of water and land caused by combustion generators.

B

Back Hauling the practice used by truckers of finding another load after their primary load has been delivered and hauling it back to the original point of departure. Back hauling can lead to contamination of food if a truck delivering toxic waste to a disposal facility (see HAZARDOUS WASTE DISPOSAL) or toxic chemicals to an industrial installation picks up a load of food for the back haul (or vice versa).

Backdraft a condition in which the draft of a flue or chimney is insufficient to carry exhaust gases to the outdoors. Wind can cause a backdraft in a chimney with an inadequate draft, especially if the chimney is cool. The condition can also occur even with adequate draft when an exhaust fan or COMBUSTION APPLIANCE is used in a tightly sealed building with no open window or other source of outdoor air. The result is that outdoor air flows down the chimney and exhaust gases are pushed into the building, a potentially dangerous situation. CARBON MONOXIDE and other toxic combustion gases can quickly build to lethal levels, and unburned fuel from gas appliances can cause an explosion hazard. See also INDOOR AIR POLLUTION, VENTILATION.

Background Radiation radiation from natural sources. Cosmic rays that penetrate the earth's atmosphere, and RADON and its decay products, which are in turn the decay products of the uranium and thorium that are found in the Earth's crust, are the sources of background radiation. The increase in low-level radiation that has resulted in some areas from the atmospheric fallout from nuclear weapons tests, emissions from nuclear power plants and nuclear accidents is also commonly referred to as background radiation. The buildup of radon gas and associated increases in radiation that can occur in a building or a mine are not considered to be background radiation. See also MEASUREMENTS, RADIOACTIVITY, URANIUM.

Bacteria single-celled organisms of the order Monera. Bacteria play an indispensable role in the breakdown of waste and in the production of soil from the organic remains of plants and animals and inorganic rock particles. They are found nearly everywhere: in water, air, soil, most food and drink and in the bodies of plants and animals. Cyanophytes (blue-green algae) are closely related to bacteria, and the line dividing the two species is vague. The fact that the same species of bacteria may take different forms adds to the difficulty of classifying bacteria by species and genera. Fossil bacteria have been found in rocks more than three billion years old.

Each bacterium consists of a single cell of protoplasm surrounded by a cell membrane. Bacteria do not have a true nucleus, and have no cellulose in their cell walls. Many bacteria (especially bacilli) are capable of moving around on their own power, often by means of hair-like flagella that lash around to push the bacteria through liquid medium. Some bacteria have elementary sensory organs including eyespots and chemical sensors that are sensitive to changes in their environment. Bacteria nourish themselves from their surrounding media and excrete waste, which may be beneficial or harmful to other organisms. Certain bacteria protect themselves by turning into a spore that is capable of withstanding much more adverse conditions than the parent bacteria. When conditions are favorable, the spore, in turn, forms a new bacterium. Bacteria reproduce themselves by simple fission or division, starting at the nucleus. Although bacteria have only one cell, some types do not separate completely after reproduction, leading to masses made up of numerous single-cell bacteria that are stuck together. Bacterial reproduction happens extremely quickly. A single bacterium can undergo division in 30 minutes, under favorable conditions, and will produce one billion progeny within 15 hours of the first division.

There are three main groups of bacteria: *bacilli*—rod-shape bacteria that may occur singly, in pairs or in long chains; *spirilla*—curved or spiral bacteria that may be shaped like a comma or a spiral; and *cocci*—spherical bacteria that may be found in long beaded chains (*streptococci*), in flat clusters (*staphylococci*) or cube-shaped masses (*sarcina*). Of the two basis types of bacteria, *saprophytes* feed on dead animal or vegetable material, and *parasites* feed on living matter. Saprophytes serve an essential function by breaking dead material down into its elements and thereby furnishing essential elements to nourish living beings. Parasites are also often essential to the functioning of living things (as with the bacteria that help break down food in an animal's stomach), but also have the potential of harming their host. AEROBIC BACTERIA require oxygen for growth, while ANAEROBIC BACTERIA thrive in the absence of oxygen. Bacteria are the active agents in processes involving fermentation. Pathogenic bacteria can cause diseases such as cholera, typhoid fever, dysentery, syphilis, tuberculosis and several forms of pneumonia. See also COMPOSTING, CRYPTOSPORIDIUM, HETEROTROPHIC BACTERIA, LEGIONELLA BACTERIA, MICROORGANISMS.

Bactericide   a PESTICIDE that is used to kill bacteria. See DISINFECTION.

Baghouse   a network of tubes or bags made of cotton, felt or synthetic fibers that filter contaminants out of an airstream. The dust cake that forms as the filter accumulates material inside the bags, which may be from 10 to 30 feet in length and 10 to 15 inches in diameter, can increase the filter's efficiency. The dust cake must be removed periodically to maintain adequate airflow. See also AIR POLLUTION TREATMENT.

Barium   a silver-white or yellowish METAL powder used in the production of other metals, PAPER and PESTICIDES. It is also used as an additive in lubricating oils; in the production of beet sugar and animal and vegetable oils; in the manufacture of pyrotechnics and explosives; in tanning and finishing leathers; as a mordant for fabrics and dyes; in electroplating, aluminum refining, and rubber manufacture; and in the production of PAINTS and enamels. Barium gets into air and water from industrial and municipal waste treatment plant discharges and from spills. Barium gets into water from natural sources. Most of the barium SALTS are either highly or moderately soluble in water. Once in water, barium is highly persistent, with a HALF-LIFE greater than 200 days.

Exposure to airborne barium can irritate the eyes, nose, throat, and lungs, and may cause coughing. Exposure to very high levels of airborne barium can cause vomiting, diarrhea, irregular heartbeat, paralysis and death.

Barium cyanide is on the Hazardous Substances List. The OCCUPATIONAL SAFETY AND HEALTH ADMINISTRATION has established a permissible exposure limit for airborne barium of 0.5 milligrams per cubic meter, averaged over an eight-hour work-shift. PUBLIC WATER SYSTEMS are required under the SAFE DRINKING WATER ACT to monitor concentrations of barium (MAXIMUM CONTAMINANT LEVEL = 2 milligram per liter). The BEST AVAILABLE TECHNOLOGY for the removal of barium from DRINKING WATER is ION EXCHANGE, lime softening or REVERSE OSMOSIS (see WATER TREATMENT). It is also cited in AMERICAN CONFERENCE OF GOVERNMENT INDUSTRIAL HYGIENISTS and Department of Transportation regulations. Businesses handling significant quantities of barium or barium compounds must disclose use and releases of the chemical under the provisions of the EMERGENCY PLANNING AND COMMUNITY RIGHT TO KNOW ACT.

Base   a compound capable of reacting with an ACID to form a SALT, either with or without the elimination of water. According to the Brønsted/Lowry system, a base is a molecule or ION that can take up a proton from an acid (i.e., a *proton acceptor*). Bases contain the *hydroxide ion* (OH–) or the *hydroxyl group* (OH), and are capable of yielding a hydroxide ion in aqueous solution. A base has a pH of more than 7, and will turn litmus paper blue. Bases are often slippery to the touch, may be corrosive (see CORROSIVE MATERIAL) and usually have a bitter taste. Lime, lye, AMMONIA, most household soaps and cleansers and caustic alkali are bases.

Basicity   see ALKALINITY

Battery   a device that stores electricity in the form of chemical energy. Most of the HEAVY METALS in HOUSEHOLD HAZARDOUS WASTE comes from discarded batteries, and the removal of batteries from the WASTE STREAM is critical in the production of REFUSE DERIVED FUEL. Discarded automotive batteries have seriously contaminated soil and water with LEAD at many Superfund sites and other smaller disposal sites, and many of the heavy metals found in LEACHATE come from discarded batteries. Batteries may contain lead, ARSENIC, zinc, cadmium, copper and MERCURY. About 3.7 billion batteries are sold annually in the United States, primarily to power household and personal items.

Primary batteries—which include carbon-zinc, alkaline, mercury cell and lithium—are used once and thrown away because the chemical reaction that generates current inside them is irreversible. Reliability, convenience, low cost and long shelf life are some of the qualities that make such disposable batteries popular.

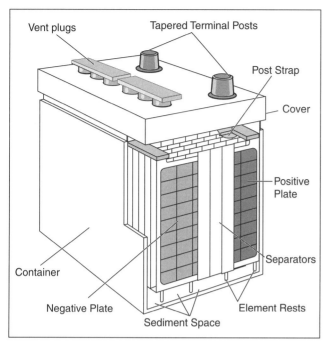

Automotive starting battery (Electrochemical Storage and Conversion: Batteries and Fuel Cells, U.S. Department of Energy)

A typical disposal C-cell battery contains more than 16,000 micrograms of mercury and other heavy metals. Alkaline-manganese batteries have not been sold in the European Community since 1993. The button batteries found in cameras, calculators, pacemakers and hearing aids often contain about 55% mercurial oxide by weight.

Secondary batteries—which include nickel-cadmium, lead-acid and nickel-hydride—are rechargeable batteries that produce electricity as the result of chemical reactions that are reversible. They are more expensive than primary batteries, and typically have one-third to one-half the usable life of a primary battery before they must be recharged. Secondary batteries tend to run out of energy suddenly, unlike a primary battery, which loses power gradually. Cadmium is what makes most secondary batteries rechargeable. Lead is also used because it is resistant to corrosion and because it accepts and yields electrons easily, and other heavy metals are present in most batteries. Far fewer rechargeable batteries enter the waste stream. The ENVIRONMENTAL PROTECTION AGENCY estimated in 1989 that nickel-cadmium batteries are the largest source of cadmium in U.S. landfills. See also AUTOMOTIVE PRODUCTS, MANGANESE, NICKEL.

Benzene a clear liquid HYDROCARBON that can be recognized by its distinctive, pleasant odor, which is detectable at concentrations of more than 12 parts per million (ppm). Benzene is a component of GASOLINE. It is used as a SOLVENT in PAINTS, inks, oils, PLASTICS, paint removers and rubber cement. It is also used to extract oil from nuts and seeds; in the manufacture of chemicals; and in the production of detergents, explosives and drugs. Benzene was first synthesized by the English chemist Michael Faraday in 1825. It is derived from coal tar and PETROLEUM. The benzene molecule is the basis for most aromatic compounds (see AROMATIC HYDROCARBON). It is extremely flammable (see FLAMMABLE LIQUID) and extremely volatile (see VOLATILITY).

Benzene is only slightly soluble in water, and tends to float to the surface and quickly evaporate, so concentrations in surface water are generally low (see WATER POLLUTION). In GROUNDWATER, however, benzene levels can build to dangerous levels. Gasoline seeping from LEAKING UNDERGROUND STORAGE TANKS, leaking pipelines, improper disposal of solvents and surface fuel spills are common ways in which groundwater becomes contaminated with benzene (see GROUNDWATER POLLUTION). Benzene gets into outdoor air by evaporating from gasoline. When other aromatic hydrocarbons found in gasoline are burned, they produce benzene, which is a component of automobile exhaust. An estimated 85% of the benzene found in outdoor air comes from gasoline. It gets into indoor air from cigarette smoke (see ENVIRONMENTAL TOBACCO SMOKE) and from the indoor use of solvents, paints, varnishes and other household products of which it is a component.

Benzene can affect health when inhaled or ingested or absorbed through the skin. It is a proven CARCINOGEN and MUTAGEN and a possible TERATOGEN. Benzene causes leukemia, anemia and other blood-related diseases in humans. Chronic exposure (see CHRONIC EFFECTS) can at first cause headaches, fatigue, anemia, loss of weight and dizziness. More prolonged exposure can lead to pallor, nosebleeds and bone-marrow damage. Acute exposure (see ACUTE EFFECTS) can cause skin irritation and drowsiness.

Benzene concentrations in public drinking water systems are regulated under the SAFE DRINKING WATER ACT (MAXIMUM CONTAMINANT LEVEL = 5 ppm). Airborne benzene is regulated under the CLEAN AIR ACT. The OCCUPATIONAL SAFETY AND HEALTH ADMINISTRATION has established a permissible exposure limit for benzene of 1 parts per million averaged over an eight-hour workshift, with 5 ppm not to be exceeded during any 10 minute period. Benzene and benzidine are on the HAZARDOUS AIR POLLUTANTS LIST, and benzene is on the Hazardous Substances List. Businesses handling significant quantities of benzene must disclose use and releases of the chemical under the provisions of the EMERGENCY PLANNING AND COMMUNITY RIGHT TO KNOW ACT. See also AIR POLLUTION, AUTOMOTIVE AIR POLLUTION, BENZENE RING.

Benzene Ring   a closed chemical structure consisting of six BENZENE molecules. The six carbon atoms in the benzene ring are arranged in a hexagon with alternating single and double covalent bonds. Each carbon atom is also bonded to one HYDROGEN atom. Molecules of the chemical family of aromatic compounds, of which the benzene ring is the basis, are formed when one or more of the hydrogen atoms in the benzene ring are replaced by another group of atoms. The new group may be another benzene ring, which may in turn have yet another benzene ring attached to it. The more complex aromatics, such as POLYCHLORINATED BIPHENYLS, are formed in this manner.

Aromatic compounds with equal proportions of identical elements often have differing characteristics. This is because the chemical groups involved are attached to the benzene ring in different patterns. For this reason, the names of aromatic compounds identify the points of attachment to the benzene ring as well as the proportion of elements found in each molecule.

Benzo-a-pyrene   a tarry, organic chemical released when GARBAGE, WOOD or FOSSIL FUELS are burned. Benzo-a-pyrene (BAP), a POLYAROMATIC HYDROCARBON, is a product of incomplete combustion, especially combustion at a temperature too low to consume FUEL completely. It is released into the atmosphere attached to SMOKE particles. Coal tars and coal-tar volatiles (see VOLATILITY) including creosote contain BAP. Asphalt contains a small amount of BAP. Pure benzo-a-pyrene is a yellow solid made up of plate- or needle-shaped crystals. It is slightly soluble in water, and binds to many forms of organic matter. It has no commercial uses.

Benzo-a-pyrene gets into outdoor air from forest fires, factory emissions, wood- and coal-fired stoves and furnaces, and gas- and diesel-powered automobiles. It gets into indoor air from the combustion of tobacco and cooking food, and from smoky outdoor air. Higher levels are found in the workplace, usually associated with the production of coke (see COAL), creosote, smoked meat and the incineration of TRASH and garbage. There is some evidence that the BAP found in ENVIRONMENTAL TOBACCO SMOKE is the reason it causes cancer. It has been shown in animal tests to be a MUTAGEN and to cause reduced body weight of offspring and to cause bronchitis. The ENVIRONMENTAL PROTECTION AGENCY has classified BAP as a probable human CARCINOGEN on the basis of epidemiological studies. PUBLIC WATER SYSTEMS are required under the SAFE DRINKING WATER ACT to monitor for benzo-a-pyrene (MAXIMUM CONTAMINANT LEVEL = 0.0002 milligram per liter). The BEST AVAILABLE TECHNOLOGY for the removal of benzo-a-pyrene from DRINKING WATER is granular ACTIVATED CARBON. The OCCUPATIONAL SAFETY AND HEALTH ADMINISTRATION (OSHA) has established a limit for coal-tar volatiles (including BAP) in workplace air of 0.2 milligrams per cubic meter (0.2 mg/m^3).

Berm   an earth barrier used to enclose a pond, landfill or other waste disposal site. Earth piled on a hillside to support a road or enclose an irrigation ditch is also called a *berm*. See also CONFINED DISPOSAL FACILITY, LANDFILL.

Beryllium   a hard, brittle, gray-white METAL used in the manufacture of electrical components, chemicals, ceramics, metal alloys and x-ray tubes. Beryllium gets into air and water from industrial emissions, and is sometimes found in the acidic RUNOFF from mines.

Beryllium is a probable human CARCINOGEN. Limited evidence suggests that it is linked to lung and bone cancer in humans, and it has been confirmed as a cause of lung and bone cancer in animals. Other CHRONIC EFFECTS include disorders of the respiratory system, heart, liver and spleen. Severe bronchitis or pneumonia can occur within two days of exposure to high concentrations. Death can result in severe cases. Exposures to high concentrations or repeated exposure to lower levels can cause scars to develop in the lungs and other body organs. In severe cases, grave disability and heart failure can result. When beryllium particles get under cuts in the skin, ulcers or lumps can develop. ACUTE EFFECTS include itching, and burning eyes, nasal discharge, tightness in the chest, cough, shortness of breath or fever. Overexposure can severely irritate the airways and lungs.

PUBLIC WATER SYSTEMS are required under the SAFE DRINKING WATER ACT to monitor concentrations of beryllium: MAXIMUM CONTAMINANT LEVEL = 0.004 milligram per liter). The OCCUPATIONAL SAFETY AND HEALTH ADMINISTRATION has established a permissible exposure limit for airborne beryllium of 0.002 mil-

1. The Benzene Molecule   2. The Phenol Molecule

3. The Benzidine Molecule

Benzene ring (*Encyclopedia of Environmental Studies*)

ligrams per cubic meter (0.0002 mg/m³) averaged over an eight-hour workshift and 0.005 milligrams per cubic meter (0.005 mg/m³), not to be exceeded during any 15-minute work period. It is also cited in NATIONAL INSTITUTE OF OCCUPATIONAL SAFETY AND HEALTH, AMERICAN CONFERENCE OF GOVERNMENT INDUSTRIAL HYGIENISTS and National Toxicology Program regulations. Beryllium is on the RTK Hazardous Substance List and the Special Health Hazard Substance List. Beryllium compounds are on the HAZARDOUS AIR POLLUTANTS LIST. Businesses handling significant quantities of beryllium or beryllium compounds must disclose use and releases of the chemical under the provisions of the EMERGENCY PLANNING AND COMMUNITY RIGHT TO KNOW ACT.

Best Available Control Technology the best AIR POLLUTION control equipment that is available for purchase when an industrial installation that will pollute outdoor air is being constructed. Major new sources of pollution are required to install the Best Available Control Technology (BACT). The cost of the equipment and its efficiency are considered when a regulatory agency determines what is the "best" technology available. BACT standards are established on a case-by-case basis. See also CLEAN AIR ACT, CLEAN WATER ACT, WATER QUALITY STANDARD.

Best Available Retrofit Technology (BART) the best air pollution control technology that can be added to an existing pollution source at a reasonable cost. See also BEST AVAILABLE CONTROL TECHNOLOGY, CLEAN AIR ACT, CLEAN WATER ACT.

Best Available Technology (BAT) the best WATER POLLUTION control equipment that is currently available for the removal of POLLUTANTS from industrial sources. The SAFE DRINKING WATER ACT requires that along with each MAXIMUM CONTAMINANT LEVEL (MCL) established for a pollutant, that the ENVIRONMENTAL PROTECTION AGENCY (EPA) also specify the best available treatment technology that is available at a reasonable cost to meet the MCL. The act requires that best available technology be a procedure or device that has proven effective in field applications, not solely under laboratory conditions. Adoption of the BAT is not mandatory, however, and any "appropriate technology" that meets the designated MCL and that has been accepted by the state water-quality agency can be employed. For some contaminants, such as microorganisms, a treatment technique rather than an MCL is specified.

Granular ACTIVATED CARBON is designated in the Safe Drinking Water Act as the best available technology for the control of synthetic organic chemicals. Only those PUBLIC WATER SYSTEMS that could demonstrate that their water source was too pure to need filtering were exempt, and all others were required to install filtration equipment by the end of 1991. Reverse osmosis is designated as the BEST AVAILABLE TECHNOLOGY for the removal of BARIUM, CADMIUM, CHROMIUM, MERCURY, NITRATES, NITRITES, RADIUM, SELENIUM, synthetic organic chemicals and URANIUM. Under the National Pollutant Discharge Elimination System (NPDES), pollution permits are to be granted only when applicants have proven that the toxicity of the effluent they plan to emit will be treated using the best available technology. See also CLEAN AIR ACT, CLEAN WATER ACT, NEW SOURCE PERFORMANCE STANDARDS, PRIORITY POLLUTANT, REVERSE OSMOSIS, WATER QUALITY STANDARD.

Best Conventional Waste-Treatment Technology the pollution-control technology for the removal of CONVENTIONAL POLLUTANTS, primarily NUTRIENTS, MICROORGANISMS, SEDIMENTS, OILS and GREASES. Best conventional waste-treatment technology (BCWTT) requirements were established under the 1977 amendments to the CLEAN WATER ACT, and are determined on a case-by-case basis. The cost/benefit ratio of installing the pollution control equipment must be considered. See also INDUSTRIAL LIQUID WASTE. Compare BEST AVAILABLE TECHNOLOGY.

Best Management Practice the best available method of controlling WATER POLLUTION from diffuse sources (see DIFFUSE POLLUTION). Best management practices are legally required on logging operations in some states, but are voluntary in most. As applied to logging, the proper construction of roads and skid trails to avoid excessive erosion, SEDIMENTATION and contaminated RUNOFF are the most important aspects of best management practices. Best management practices, which can also be applied to farming, mining and other resource harvesting techniques, are the equivalent for diffuse pollution sources of the best available technology standards applied to point sources of pollution.

Best Practicable Technology the best practical technology available for the control of industrial point-source water pollution. Publicly owned sewage treatment plants must conform to a modified version of best practicable technology (BPT). Cost/benefits analysis and the age and condition of the polluting installation are considered when determining the BPT to be applied. BPT standards are normally set by surveying the average pollution-control technology in place at similar installations in the same region. BEST AVAILABLE TECHNOLOGY standards are stricter than BPT requirements. Industrial polluters have argued that "practicable" technology is meaningless where receiving waters are already very polluted or are very large in relation to the volume of effluent produced, but both the ENVIRONMENTAL PROTECTION AGENCY and the courts have rejected this argument. See also BEST CONVENTIONAL WASTE-TREATMENT TECHNOLOGY.

Beta Particles   negatively charged electrons that are emitted by the radioactive decay of certain nuclides (see RADIOACTIVITY, RADIONUCLIDE) including RADIUM 228. Beta particles travel near the speed of light and have about 100 times the penetrating ability of ALPHA PARTICLES. Beta-emitting materials found in surface water are primarily the result of the atmospheric testing of nuclear weapons. GROUNDWATER contaminated with beta particles can be found in areas where underground nuclear testing has occurred including Rio Blanco county in Colorado, where tests involving the detonation of nuclear bombs to stimulate natural gas production were carried out during the 1970s. PUBLIC WATER SYSTEMS are required under the SAFE DRINKING WATER ACT to monitor concentrations of beta particles and photon emitters (MAXIMUM CONTAMINANT LEVEL = 4 millirem). Compare GAMMA RAY, X-RAY. See also GROUNDWATER POLLUTION, NUCLEAR REACTOR.

Bioaccumulation   a process through which small concentrations of an environmental toxin can build to high concentrations of the contaminant in the tissues of plants and animals. Bioaccumulation is also called *biomagnification* and *bioconcentration*. Among the factors that lead to bioaccumulation are the fact that many chemically stable toxins have an affinity for fatty tissue. Concentrations of the compound can build to toxic levels from regular exposure very low concentrations of a toxin experienced over a long period of time. The concentrating effect of the food chain is also a form of bioaccumulation. One example of this process starts when single-celled organisms such as plankton make toxins from polluted water a part of their tissue. In the tissue of small fish that routinely feed on the plankton, the concentration of the toxin is even higher. When thousands of the smaller fish are eaten by a larger fish, such as a salmon, even higher concentrations of the toxin result. And when a human or a bear feasts on the fish, the accumulated burden of toxins can become [yet another step higher and] potentially more lethal in their tissue. See also BIOAVAILABILITY, CHLORINATED HYDROCARBON, DIOXIN, LEAD, PERSISTENT COMPOUND, POLYCHLORINATED BIPHENYLS (PCBs), TOXAPHENE.

Bioavailability   the ease with which a compound is taken up by living organisms. Bioavailability is as important as toxicity in determining the hazard posed by a particular toxin. An environmental toxin with a high bioavailability is more likely to become part of the food chain, and can therefore be more hazardous than a more toxic compound with a low bioavailability. In a plant nutrient, a high degree of bioavailability is an asset. Nitrates in the soil, for instance, are of much more use to most plants than is nitrogen in its natural, gaseous form because of the nitrates high bioavailability. See also BIOACCUMULATION, HAZARDOUS MATERIAL, PERSISTENT COMPOUND.

Biochemical Oxygen Demand (BOD)   the demand for oxygen caused by the decomposition of biodegradable impurities in wastewater (see BIODEGRADABILITY, DISSOLVED OXYGEN)—also called *biological oxygen demand*. The METABOLISM of aquatic microorganisms is responsible for most of the oxygen demand created by wastewater, but chemical oxidants also consume dissolved oxygen (see CHEMICAL MANUFACTURING WASTE). Heat and nutrients in effluent from sewage treatment plants and industrial facilities (see INDUSTRIAL LIQUID WASTE) can also fuel increased growth of algae and other MICROORGANISMS in water and therefor cause an increase in biochemical oxygen demand which is not a specific pollutant, but rather a measure of the cumulative effects of several contaminants. A high BOD level indicates that water quality has been degraded, but cannot determine the source of the degradation. Water with a high BOD generally contains numerous microorganisms and is receiving excessive nutrients, conditions associated with EUTROPHICATION. BOD increases as heat increases because the metabolism of all aquatic life speeds up in warmer water. An 18°F increase in temperature—not uncommon at a large power plant—is enough to double the metabolism of aquatic organisms. SEDIMENTATION can cause increases in water temperature and therefore in BOD. A relatively small amount of wastewater with a very high BOD reading will generally have less effect on the health of an aquatic ecosystem than a relatively small amount of BOD in a very large amount of wastewater. The microorganisms responsible for biochemical oxygen demand are very sensitive to toxic chemicals, so BOD testing of wastewater containing many toxics will generally give an artificially low reading, since many of the microbes present will die from exposure to the toxins while the test is in progress. Many of the suspended solids and other materials responsible for BOD are removed from sewage and industrial wastewater during primary sedimentation (see SEWAGE TREATMENT).

Testing for BOD

Two or more one-liter bottles are filled with the water to be tested, and the dissolved oxygen content of water in the first bottle is tested. The second bottle is allowed to stand for a specified period, usually five days, and its dissolved oxygen level is then measured. BOD levels of additional water samples are measured at regular intervals. Test results are expressed as the parts per million (ppm) of dissolved oxygen consumed in a sample that stood for a given number of days. Water that contained 8 ppm dissolved oxygen at the beginning of the test period and 5 ppm after standing ten days would be denoted 3 ppm BOD_{10} because 3 ppm of the dissolved oxygen in the water had been consumed over the test period. Samples are sometimes allowed to stand for a very long period of time, until all biological activity in the sample has stopped, and such tests are denoted BOD_{ULT}. The BOD of a normal stream is in the range of 1 to 2 ppm, while untreated sewage normally has a BOD between 100 and 400 ppm, which is reduced to about 10 to 30 ppm in the sewage treatment process.

Congress passed legislation in 1987 requiring that primary treatment remove 30% of the BOD present at sewage treatment plants with no secondary treatment facilities. The ENVIRONMENTAL PROTECTION AGENCY proposed a regulation in 1991, not yet implemented, that included the 30% reduction. The standard was easy for cities with large industrial polluters on the collection system to meet, but smaller cities with more diluted sewage—many with BOD concentrations in effluent outfalls well below federal standards—found compliance much more difficult. Some allowed increased rates of BOD emissions into sewers in anticipation of the required reductions.

See also AEROBIC STABILIZER, CHEMICAL OXYGEN DEMAND, EFFLUENT STANDARD, FLOCCULATION, SEPTIC SYSTEM.

Biodegradability the ability of a substance to be broken down by microbial action once released into the environment. The molecules in a biodegradable detergent, for instance, are converted to alcohols and simple hydrocarbons by the metabolic action of microorganisms found in the soil and water. Substances such as CHLORINATED HYDROCARBONS and METALS such as MERCURY and LEAD that are not biodegradable are especially threatening as pollutants because they can accumulate in sediments and living tissue (see BIOACCUMULATION, PERSISTENT COMPOUND). See also BIOLOGICAL TREATMENT, BIOREMEDIATION. Compare GEODEGRADABILITY.

Biodegradable Plastics plastic materials that will break down more quickly than most plastics when exposed to sunlight and weathering. Plastics are rendered biodegradable by the addition of a substance, usually corn starch, that can be attacked by MICROORGANISMS. The addition of corn starch, however, weakens the plastic enough that extra resin must be added if the biodegradable product is to be as strong as the regular plastic version. Studies by the Garbage Project have shown that neither biodegradable or regular plastic products go through much decomposition once buried out of the sunlight in a landfill. And even if the biodegradable product breaks down as it was designed to do, it simply breaks into smaller pieces and loses none of its volume. Some experts have expressed concern that the fact that biodegradable plastics may eventually break down could actually release more toxins than would be released by normal plastic, and that the toxins, once released, would become potential sources of air and WATER POLLUTION (see AIR POLLUTION, GROUNDWATER POLLUTION). Biodegradable plastics cannot be recycled because the corn starch they contain is seen as a contaminant to recyclers. Although biodegradable plastics enjoyed a short period of popularity during the 1980s, when they were used primarily as a tool to market products such as disposable diapers, the above drawbacks probably means their use will probable never become widespread. See also PLASTICS RECYCLING, RECYCLING.

Biogas any GAS that is a BYPRODUCT of the METABOLISM of MICROORGANISMS (see METHANE).

Biological Aerosols PARTICULATES that have a biological origin. Pollens, spores and microbial waste are examples of biological AEROSOLS. The larger aerosols only are airborne when there are sufficient air currents to support them, while the smallest can remain aloft indefinitely. See also INDOOR AIR POLLUTION, MICROORGANISMS.

Biological Oxygen Demand (see BIOCHEMICAL OXYGEN DEMAND)

Biological Treatment treatment of WASTEWATER and DRINKING WATER that relies on the METABOLISM of MICROORGANISMS (see HAZARDOUS WASTE TREATMENT, SEWAGE TREATMENT, WATER TREATMENT)—also called *biodegradation* or *biotransformation*. Aerobic organisms require oxygen for metabolism, while anaerobic organisms do not (see AEROBIC BACTERIA, ANAEROBIC BACTERIA). *Primary biodegradation* (*biotransformation*) alters the structure of a compound so that its physical and chemical properties are changed. *Secondary biodegradation* (*mineralization*) breaks the substance completely down, with CARBON DIOXIDE and WATER as principal waste products. Microorganisms, primarily aerobic bacteria, convert and metabolize dissolved and colloidal materials (see COLLOIDAL SOLIDS) in SEWAGE during SEC-ONDARY TREATMENT (see FLOCCULATION, TURBIDITY), and biological treatment is also an essential element in the neutralization of HAZARDOUS WASTE (see BIOREMEDIATION). Soil microorganisms can often break down toxins with which they come in contact, with the temperature, permeability and acidity of the soil affecting the degree to which they neutralize the toxins.

A technique called *land farming* is used to remove heavy petroleum oils POLYCHLORINATED BIPHENYLS (PCBs) and other toxins from soils. The soil to be treated is excavated and placed on a liner to prevent the spread of contamination, and fertilizer, carbon and oxygen are added to facilitate microbial action. Polluted water (and sometimes soil) can be purified in a *biotower* or *reactor vessel*, which is an enclosed chamber in which temperature, pH and nutrients are carefully regulated to favor the metabolism of bacteria that are fixed to activated carbon or biocarriers.

Bioremediation   the neutralization of hazardous and toxic substances in air, water or soil through the metabolic processes of MICROORGANISMS, primarily AEROBIC BACTERIA. Naturally occurring microbes are continually breaking down toxins, but the chemical diversity of petrochemicals usually makes it impossible for them to neutralize all the toxins with which they are confronted. New strains of microorganisms capable of consuming a specific toxin are developed in the laboratory using recombinant DNA and other techniques. By seeding a polluted aquifer or contaminated soil with some of the designer microbes, more complete neutralization of toxic compounds is possible. Bioremediation has been successfully used to clean up groundwater contaminated with gasoline from leaking storage tanks and surface leaks and spills—attacking the contaminants in the groundwater *in situ* by injecting the microorganisms directly into the aquifer—and to clean up oil released in maritime accidents. Microbial action doesn't break down petroleum quickly enough to prevent widespread environmental damage, but oils that spill or leak into oceans are eventually broken down by natural processes, or they would quickly become completely covered with oil slicks. The nature of the contaminated soil or body of water, the type and variety of toxins involved and climate influence how long it will take for neutralization. See also HAZARDOUS WASTE TREATMENT.

Black Water   the water in sewage that carries FECES, urine and other substances that are strongly active biologically. Black water is primarily the EFFLUENT from TOILETS and GARBAGE DISPOSALS. See also SEWAGE TREATMENT. Compare GRAY WATER.

Blue-Green Algae   a phylum of more than 7,000 unicellular organisms called *cyanophytes*. Blue-green algae contain chlorophyll, and have many characteristics of true ALGAE, with which they were once classified, but lack a complete nucleus and have no cellulose in their cell walls, and are therefore usually now classified as BACTERIA (and are sometimes called *cyanobacteria* or *blue-green bacteria*). Some waterborne cyanophytes grow long filaments, and are responsible for *algal bloom* (also called *bacterial bloom*), a phenomenon caused by the explosive growth of cyanophytes responding to optimum water temperature and nutrient concentration. Algal bloom can clog waterways, dye water green and kill other forms of aquatic life (see EUTROPHICATION).

Boiler Slag   impurities that are removed during the smelting of iron ore, most of which float to the surface of the molten ore. See also STEEL.

Bones   the hard tissue that makes up the skeleton of most vertebrates. The inorganic compounds in bone tissue, primarily calcium phosphate, make them hard and rigid, and make them less vulnerable to breakdown by microorganisms. Antlers, tusks and teeth are made up of bony material. The bones of prehistoric humans are studied by archaeologists to learn more about ancient cultures. Animal bones were used by early man as tools. Alpha radiation, lead and certain toxic chemicals tend to accumulate in bones. Red blood cells are produced by bone marrow. Toxins such as TOLUENE and XYLENE damage bone marrow and thereby reduce the concentration of red blood cells in the bloodstream, which can lead to anemia. Medical x-rays are used to inspect bones for cracks or breaks. See also ARSENIC, ARTIFACT, METHOXYCHLOR, PHOSPHOROUS, VINYL CHLORIDE.

Bottle Bill   an ordinance requiring a deposit on cans and bottles containing beverages. The deposit, often about ten cents per bottle, is intended to encourage the consumer to return the bottles. Support for efforts to pick up roadside litter are also often part of such legislation, which in many cases is statewide. In Oregon, where the first bottle bill was passed in 1972, officials report that 90% of the beer and pop containers on which deposits are placed are returned and that roadside litter has been reduced by 80% since the legislation was passed. Groups backed by the beverage and bottling industries launch vigorous campaigns against bottle bills whenever they are introduced, often citing the "failure" of the bottle legislation in Oregon. The eight other states that had enacted bottle laws as of 1988 are Iowa, Michigan, Massachusetts, Vermont, Maine, Connecticut, New York and Delaware. A

National Beverage Container Reuse and Recycling Act was sponsored by 60 members of Congress in 1992, but it was not adopted. See also GLASS; NO DEPOSIT, NO RETURN; ONE-WAY PACKAGING RECYCLING.

Bottom Ash ash and hot cinders that are heavy enough that turbulence in the combustion chamber does not carry them up the exhaust stack (as opposed to FLY ASH, which either is ejected into the atmosphere or is removed from effluent with pollution control equipment) (see AIR POLLUTION TREATMENT). Bottom ash is normally not as toxic as fly ash, which often must be disposed of as hazardous waste. Relatively high concentrations of HEAVY METALS such as LEAD and CADMIUM can sometimes be found in bottom ash, however. Bottom ash is generally cooled with water in a quench tank, where it is mixed with fly ash and the resulting slurry is funneled to a truck for disposal. The ENVIRONMENTAL PROTECTION AGENCY estimates that a little more than one-third of the ash from U.S. incinerators is dumped in an ASHFILL, and that a little less than one-fifth is dumped in a MUNICIPAL SOLID WASTE landfill, with the fate of the remaining one-half unknown. See also COAL, COMBUSTION.

Breeder Reactor a NUCLEAR REACTOR that turns non-fissionable URANIUM or thorium into PLUTONIUM while producing heat that is used to generate electricity. A breeder reactor creates more fuel than it burns. When a fissionable nucleus splits, two or more neutrons are released. One of the free neutrons must hit another atom and cause FISSION to maintain the nuclear chain reaction, but the rest of the neutrons can create nuclear fuel from nonfissionable uranium (U^{238}) by attaching themselves to the atom's nucleus to create U^{239}, which quickly decays (within about a month) to form Pu^{239}, plutonium. Fissionable uranium (U^{235}) typically makes up less than 1% of the uranium found in uranium ore, and the rest is U^{238}. The same process takes place in a conventional fission reactor, with almost one U^{239} NUCLIDE being created for every two fissionable nuclei destroyed. In a breeder reactor, four fissionable nuclei are created for every three that are destroyed. The amount of time required for a breeder reactor to produce enough fissionable material to fuel another reactor of a similar size is called its *doubling time*. An advanced breeder reactor is expected to have a doubling time of 10 to 15 years. Water can not be used as a coolant with the "fast" or high-energy neutrons used in a breeder reactor, and molten sodium is used instead. The fuel assembly and control rods are similar to those used in a conventional reactor, but surrounding them is a blanket of U^{238} in which plutonium is formed. Because the liquid sodium that carries heat

away from the reactor becomes radioactive after being exposed to intense neutron bombardment in the reactor core, the heat must be transferred to a second cooling loop in a heat exchanger to isolate the radiation. The secondary cooling loop carries the heat to a steam generator, and the steam, in turn, is used to drive an electrical generator.

Plans for plutonium-fueled fast-breeder reactors are on hold due to high costs and failures, causing problems for nations with REPROCESSING PLANTS. The only large-scale fast breeder in operation is France's 1,200 megawatt Superphénix, which cost three times as much to build as a standard light-water reactor. The reactor has had so many technical problems that, as of the mid-1990s, it has been online only about 40% of the time, and the French government is reportedly considering shutting it down. A prototype breeder reactor in India has not operated for more than a few minutes at a time. See also RADIOACTIVE WASTE, RADIOACTIVITY.

Brine water with a high concentration of salt. Large quantities of brine are brought to the surface in petroleum and natural gas wells, and disposing of it has long been a problem for the industry. Brine was once pumped into shallow ponds designed to allow the liquid to seep into the ground, but the serious contamination of surface AQUIFERS has largely ended the practice. The first injection wells were used by the petroleum industry around the turn of the twentieth century to dispose of brine, and an estimated 40,000 injection wells have been used for the purpose since that time. Brine often contains heavy metals and other contaminants, and is a major source of GROUNDWATER POLLUTION. Groundwater that has been isolated in a deep aquifer for a long period of time usually has a high salt content, and the presence of such brine is one sign of a good site for DEEPWELL DISPOSAL. Brine sludges from the production of chemicals are often full of toxic compounds that often must be disposed of as HAZARDOUS WASTE.

Browning-Ferris Industries a U.S.-based international waste-management firm. Browning-Ferris Industries (BFI) collects garbage from more than 4.5 million homes and half a million businesses, operates more than 100 landfills worldwide and is a leading recycler, with more than a million customers. It has developed the medical waste-disposal market and has been trying to increase its share of the hazardous waste business, which in the mid-1990s dominated by Chemical Waste Management (see WMX TECHNOLOGIES). When William D. Ruckelshaus, twice director of the ENVIRONMENTAL PROTECTION AGENCY, signed on as BFI's chief executive in 1988, it was seen as a major coup for the firm. While Ruckelshaus has improved the organization's spotty environmental record, BFI's highly decentralized structure in which local managers run their operations almost like independent businesses (which most were before being acquired by BFI) has made improvement difficult.

In addition to hauling MUNICIPAL SOLID WASTE, Browning-Ferris owns and operates an ASBESTOS removal company and plants to reclaim fuels from solvents. BFI entered into a joint venture in 1983 with Air Products and Chemicals to design, build, own and operate municipal garbage incinerators (see AMERICAN REF-FUEL). BFI has also branched into street sweeping, paper shredding and renting portable toilets. Address: Browning Ferris Industries, 757 N Eldridge, Houston, TX 77079. Phone: 713-870-8100.

Bubble an entire industrial plant or group of plants that are treated as a single POINT SOURCE of AIR POLLUTION for regulatory purposes. All the smokestacks, vents, evaporating ponds and other sources of contaminants are treated as if they are under a giant bubble with only one exhaust opening to the atmosphere. The idea is to control total emissions from an industrial complex while giving plant managers the latitude to decide the rate of emissions from the various air pollution sources within the complex. The manager might choose, for instance, to cut emissions from a smoke stack for which pollution-control equipment (see AIR POLLUTION TREATMENT) is readily and economically available, while allowing the concentration of contaminants from a stack that would require more expensive and less proven technology to continue at a rate higher than the overall standard. A "new-source, old-source bubble" allows a firm to install air pollution control equipment in a new installation that is less than state of the art on the provision that the company cut emissions from an existing source in the same AIRSHED enough to offset the additional pollution generated. A "mobile bubble" averages the emissions from all the motor vehicles in an airshed and allows some vehicles to emit more pollutants than others provided that total pollutant load generated by vehicles operating within the bubble meets the standard. See also AMBIENT AIR QUALITY, CLEAN AIR ACT.

Building Materials materials that are used to construct a building. A typical building contains products made of CONCRETE, GLASS, GYPSUM, METALS, PAINT, PAPER, PLASTICS and WOOD. Building materials include HAZARDOUS MATERIALS such as glues, paints and SOLVENTS, and toxic materials, including LEAD and ASBESTOS, that were used in building materials in the past. Dust or smoke from many building materials created during construction can cause INDOOR AIR POLLUTION. The environmental cost of building materials includes resource depletion (see RESOURCE DEPLETION COST), energy use and pollution (see AIR POLLUTION, WATER POLLUTION) associated with the conversion of a natural resource into a finished building product including the effects of logging, mining, milling and manufacturing and transportation. Building materials are an important but little researched component of MUNICIPAL SOLID WASTE (see CONSTRUCTION WASTES, DEMOLITION WASTES). See also APPLIANCES, CARDBOARD, FURNITURE, INDUSTRIAL SOLID WASTE, PACKAGING, SOLID WASTE.

Byproduct waste material created by any process. Any physical, chemical or biological process produces byproducts, which may come in the form of heat (see THERMAL POLLUTION) NOISE or light or of materials in solid, liquid or gaseous form (see GASEOUS WASTE, LIQUID WASTE, SOLID WASTE). The byproducts of bringing an automobile to a stop (removing its physical speed) include heat, which is removed from the brakes through radiation and convection; a few minute solid particles (see PARTICULATE) that are scoured from the brakes and tires (and maybe some SMOKE in a really hard stop); and noise that can range from a barely discernible scraping sound to a loud squealing. The tires and brakes must be replaced eventually, and the discarded brake parts and tires are also partially a byproduct of the many stops they have been impelled to go through while on the automobile. The byproducts of bringing an automobile to a stop without the use of brakes can be considerably more costly, and is the reason God created auto insurance.

Virtually all the nutrients used by life forms in the biosphere are the byproducts of the METABOLISM of some other life form. The BIOLOGICAL TREATMENT of wastewater makes use of the metabolic action of BACTERIA and other MICROORGANISMS to break down NUTRIENTS and toxins (see FLOCCULATION, HAZARDOUS WASTE TREATMENT, SECONDARY TREATMENT, SEWAGE TREATMENT, WATER TREATMENT). Many hazardous byproducts of industrial processes are reused or recycled as a source of energy, feedstock or process chemicals for another industrial process. The byproducts of all kinds of processes are found in GARBAGE, which can be converted into energy (see REFUSE DERIVED FUEL, WASTE-TO-ENERGY PLANT) and much of which can be reused or recycled (see RECYCLING, REUSE).

C

Cadmium an ELEMENT that may occur as a bluish METAL or grayish powder. Cadmium is used in the electroplating of other metals, in batteries, pigments, in the production of metal alloys, CORROSION inhibitors, chemicals, PLASTICS, NUCLEAR REACTOR fuel rods, photoelectric cells and nickel-cadmium electrical storage batteries and as a catalyst. Approximately 1,700 metric tons of cadmium were produced by U.S. manufacturers during 1993. About 66% of the cadmium used by U.S. industry is imported. It can get into indoor air as the result of welding, brazing, soldering, plating, cutting, grinding and metallizing operations. It gets into outdoor air and water primarily from industrial effluents and landfill leaching. Cadmium is found in zinc, COPPER and LEAD ores. Cigarette smoke also contains traces of cadmium (see ENVIRONMENTAL TOBACCO SMOKE). It is difficult for the body to eliminate cadmium, so concentrations tend to increase with continued exposure. Cadmium's toxicity in fresh water is a function of the water's hardness—the harder the water, the lower the toxicity. Cadmium is slightly soluble in water. Up to 1 milligram will mix with a liter of water. Once it has contaminated water, cadmium is highly persistent, with a HALF-LIFE of greater than 200 days. Cadmium is used as a stabilizer in PVC pipe and is often present as an impurity along with the zinc used in galvanized STEEL pipe. Corrosion of such pipes is its primary source in drinking water.

Heating or grinding materials that contain cadmium can cause a flu-like illness with chills, headache or fever. Exposure to high concentrations of airborne cadmium can cause nausea, salivation, shortness of breath, chest pain, cough, a buildup of fluid in the lungs, vomiting, cramps and diarrhea. Symptoms may not manifest for four to eight hours after exposure. CHRONIC EFFECTS include emphysema and lung scarring; eventually death can result. Repeated exposure to lower concentrations can cause permanent kidney damage and kidney stones, emphysema, anemia and loss of smell. Cadmium (especially cadmium oxide) is a probable human CARCINOGEN. There is evidence that it is linked to prostate and kidney cancer in humans and it has been shown to cause lung and testes cancer in animals. Cadmium may also damage human testes and may affect the female reproductive cycle.

The OCCUPATIONAL SAFETY AND HEALTH ADMINISTRATION has established a permissible exposure limit for airborne cadmium of 0.2 milligrams per cubic meter averaged over an eight-hour workshift and 0.6 mg/m^3, not to be exceeded during any 15-minute work period. Cadmium is also cited in NATIONAL INSTITUTE FOR OCCUPATIONAL SAFETY AND HEALTH and AMERICAN CONFERENCE OF GOVERNMENT INDUSTRIAL HYGIENISTS regulations. Cadmium compounds are on the Hazardous Substances List, the Special Health Hazard Substance List and the HAZARDOUS AIR POLLUTANTS LIST. PUBLIC WATER SYSTEMS are required under the SAFE DRINKING WATER ACT to monitor concentrations of cadmium (MAXIMUM CONTAMINANT LEVEL = 0.005 milligrams per liter). The BEST AVAILABLE TECHNOLOGY for the removal of cadmium from DRINKING WATER is coagulation-filtration, lime softening, ION EXCHANGE or REVERSE OSMOSIS (see WATER TREATMENT). Businesses handling significant quantities of cadmium or cadmium compounds must disclose use and releases of the chemical under the provisions of the EMERGENCY PLANNING AND COMMUNITY RIGHT TO KNOW ACT.

California Air Resources Board an 11-member board appointed by the governor. The Air Resources Board has 900 employees and a $100-million annual budget. The board approved a plan in 1989 calling for the reformulation of aerosol sprays, stricter controls on industrial emissions, the introduction of cleaner fuels and cleaner-burning engines. The board adopted the CALIFORNIA CLEAN AIR ACT in 1990 to become the first government body in the world to require that electric vehicles be mass produced to improve air quality.

California Clean Air Act a law passed by the California Legislature in 1988 intended to reduce smog in the state's urban areas. The California Clean Air Act (CCAA) imposes the strictest vehicle-emission standards in the world. Adopted by the CALIFORNIA AIR RESOURCES BOARD in September, 1990, just a month before passage of the federal CLEAN AIR ACT (CAA), the CCAA had a major impact on the federal legislation. California was, at the time, the only state with auto emissions standards independent of those imposed by the federal government. The federal CAA includes a provision that allows states to impose stricter emission standards than those it prescribes primarily to accommodate the CCAA. During 1992, nine northeastern states and the District of Columbia moved to adopt the California standards in preference to the weaker CAA standards.

With an estimated 22 million vehicles, automotive emissions are the biggest single source of AIR POLLUTION in California (see AUTOMOTIVE AIR POLLUTION), and the state has endured the longest history of urban smog. Although the state already had the nation's strictest tailpipe emission standards before passage of the CCAA, 53% of the HYDROCARBONS and NITROGEN OXIDES and 82% of CARBON MONOXIDE in the Los Angeles basin still came from vehicles. Under the act, at least 10% of the vehicles sold new from 1994 through 1996 had to meet a standard of 0.125 grams of hydrocarbons emitted per mile (as compared to the federal standard at the time of 0.41 grams per mile). In 1997, 25% of the vehicles sold in the state were to be low-emission vehicles (LEV) that emit no more than 0.075 grams per mile (gpm) and 0.2 gpm of nitrogen oxides. By the year 2000, 96% of the new vehicles sold in the state are required to meet the LEV emission standard, with the remaining 4% consisting of even cleaner running vehicles.

Internal combustion engines fueled with propane, compressed natural gas, a gasoline/methanol blend and cars burning reformulated gasoline in engines with new, electrically heated CATALYTIC CONVERTERS are expected to be capable of meeting the California LEV standards. Zero-emission vehicles must equal 2% of all new car sales by 1998, and sales must be up to 10% of all new vehicles by 2003. Vehicles with electric motors are the only zero-emission vehicles now in existence. The CCAA includes measures designed to reduce the number of single-occupant cars on the state's roads. Companies with more than 100 employees that commute to work are required by the act to facilitate car pooling and van-pooling. The act requires that all the state's vehicles must have an emissions test performed regularly at one of the 9,000 commercial garages in the state that have been certified to conduct testing. The state is currently locked in a controversy with the ENVIRONMENTAL PROTECTION AGENCY (EPA) over that requirement. The EPA claims that state-run emissions certification centers would be twice as effective at controlling emission as the commercial garages, and has threatened to cut off $700 million in highway funds if the state does not establish separate testing centers. The state says that compliance with vehicle emission standards would be no better if the state owned and operated test centers than it is under the current system, and that it cannot afford to build separate centers in any case.

Cap an impervious cover used to permanently cover a full landfill or a hazardous waste site. A typical cap generally consists of two or more layers of clay, at least one plastic liner and a layer of topsoil planted with grass and other vegetation. Runoff is prevented from leaking into a closed landfill or hazardous waste site by the cap, which could lead to the creation of toxic LEACHATE that could cause water pollution. Except for the network of vent pipes sticking up through the turf, the top of many old landfills look like a sculpted hill.

Carbamate a group of PESTICIDES made up of carbon, HYDROGEN, oxygen and NITROGEN. ALDICARB is the best known of the carbamate pesticides.

Carbon Dioxide a colorless, odorless GAS that makes up about 0.03% of the ATMOSPHERE. Carbon dioxide exists either as a gas or as a solid called *dry ice*. It is produced naturally by plant and animal metabolism, by the decomposition or combustion of materials that contain carbon and by the weathering of rocks such as limestone that contain carbon. CO_2 is also produced when carbon-containing materials are burnt in the presence of oxygen. The use of WOOD, COAL, OIL and GASOLINE as a FUEL produces most manmade CO_2. The production of cement, in which carbon-bearing minerals are heated, also produces the gas. The carbohydrates created during photosynthesis use atmospheric CO_2 and give off oxygen, while plant metabolism consumes oxygen and gives off CO_2. Extra carbon dioxide is sometimes added to greenhouse air to promote plant growth by facilitating photosynthesis, usually by allowing dry ice to "melt" into a gas. Levels of the gas in indoor air are sometimes monitored to determine if VENTILATION inside a building is adequate. Carbon dioxide injected into carbonated beverages gives them their fizz. An atmosphere of pure carbon dioxide will not generally support combustion, and the gas is used in fire extinguishers for this reason. Magnesium and a few other substances will, however, burn in such an atmosphere if they are first ignited in the presence of oxygen.

Carbon dioxide is the principal greenhouse gas, making up an estimated 54% of the total (see GLOBAL WARMING). This has led to discussion about reducing CO_2 emissions, but doing so is difficult because the use of fossil fuels is the backbone of the world economy. Electric power plants, motor vehicles, industrial smokestacks, wood stoves and furnaces all emit the gas, and this has lead to dramatic increases in the concentration of CO_2 in the atmosphere, beginning with the Industrial Revolution in about 1850. The concentration of atmospheric CO_2 has increased from about 280 parts per million (ppm) in 1800 to 356 ppm in 1993 and is expected to increase another 15% by the year 2000. In industrialized nations, an average of five tons of CO_2 are emitted each year per resident, while only .2 to .6 tons per capita are emitted in developing countries. The gap is narrowing, however, since CO_2 production is increasing at about 6% per year in developing countries compared to a 1% rate of growth in Western Europe and North America.

World economic conditions are directly linked to the production of carbon dioxide. Emissions have increased at about 4% per year from 1860 to 1970, with dips for the world wars and the Great Depression. The 1973 oil embargo halved the rate of increase. CO_2 emissions remained steady at 5.3 billion tons per year from 1979 to 1985, primarily as the result of energy conservation efforts, and averaged 6 billion tons a year from 1989 to 1993. The pause in emission growth was the result of a recession in industrial countries and the economic breakdown of the Soviet Union, where 26% less CO_2 was emitted in 1993 than in 1988. Big increases in emissions in developing nations such as India, China and South Korea during the period kept worldwide emission steady. An estimated 2.4 billion tons of carbon dioxide was emitted through the combustion of petroleum products during 1993, about 40% of the total emitted by the combustion of fossil fuels. Other sources related to human development include the flatulence of livestock, responsible for an estimated 15% of atmospheric CO_2, and deforestation, which produced an estimated one to two billion tons of the gas as the result of burning and decomposition. About one-third of the carbon dioxide being released each year into the atmosphere does not stay there, and scientists are hard pressed to say exactly why. One theory is that the production of ALGAE and calcium carbamate shells in the oceans is responsible.

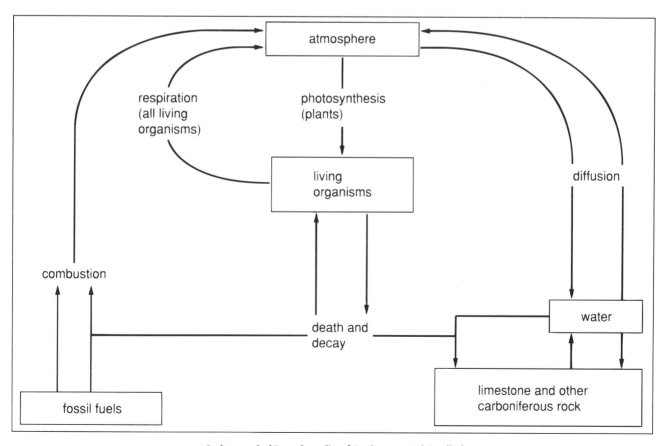

Carbon cycle (*Encyclopedia of Environmental Studies*)

Carbon Monoxide a colorless, odorless gas formed by the incomplete combustion of materials that contain carbon. Carbon monoxide (CO) is formed when there is not enough oxygen to allow complete combustion (as with a fire in an enclosed space), when a carbon-based fuel passes through a combustion chamber too quickly to allow a complete burn (as with GASOLINE in an INTERNAL COMBUSTION ENGINE) and when a fuel is burned at too high a temperature. It is one of the five CRITERIA POLLUTANTS named in the CLEAN AIR ACT. An estimated 50% of atmospheric carbon monoxide emissions comes from motor vehicles. Carbon monoxide is an important industrial chemical because it combines readily with other substances. It reacts with members of the IRON family to form nickel and iron carbonyls, a reaction that is used in the production of iron. It combines with HYDROGEN to form METHANOL. Carbon monoxide can react with CHLORINE under the influence of heat or light to form PHOSGENE, an extremely poisonous gas that often forms when PLASTICS and other synthetic materials burn in confined, unventilated spaces. Industrial carbon monoxide is produced by forcing air through layers of hot COAL and COKE.

Carbon monoxide poisoning kills about 1,500 people a year and makes another 10,000 sick, making it the most common cause of death from poisoning, according to a report by the Journal of the American Medical Association. Expectant mothers, infants, the elderly and people with respiratory or coronary problems are especially susceptible to the effects of carbon monoxide. The fact that carbon monoxide is invisible and has no odor and the relatively mild symptoms of carbon monoxide poisoning—a mild headache and mental fatigue, which can quickly be followed by drowsiness, unconsciousness and death—make it especially dangerous. Levels as low as .001% can cause symptoms of carbon monoxide poisoning, and 2% is usually fatal. Carbon monoxide has affinity for the hemoglobin in blood, where it replaces oxygen, interfering with the bloodstream's ability to carry oxygen to the cells.

Many people succumb to accidental carbon monoxide poisoning. A faulty heating system—a lantern or a stove or a portable space heater used inside a tent or a camper without adequate ventilation, a faulty exhaust system that releases fumes inside a vehicle, an automobile or lawnmower engine run inside a closed garage, or ice hockey fans sickened by the fumes from the Zamboni machine—the causes of fatal carbon monoxide poisoning are common and recurring. Regularly inhaling lower levels of CO over a longer period of time can lead to flu-like symptoms.

Improving the efficiency with which motor vehicle engines burn fuel is the most important way of reducing carbon monoxide levels in outdoor air. Alternative fuels (see GASOHOL), more efficient engines (see INTERNAL COMBUSTION ENGINE), electric vehicles, carpooling and improved mass transit are all ways of combating a CO problem. In the home, keep the furnace and any other combustion appliances in good working order. Make sure there is adequate ventilation when any combustion appliance is used indoors and use caution in the garage or any other enclosed space where a motor is running. Nine parts per million is the federal limit for carbon monoxide. Carbon monoxide can be detected by chemical action in detector tubes and badges and by infrared spectrometry (see AIR POLLUTION MEASUREMENT, INDUSTRIAL AIR POLLUTION MEASUREMENT). See also CLEAN AIR ACT, POLLUTANT STANDARD INDEX.

Carbon Tetrachloride a clear, colorless, nonflammable liquid with an ether-like odor. Carbon tetrachloride is used to put out fires, as a SOLVENT, in refrigerants and aerosols, for dry-cleaning clothes and in the production of PESTICIDES. It can get into air and water from industrial effluents, municipal treatment plant discharges and spills. Carbon tetrachloride is moderately water soluble, and is nonpersistent in water, with a HALF-LIFE of less than two days.

The acute health effects (see ACUTE EFFECTS) of exposure to carbon tetrachloride include eye irritation, dizziness and light headedness, which can lead rapidly to unconsciousness and death. It can make the heart beat irregularly or stop. Carbon tetrachloride can also damage the liver and kidneys enough to cause death. Chronic health effects (see CHRONIC EFFECTS) can include severe liver and kidney damage, and thickening and cracking of the skin. Carbon tetrachloride has been shown to cause liver cancer in animals, and there is some evidence that it causes liver cancer in humans. There is limited evidence that it may be a human TERATOGEN. Carbon tetrachloride can enter the body by inhalation or by passing through the skin.

PUBLIC WATER SYSTEMS are required under the SAFE DRINKING WATER ACT to monitor concentrations of carbon tetrachloride (MAXIMUM CONTAMINANT LEVEL = 0.005 milligrams per liter). The OCCUPATIONAL SAFETY AND HEALTH ADMINISTRATION has established a permissible exposure limit for airborne carbon tetrachloride of 10 parts per million (ppm) averaged over an eight-hour workshift, with exposure to concentrations of between 25 and 200 parts per million allowed for a maximum of five minutes in any four hours. Carbon tetrachloride is also cited in NATIONAL INSTITUTE OF OCCUPATIONAL SAFETY AND HEALTH, AMERICAN CONFERENCE OF GOVERNMENT INDUSTRIAL HYGIENISTS and Department of Transportation regulations. It is on the Hazardous Substance List, the HAZARDOUS AIR POLLUTANTS LIST and the Special Health Hazard Substance

List. Businesses handling significant quantities of carbon tetrachloride must disclose use and releases of the chemical under the provisions of the EMERGENCY PLANNING AND COMMUNITY RIGHT TO KNOW ACT.

Carcinogen a substance that causes or promotes the development of cancer. Although the causes of cancer are not known with certainty, a handful of toxins have been so frequently associated with the disease that they are considered to be carcinogens. Many carcinogens have little or no ACUTE EFFECTS. The first symptoms of cancer resulting from exposure typically will not manifest for decades. This long period of latency makes it virtually impossible to link a specific cancer with a particular carcinogen. Many carcinogens, since they have the ability to enter a cell and cause damage, are also MUTAGENS and TERATOGENS. In large doses, a carcinogen kills cells. In smaller doses, a carcinogenic substance causes cells to mutate or to undergo a permanent physical, usually genetic, change. When a mutated cell divides, the resulting cells copy its abnormalities. Many scientists believe there is no safe level of exposure to a carcinogen. The Delaney Clause of the Federal Food, Drug and Cosmetic Act of 1958 bans the use of any amount of a substance that is a proven carcinogen in animals or humans as a food additive.

Epidemiological studies have produced much of what is known about carcinogens. Such studies have shown that people that routinely work with substances such as ASBESTOS, URANIUM, aniline dyes, ARSENIC, VINYL CHLORIDE, ethylene oxide, BENZENE, benzidine dyes, BERYLLIUM, CADMIUM, CHROMIUM, PETROLEUM distillates (see DISTILLATION), FORMALDEHYDE and various PESTICIDES are statistically more likely to get cancer. Tests in which animals are exposed to suspected carcinogens and are then closely observed for signs of disease have also contributed to our knowledge of cancer. Substances that cause cancer in such laboratory experiments are termed possible human carcinogens.

There are three main classes of carcinogens: chemicals, radiation, and viruses. Epidemiologists estimate that 80% of cancers are caused by exposure to carcinogens. It is also possible to inherit a tendency to develop cancer. Scientists estimate that viruses may at least contribute to the development of tumors in about 5% of cases. As of 1990, only about 5 percent of cancers could be linked to environmental POLLUTION (see AIR POLLUTION, WATER POLLUTION) or occupational exposure to carcinogens. Almost all of the 600,000 cases of non-melanoma skin cancer reported each year in the United States are considered to be related to the ultraviolet rays of the sun. Recent epidemiological studies show that exposure to the sun is also a major factor in the development of melanoma, a much more serious skin cancer.

Smoking
Lung cancer resulting from cigarette smoking (see ENVIRONMENTAL TOBACCO SMOKE) is the leading cause of cancer deaths in industrial nations, according to a report released by the New York Academy of Sciences, causing an estimated 40% of cancer mortality among men and 30% among women. In the United States, it is thought that cigarette smoking initiates nearly half the cancer cases resulting from environmental toxins, and that smoking is related to about 30% of all cancer deaths. The American Cancer Society estimates that cigarette smoking causes 85% of lung cancer cases among men and 75% among women. Smoking is also plays a role in cancers of the mouth, pharynx, larynx, esophagus and urinary bladder. Excessive drinking of alcohol, especially when accompanied by any type of tobacco use, also increases a person's risk of cancers of the mouth, larynx, throat, liver, and esophagus.

Cardboard a heavy grade of paper that is used in packaging, building supplies, the production of matches, hardbound book covers and numerous other products. Corrugated cardboard is made by gluing three or more layers of heavy brown kraft paper (see PAPER) together with smooth inner and outer surfaces and a corrugated inner layer. Cardboard is used in every imaginable kind of packaging, from boxes for breakfast cereal and egg cartons to shipping cartons for refrigerators. Composite packaging composed of cardboard and plastic is used to package products that are to be sold in retail stores. Higher grades of bleached, single-layer cardboard are produced for use as poster board and picture frames. The market for recycled old corrugated containers took off in 1994 due to a lack of wood chips to be converted into kraft paper on the west coast. Kraft paper bags can be recycled along with old corrugated containers. Mixed residential paper, a low-cost or no-cost recycled material, is used in the production of lower grades of cardboard.

Catalyst a substance that causes or accelerates a reaction between two or more other substances called *reactants*. The catalyst's properties are not changed as a result of the reaction it initiates. The catalytic action of enzymes plays a central role in metabolism, and catalysts also widely used in industrial processes. A *heterogeneous catalyst* has an adsorptive surface (see ADSORPTION) that holds the reactants in contact with one another long enough for the chemical reaction to take place (see CATALYTIC CONVERTER). In a *homogenous catalyst*, the reactants and the catalyst are in the same physical form (liquid, solid or gas).

Catalytic Converter   a device mounted just behind the exhaust manifold on an internal combustion engine that uses a heterogeneous catalyst to reduce the vehicle's emission of air pollutants (see AUTOMOTIVE AIR POLLUTION). Ceramic beads coated with the METALS platinum and palladium are at the heart of most catalytic converters. Exhaust gases hot from the engines' exhaust valves are forced through the beads. Most of the CARBON MONOXIDE and HYDROCARBONS in the exhaust stream react with the metal catalysts to produce CARBON DIOXIDE and water vapor. Reports of suicide attempts that have failed because of too little carbon monoxide getting past an automobile's exhaust are becoming more common. NITROGEN OXIDES (NO_x) are a byproduct of the catalytic action. Traces of sulfur in fuel can cause a converter to emit HYDROGEN SULFIDE—the source of the rotten-egg smell in an area where many car engines are running—and sulfur trioxide (see SULFUR DIOXIDE), which results in the emission of a fine mist of sulfuric acid. The adsorptive surfaces in a catalytic converter are easily clogged by impurities in gasoline, especially by LEAD. Unleaded fuel refined specifically for use with an exhaust-system catalyst must be used, and the converter must be periodically regenerated or replaced.

The CALIFORNIA AIR RESOURCES BOARD has directed that new automobiles sold in the state must emit no more than 0.125 grams of hydrocarbons. The first of a new generation of low-emission vehicles, which are defined under the California regulations as emitting less than 0.075 grams per mile (gpm) and 0.2 gpm of nitrogen oxide, must be available by 1997, according to the regulations. Catalytic converters, cleaner fuels and more efficient engines will be the principal ways in which auto makers can fulfill the standards. New electrically-heated converters capable of reducing hydrocarbon emissions to 0.03 gpm are coming on the market. The converters use an electric heating element to maintain the high temperature necessary for the catalyst to perform properly before the engine reaches normal operating temperature. The federal hydrocarbon standard, which is being phased in between 1993 and 1998, is 0.25 grams per mile. See also AIR POLLUTION TREATMENT, CALIFORNIA CLEAN AIR ACT.

Celebrity Garbage   trash from the households of famous people. The trash can contains a wealth of information about the life that goes on within a household, which is why police routinely go through the trash cans at the scene of a crime and why an archaeologist sifts through a MIDDEN (see ARCHAEOLOGY) or a modern garbage dump (see GARBAGE PROJECT). There is ample public interest on what goes on behind the closed doors of a celebrity's household, and more than

one writer has performed basic research for an article or book about a celebrity by going through their garbage. See also GARBOLOGY.

Cellulose   the principal material found in the cell walls of all vegetable matter. Cellulose forms the skeleton of plants and a protective coating around living plant protoplasm, and is found in the cell walls of some algae. More than one-third of all vegetable matter is cellulose. There are several closely related kinds of cellulose, all with extremely complex molecular structures. In general, woody, fatty and gummy substances in plants contain cellulose. MICROORGANISMS in the digestive tract of hoofed animals are capable of breaking down the cellulose found in grass and other plants into a form the animal can assimilate for nutrition. Cellulose in the form of discarded wood, paper, fabrics and synthetic materials makes up a large but undetermined amount of the material in a typical landfill, and is a principal source of energy in an incinerator or refuse-derived-fuel plant.

Cellulose and wood and paper products that contain cellulose are derived from wood, cotton, linen rags, hemp, flax and a large variety of other vegetative sources. Most solvents can not dissolve cellulose, and pure cellulose can be produced by using solvents such as water, alcohol, ether, dilute alkalis and acids to remove impurities. Wood and paper products are principally cellulose, and it is used in the production of paper, adhesives, synthetic fabrics, explosives and plastics. Sulfuric acid reacts with cellulose to form the soluble starches glucose, used as a food sweetener, and amyloyd, a starch used to coat parchment paper. By treating cellulose with an alkali and then exposing it to carbon disulfide, it dissolves to become cellulose xanthate, the raw material from which rayon and cellophane are made. Cellulose acetates are made by treating cellulose with acetic anhydride. They are used as glazings, in the manufacture of safety glass and as a molding material in the production of a variety of plastic products. Cellulose ethers are used in the production of adhesives, soaps and synthetic resins. Cellulose nitrates are a series of flammable and explosive compounds formed by mixing cellulose with a mixture of sulfuric acid and nitric acid. The celluloid that was once used to produce film is made by mixing cellulose nitrate with a plasticizer such as camphor.

Center for Safety in the Arts   a clearinghouse for information on health hazards associated with the visual and performing arts. The Center's Art Hazard Information Center answers artists' questions about art-related health problems, makes referrals to physicians with occupational health experience and publishes a newsletter and fact sheets on hazards associated with the arts. The Center for Safety in the Arts is partially supported by grants from the National Endowment for the Arts, New York State agencies and several arts-related unions. Address: 5 Beekman St., Suite 820, New York, NY 10038. Phone: 212-227-6220.

Center for Plastics Recycling Research   see PLASTICS RECYCLING FOUNDATION

Center for Waste Reduction Technologies   an industry-based consortium composed of chemical manufacturers, engineering and construction companies, consultants, government laboratories and government agencies that sponsors and coordinates research and development into new and innovative waste-reduction technologies in collaboration with industry, government and academia. The Center for Waste Reduction Technologies (CWRT), founded in 1991, also disseminates information about waste-product reduction for use in undergraduate curriculums, industrial planning and government rule making. The CWRT is now working on a computerized evaluation tool called the Clean Process Advisory System (CPAS) in cooperation with researchers from the National Center for Manufacturing Sciences and the Center for Clean Industrial and Treatment Technologies. The CPAS will enable the designers of industrial facilities to explore options that will prevent pollution through modeling the operation of the facility while it is still in the design phase. Address: Center for Waste Reduction Technologies, American Institute of Chemical Engineers, 345 East 47th Street, New York, NY 10017. Phone: 212-705-7462.

Centers for Disease Control and Prevention   see PUBLIC HEALTH SERVICE

Centrifugal Cleaner   see CYCLONE

CFCs   see CHLOROFLUOROCARBONS

Chemical   a substance that is obtained by a chemical process, that is used in the production of chemicals or used to produce a particular chemical effect. The chemical elements are the building blocks of matter, and chemical reactions are at the heart of photosynthesis in plants and of METABOLISM and associated processes such as DECAY, PUTREFACTION and the production of biological wastes. Chemicals are chemically active, as opposed to INERT MATERIALS, which do not enter into chemical reactions. The chemical properties of a substance can be revealed by the way in which the material reacts to exposure to various wavelengths of electromagnetic

radiation and to other chemicals. Organic chemicals are compounds that contain carbon. There are two million recognized chemical compounds, and 30,000 chemicals are now used in commerce. Chemical sales in the United States exceed $100 billion per year, and represent 6% of gross national product. Many synthetic chemicals are based on petroleum (see PETROCHEMICAL).

The production, use and disposal of chemicals is the source of most environmental pollution (see AIR POLLUTION, WATER POLLUTION) and HAZARDOUS WASTE. A *fine chemical* is produced and handled in relatively small amounts, and usually occurs in a relatively pure form. Photographic chemicals, pharmaceuticals and perfumes are fine chemicals. A *heavy chemical* is produced in large quantities (often a ton or more at a time), in a relatively unrefined form. The acids, alkalis and salts used in industrial processes (see PROCESS CHEMICAL) are heavy chemicals. See also BIOCHEMICAL OXYGEN DEMAND, CHEMICAL MANUFACTURING WASTE, CHEMICAL OXYGEN DEMAND.

Chemical Manufacturing Waste   the byproducts of the manufacturing processes that produce chemicals. More than 700 chemical manufacturing facilities produced 218 of the total of 275 trillion tons of HAZARDOUS WASTE reported to the ENVIRONMENTAL PROTECTION AGENCY (EPA) in 1986. Two million chemical compounds are now recognized, and 30,000 chemicals are currently used commercially (see PROCESS CHEMICAL). More than one hundred petrochemical plants are located along the Mississippi between New Orleans and Baton Rouge, and the lower Mississippi River receives treated effluent from all of them. Toxic process chemicals are regulated under the Toxic Substances Control Act, and hazardous chemicals in waste are managed under the RESOURCE CONSERVATION AND RECOVERY ACT and the COMPREHENSIVE ENVIRONMENTAL RESPONSE COMPENSATION AND LIABILITY ACT. Chemical waste can be extremely long-lived if it is not biodegradable (see BIODEGRADABILITY). POLYCHLORINATED BIPHENYLS (PCBs), for instance, are still found in the Great Lakes and the Hudson River even though they were banned in the TOXIC SUBSTANCES CONTROL ACT (TSCA) of 1976.

Waste associated with chemical production includes still bottom, a thick, tar-like substance created in the production of organic chemicals and other organic and oily residuals of petroleum products; spent catalysts, solvents and organic solutions produced in the manufacture of pharmaceuticals, rubber, plastics and organic chemicals; organic sludges generated in the treatment of industrial wastewater (see INDUSTRIAL LIQUID WASTE, pesticide manufacturing wastes); leaks and spills of the chemicals used in industrial processes and materials contaminated in cleaning them up; and ASH from toxic-waste incinerators (see FLY ASH, INCINERATION, TOXIC- ITY CHARACTERISTIC LEACHING PROCEDURE). Industrial liquid waste must generally be stabilized through pre- treatment before it can be emptied into a municipal sewer system. See also INDUSTRIAL SOLID WASTE.

Chemical Oxygen Demand   sometimes used rather than BIOCHEMICAL OXYGEN DEMAND (BOD) to assess the amount of nutrients and ORGANIC MATERIAL in wastewater. Chemical oxygen demand (COD) is mea- sured by adding an oxidizing agent (see OXIDATION) such as potassium dichromate to a water sample, then boiling the sample until all oxidation has ceased. The amount of the oxidant remaining in the sample is then measured, and the COD is calculated on the basis of the amount of oxidant consumed. For a water sample containing primarily biodegradable contaminants and few chemicals, the COD is roughly equivalent to BOD, and it can be determined much more quickly.

Chemical Waste Management   the largest hazardous- waste disposal company in the United States. Chemical Waste Management owns and operates six hazardous waste landfills and 19 hazardous waste treatment, resource recovery or disposal facilities. CWM's sub- sidiary, Chem-Nuclear, is the only private provider of low-level nuclear waste disposal east of the Mississippi. A related company, CWM Technologies, develops new technologies for the recycling, treatment, neutralization and disposal of hazardous waste. CWM is a subsidiary of WMX Technologies. Address: 3001 Butterfield Rd, Oakbrook, IL 60521. Phone: 708-218-1500.

Chemosterilzation   see DISINFECTION

Chernobyl   a Russian nuclear power facility that suf- fered a catastrophic explosion and core meltdown in April, 1986. The Chernobyl complex, located about 80 miles northwest of Kiev in Ukraine on the Pripyat River, consisted of four operating reactors at the time of the disaster, with two more under construction. The installation would have become the world's largest nuclear power plant—capable of providing electricity to 7 million people—if the accident had not occurred. Two of the Chernobyl reactors were still being run by the Ukrainian government as of 1996, despite the protests of the United States and the European Union and their efforts to raise funds to replace them. The RBMK-1000 reactors used at Chernobyl are very differ- ent from the commercial reactors found in the west in that they use graphite as a moderator (see NUCLEAR REACTOR for description). The lack of a containment structure around the building housing the reactor core, however, is the most notable difference. Such a struc-

ture might have been able to withstand the explosions and flying debris unleashed in the reactor core by the accident. To make matters worse, the roof of the build- ing was coated with bitumen, which is so flammable that it was banned for use on other buildings in the Soviet Union.

The chain of events that led to the meltdown was ini- tiated by a test of equipment intended to stabilize the electricity output of the power turbines as they lost their inertia and slowly coasted to a stop as the result of a power outage. It takes three minutes for the diesel backup generators to produce sufficient power to run the 5,500-kilowatt pumps that circulate coolant through the reactor core, more than can be supplied by battery storage. Because the core of the RBMK-1000 reactor cannot operate without the circulation of coolant for more than a minute or so without risking serious damage, the ability to run the pumps until the diesel generators come on line is critical. One of the two steam turbines driven by the Number Four reactor was shut down. The output of the reactor was to be reduced to about 800 megawatt (MW) while the test was underway—about half the output required to run one of the turbine generators—because if it were shut down much below this level, neutron-absorbing xenon-135 would be produced. Only after the xenon had been purged from the core by slowly increasing the rate of the nuclear reaction over a period of several days could the reactor be returned to full power.

In powering down the reactor, the control rods were accidentally lowered too far, and output of the reactor was reduced to less than 30 MW. The operators then raised some of the control rods in an attempt to increase the reactor's production of steam; they also increased the flow of cooling water through the core, thus reducing the reactor's ability to produce steam. The thermal output of the reactor rose to about 200 MW—still a level insufficient for the test—after the additional control rods were withdrawn. The opera- tors decided to attempt to boost the energy output by withdrawing more control rods from the core. Safety systems would have normally shut down the reactor at this point, but they had been disabled for the duration of the test. The flow of steam to the turbine was cut to start the experiment. Inertial energy from the turbine was to provide power to run four coolant circulation pumps, even though the turbine was spinning too slowly to accomplish the task.

The cutoff of steam to the turbine and the reduction of the flow of cooling water through the reactor at this critical moment caused steam voids to form quickly throughout the reactor core. The power output of the reactor began to rise, and the operators realized the reaction was going out of control. They pushed the emergency shutdown button, but it was too late. The surge of heat inside the reactor core had already caused so much damage that the control rods jammed before they could slow the reaction. An explosion caused by water flashing to steam when it hit hot core fragments on the reactor floor was quickly followed by a larger hydrogen blast that rocked the building, throwing molten pieces of the core through the roof and onto the roof of the adjoining machinery building and onto the grounds. A fire sprang up immediately on what remained of the roof of the Number Four reactor building and on the adjacent machine building. The heroic efforts of firemen—none of whom knew that they were being exposed to lethal levels of radiation—were all that kept the fire from spreading to the other three operating reactors at the Chernobyl complex. An estimated 50 tons of nuclear fuel were vaporized and injected into the atmosphere by the disaster—at least 100 times the radiation released by the atomic bomb that leveled Hiroshima—and an additional 70 tons of radioactive fuel were dispersed around the site. Another 50 tons of nuclear fuel remained in the reactor vault along with an estimated 800 tons of the 1,700 tons of graphite from the core. The explosions opened a path for airflow through the core, causing the hot graphite to ignite. Firemen on the roof were exposed to radiation from airborne particles and hot debris from the core, and many would eventually die from their exposure. All the firemen who fought the graphite fire in the reactor core, which was to burn for more than a week, soon died of acute radiation poisoning. More than 5,000 metric tons of a mixture of sand, clay, lead, boron carbide and dolomite was dumped on the remains of the core over the next six days in an effort to smother the graphite fire. The sorties stopped when heat from fission and the continuing combustion of graphite under the growing pile of sand caused the temperature of the core materials to build. A slow meltdown of many of the remaining core materials led, over the next few days, to releases of radioactivity on the scale of those caused by the original explosion.

Wind and precipitation distributed fallout from Chernobyl—primarily radioactive cesium 137 and strontium 90—over 50,000 square miles of prime farmland across Ukraine, Belarus, Russia and other northern European countries. The estimated 50 million curies of radiation that were released into the atmosphere at Chernobyl dwarfs the 1979 accident at THREE MILE ISLAND, where only about 50 curies were released. Pripyat, a town of 45,000 just north of Chernobyl, was permanently evacuated as were numerous smaller towns in the region. A massive STEEL-and-CONCRETE "sarcophagus" more than 10 stories high and 59 feet

thick in places has been built over the hot remains of reactor No. 4. Of the 100,000 people who built the vault in 1986, several thousand died from radiation poisoning. Cancer rates in Ukraine, as well as in bordering Belarus and Russia, have far exceeded predictions. According to research published in the *British Medical Journal*, the rate of thyroid cancer in a region north of the Chernobyl nuclear plant is nearly 200 times higher than normal. Fallout from Chernobyl circled the globe, sparking renewed international concern over the safety of nuclear power.

Chiffonier   see RAG PICKER

Chlordane   a chlorinated-hydrocarbon INSECTICIDE (see CHLORINATED HYDROCARBON) manufactured by Velsicol Corporation. Pure chlordane is a colorless-to-amber, odorless, thick liquid, but the commercial product has a CHLORINE-like odor. Chlordane was the leading household insecticide in the United States during the 1960s and early 1970s. It was banned for use on food crops in 1975. It was used for termite control during the next decade, but evidence that air inside homes treated with chlordane or the related chemical, heptachlor, became tainted with the PESTICIDES led to a complete ban in 1987. Because of its widespread general use in the past, its persistence and present use of existing stocks, chlordane enters the environment in agricultural and residential RUNOFF and can be found in indoor air of houses in which it was used to treat termites. Chlordane is slightly soluble in water. Once it gets into water, it is highly persistent, with a HALF-LIFE of greater than 200 days.

Chlordane is a possible human CARCINOGEN that has been shown to cause liver cancer in animals. It is a possible human TERATOGEN, and may decrease fertility in males and females. Chlordane may also damage the liver and kidneys. An acne-like rash may appear following skin contact.

PUBLIC WATER SYSTEMS are required under the SAFE DRINKING WATER ACT to monitor concentrations of chlordane (MAXIMUM CONTAMINANT LEVEL = 0.002 milligrams per liter). The BEST AVAILABLE TECHNOLOGY for the removal of chlordane is granular ACTIVATED CARBON. Chlordane is regulated by the OCCUPATIONAL SAFETY AND HEALTH ADMINISTRATION and cited by AMERICAN CONFERENCE OF GOVERNMENT INDUSTRIAL HYGIENISTS. It is on the Hazardous Substance List, the Hazardous Air Pollutants List and the Special Health Hazard Substance List. Businesses handling significant quantities of chlordane must disclose use and releases of the chemical under the provisions of the EMERGENCY PLANNING AND COMMUNITY RIGHT TO KNOW ACT.

Chlorinated Dioxin   see DIOXIN

Chlorinated Hydrocarbon   a HYDROCARBON in which one or more of the HYDROGEN atoms has been replaced by a CHLORINE atom. Most chlorinated hydrocarbons are solids or thick fluids. Chlorinated hydrocarbons are used in lubricating fluids and, because of their low electrical conductivity, as insulating fluids in transformers and other electrical equipment. Although they are of relatively low acute toxicity to mammals (see ACUTE EFFECTS), many chlorinated hydrocarbons are of high acute toxicity to invertebrates and are the active ingredient in some PESTICIDES. They are a PERSISTENT COMPOUND that can build to lethal levels in living tissue over a period of months or years (see BIOACCUMULATION). Chlorinated hydrocarbons are carcinogens and mutagens, so their use is tightly regulated or banned. See also CHLORDANE, HEPTACHLOR, ORGANOCHLORIDE, POLYCHLORINATED BIPHENYLS, TOXAPHENE.

Chlorination   see CHLORINE, DISINFECTION

Chlorine   a dense, greenish-yellow gas with an acrid, unpleasant odor, or a liquid solution. Chlorine is the fourteenth most common ELEMENT on the earth, and is one of the most reactive (see REACTIVITY) substances known. It does not exist naturally in its pure state, but the chlorine atom is the basis of countless other compounds. The oxidizing abilities of chlorine are the basis of its use both as a PESTICIDE and DISINFECTANT and as an industrial process chemical. Chlorine is used in the production of SOLVENTS, industrial chemicals, bleaches, synthetic rubber and PLASTICS. Many chlorine compounds are extremely toxic. Chlorine dioxide is used in most U.S. public drinking water systems that use surface water or groundwater from shallow aquifers as a disinfectant. Chlorine reacts with POLYCYCLIC ORGANIC MATTER to form TRIHALOMETHANES, a family of toxic compounds that includes CHLOROFORM. Chlorine compounds are the active ingredient in INSECTICIDES, HERBICIDES, FUNGICIDES, RODENTICIDES and disinfectants (see CHLORINATED HYDROCARBON) and in household cleansers, bleach, DETERGENTS and dish soaps. Sodium chloride, or table SALT, is an essential dietary element. Most chlorine is produced commercially by passing an electric current through salt water.

Chlorine gas gets into the atmosphere from industrial emissions, accidents and spills. It can get into indoor air from household products and solvents containing chlorine compounds. DIOXIN and chloroform are sometimes produced when chlorine-base cleansers are used. Chlorine is highly soluble in water, is non-persistent with a HALF-LIFE of less than two days.

Chlorine gets into water from SEWAGE TREATMENT plants, which use it as a disinfectant, from industrial effluents and from spills.

The ACUTE EFFECTS of exposure to moderate levels of chlorine gas may include irritation of the eyes, nose and throat, and chest pain. Higher concentrations can burn the lungs and cause a buildup of fluid in the lungs and death. Contact can severely burn the eyes and skin. Repeated exposures or a single high exposure can permanently damage the lungs, and can also damage the teeth and cause a skin rash. Exposure to a concentration of 25 parts per million (ppm) chlorine is immediately dangerous to life and health. The human nose can detect chlorine at a concentration of about 0.31 ppm. There has been a movement among environmentalists during the 1990s to outlaw most or all uses of chlorinated compounds because of the health threat associated with many such chemicals. There is no more basic chemical than chlorine, however, and changing over to processes that do not rely on chlorine would be an undertaking of mind-boggling proportions (and expense).

The OCCUPATIONAL SAFETY AND HEALTH ADMINISTRATION has established a permissible exposure limit for airborne chlorine of 1 ppm, not to be exceeded at any time. It is also cited in AMERICAN CONFERENCE OF GOVERNMENT INDUSTRIAL HYGIENISTS, NATIONAL INSTITUTE OF OCCUPATIONAL SAFETY AND HEALTH, ENVIRONMENTAL PROTECTION AGENCY and Department of Transportation regulations. PUBLIC WATER SYSTEMS are required under the SAFE DRINKING WATER ACT to monitor concentrations of chlorine. The proposed MAXIMUM DISINFECTANT RESIDUAL LEVEL (MDRL) for chlorine and chloramine is 4 milligrams per liter, the MDRL for chlorine dioxide is 0.8 mg/L. Chlorine is on the Hazardous Substance List and the HAZARDOUS AIR POLLUTANTS LIST. Businesses handling significant quantities of chlorine or chlorine dioxide must disclose use and releases of the chemical under the provisions of the EMERGENCY PLANNING AND COMMUNITY RIGHT TO KNOW ACT.

Chlorofluorocarbons (CFCs)   a group of chemical substances including chlorofluoromethane and similar compounds of CHLORINE, fluorine and carbon that have been used as SOLVENTS, especially in the electronics industry; as refrigerants, heat-transfer fluids such as Freon in home and automobile air conditioners and refrigerators; and as propellants for aerosol products. Chlorofluorocarbons (CFCs) are nonpoisonous and chemically inert gases that are so durable that they can stay airborne for years after release—even though they are four to eight times heavier than air. Atmospheric turbulence cannot only keep them aloft, but eventually can carry them into the stratosphere.

Once in the stratosphere, the CFC molecules find their match: they are torn apart when struck by ultraviolet rays that have almost the energy they had when emitted from the sun. As the CFCs are broken down, chlorine, which can in turn destroy ozone molecules, is released. Many experts link the decade-long decline in STRATOSPHERIC OZONE readings with the presence of chlorine from CFCs. Hydrofluorocarbons or HFCs, have been widely adopted as a CFC substitute, although concern that they may also damage the ozone layer has been expressed.

Chloroform   a clear, colorless liquid with a pleasant, sweet odor. Chloroform—also called *trichloromethane*—is nonflammable and is heavier than water. It is used as a SOLVENT and as a cleansing agent; in making dyes, drugs, and PESTICIDES; and in fire extinguishers to lower the freezing point of CARBON TETRACHLORIDE. It was widely employed as an anesthetic in the past. Chloroform gets into air and water from industrial effluents, municipal waste treatment plant discharges and spills. Chloroform is a TRIHALOMETHANE (THM), which may be formed when drinking water is treated with chlorine. Chloroform in domestic water can evaporate into indoor air when it is sprayed from a faucet or shower head or is boiled or agitated. Chloroform is highly soluble in water. Concentrations of 1,000 mg and more will mix with a liter of water. Chloroform is nonpersistent in water, with a HALF-LIFE of less than two days.

Acute health effects (see ACUTE EFFECTS) include dizziness, lightheadedness, nausea, confusion and headache. Exposure can cause the heart to beat irregularly or to stop, resulting in death. Contact can irritate the skin, causing a rash or burning sensation on contact. Liquid chloroform can cause severe eye burns. Exposure to chloroform vapor can irritate the nose and throat. Chloroform's most serious CHRONIC EFFECT is that it is a probable CARCINOGEN in humans. It has been demonstrated to cause liver, kidney and thyroid cancer in animals. There is limited evidence that chloroform is a TERATOGEN in animals, and it is therefore considered a possible teratogen in humans. Chloroform can damage the liver. Repeated skin contact with the liquid may produce skin drying and cracking. Repeated exposure may affect the kidneys and nervous system.

The OCCUPATIONAL SAFETY AND HEALTH ADMINISTRATION has established a permissible exposure limit for airborne chloroform of 50 parts per million, not to be exceeded at any time. Chloroform is also cited in AMERICAN CONFERENCE OF GOVERNMENT INDUSTRIAL HYGIENISTS, National Toxicology Program, Department of Transportation and NATIONAL INSTITUTE FOR OCCUPATIONAL SAFETY AND HEALTH regulations. Chloroform

is on the Hazardous Substance List, the HAZARDOUS AIR POLLUTANTS LIST and the Special Health Hazard Substance List. PUBLIC WATER SYSTEMS are required under the SAFE DRINKING WATER ACT to monitor for chloroform and other THMs. Businesses handling significant quantities of chloroform must disclose use and releases of the chemical under the provisions of the EMERGENCY PLANNING AND COMMUNITY RIGHT TO KNOW ACT. See also INDOOR AIR POLLUTION.

Chromium   a steel-gray, lustrous METAL often used in powder form. Chromium is used in photography, in the production of stainless STEEL and other metal alloys, in the chrome plating of other metals and in a wide variety of industrial processes. Approximately 113,000 metric tons of chromium were produced by U.S. manufacturers in 1993. More than 80% of the chromium used by U.S. industry is imported. It gets into air and water from industrial and municipal effluent. Chromium can get into indoor air from welding, brazing, soldering, plating, cutting and metallizing. Chromium and its SALTS range from low to high in solubility. Chromium is highly persistent in water, with a HALF-LIFE of more than 200 days. Chromium is more toxic in soft water than in hard water.

Chromium is a CARCINOGEN and MUTAGEN. It has been linked to lung and throat cancer in humans. Acute health effects (see ACUTE EFFECTS) include "metal fume fever," a flu-like illness with chills, aches, cough and fever that lasts for about 24 hours after inhaling chromium fumes. Chromium metal ore has been reported to cause allergy. Chromium particles can irritate the eyes. Exposure to high levels of *hexavalent chromium*, the most toxic of the chromium compounds, causes ulcers, respiratory disorders and skin irritation in humans. *Trivalent chromium* is an essential dietary trace mineral.

The OCCUPATIONAL SAFETY AND HEALTH ADMINISTRATION has established a permissible exposure limit for airborne chromium of 1 milligram per cubic meter, averaged over an eight-hour work shift. It is also cited in AMERICAN CONFERENCE OF GOVERNMENT INDUSTRIAL HYGIENISTS and National Toxicology Program regulations. Chromium is on the Hazardous Substance List and the Special Health Hazard Substance List. Chromium compounds are on the HAZARDOUS AIR POLLUTANTS LIST. PUBLIC WATER SYSTEMS are required under the SAFE DRINKING WATER ACT to monitor concentrations of chromium (MAXIMUM CONTAMINANT LEVEL = 0.1 milligrams per liter for total chromium). The BEST AVAILABLE TECHNOLOGY for the removal of chromium from DRINKING WATER is coagulation-filtration, ION EXCHANGE, lime softening (chromium III only) or REVERSE OSMOSIS (see

WATER TREATMENT). Businesses handling significant quantities of chromium or chromium compounds must disclose use and releases of the chemical under the provisions of the EMERGENCY PLANNING AND COMMUNITY RIGHT TO KNOW ACT.

Chronic Effects   symptoms of exposure to a toxin that may occur months or years after the fact. Chronic effects indeed may not become manifest until decades after exposure, but once symptoms appear, they tend to be long-lasting or permanent. Either a single high-level exposure or long-term low-level exposure to a toxin can cause chronic health effects. The chronic TOXICITY of a substance can, in many cases, cause a far more serious and debilitating disease than the acute toxicity (see ACUTE EFFECTS) of the same substance. Substances that can cause serious chronic conditions such as heart and circulatory disorders, cancer and reproductive problems frequently have no discernible acute effects at all. Many MUTAGENS similarly have little or no acute effects, but may cause genetic abnormalities that will last for generations.

It can be next to impossible to determine a substance's chronic effects. To start with, basic research simply has not been done on the acute or chronic health effects of tens of thousands of chemicals in common use today. Even less data is available on the health effects of multiple contaminants, which in many cases have a synergistic effect. The extreme mobility of modern society, the high number of potential toxins to which an individual may be exposed, the length of time between exposure and the emergence of symptoms and the fact that exposure most likely will occur without the victim's knowledge compound the difficult of accurately diagnosing chronic effects.

Cigarette Smoke   see ENVIRONMENTAL TOBACCO SMOKE

Citizens' Clearinghouse for Hazardous Waste   a citizens' group that acts as a repository for information on HAZARDOUS WASTE and environmental toxins and as a supporter of local groups concerned with pollution. The Citizens' Clearinghouse for Hazardous Waste (CCHW) approaches pollution as more of an environmental justice issue than do mainstream environmentalists, asserting that neighborhoods composed primarily of low-income people and minorities are more likely to suffer contamination. The CCHW campaign to put a stop to the land disposal of hazardous waste might be considered as an instance of the "not-in-my-backyard" syndrome (see NIMBY), but it is called "stopping up the toilet" by movement insiders. The CCHW and the local groups it works with have the philosophy that industrial polluters will be forced to find safer disposal alternatives for hazardous materials if their attempts at siting dumps and incinerators to dispose of hazardous waste are thwarted. The CCHW played a major role in the MCTOXICS CAMPAIGN in which McDonald's Corporation decided to discontinue the use of POLYSTYRENE foam "clamshells" as a container for hamburgers. Address: Citizen's Clearinghouse for Hazardous Waste, Box 6806, Falls Church, VA 22040. Phone: 703-237-2249.

Clark Fork River   a river that flows through western Montana and northern Idaho. The upper stretches of the Clark Fork were contaminated by years of copper mining starting in the early 1800s at one of the river's headwaters in Butte, Montana. Smelter waste containing CYANIDE, COPPER, ARSENIC, CADMIUM, LEAD and zinc was dumped into the river for years, creating major pockets of toxic sediment in the river and its flood plain at least as far west as Missoula, 130 miles downstream. The nation's largest Superfund site now stretches from the Butte area to the Milltown Reservoir, just upstream of Missoula, comprising the entire river and its flood plain with a total of 25 individual HAZARDOUS WASTE sites in its drainage. The reservoir, constructed by Montana Power Company in 1907, contains an estimated 6.5 million tons of toxic sediment, which has caused GROUNDWATER contamination in nearby Bonner.

The birthplace of the Clark Fork is just below the Warm Springs settling ponds, which contain an estimated 19 million cubic yards of heavily contaminated sediment. Silver Bow Creek, which drains the high valley in which Butte is located, today wanders through a wasteland of barren century-old mill TAILINGS. The Berkeley Pit, a 2,200-foot deep open pit mine (see STRIP MINE), has been slowly filling with water since the pumps that kept it clear were shut off in 1983, and is expected to fill to the point that it will start to flood local aquifers with extremely toxic water by 1996. At the copper smelter in nearby Anaconda, an estimated 185 million cubic yards of heavy-metal laced tailings, 27 million cubic yards of furnace slag and 300,000 cubic yards of arsenic-rich flue dust (see AERIAL DEPOSITION) have contaminated most of the soils, surface water and groundwater. Well over $50 million has been spent on remediation of the pollution in Butte, Anaconda and the Clark Fork, primarily for planning. Atlantic Richfield Corporation bought the Butte mines from Anaconda Corporation in 1977, and inherited responsibility for cleaning up the mess. ARCO in turn sold the mines in 1983, but retained legal responsibility for the cleanup in the sale agreement.

Clean Air Act several related pieces of federal legislation designed to improve and protect outdoor air quality. The original Clean Air Act (CAA) became law in 1963, but amendments passed in 1966, 1970, 1977 and 1990 have largely rewritten the original law. (The original act together with all its amendments is referred to as the Clean Air Act in this document.) The act has two primary goals: to prevent the further deterioration of U.S. air quality, and to reduce or eliminate certain air pollutants that have been identified as being hazardous to humans and to the environment. The CAA has four primary goals: attainment and maintenance of air quality standards; motor vehicles and alternative fuels; toxic air pollutants; and acid precipitation. POLLUTION permit and enforcement specifics and provisions for phasing out chemicals that damage the ozone layer (see STRATOSPHERIC OZONE) are also included. A deadline of December 1987 was set in the act for cities to meet federal air-quality standards, but more than 60 urban areas were unable to comply.

The Clean Air Act Amendments of 1990 were passed after nearly a decade of often intense debate among politicians, industrial lobbyists and environmentalists. Industry feared the amendments would cost too much and put too many liabilities on polluters, while environmentalists argued the provisions did not go far enough to protect public health and the environment. After passage of the CAA, the Bush administration was charged by environmentalists with trying to postpone and water down the act by gumming up the administrative process. During 1991, eight northeastern states—after growing impatient with the slow pace of act implementation—passed their own restrictions on NITROGEN OXIDE emissions from oil-, coal-, and natural-gas-fired electrical generating stations. In May, 1992, the SIERRA CLUB sued the ENVIRONMENTAL PROTECTION AGENCY (EPA) for failing to meet deadlines imposed under the legislation.

The act labels regions that violate AIR POLLUTION standards "non-attainment areas." Under the law, tighter restrictions on air pollution sources are to be applied in such areas. Pollution in a non-attainment area is classified as marginal, moderate, serious, severe or extreme, depending on the concentration and toxicity of the contaminants present. The aggressiveness of the steps taken to combat pollution under the CAA increases with the degree of contamination of an area's air. A system of emission allowances was established by the 1990 Amendments. A business that emits less pollution than allowed under CAA limits is under the rules to sell "pollution credits" to another company that is in violation of federal air pollution guidelines but that cannot currently afford the investment required to rectify the problem. In the first auction of pollution allowances in Chicago in March 1993, $21 million worth of credits were sold to utilities, brokerage firms and environmental groups, which intended to retire the pollution allowance permanently rather than sell it to a polluter.

The EPA is charged with overall implementation of the CAA, with the responsibility for monitoring and local enforcement generally delegated to state agencies. The act says that each state must develop a CAA implementation plan and identify areas not in compliance with NATIONAL AMBIENT AIR QUALITY STANDARDS. If the EPA finds that a state's plan meets the act's requirements, the state is left to deal with individual air polluters however it sees fit. The EPA has the power to impose sanctions on states that fail to meet the requirements of the act, including the loss of most federal highway funds.

Motor Vehicles and Alternative Fuels

Since automobile exhaust is a primary ingredient of all urban smog, the act includes provisions intended to clean up vehicle emissions. Cities that exceed federal limits for ozone, which is produced when volatile organic compounds (VOCs) from vehicles and other sources react with atmospheric gases and sunlight, or for CARBON MONOXIDE, which is the result of incomplete combustion of gasoline and other fuels, must use special fuels during the pollution season. The 1970 CAA called for cars that could run on lead-free fuels by 1975, and initiated the phase-out of leaded gasoline.

The 1990 CAA amendments include tighter emission standards for new cars, trucks and urban buses. Regular emission testing of existing automobiles can also be mandated in areas that violate ozone or carbon monoxide limits. Independent auto tune-up shops were expected to reap a windfall of $2 to $5 billion annually in repairs and testing as a result of the rules' passage. The amendments authorize the stricter vehicle emission standards passed by the CALIFORNIA AIR RESOURCES BOARD a month before CAA passage (see CALIFORNIA CLEAN AIR ACT). During 1992, nine northeastern states and the District of Columbia moved to adopt the California standards in preference to the weaker CAA emissions codes.

The 1990 amendments ordered that cleaner-burning reformulated gasoline be ready for use by the summer of 1995 in nine areas that violate federal ozone limits. The gasoline was required to reduce VOCs and other hazardous emissions an average of 15% below levels generated by fuels in use in 1990. The amendments also ordered the use of oxygenated fuels during high pollution seasons in areas that fail federal carbon monoxide standards. Oxygenated gasoline contains more oxygen because ETHANOL, METHANOL or the methanol derivative, METHYL-TERTIARYBUTYLETHER (MTBE), has been added. The presence of the extra oxygen leads to cleaner combustion and reduced carbon monoxide emissions. The amendments also limited the SULFUR content of DIESEL fuel to less than .05% by weight.

Fumes that evaporate from gasoline when it is being pumped into a tank are a major source of urban air pollution. The amendments address fumes from fuel tanks of all sizes, from fuel wholesalers bulk storage and transport trucks to the neighborhood gas station to vehicles being fueled there. They call for the development of gasoline formulations with a lower volatility for use during the ozone pollution season. Onboard vapor recovery systems, which would be capable of recovering at least 95% of the gases that evaporate while refueling, are required by the legislation as an option on cars and light-duty trucks. Fuel dealers and service stations are required to install systems to recover the vapors released when tanks are filled.

Toxic Aid Pollutants

The 1990 amendments list 189 hazardous air pollutants that must be regulated. The amendments require that the maximum achievable control technology (MACT) be used to control emissions of the contaminants on the list. The cost of the pollution-control technology, its environmental impacts (in areas other than air pollution) and its energy requirements are all to be considered when specifying a MACT. Technological pollution controls (such as SCRUBBERS), changes in industrial processes, substitution of process materials and special operator training are all possible forms of MCAT. The act requires that the EPA establish emission standards that "provide an ample margin of safety to protect the public health."

Acid Rain

To reduce emission of the precursors of acid precipitation, sulfur dioxide and nitrogen oxides, the 1990 amendments offer incentives for clean-coal technology. Included are measures that remove impurities from coal before it is burned, that improve the efficiency of coal combustion or that remove contaminants from exhaust gas. The Department of Energy, using funds provided by the EPA, is authorized to spend up to $2.5 billion on a series of clean coal technology demonstration projects. Government funds will cover at least 20% of the project's cost.

As a result of the legislation, sales of low-sulfur western coal has improved. Utilities that burn low-sulfur coal often meet CAA sulfur dioxide limits without installing any further flue-gas cleaning equipment. Technology that removes sulfur and other impurities

from coal has also received a boost. Sales of high efficiency coal combustion systems such as fluidized-bed incinerators have also improved as a result of the act.

CFC-Ozone Layer

The 1990 amendments call for the phase-out of all compounds known or strongly expected to cause deterioration of the stratospheric ozone layer including CFCS, HCFCS, HALONS, CARBON TETRACHLORIDE and METHYL CHLOROFORM. See also HYDROCARBON, LEAD, NATIONAL EMISSIONS STANDARDS FOR HAZARDOUS AIR POLLUTANTS, PREVENTION OF SIGNIFICANT DETERIORATION, REASONABLY AVAILABLE CONTROL TECHNOLOGY.

Clean Water Act   a federal law and its amendments that govern the pollution of surface water. First enacted in 1972 as the Water Pollution Control Act (WPCA), the law is intended to "restore and maintain the chemical, physical and biological integrity of the nation's waters," to make all navigable waters in the United States clean enough for fishing and swimming by 1983 and to end the discharge of pollutants into such waters by 1985 (see ZERO DISCHARGE). The act calls for the ENVIRONMENTAL PROTECTION AGENCY (EPA) to establish water quality standards, to develop standards governing the concentration of toxins in effluent (see EFFLUENT STANDARD) based on the best water purification technology available (see BEST CONVENTIONAL WASTE-TREATMENT TECHNOLOGY) and to establish the NATIONAL POLLUTANT DISCHARGE ELIMINATION SYSTEM (NPDES), the system under which permits are issued to businesses that discharge pollutants into surface water. Publicly owned SEWAGE TREATMENT plants are required under the act to install at least secondary treatment equipment by 1983 (later revised to 1988), and federal funding for such upgrades was increased from 50 to 75%. The act initially authorized $18 billion in construction grants from 1972 through 1976. The EPA is required under the act to perform a biennial needs survey, an assessment of the cost of bringing effluent from publicly owned WASTEWATER treatment plants up to the standards specified under the act.

The U.S. Ninth Circuit Court of Appeals ruled in 1993 that violations of the terms of permits issued under National Pollution Discharge Elimination System cannot be enforced by citizens' groups under the Clean Water Act. The Supreme Court, on the other hand, has ruled that citizen's groups *do* have authority to sue to ensure that the CLEAN AIR ACT (CAA) is effective in protecting water quality. The appeals court has granted a petition for rehearing. The case questioned whether the Northwest Environmental Advocates had grounds to sue the city of Portland, Oregon, for violating its NPDES permit by occasionally dumping untreated effluent into the Columbia River from its combined storm water/sewage system. An estimated 800 to 1,000 U.S. communities have such systems.

History

The WPCA combined two separate streams of federal regulation that had existed before that time: the Federal Water Pollution Control Act of 1948 (which made federal money available to cities for the construction of treatment facilities for sewage effluent that was dumped into navigable waters with the primary goal of decreasing the health risks associated with such waterways) and the Oil Pollution Act of 1924 (which prohibited the dumping of oil in coastal waters with the primary goal of protecting the environment). The two lines of legislation were first joined in the Water Quality Improvement Act of 1970, which in turn led to the complete overhaul of federal water quality law that resulted in the WPCA.

Amendments

The Clean Water Act of 1977 was the first major amendment of the WPCA. The federal share of funding for the construction of municipal SEWAGE TREAT-MENT plants was increased in the act to 85% for projects that used innovative and alternative technologies, and the total level of federal funding for sewage treatment facilities was increased to $24.5 million through 1981. The Municipal Wastewater Treatment Construction Grant Amendments of 1981 reduced the federal share of construction costs from 75% to 55% for conventional sewage treatment facilities and from 85% to 75% for innovative technologies, and authorized $2.4 billion in construction grants from 1982 to 1985. The Clean Water Act Amendments of 1987 called for the complete transfer of pollution control programs to the states by 1994, and started the phase-out of federal construction grants, which were to be replaced by state revolving funds, which loan money to states to help with the construction of water purification facilities. The amendments called for state governments to develop programs to combat "non-point pollution" (see DIFFUSE POLLUTION). The EPA was authorized to spend $8.4 billion between 1989 and 1994 on grants under the program, at which time federal funding for sewage treatment facilities ended.

Performance

The EPA has missed most of the deadlines established in the WPCA and subsequent amendments, and has not established federal guidelines for effluent issuing from most of the 75,000 industrial facilities that dump effluent directly into surface water (see INDUSTRIAL

LIQUID WASTE, MIXING ZONE). The agency has been the target of numerous lawsuits by environmental groups over missed deadlines and lax enforcement of the Clean Water Act. Agency officials say they are doing all they can to fulfill the mandates of the CWA and other environmental legislation given their funding and staff.

Some of the most obvious contamination of surface water was cleared up by the act. The frequency with which untreated sewage and industrial waste finds its way into a watercourse has been reduced as a result of the act, and the concentration of some pollutants has been reduced as a result, but the act has fallen far short of the original goal of cleaning up surface water by 1985. The EPA estimates that 30% to 40% of American waters do not meet the Clean Water Act standards, largely because of polluted runoff and other diffuse sources of pollution. Federal standards have yet to be imposed on the majority of industrial water polluters, and entire classes of water polluters—including hazardous-waste treatment facilities; industrial laundries; hospitals; businesses that recycle solvents, automotive oil and barrels; and facilities where transportation equipment is cleaned—have altogether escaped federal regulation. Runoff from feed lots and pastures and spills from COMBINED SEWER SYSTEMS still routinely spill raw sewage into lakes and rivers. The CWA is currently up for renewal, with the odds favoring a weakened act. See also NEW SOURCE PERFORMANCE STANDARDS, PRETREATMENT STANDARDS, PRIORITY POL-LUTANT, SAFE DRINKING WATER ACT.

Coal   a rock composed of ancient vegetable matter that has been modified by heat and pressure to contain a high proportion of elemental carbon. Coal has been widely used as a FUEL since the advent of the Industrial Revolution in the 19th century, and currently provides a little less than one-quarter of the world's energy needs—down from about one-half in the 1950s. Oil, natural gas and electricity now perform many of the tasks once done by coal. Almost one billion short tons of coal were produced by U.S. manufacturers in 1991. A century ago, coal heated homes and businesses, and fueled the trains that made western expansion possible, the STEEL mills that made the train tracks and the ships that plied the nation's ports. Today, most of the coal produced is burned to generate electric power (electric utilities used 87% of 1992 coal production—up from 17% in 1949). The consumption of coal has leveled off in industrialized nations at about 820 million tons of oil equivalent (MTOE) since the late 1980s due to world-wide recession and competition from nuclear power and natural gas for the production of electricity. Coal use in the former Soviet Union declined 35% between

1988 and 1992, dropping to 410 MTOE, while consumption in current U.S. coal production is about 830 million tons per year. Coal is the fuel of choice in developing nations, on the other hand, where it is seen as an advance from the use of WOOD and dung. During the last decade the use of coal in developing countries has doubled every fifteen years. Three-fourths of China's commercial energy is produced by burning coal. World coal use was a little more than two billion tons of oil equivalent per year from 1985 to 1994. World reserves of coal are estimated to be about 786 billion tons, with about 28% of the total found in the United States.

Coal production is measured in tons of oil equivalent because the energy content of different ranks and grades of coal varies so widely depending on the origins of the deposit. Coal from mines in northwest Colorado, for instance, produces about 11,400 British thermal units (Btu) when it is burned, about 2,000 Btu per pound more than coal produced in nearby Wyoming. The Wyoming coal, on the other hand, is available in large deposits that are near the surface for about $3 a ton, while the Colorado coal must be mined and costs $17 a ton at the mine.

How Coal Is Formed

Peat is a moist organic muck that is formed through the anaerobic decomposition (see ANAEROBIC BACTERIA) of vegetation that has fallen to the bottom of a bog or swamp. It contains about 80% water. The golden age of peat was the Carboniferous era (approximately 270 to 350 million years ago), a time of luxuriant vegetation growing in vast swamplands and an era during which vast deposits of peat were laid down. Erosional forces since that time covered many of the peat deposits with layers of sand and silt, which later solidified into sandstone and shale, which in turn were buried deeply enough to lithify. The pressure exerted by overlaying rock formations transforms peat into a low grade of coal called *lignite*, which contains about 40% water, 35% elemental carbon along with a large number of complex HYDROCARBONS. Geothermal heat can convert deeply buried lignite into *bituminous* coal, which contains about 50% elemental carbon, 45% hydrocarbons and 5% water. Further heating and tectonic pressure can drive off most of the remaining hydrocarbons to form *anthracite* coal, which is about 95% carbon and 5% water. Only about 2% of U.S. coal reserves are anthracite, found primarily in one small area in central Pennsylvania. Coal is classified by rank (lignite, bituminous, anthracite and a number of intermediate developmental steps) and by *grade*, which reflects the number of impurities the coal contains.

A little more than half of U.S. coal reserves is low-grade, high-sulfur bituminous coal found in the Midwest and the Appalachians. The SULFUR DIOXIDE resulting from the use of such coal is one of the principal causes of ACID RAIN. Ohio coal typically contains about 3.5% sulfur. Most of the other half of U.S. reserves are lignite and sub-bituminous coal (higher in rank than lignite but lower than bituminous coal) located under the western High Plains and the Arctic coastal plain of Alaska. The coal has a low energy content and a low sulfur content. The use of low sulfur coal from mines in the western United States is one way utilities can meet the sulfur emission standards of the 1990 amendments to the CLEAN AIR ACT. By paying more for coal, the utilities can avoid an investment for a scrubber that can cost $30 million over the life of a power plant.

Environmental Effects of Coal

The production and combustion of coal is a major source of air and water pollution (see AIR POLLUTION, GROUNDWATER POLLUTION, WATER POLLUTION). The mining, transportation and combustion of coal all carry environmental consequences. The burned-out wasteland left behind in many parts of Appalachia and the West bear testimony to the social and environmental consequences of coal mining. The destruction of local aquifers often accompanies strip mining on the plains because in many areas coal seams are the arteries through which groundwater seeps. Toxins from the combustion of coal played a major role in the death of 4,000 people in London during a period of high pollutant concentration in 1952 (see SMOG), and similar tragedies have occurred elsewhere when weather conditions caused air to stagnate in an area where significant amounts of coal were being burned (see ATMOSPHERIC INVERSION).

Smelters and other major consumers of coal have been forced in the past to build tall smokestacks to carry air pollution away from their community as the result of an air-quality tragedy. More recently, utilities have built smokestacks that are 400 to more than 1,000 feet high so they will be able to meet local air quality standards without buying air pollution equipment that can cost hundreds of millions of dollars. The net effect has been that better air quality near the installation has been traded for poorer air quality in a large area often extending hundreds of miles downwind. Sulfur dioxide from Midwestern power plants, for instance causes acid rain and air pollution from the Appalachians to northeastern Canada. The CARBON DIOXIDE that is released when coal is burned is also of concern, if the catastrophic predictions that accompany the GLOBAL WARMING theory are to be taken seriously. The disposal of the ASH resulting from coal combustion, which is sometimes laced with toxins, is also a concern.

State-of-the-art pollution control equipment, though expensive, is capable of removing all but 0.1% of the coal ash and 5% of the sulfur dioxide from the exhaust gases emitted from an electric power plant. Research into cleaner ways of burning coal with a high sulfur content is underway. Removing most of the sulfur from the coal before combustion and mixing limestone with high-sulfur coal to absorb the sulfur dioxide as it is produced are among the techniques being explored. See also DEPARTMENT OF ENERGY, SLURRY.

Cobalt an odorless, metallic silver-gray solid or small particles. Cobalt exists in the form of various SALTS. Cobalt and its salts are used in nuclear medicine; in the semiconductor and electroplating industry; as a foam stabilizer in beer; in vitamin B_{12} manufacture; as a drier for lacquers, varnishes, and paints; and as a catalyst for organic chemical reactions. It is used in STEEL alloys, jet engines and cemented carbide abrasives and tools. Cobalt is a natural ELEMENT found in certain ores. Cobalt has several radioactive ISOTOPES, including Cobalt 60, a beta and gamma emitter used in radiation therapy, level gauges and research (see RADIOACTIVITY). Cobalt gets into air and water from industrial and municipal discharges and from spills. Water solubility of cobalt and its salts range from highly soluble to practically insoluble. Cobalt salts are generally highly persistent in water, with a HALF-LIFE of more than 200 days.

Cobalt can affect health when inhaled. Exposure causes irritation of eyes, nose, throat and lungs. High levels may cause a potentially fatal buildup of fluid in the lungs. Cobalt DUST can irritate the skin, causing a rash or a burning sensation on contact. A skin allergy can develop after repeated exposure. A severe allergic lung reaction with coughing, wheezing, chest pain and shortness of breath can also result from repeated exposure, can cause scarring of the lungs (fibrosis), which can also be fatal. Exposure to cobalt can also damage the heart. Cobalt is a possible human MUTAGEN and CARCINOGEN. Long term exposure may damage the thyroid and liver. Trace amounts of cobalt in the diet are essential to human health.

The OCCUPATIONAL SAFETY AND HEALTH ADMINISTRATION has established a permissible exposure limit for airborne cobalt of 0.05 milligrams per cubic meter, averaged over an eight-hour workshift. Cobalt is also cited in AMERICAN CONFERENCE OF GOVERNMENT INDUSTRIAL HYGIENISTS, NATIONAL INSTITUTE FOR OCCUPATIONAL SAFETY AND HEALTH and ENVIRONMENTAL PROTECTION AGENCY regulations. Radioactive isotopes of cobalt are regulated by the Nuclear Regulatory Commission. Cobalt is on the Hazardous Substance List. Cobalt compounds are on the HAZARDOUS AIR POLLUTANTS LIST.

Businesses handling significant quantities of cobalt or cobalt compounds must disclose use and releases of the chemical under the provisions of the EMERGENCY PLANNING AND COMMUNITY RIGHT TO KNOW ACT.

Cogeneration energy generation from what would otherwise be a waste product—originally defined as any process that produced two different usable forms of energy such as heat and electricity. A WASTE-TO-ENERGY PLANT that converts garbage to heat through combustion or that produces refuse derived fuel to drive an industrial process is practicing cogeneration. The PUBLIC UTILITIES REGULATORY POLICY ACT requires a preference for energy produced at cogeneration facilities.

Coke a hard, porous residue left after the destructive DISTILLATION of COAL. It was first produced as a byproduct of the illuminating-gas manufacturing process. Because of its high heat content, about 13,800 Btu per pound, coke is often used as a FUEL in industrial processes, especially in the STEEL industry. Coke is about 92% carbon, with ash composing most of the rest of its mass. It is produced commercially by heating coal to nearly 2,000°F in an oven for approximately 18 hours. Volatile hydrocarbons including BENZENE are driven off during the process, and coal tar is condensed by exposing the exhaust gases to water. The volatile gases are scrubbed with water to remove AMMONIA and with oil to remove benzene. The red-hot coke is then removed from the oven and cooled with water, and the still-hot oven is again loaded with coal. Coke oven emissions are on the HAZARDOUS AIR POLLUTANTS LIST (as a group).

Coliform Bacteria bacteria found in surface water and shallow aquifers that are associated with mammalian feces. PUBLIC WATER SYSTEMS are required under the SAFE DRINKING WATER ACT to monitor for total coliforms, FECAL COLIFORM BACTERIA and E. Coli. The MAXIMUM CONTAMINANT LEVEL GOAL for coliforms is zero. For a water utility that takes 40 or more samples per month, no more that 5% of the samples taken may test positive for coliforms. For a utility that takes fewer than 40 samples per month, no more than one positive sample is allowed. To avoid the requirement that water systems drawing from surface sources install filtration equipment, a utility must demonstrate that the total coliform bacteria count in its drinking water source is 0.2 milliliters or less in at least 90% of the samples taken.

Colloidal Solids solid materials that are small enough to remain in suspension in a liquid indefinitely (see SUSPENDED SOLIDS, TURBIDITY). The solids in a colloidal suspension will not settle out and can not be removed with a normal water filter. A colloidal particle remains in suspension because it is light enough that the force exerted by the water molecules that are constantly bombarding it from all sides is sufficient to offset the pull of gravity. A REVERSE OSMOSIS membrane and DISTILLATION will both remove colloids. A suspension of colloids is uniform throughout, and has a normal boiling and freezing point. Milk, blood and sap are common colloidal suspensions. Colloidal suspensions of a liquid in a gas are also possible (see SUSPENSION).

Color   low concentrations of lignins and tannins can give water color, and their reaction with other substances can intensify the effect. Tannins react with iron to form iron-tannate, which is commonly used to make blue-black ink. The *true color* of a water sample is the result of substances held in SOLUTION, while *apparent color* is the result of materials held in SUSPENSION (see COLLOIDAL SOLIDS, SUSPENDED SOLIDS). See also EUTROPHICATION, TURBIDITY, WATER POLLUTION.

Combined Sewer System   a municipal SEWAGE COLLECTION SYSTEM in which storm water runoff flows through the same mains as SEWAGE. Many of the nation's older cities have combined sewer systems that were installed at a time when the sole function of both STORM SEWER and sanitary sewer systems was to transport waste to the nearest watercourse or the ocean for disposal without treatment. Accumulated grit, leaves, street sweepings and other debris must be removed periodically from catch basins in most combined sewer systems. When sewage treatment facilities are added to such collection systems, municipalities are confronted with an inherent problem: runoff entering during periods of heavy rain or quickly melting snow (see STORM RUNOFF) can boost the total volume of wastewater a hundredfold, and it is impracticable to build a treatment plant that can process peak flows. To prevent damage to the sewage treatment plant at such times, overflow points were constructed through which water drains out of the system into a nearby watercourse. Storm water and sewage are thoroughly mixed inside a combined sewer system, however, and such combined sewer overflows normally entails at least occasional discharge of raw sewage into waterways.

According to the 1990 Needs Survey, a biennial assessment of the cost to community sewer systems of compliance with the CLEAN WATER ACT, it would cost at least $3.2 billion to improve the quality of effluent from combined sewer overflows to Clean Water Act standards, although the estimate is probably low,

since only 154 of the 1,000 that are affected under the 1989 National CSO Strategy were examined. The strategy requires at least some form of treatment in all combined systems and compliance with state water quality standards. Combined systems will typically have to install a treatment plant on the storm water outflow and a storage reservoir capable of capturing overflows for later treatment. At least half the cost of installing a storm water treatment system goes for treatment of some of the heaviest anticipated flows of WASTEWATER.

Combustion   the rapid OXIDATION of a substance with the simultaneous release of heat and, usually, light. The combustion of common fuels involves their chemical combination with atmospheric oxygen and the resultant release of CARBON MONOXIDE, CARBON DIOXIDE and water. A variety of other gases such as SULFUR DIOXIDE may also be released by combustion, depending on the nature of the fuel. The rapid oxidation that characterizes combustion may also be supported by substances other than oxygen, including nitric acid, certain perchlorates, CHLORINE and FLUORINE. Combustion is used to break down hazardous waste in an incinerator (see HAZARDOUS WASTE DISPOSAL, MEDICAL WASTE INCINERATOR), and to convert the energy in garbage to heat or fuel in a trash-to-energy plant. The *combustion point* of material is the point at which it will support open flames. *Spontaneous combustion* occurs when the heat produced by the decay of organic material, usually as the result of the metabolic action of AEROBIC BACTERIA, is not carried away as quickly as it is produced, and heat eventually builds to the material's combustion point. It occurs primarily in the middle of a large mass of material such as a hay stack, where heat from the decomposition of organic materials in the presence of water is prevented from escaping by a thick blanket of dry, insulating hay. Piles of coal, sawdust and wood chips are also capable of supporting spontaneous combustion. Even a pile of oil-soaked STEEL turnings from a lathe have been reported to have ignited because of spontaneous combustion and burned with a heat so intense that the pile was fused together into a nearly indestructible mass. See also FIRE, COMBUSTION AIR, COMBUSTION BYPRODUCTS.

Combustion Air   the air consumed when FUEL is burned. Inadequate combustion air in a tightly sealed building can lead to a BACKDRAFT. Providing a source of outdoor air for wood stoves, fireplaces, furnaces and other combustion appliances that require lots of combustion air is essential in a tight building. See also COMBUSTION, INDOOR POLLUTION.

Combustion Appliances household appliances that burn a fuel. Gas stoves and ovens, portable space heaters, furnaces and wood stoves are the principal combustion appliances. The improper operation of combustion appliances can lead to pollution of indoor air (see INDOOR AIR POLLUTION) with toxins including BENZO-A-PYRENE, CARBON MONOXIDE, NITROGEN OXIDES, and PARTICULATES. See also BACKDRAFT, COMBUSTION AIR, COMBUSTION BYPRODUCTS.

Combustion Byproducts gaseous, liquid or solid materials that are produced incidentally as a result of the COMBUSTION of WOOD, COAL and PETROLEUM products to obtain heat or light. Common gaseous combustion byproducts include CARBON MONOXIDE, CARBON DIOXIDE, NITROGEN OXIDES, SULFUR DIOXIDE, polynuclear AROMATIC HYDROCARBONS including BENZO-A-PYRENE and a variety of other HYDROCARBONS. PARTICULATES released include DUST, ASH, SMOKE, soot and liquid droplets. Ash is the solid byproduct of combustion. See also COMBUSTION AIR.

Community Right to Know Act a provision of the Superfund Amendments and Re-authorization Act of 1986 (see COMPREHENSIVE ENVIRONMENTAL RESPONSE COMPENSATION AND LIABILITY ACT) that requires businesses handling hazardous materials to report the presence of the materials to local authorities and to help develop local emergency plans to be followed in the event of an accidental release. See also TOXICS RELEASE INVENTORY.

Compaction the reduction of the volume of solid waste through crushing. A COMPACTION TRUCK, a TRASH COMPACTOR and a foot crushing an aluminum can on the garage floor all put the principle of compaction into action.

Compaction Truck a truck that crushes garbage with a large hydraulic press. A compaction truck increases the density of the trash and thereby the capacity of the truck. See also GARBAGE TRUCK, SOLID WASTE DISPOSAL.

Composite Packaging a packaging system that uses two or more types of materials such as paper and plastic or two different types of plastics. Composite packaging is rarely recyclable (see RECYCLING). Paper and cardboard sheet laminated to metal foil or plastic sheet is virtually impossible to recycle with the technology available today.

Composting the decay of ORGANIC MATERIAL such as yard waste, manure and wood by AEROBIC BACTERIA. Composting completes the cycle of birth, growth and death by decaying substances that were once alive and returning nutrients and organic material to the soil. It is also a way of reducing the bulk of solid waste such as food, grass clippings, leaves, garden waste, sewage sludge, wood, PAPER, CARDBOARD and nonsynthetic textiles that make up a large part of the volume of GARBAGE. Well established in Europe, where limited space for landfills has forced municipalities to limit the volume of their solid waste for decades, large-scale composting is just getting started in the United States, with a few hundred installations planned or operating. Composting can be an especially effective way of reducing the bulk of garbage that must be composted in warmer parts of the country where plant growth and yard work occur year-round. Archaeologists have unearthed what look like composting pits that were used about 4,000 years ago at Knossos in Crete, and the composting of agricultural and domestic waste has been an integral part of the waste-disposal/nutrient cycle of most cultures since that time. Only since the advent of the petroleum age have cultures had the ability to produce fertilizer from other sources (see AMMONIA, HYDROGEN CHLORIDE, NITRATE, NITROGEN, PHOSPHATE, PHOSPHOROUS, PETROLEUM).

The aerobic bacteria that break down the materials in a compost pile work best if the material to be composted is chopped up and kept wet and the pile is exposed to oxygen. Almost no natural composting occurs in a LANDFILL because wastes are generally dry and tightly compacted, and oxygen cannot penetrate far below the surface. Composting costs less than INCINERATION and sometimes less than landfill, depending on the method used. The thousands of cubic yards of compost produced annually by a large composting operation is often used as topsoil, mulch, soil nutrient for municipal parks and golf courses and as landfill cover, or may be sold or given away to local residents. Even the feces from the municipal zoo can be composted (see ZOODOO). CARBON DIOXIDE and water vapor are the principal byproducts of composting.

With any composting method, the presence in garbage of materials that will not compost—including PLASTICS, METAL, GLASS, wire and rocks—is a problem, so the first step is to sort out such materials. Toxic substances, especially chlorine-based compounds, can cause the release of toxic gases as the compost decomposes. It can be difficult or impossible to remove such substances, so preventing their being thrown away with other trash is important. *Mechanical composting* takes place inside drums, silos and tunnel reactors that pulverize, turn and aerate compost (see AERATION), and is fairly common in Europe. Compost can be produced in five to ten days using mechanical methods, although the curing and stabilizing process that occurs after compost comes out of a mechanical composter may take another two to three weeks. A mechanical composting installation takes less space and less time and produces fewer odors than other methods, but requires a much larger capital investment. Urban areas are best suited for such an installation. *Windrow composting* takes place outdoors in long rows of compost with lanes between. The compost is aerated either via piping and blowers (the high-rate method, which takes two to four weeks) or by mechanical mixing (the low-rate method, which takes five to ten weeks). Windrow composting requires more than 100 acres of land for an operation that produces 10,000 cubic yards of compost a year. *Cocomposting* involves mixing sewage sludge and household waste at a fixed proportion. The compost is remixed and shifted regularly, and natural microbial aerobic fermentation changes the material into natural additives or "bugs." With cocomposting, heavy metals and toxic materials dumped into sewers by industry can cause problems. The heavy metals in inks and pigments, and pesticides in yard waste, can also contaminate compost. Siting a large-scale composting operation can prove nearly as difficult as siting a landfill (see NIMBY). Outdoor operations can involve dust and odors, and any composting installation will involve increased traffic.

The California Integrated Waste Management Act, which requires cities and counties to eliminate one-quarter of the garbage headed for landfills by the end of 1995 and eliminate half by the year 2000, is the toughest of numerous pieces of state laws that call for RECYCLING and composting. Yard waste has been banned from landfills in at least ten U.S. states to encourage small-scale composting. As a result, an increasing number of municipalities are looking to composting as a way of reducing the volume of organic materials going to their landfill (see SOURCE REDUCTION). In Fremont, California, residents are being charged $5 per month for the removal of yard waste unless they have no yard or dispose of all their grass and shrub clippings in a compost pile. The Fetzer winery in California is composting pomace from its winemaking operations, reconditioning old wine barrels and recycling other materials used in winemaking to landfill dump fees by 70% at a savings of more than $20,000 per year. The state of California plans to spend $355,000 on composting demonstration projects by the end of 1996. Among the first projects funded are large-scale composting operations that will consume yard trimmings, vegetable-processing waste, manure, spoiled hay and waxed cardboard. The Marina landfill in Monterey County will compost yard trimmings, which account for 15% to 25% of waste volume in the state. A dozen rows of compost that are 12 feet high, 15 feet wide and 250 feet long are situated on 10 acres at the dump. Workers regularly test the composting materials for METALS, CHLOROFORM and salmonella. See also GLOBAL WARMING, MANDATORY RECYCLING, MCTOXICS CAMPAIGN. Compare ANAEROBIC DIGESTER.

Composting Toilet   a toilet that uses microbial action to break down feces and urine in human waste. In a typical installation, wastes from the kitchen and toilet are collected in a large tank made of fiberglass or CONCRETE. AEROBIC BACTERIA and sometimes worms in the tank break down wastes, which are agitated occasionally. A composting toilet in a typical household produces about twenty gallons of humus per year. GRAY WATER from sinks, tubs and appliances is typically routed to a separate treatment system and drainfield. No WASTEWATER is produced by the toilet. See also SEPTIC TANK, SEWAGE TREATMENT.

Compound   a substance with more than one ELEMENT chemically bonded in its molecule. The proportion of the elements contained in a compound remains the same throughout the compound and wherever the compound is found. A compound is different from a *mixture*, a substance containing two or more elements or compounds that are not chemically bonded to each other.

Comprehensive Environmental Response Compensation and Liability Act   (CERCLA or Superfund) a federal statute signed into law in 1980 intended to identify, clean up and assess liability for HAZARDOUS WASTE sites that have contaminated air, water or soil. A Hazardous Substances Response Trust Fund (the Superfund Trust Fund)—administered by the ENVIRONMENTAL PROTECTION AGENCY (EPA) and supported by taxes on oil, industrial chemicals, fuels and corporate income—was established under the act. The law's "retroactive liability" provision holds polluters (or their legal successors) financially responsible for cleanup, even in instances where hazardous materials that were disposed of legally later cause contamination. The Superfund is used only to finance cleanups where the party responsible for the contamination cannot be charged with the cost, and to pay for actions taken under the Emergency Response Program, which pays for quick, relatively inexpensive responses that address the immediate threat posed by hazardous substance at a waste-disposal site. Emergency cleanups are undertaken if the threat to human health or of serious environmental contamination is deemed too high without immediate corrective action. A public drinking water supply threatened with contamination by hazardous wastes, toxic substances leaking from tanks or drums or unstable wastes that may explode or cause extensive air pollution if not contained are the kinds of situations typically dealt with by the Emergency Response Fund. The Emergency Response Program can furnish bottled water if a supply of drinking water is contaminated by hazardous waste, and can provide temporary or permanent housing for people whose health is threatened by proximity to hazardous materials. The EPA has conducted more than 3,000 such actions since 1980, with the U.S. Coast Guard in charge of some of the on-site work. More than 49 million people live within three miles of a site where an emergency response has been undertaken, and 850,000 live within 500 yards.

By mid-1993, the Superfund program had cost $15 billion. A total of 161 sites had been cleaned up, but another 1,256 remained on the NATIONAL PRIORITIES LIST of contaminated sites. More than one-third of the hazardous waste removed from Superfund sites goes to a facility operated by Chemical Waste Management in Emelle, Alabama (see HAZARDOUS WASTE DISPOSAL). Contaminated soil and stabilized hazardous materials from Superfund sites is a major source of hazardous waste. Congressional debate over the renewal of CERCLA began in 1995 with a move to remove or tone down the law's strict retroactive liability provisions gaining considerable support. The ten largest U.S. military contractors are responsible for more than 200 of the nation's more than 1,200 Superfund sites.

The trust fund, which initially amounted to $1.6 billion, was increased to $9 billion when Congress passed the Superfund Amendments and Reauthorization Act (SARA) in 1986. Frustration in Congress with the slow pace of the cleanup program led to a requirement that rehabilitation be started on 375 sites by 1991. State agencies were given the authority in the amendments to prosecute owners of toxic waste sites that pollute air, soil or water. The EMERGENCY PLANNING AND COMMUNITY RIGHT TO KNOW ACT, which requires industrial users of toxic chemicals to notify local governments about the use and accidental release of certain toxic materials, was also a part of the SARA package (see TOXICS RELEASE INVENTORY).

Landfills and CERCLA

More than 225 municipal landfills in the U.S. have been declared Superfund sites. The EPA adopted an "Interim Municipal Settlement Policy" in 1989 in which the agency agreed not to pursue cities that had dumped only MUNICIPAL SOLID WASTE (MSW) for cleanup costs associated with the sites. CERCLA does, however, allow the parties that are found responsible for contamination at a Superfund site to sue others that disposed of hazardous materials at the site to recover part of their costs. Highly toxic industrial waste

dumped along with MSW is responsible for most of the environmental contamination at the majority of the landfills that have been listed as cleanup sites, but industrial polluters have nevertheless successfully been taking municipalities to court to recover part of their cleanup costs. Although the concentration of hazardous materials in MSW is extremely small compared to that found in industrial waste, litigants have demonstrated in court that MSW does cause pollution (see HOUSEHOLD HAZARDOUS WASTE).

A 1993 ruling in California district court held that 14 southern cities were responsible for toxic materials in commercial and household garbage that was hauled to the Operating Industries landfill in Monterey Park, now a 190-acre Superfund site. The U.S. Appeals Court in Washington, DC, ruled in 1992 that insurance companies must pay claims made by industrial policyholders held responsible for the cleanup of damage caused by toxic waste. A 1994 Supreme Court decision denied the right of those who clean up their environmental contamination to recover legal fees they incur in getting other polluters to help pay for the cleanup.

A *Hazard Ranking System* is used to determine which sites should be listed on the National Priorities List of disposal sites to be cleaned up. The system is based on the Environmental Protection Agency's assessment of the threat to public health and environment. A site may also be put on the National Priorities List if the Agency for Toxic Substances and Disease Registry issues a health advisory for the site or if a site is chosen as a state's top priority cleanup site.

A *Technical Assistance Grant Program* established under CERCLA provides grants of up to $50,000 per site to allow citizen's groups to hire technical experts to help them through the maze of paperwork, extremely technical (and often tedious) data and testimony surrounding a Superfund site. The *National Response Center Hotline* (800-424-8802) is a 24-hour hotline that accepts calls about potential hazardous waste sites. For descriptions of specific superfund cleanup sites, see CLARK FORK RIVER, HANFORD NUCLEAR RESERVATION, KENILWORTH DUMP, LOVE CANAL, OAKRIDGE NATIONAL LABORATORIES, ROCKY FLATS, SAVANNAH RIVER, URAVAN URANIUM MILL, VALLEY OF GEHENNA, WOBURN.

Concentration   the amount of a CONTAMINANT found in a given volume of air or WATER. See AIR POLLUTION, GROUNDWATER POLLUTION, POLLUTION, TOXICITY, WATER POLLUTION.

Concrete   a BUILDING MATERIAL made by combining cement with water and a mineral aggregate, usually washed sand and gravel. Wet concrete is poured or pumped into forms, and has a very hard, rock-like surface when it sets. It normally must be kept moist for a minimum of three to five days before it is strong enough for the forms to be removed, although the process can be accelerated chemically through the addition of salts, usually calcium chloride, or through the use of steam. Gravel is added to concrete aggregate for economy. Concrete is used to make roads, bridges, retaining walls, buildings, swimming pools, boat hulls, sewer pipe, septic tanks, piers and concrete blocks, curbs and bricks. Concrete is very hard, but it has a relatively low tensile strength and must be reinforced with steel mesh or rods. Frost and corrosive materials can splay the surface of concrete, and it can crack, but the material is in general comparable to a rock in its durability and lifetime. For this reason, the disposal of concrete can be a problem. Used concrete is sometimes crushed and used as aggregate in new concrete, and demolished concrete is used as fill material. Chunks of concrete are sometimes reused as stepping stones and other items, but a large yet undocumented amount ends up at landfills.

Cement, the most important ingredient in concrete, is a gray, powdery mixture. Hydration, a chemical reaction between water and cement, is the basis of cement's ability to bind the materials used in concrete together. FLY ASH from air pollution control devices has some of the same binding abilities, and it is seeing increased use in the production of concrete. The raw materials used in cement production include limestone, cement rock, oyster and coquina shells, marl (a sand, silt or clay with a high concentration of calcium compounds), shale, sand, iron ore and clay—sources of lime, silica, aluminum and iron. The raw ingredients are pulverized and mixed together in proportions depending on the desired finished product. The mixture is then fired in a kiln at about 2,700°F, and formed into *clinkers*. The clinkers are mixed with a small amount of gypsum, which controls the rate of hydration in the finished cement, and the mixture is ground up into a fine powder called *portland cement*. Portland cement is based on an 1824 British patent on the manufacture of cement. Cement's color resembled Portland stone, which was widely used in building construction in England at the time, so the inventor added the name to his cement. Cements are available in a variety of formulations designed for particular applications where qualities such as fast curing or resistance to sulfates is important. Slow-curing concrete must be used to minimize the buildup of heat as a byproduct of hydration in applications such as dams or the footings of a skyscraper where a large mass of concrete is poured. More than 80 million short tons of cement were produced by U.S. companies during 1993.

Many cement kilns are reducing their fuel consumption by burning used tires. The steel in radial tires, when burned enhances the chemical process that creates cement. A similar effect can be obtained by using certain kinds of hazardous waste as fuel, but the legal and political hurdles of licensing a cement kiln as a hazardous waste incinerator can be substantial (see NIMBY). Dust from cement kilns is specifically exempt from regulation under the RESOURCE CONSERVATION AND RECOVERY ACT.

Condensation the conversion of a substance from a gas to a liquid or solid as the result of cooling. The condensation and evaporation of water are at the heart of the HYDROLOGIC CYCLE. When a vapor condenses, the latent heat of vaporization is released. When water vapor condenses to form precipitation during the hydrologic cycle, the latent heat that is released plays a key role in carrying warmth from the oceans over the continents. *Condensate* forms on surfaces that are cooler than air's *dew point*, the temperature at which water vapor begins to change into a liquid. The condensate that can form on cool windows and walls in buildings can damage building materials and support microbial growth that can lead to INDOOR AIR POLLUTION. *Transpiration* is the evaporative loss of water vapor from leaves, needles and other parts of plants and trees. An estimated one million gallons of water passes into the atmosphere from an acre of hardwood forest each year.

Conductivity the ability of a solid, liquid or gaseous material to conduct heat or electricity. *Thermal conductivity* (k), is defined as the quantity of heat measured in British thermal units (Btus) per hour that will pass through one square foot of a given material that is one inch thick if there is a temperature difference of 1∞F between the two sides of the material. *Conductance* (C) is also a measure of the heat flow in Btus per square foot through one square foot of a material with a 1∞F temperature difference, but it is used to measure a material's total resistance to heat flow, not just that of a one-inch-thick section. Dividing a material's conductance by its thickness yields its k-value (C = k/X). *Resistance*, which is the reciprocal of conductance (R = 1/C = X/k), is commonly used to express the heat flow of insulation in terms of its *R-value*. R-11 fiberglass insulation, for example, is 3 1/2 inches thick, has an R-value of 11, a conductance of 1/R = 1/11 or about .09 Btu/hour/square foot. A material's R-value gets larger as its resistance to heat flow increases, while its conductance gets lower.

Electrical conductivity, which is expressed in *ohms*, is the inverse of electrical RESISTIVITY, which is expressed

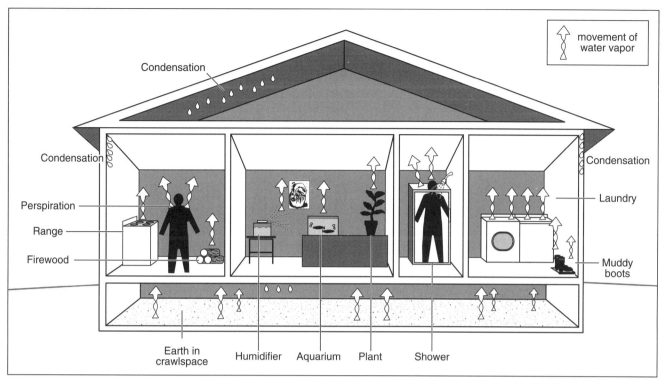

movement of
water vapor

Condensation

Condensation

Perspiration

Range

Firewood

Condensation

Laundry

Muddy
boots

Earth in
crawlspace   Humidifier   Aquarium   Plant   Shower

Sources of water vapor (*Residential Energy*)

in *mhos*—the inverse of ohms. Electrons that are free to move within a metal are said to be in the *conductivity band*. The electrical conductivity of a solution can be uses to measure the concentration of ions it contains. Pure water does not conduct electricity, nor do solutions of electrically inert substances such as sugar or alcohol in water. Solutions containing electrovalent compounds (IONS) such as COPPER, sodium chloride, hydrogen chloride or potassium nitrate conduct electricity, and are called *electrolytes*. By measuring the electrical current (in *micromhos* or *μmhos*) that will pass between two electrodes through a sample of the water (or other liquid), the purity of the water can be ascertained. About 10 μmhos per centimeter will pass through pure water, while highly mineralized water may conduct more than 40,000 μmhos. See MEASUREMENTS for a complete treatment of the units of measurement used for electricity and heat.

Cone of Depression   a depression in the surface of the WATER TABLE that forms when a pump removes water from an AQUIFER more quickly than it is replenished.

Confined Disposal Facility   an enclosure for the isolation of contaminated dredge spoil. In *diked disposal*, a dike is built near the area to be dredged with large boulders lined inside with a layer of sand. Dredge

spoil is placed inside the enclosure, and most suspended contaminants are filtered out of the water that seeps out through the sand. Dissolved contaminants and smaller suspended particles can usually escape from a diked disposal area with relative ease, however. In *upland disposal*, dredge spoil is transported, usually by truck or railroad car, but occasionally on a conveyor belt to a landfill. See also INTERNAL LOADING.

Construction Wastes   materials that are thrown away during the construction of new buildings and during the renovation of existing structures. Construction wastes and DEMOLITION WASTES typically make up about 12% of the volume of a typical landfill's contents, according to estimates by the Garbage Project. Construction and demolition wastes are intimately mixed in the dumpster on a typical remodeling site with PAPER, CARDBOARD and plastic PACKAGING from building materials; scraps of WOOD, METAL, electrical wire and insulation; and bent nails, broken GLASS and assorted grit, plaster and GYPSUM wallboard, old plumbing fixtures, broken-down windows and other demolition debris. An average American accounts for the use of 540 tons of construction materials in his or her lifetime, according to estimates by the WORLD-WATCH INSTITUTE. Less is known about construction wastes than other waste categories because of the wide

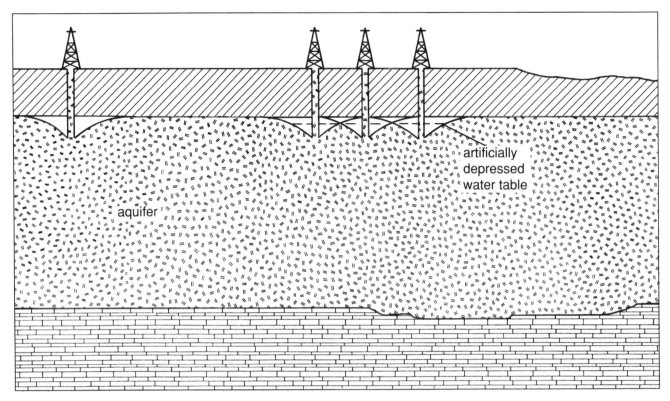

Cones of depression (*Residential Energy*)

range of materials that can be involved and because there are no trade associations to take an interest. See also BUILDING MATERIALS.

Consumerism a condition in which the act of buying things becomes an end in itself. Consumerism and POPULATION GROWTH are the principal causes of an increasingly tight supply of natural resources and an increasing rate of global environmental degradation. The consumption of materials and energy in industrialized nations—which comprise about one-fifth of the world's population, but consume about four-fifths of world production of commodities such as ALUMINUM, IRON, STEEL, PAPER and WOOD—has risen dramatically during the latter half of the twentieth century. Referring to people as "consumers" implies that their most important function is buying things, using them up and buying again, and under the economic formula predominant in most industrialized nations, this description is accurate. To consume means "to destroy or to do away with completely," and, in keeping with this definition, the consumer lifestyle accelerates the conversion of raw materials into waste products. The idea, however, is to consume, not to consummate—and the consumer is faced with a constant "need" to buy. Once the basics have been

acquired, replacing the original purchases with newer, better, things can become a self-perpetuating condition, one that is highly desirable for the manufacturers and purveyors of goods. Consumerism is a FORMATION PROCESS.

The process of consumerism are established early. A toddler is confronted with a barrage of advertisements that start as soon as he or she is capable of comprehending that they can actually possess the toy that all the kids are having so much fun with in the television commercial. The creation through advertisements of largely artificial "needs" continues as children grow into adults. Radio, television, ads in print media, billboards, match books, bus benches, subway walls—a commercial message can be seen practically anywhere. Each product looks desirable—even indispensable—in its advertisement, but the looks are too often only skin deep. The consumer frequently finds that more effort has gone into selling the product than has been devoted to the production of high-quality, durable goods or clever design. In fact, products such as kids' toys and games, automobiles and clothing are typically sold with the idea that they will soon be discarded (see PLANNED OBSOLESCENCE). Cheap materials and poor design and workmanship too often mean that the toys, tools, clothing and countless other modern products look and sound great in an ad, but are quickly used up, broken down and thrown away. The old saying, "the bitterness of poor quality lingers long after the sweetness of low price," seems to apply.

Consumerism and the associated waste of materials is not new. During the Classic phase of the Maya, about 300 A.D., what is referred to by archaeologists as a "cult of *conspicuous consumption*" flourished, as it has among other early peoples. Ornate temples, extravagant ceremonial clothing and lavish burials typified the era. Such practices ended with the advent of the Decadent Mayan Era, when hard times necessitated a switch to REUSE, RECYCLING and a spare life. The robust production of trash generally means economic prosperity, and the United States and other modern industrial cultures are still in their classic phase, in terms of garbage production.

Maintaining the vitality of the economy while voluntarily adopting "decadent" techniques of resource use such as recycling and reuse is seen by many as a way of attaining sustainable resource use and of thereby avoiding an involuntary decadent period forced by the scarcity of resources (see SUSTAINABLE SOCIETY). This would mean replacing consumerism with "conservatism," which through durability, repairability, recycling and reuse would maximize the amount of use a resource gets between its source as a natural resource and its deposition in a landfill. Rather than accumulating things largely for the sake of buying them, as is the case with consumerism, the question would be just how many things are required for an acceptable quality of life. Voluntary simplicity— adopting a lifestyle that has a minimal environmental impact—would be the word of the day, and the economic formula under which success is defined by the amount of VIRGIN MATERIALS extracted and the amount of waste produced would, itself, be discarded. See also CONVENIENCE PACKAGING, MATERIALISM.

Contaminant any substance that is found in an environment where it does not belong. When a grain of sand finds its way with the lettuce from farm to salad, the sand has become a contaminant somewhere along the way. Contaminants often render the materials in which they are found such as air or WATER, unfit for its intended use. Contaminants are not always bad, however—the "contaminants" found in honey or a fine wine give it its distinctive flavor, for instance. A *micro contaminant* is a pollutant that poses a significant health threat even at very low concentrations. Complex ORGANIC COMPOUNDS such as DDT, PCBs (see POLYCHLORINATED BIPHENYLS) and DIOXIN are the most common micro contaminants. Contaminants are

notnecessarily harmful, and may actually be beneficial. The distinctive flavor of different types of honey, for instance, is caused by foreign elements that vary from batch to batch, depending on where the bees obtained nectar, and the mineral contaminants in water similarly give it its flavor. Compare POLLUTANT. See AIR POLLUTION, GROUNDWATER POLLUTION, POLLUTION, TOXICITY, WATER POLLUTION.

Convenience Packaging ONE-WAY PACKAGING that is designed principally for the convenience of the person that buys the product. Small portions of food and drink packaged in individual plastic containers and FAST-FOOD PACKAGING are examples of convenience packaging. Since one of the principal conveniences of small packages of food and drink are that they can be taken anywhere, discarded paper and plastic wrappers and containers and little cardboard carry boxes are a principal component of LITTER. See also CONSUMERISM, PACKAGING).

Conventional Pollutant a POLLUTANT that is found in normal household SEWAGE. A conventional pollutant contaminates a body of water by changing its trophic level or dissolved-oxygen balance (see BIOCHEMICAL OXYGEN DEMAND, EUTROPHICATION) or otherwise physically altering its character. NUTRIENTS, especially PHOSPHOROUS and NITROGEN, BACTERIA and other MICROORGANISMS, SILT and other SEDIMENTS, OILS and GREASES are the principal conventional pollutants. Toxic substances are not conventional pollutants.

Conventional Toxic Substance an inorganic or simple ORGANIC COMPOUND that is toxic (see TOXICITY). ARSENIC, CYANIDE, AMMONIA, boric acid, CHLORINE and other HALOGENS and PHENOL are all conventional toxic substances.

Copper a soft, lustrous METAL used in the production of electrical and plumbing equipment, building materials, jewelry, coins and many other products. It is valued because of its resistance to corrosion and weathering, because it is a good conductor of electricity and is ductile (capable of being drawn into a wire). About 1.8 million metric tons of copper are produced each year in the United States for use in the production of copper electrical wire, electronics equipment, roofing, pots and pans and many other products. It is also used in chemical and pharmaceutical industries and in the production of parts for machinery. IRON is the only metal that is more widely used. Because of its value, copper is often recycled, although old copper pipe and wire often end up in landfills along with other DEMOLITION WASTES. The mining and smelting of copper are major sources of air and water pollution and energy consumption. Toxic copper fumes and vapors can be created through welding, brazing, soldering, plating, cutting, and metalizing involving the metal. Copper is found in road dust. Galvanic CORROSION can occur when buried copper and galvanized STEEL pipes pass to close to one another. Copper freshly exposed to the atmosphere has a bright surface with hues of brown, red and yellow. A thin film called a *patina* quickly starts to form on copper that is exposed to the elements. It is made up of copper salts created by oxidation, and can stain the exposed surface brown, black or green or combinations of the three, depending on how rainy the climate is and the makeup of the rain.

Copper is undoubtedly the first metal used by humans. Naturally occurring *native copper* is fairly widely distributed, and the soft metal is easily worked, even without heating. Copper jewelry and utensils are found at prehistoric archaeological sites around the world, and such artifacts are especially common near Lake Superior, on the British Isles, and on Cyprus. Copper received its name from the Latin name for Cyprus, "cuprum," because the Romans smelted the metal there. By melting copper and brass together, early man was able to cast tools and weapons made of bronze, the metal of the Bronze Age.

Health

Copper itself is not toxic, but it forms compounds that are toxic in the environment. Exposure to copper dust or fumes can irritate the eyes, nose and throat, and copper particles can damage the eyes, impairing vision and even causing blindness. Copper fumes can cause "metal fume fever." Symptoms include a metallic taste in the mouth, chills, fever, aches, cough and chest tightness. Exposure to copper can also cause a skin allergy, and repeated exposures can cause thickening of the skin. Skin and hair can take on a greenish color with frequent exposure to copper. Low concentrations of copper are commonly found in water, and low levels of copper are an essential dietary element for both plants and animals.

Water Pollution

Copper pipe and fixtures can contaminate drinking water if water is aggressive (see AGGRESSIVE WATER). Copper compounds have high acute and chronic toxicity to aquatic life. The toxicity of copper compounds to aquatic life varies according to the mineral content and PH of the water. The concentration of copper found in fish tissues in water polluted with copper is generally much higher than the average concentration of copper in the water from which the fish was taken (see BIOACCUMULATION).

Mining

Mine and mill tailings and air pollution generated in copper smelting are major sources of air pollution. Water and soil polluted by copper mining and smelting that occurred a century ago are still causing serious environmental problems in many old copper-mining communities (see CLARK FORK RIVER). Copper is smelted principally from chalcopyrite ($CuFeS_2$), chalcocite (Cu_2S) and cuprite (Cu_2O). Copper smelters can produce massive quantities of sulfur dioxide, which is produced as a byproduct when the *copper sulfide* and *copper oxide* produced by smelting are reduced to pure copper by heating. Smelters also frequently emit heavy metals such as ARSENIC and LEAD that can contaminate and even sterilize soils in the surrounding area through DUSTFALL. Pyrite in tailings from copper mines and smelters is a source of acidic runoff, and cadmium, lead, zinc and arsenic are also picked up by runoff that passes through tailings associated with copper production. The roasting of copper ore was once accomplished by open burning in the early days of Butte, Montana, until the severe health problems suffered by city residents put an end to the practice (see OPEN BURNING).

Copper is used in the production of pesticides and wood preservatives. *Bordeaux mixture*, a mixture of copper sulfate and calcium hydroxide, was once widely used to control fungal growths on agricultural crops and landscape plants, but its use has been largely discontinued due to its toxicity. PUBLIC WATER SYSTEMS are required under the SAFE DRINKING WATER ACT to monitor concentrations of copper. Treatment technology rather than a MAXIMUM CONTAMINANT LEVEL is specified for copper. The concentration of copper is measured at the user's tap, since the metal can be picked by aggressive finished water as it passes through the distribution system and the household plumbing system (action level = 1.3 milligrams per liter). Acidic drinking water must be neutralized under the rules to make it less corrosive, and the public must be educated about ways of reducing exposure. The OSHA permissible exposure limit for airborne copper dusts and mists is 1.0 milligrams per cubic meter, and for copper fumes is 0.1 milligrams per cubic meter averaged over an 8-hour workshift. Copper is also cite in AMERICAN CONFERENCE OF GOVERNMENTAL INDUSTRIAL HYGIENISTS guidelines.

Corrosion the process of chemical or electrochemical destruction of one substance by another. While corrosion chemically changes one material into another, mechanical processes such as abrasion and erosion physically alter a substance by breaking off small pieces without altering the material chemically. The new material created by corrosion is generally weaker than the substance from which it is derived, although certain corrosion products (such as aluminum oxide) are actually harder than the parent material.

There are two kinds of corrosion: *chemical corrosion* and *electrolytic corrosion*. Chemical corrosion is characterized by a chemical reaction between a corroding agent and a corrosive material in which one or more new substances substantially different from the parent materials is created. The formation of rust on an IRON surface is a familiar example—a reaction of oxygen and iron in the presence of water that yields hydrated iron oxides. Electrolytic corrosion involves the passage of an electric current through a METAL or other conductive material (the *electrode*) that is immersed in a conductive *electrolyte* such as salt water or damp soil. The electric current flows into the conductor at the *cathode*, and flows out at the *anode*, removing molecules of the material in the process. The dislodged molecules are free ions, which are permanently separate from the parent material. *Stray-current* electrolytic corrosion occurs when direct current electricity leaks out of an external source such as a frayed power cord through an electrolyte and through an electrode. Alternating current electricity does not cause electrolytic corrosion because the cathode and the anode switch places with each cycle so that any molecules dislodged by the electric current are immediately re-deposited. *Galvanic corrosion* requires a difference in electrical potential between a metal (such as a buried pipe) and the surrounding electrolyte (such as moist soil). The difference in potential generates a weak electrical current. Differences in soil chemistry or water content between two ends of a buried metal pipe or the presence of certain soil bacteria can establish the conditions necessary for an electrolytic reaction to take place. Electrolytic flow can also be established when two metals with differing electrical properties are buried in close proximity to one another in a soil capable of acting as an electrolyte. Galvanic corrosion can occur, for instance, when buried copper and galvanized STEEL pipes pass too close to one another.

Corrosion Protection

Coating susceptible surfaces with an inert material and making sure all components of a system (such as a LEACHATE COLLECTION SYSTEM) that will come in contact with corrosive materials are corrosion resistant can prevent chemical corrosion. Stray current corrosion can be stopped by adequately insulating the source of current or by insulating the electrolyte from electric current from any source. Galvanic corrosion can be controlled by using backfill around underground pipes or buildings that is of similar composition and acidity and by avoiding differences in soil moisture content once the backfill is in place. A *sacrificial anode* can protect any potential electrode—even a conductive metal immersed in a good electrolyte such as seawater from the effects of all types of galvanic corrosion. If a rod of magnesium or other metal from the low end of the galvanic series is connected electrically to the electrode to be protected, it becomes the anode of the galvanic system, and thereby all the galvanic corrosion is experienced by the sacrificial material. See also COPPER, GALVANIC SERIES, LEAD, RESISTIVITY.

Corrosive Material   A material that can chemically or electrochemically alter other substances. Corrosive materials are generally liquids that will corrode normal materials such as the STEEL, ALUMINUM and RUBBER used in tank trucks, pipelines and storage vessels. Most corrosive materials contain strong ACIDS or alkalis that are capable of causing external or internal tissue damage referred to as burns. Ingestion of a significant amount of a corrosive material can cause vomiting, with blood often appearing in the vomitus. Hydrochloric acid, carbolic acid, bichloride of mercury and AMMONIA are commonly encountered corrosive materials. See also CORROSION, RESISTIVITY.

Corrosivity   the ability of a material to cause CORROSION. Compare REACTIVITY, RESISTIVITY. See also CORROSIVE MATERIAL.

Cosmic Radiation   see BACKGROUND RADIATION

Council on Packaging in the Environment   a national coalition of manufacturers and users of packaging materials, suppliers, retailers, trade associations and recyclers. The Council was formed in 1986 to implement programs that "integrate the concepts of environmental and economic sustainability in the design, manufacture and use, and disposal of packaging." The Council on Packaging in the Environment (COPE) conducts surveys of consumer awareness and attitudes regarding packaging, and makes information on packaging available to the media and the public. COPE publishes short reviews of issues relating to packaging called *Backgrounders*. COPE members include Coca Cola, Colgate-Palmolive, Dow Chemical U.S.A., Nabisco Foods Group and the STEEL RECYCLING INSTITUTE. For further information, contact COPE at 1255 Twenty-third Street, NW, Washington, DC 20037. Phone: 202-331-0099.

Cremation   the reduction of dead human and animal bodies to ashes in a specially designed furnace. The remaining ashes are generally preserved in an urn. Cremation on a funeral pyre was the normal way of disposing of human remains in many early cultures. The Jewish belief that cremation was a desecration of an act of God and the Christian belief that the destruction of dead bodies would prevent their resurrection ended the practice in countries where the religions were widely practiced. Concern about the spread of contagious disease, and a lack of space for graveyards in urban areas revived the practice in more recent times. The first crematorium in the United States was built in Washington, Pennsylvania in 1876. In addition to reducing the demand for burial sites, cremation avoids the pollution that sometimes contaminates water under graveyards. Largely because of a shortage of space for graveyards, about 70% of British dead are now cremated rather than buried—a ratio higher than most other countries with the exception of Sweden and Japan. See also GRAVEYARD.

Criteria Pollutant   any of the seven POLLUTANTS designated in the CLEAN AIR ACT amendments of 1970. The criteria pollutants are TOTAL SUSPENDED PARTICULATES, SULFUR DIOXIDE, CARBON MONOXIDE, nitrogen dioxide, OZONE, HYDROCARBONS and LEAD. They are called criteria pollutants because they are governed according to criteria established by the ENVIRONMENTAL PROTECTION AGENCY rather than by specific standards governing the concentration of a contaminant. The legislation called for *primary standards* that are designed to protect human health and are supposed to be strict enough to prevent a significant number of exposed individuals from suffering health effects as a result. Stricter *secondary standards* are designed to protect the health of wildlife, plants and other living organisms and to prevent damage to property, clothing and intangibles such as visibility—factors that have only an indirect effect on human life. Although the CLEAN AIR ACT does not allow the consideration of the economic impacts of compliance as a factor in setting standards for criteria pollutants, state governments, which are responsible for on-the-ground enforcement of the act, can make the cost of compliance part of the state pollution control strategy. See also POLLUTANT STANDARD INDEX.

Critical Mass   see FISSION

Crocidolite   see ASBESTOS

Cross-Media Pollution   pollution that is caused by the cleanup of another form of pollution. Toxic emissions from an incinerator that is burning hazardous waste and WATER POLLUTION caused by leaks from a landfill that contains hazardous waste generated by pollution control equipment are examples of cross-media pollution. The lines between solid, liquid and gaseous waste is far from clear.

Solid particles dissolved or suspended in wastewater, for instance, are LIQUID WASTE until they are removed by pollution-control equipment, when they become SOLID WASTE (although solid waste such as SEWAGE SLUDGE may be more than 90% water). The VOLATILE ORGANIC COMPOUNDS (VOCs) that are often found in a polluted river similarly are liquid waste until they evaporate, when they become gaseous waste. The evaporation of VOCs in a badly polluted river can be a major source, which is why tourists are cautioned about hanging out for too long in the polluted mists that swirl around Niagara Falls.

An example of cross-media pollution occurred recently in Denver, Colorado, where METHYL-TERTIARYBUTYLETHER (MTBE) has been showing up consistently in tests of ground water and precipitation. MTBE is added to oxygenated gasoline sold in the city in the winter because the city does not meet air quality standards stipulated in the CLEAN AIR ACT. It is not known whether the MTBE is getting into water from leaking underground storage tanks, from surface runoff or through a combination of the two. See also AIR POLLUTION TREATMENT.

Crude Oil   see PETROLEUM

Cryptosporidium   a pathogenic water-borne parasite sometimes found in DRINKING WATER. Cryptosporidium in drinking water has been responsible for outbreaks of gastroenteritis, a flu-like illness. More than 20,000 cases of gastroenteritis were reported in Carrollton, Georgia, in 1987, and 400,000 cases and 100 deaths occurred in Milwaukee, Wisconsin in 1993. Cryptosporidium is a protozoan that is found in the intestinal tracts of animals. The ENVIRONMENTAL PROTECTION AGENCY has proposed that cryptosporidium be monitored as part of the enhanced surface water treatment rule. The MAXIMUM CONTAMINANT LEVEL goal for cryptosporidium is 0. See also DISINFECTION, GIARDIA, LEGIONELLA BACTERIA, MICROORGANISMS, TURBIDITY.

Curie   see RADIATION MEASUREMENT

Cut-Off Wall   a barrier usually composed of concrete, a plastic membrane, clay or a combination of materials that intercepts the flow of polluted groundwater (see GROUNDWATER POLLUTION, GROUNDWATER TREATMENT). A trench that goes to the bottom of the affected AQUIFER must first be excavated before such a cut-off wall can be installed. Cut-off walls can also be established by (a) freezing soil and groundwater through the use of cryogenics. See also INTERCEPT WELL.

Cyanide a compound of *cyanogen*, or a salt or ester of *hydrocyanic acid*. Cyanogen can be produced by heating silver cyanide salts, and can undergo basic hydrolysis to yield cyanide and cyanate ions. The cyanate ions can, in turn, unite with hydrogen ions to form the poisonous gas *hydrogen cyanide*, which forms hydrocyanic acid when in solution. PUBLIC WATER SYSTEMS are required under the SAFE DRINKING WATER ACT to monitor concentrations of cyanide (MAXIMUM CONTAMINANT LEVEL = 0.2 milligrams per liter).

Cyanophyte see BLUE-GREEN ALGAE

Cyclone an air pollution treatment device—also called a *dry centrifugal collector*—in which contaminants are removed from gaseous industrial effluent (see INDUSTRIAL AIR POLLUTION) through centrifugal force. Exhaust gases create a vortex as they flow through a tapered, cylindrical chamber, and centrifugal force throws particles to the outside of the collector.

The particles then settle to the bottom of the collector for removal. A centrifugal collector can remove up to 70 percent of particles 5 microns and larger in diameter from an airstream. Centrifugal collectors are generally used as primary collectors for FLY ASH, WOOD, DUST and PLASTICS dust. A *spray chamber scrubber*, usually water, is sprayed into the top of the chamber, and runs down the vessel's walls, absorbing contaminants from the airstream as it goes. The contaminants are then removed and wastewater is either reused or disposed of.

D

Dander tiny scales of human or animal skin and pieces of hair and feathers that are light enough to become airborne. Many allergy sufferers are sensitive to airborne dander. Dander is a common component of DUST. See also MICROORGANISMS, PARTICULATE.

Debris the remains of something broken down—an accumulation of loose pieces of rock—or the waste sand and gravel produced by hydraulic mining. Organic waste from dead or damaged tissue is also often called debris as is roadside LITTER and scattered TRASH. Compare GARBAGE, REFUSE.

Decay the decomposition of organic material as the result of the action of AEROBIC BACTERIA. Decay oxidizes the products of PUTREFACTION, eliminating offensive odors. Decay processes are an essential part of the recycling of resources from dead organisms into nutrients for the nourishment of living things. Radioactive decay products are ALPHA PARTICLES, BETA PARTICLES and GAMMA RAYS released when one ELEMENT or ISOTOPE is transformed into another through the partial disintegration of the atomic nucleus. They are produced by the mining and milling of radioactive fuels, the operation of nuclear reactors and the explosion of nuclear weapons. See also COMPOSTING.

Decay Product see REACTOR WASTE

Deepwell Disposal the disposal of industrial liquid waste by pumping it into AQUIFERS that may be anywhere from 1,000 to 20,000 feet below the surface. Used since about the beginning of the twentieth century by the petroleum industry to dispose of drilling BRINE, deepwell disposal was adopted by the chemical industry in the 1950s as a way of disposing of hazardous chemical manufacturing byproducts. More than half of the liquid HAZARDOUS WASTE produced in the United States is now disposed of in approximately 600 operating injection wells. Aquifers that contain highly saline water (with more than 10,000 parts per million total dissolved solids) are considered to be ideal for deepwell disposal, because their high degree of salinity generally indicates that they are isolated from other aquifers. The ENVIRONMENTAL PROTECTION AGENCY requires that the geologic formation into which wastes are injected be at least a quarter mile below any aquifers that furnish DRINKING WATER and that it be separated from any such aquifers by an impermeable layer of rock.

Because the contents of an injection well are under pressure, the seal on the well itself and the impermeable layer enclosing the underground structure into which wastes are being injected must be very tight. Cracks in the underground strata or an unmapped old well can be opened up by the pressure, allowing wastes to flow out of the disposal area. Both the injection well and the impermeable formation that isolates the disposal area must be composed of materials that can not be broken down chemically by the wastes being disposed of. If the underground disposal area should be breached by another well, a toxic geyser could result, so the location of deepwell disposal sites must be adequately mapped and known to other potential well drillers.

In the 1960s, an injection well near Denver, Colorado, caused a series of minor earthquakes, apparently because it lubricated a buried geologic fault, allowing it to move more easily. Deepwell disposal is illegal in several states including Alabama, Florida, New York, and New Jersey.

Demographic Marker an object that can be analyzed by an archaeologist to learn something about human behavior in the society that produced it. Demographic markers are used in archaeology to make suppositions about life in ancient cultures and in GARBOLOGY to trace behavior patterns in modern cultures. An *exclusive demographic marker* gives information about a specific sector of the population. The number of babies in a household or a community can be estimated by counting disposable diapers thrown in the garbage can or the landfill, for instance, and the number of discarded toys and toy packaging can similarly be counted to estimate the number of children in a given population.

See also FORMATION PROCESS, MIDDEN, PULL-TAB TYPOLOGY, STRATIGRAPHIC LAYER.

Demolition Wastes   waste materials produced by the demolition or renovation of buildings, bridges, streets and sidewalks and other structures. Demolition wastes typically contain reusable items such as scraps of new building materials, used lumber (see WOOD WASTE), old windows, doors, serviceable plumbing fixtures and recyclable items such as ALUMINUM, STEEL, WOOD and CARDBOARD that frequently become municipal solid waste and are buried at a landfill. Hazardous materials such as LEAD and ASBESTOS are often found in demolition wastes. The finest fraction of the solid waste created by demolition comes in the form of DUST, which can cause heavy short-term contamination of air in the neighborhood, and paint chips, which can cause long-term lead contamination in soil. Lead dust can also become trapped in cracks and carpets and cause indoor air pollution in the future. Dismantled nuclear facilities are a form of demolition wastes that are especially difficult to throw away (see RADIOACTIVE WASTE DISPOSAL). Asphalt shingles can be melted down along with old asphalt pavement to make new pavement or a dust-proofing material (see ASPHALT). See also BUILDING MATERIALS, CONSTRUCTION WASTES.

Department of Energy   the federal agency responsible for the coordination and administration of the energy-related functions of the federal government. The responsibilities of the Department Of Energy (DOE) include:

- marketing electricity generated at federal facilities;
- long-term energy research and development;
- energy conservation programs;
- the compilation and publication of energy statistics.

The Department of Energy Organization Act of 1977 put the energy-related components of the Energy Research and Development Administration, the Federal Energy Administration, the Federal Power Commission and the Alaska, Bonneville, Southeastern and Southwestern Power Administrations under the control of the new agency. The power marketing functions of the Bureau of Reclamation were also moved to the DOE along with a variety of smaller energy programs from other federal agencies.

The Federal Energy Commission is an independent five-member commission within the DOE. The Assistant Secretary For Energy-Conservation And Renewable Energy Programs is responsible for DOE efforts to stimulate the development of renewable energy sources including solar, biomass, wind, geothermal and alcohol fuels, and for programs aimed at reducing the waste of electricity and fossil fuels by increasing the efficiency of automobiles, power plants and other processes. The Secretary also administers programs that support state energy planning, low-income energy assistance and weatherization and energy conservation in schools, hospitals and public buildings. The DOE Office of Environment, Safety and Health oversees DOE operations to see that they conform with federal regulations and policies governing occupational safety and health, the environment and national security. The office carries out legal functions mandated in the PRICE-ANDERSON ACT.

DOE nuclear programs are managed by:

- The Office of Nuclear Energy, which administers research-and-development in nuclear FISSION. The office is also responsible for the treatment and stabilization of radioactive waste under the Remedial Action Program, which also addresses the cleanup of nuclear sites operated under DOE authority.
- The Office of Civilian Radioactive Waste Management, responsible for the management of federal programs and research relating to the disposal of high-level radioactive waste and spent nuclear fuel. It was established by the Nuclear Waste Policy Act of 1982.
- The Office of Defense Programs, which manages the storage and disposal of nuclear waste from weapons production and nuclear vessels. It also is responsible for research into inertial confinement fusion.

Other DOE departments with duties relating to environmental contamination and waste disposal include:

- The Office of Environmental Restoration and Waste Management, which oversees programs involving the cleanup of contaminated sites and the storage and disposal of HAZARDOUS WASTE.
- The Office of Energy Research, which administers programs associated with fusion energy.
- The Office of Science Education and Technical Information, which coordinates the publication of information gained through DOE research and development programs.

The DOE Clean Coal Technology Program (see ACID RAIN) is administered by the Assistant Secretary for Fossil Energy. The Conservation and Renewable Energy Inquiry and Referral Service provides information on energy conservation and renewable energy. (Address: CAREIRS, PO Box 3048, Merrifield, VA 22116; telephone: 800-623-2929.) The Energy Efficiency and Renewable Energy Clearinghouse is a computer database that contains information on recycling, WASTE-TO-ENERGY PLANTS, energy conservation and renewable energy. (Telephone: 800-523-2929 or 800-273-2957 for the hearing impaired). DOE's Energy Information Administration (EIA) is a source of up-to-date information on energy use in the United States. (Address: Energy Information Administration, National Energy Information Center, U.S. Department of Energy, Washington, DC 20585; telephone: 202-586-8800 for more information.)

Depressurization   the creation of a slight vacuum inside a tightly sealed building through the use of ventilators and exhaust flues (which remove air from the building), or the use of combustion equipment (which consumes oxygen, thereby reducing the amount of air inside the building) without the introduction of outdoor air to replace it. Depressurization can increase the flow of RADON through the cracks in a foundation into a building, and, in extreme cases, result in lethal BACK-DRAFT. It is used inside the containment structure of a nuclear reactor to assure that any air movement is into, not out of, the building, so that in the event of a release of radioactive gases inside the building, these gases will not escape into the environment. See also INDOOR AIR POLLUTION, VENTILATION.

Desalination   the removal of salts from a liquid, usually WATER. More than 7,500 desalination plants are now in operation in more than 120 countries producing a combined 3.5 billion gallons of fresh water per day. Saudi Arabia alone produces nearly one billion gallons of potable water per day from seawater at desalination plants on the Red Sea and the Persian Gulf. Northern Africa, the Middle East and the Caribbean have most of the world's desalination plants, although plants are in operation or under construction to purify seawater in coastal California cities and brackish GROUNDWATER in Florida. The United States is required under a treaty with Mexico to install a desalination plant on the Colorado River that will reduce its salt content before it flows into Mexico.

Thermal processes remove contaminants by boiling or freezing the water being processed, leaving salts and other contaminants behind in the process. In DISTILLATION, water is boiled and the resulting steam recondensed to produce potable water. In FREEZE DISTILLATION, the brine is chilled to below 32°F (0°C), causing the water to freeze and separate from other elements. With REVERSE OSMOSIS, water is forced

through a single membrane with microscopic pores, while in ELECTRODIALYSIS, multiple membranes are employed, and salt ions are driven toward electrodes as they pass through selective membranes. Both reverse osmosis and dialysis produce a stream of purified water and a wastestream of BRINE. In ION EXCHANGE, less harmful and/or more controllable ions of elements such as sodium are exchanged for the contaminating ions. Salts, minerals and radioactive materials can be removed from water via ion exchange. The process produces a potentially toxic brine.

Desertification   the gradual loss of organic materials, fertility and water in a soil that can lead to a permanent loss of the soils ability to grow plants. Desertification is, by definition, caused by human activities and not by climatic change. Overgrazing, the lowering of the water table due to overpumping and other destructive agricultural practices that put a burden on soil moisture are its causes. The number of MICROORGANISMS in a soil that is undergoing desertification slowly decreases until the land is no longer capable of supporting much more than drought-resistant plants such as thistles and cactus, and the soil can decline so far that it will no longer support even these. An estimated 11% of the world's land, 4.6 million square miles, has suffered moderate to severe desertification since 1945. Farmers abandon an estimated 27,000 square miles each year because the land has become too depleted to farm at a profit. An estimated 225 million acres in the American Southwest are affected by desertification.

Destructor   a name given to garbage incinerators in the days when INCINERATION was principally a means of reducing the volume of solid waste. The first garbage incinerator, which was called a destructor, was opened in Nottingham, England in 1874. Destructors—also called crematories—were mass-burn incinerators whose sole function was the incineration of waste, in contrast to installations that produced energy and resource-recovery operations such as the one instituted by New York City around the turn of the century (see George E. WARING Jr.).

Eco Logic, Inc., which has headquarters in Rockwood, Ontario, has developed a device called a destructor that neutralizes toxic chemicals such as POLYCHLORINATED BIPHENYLS (PCBs), POLYAROMATIC HYDROCARBONS (PAHs), chlorinated dioxins and chlorinated solvents through a process called *reductive dechlorination* (see HAZARDOUS WASTE TREATMENT). The process utilizes heat and hydrogen, rather than oxygen, to break down organic molecules into water, carbon dioxide, salts and methane gas. The only atmospheric emissions from the destructor is from its natural-gas-fueled power boiler.

Detector Tube   a transparent cylinder that contains chemicals that change color or extend a stain in proportion to the amount of a specific airborne contaminant that passes through the tube. An air sample is taken by drawing a known quantity of air through the tube with a hand pump. Detector tubes are used as a screening technique, and are not highly accurate. The degree of color change can be measured with a photometer. A detector tube may react only to a single airborne chemical such as CARBON MONOXIDE, or may be designed to sense the presence of an entire category of compounds such as the CARBAMATES.

Detergent   a compound that breaks down OIL and GREASE into microscopic droplets small enough to be carried away by WATER. Dirt on clothing and household dust and grime are commonly mixed with oils and greases, so detergents must be used to remove them. The discharge of nutrients in detergents based on *trisodium phosphate* has led to EUTROPHICATION and bans on phosphate-based detergents in many areas. Phosphate replacements can also cause pollution, however, and one, nitrilotriacetic acid, had to be abandoned when it was discovered that it was a MUTAGEN and that it formed compounds with HEAVY METALS that made them more easily assimilated by living beings (see BIOAVAILABILITY). ABS detergents are based on alkylbenzenesulfonate, a HYDROCARBON that is not readily biodegradable (see BIODEGRADABILITY). Durable masses of foam can build up on lakes and streams that receive effluent containing ABS-based detergents, usually originating from sewage treatment plant outfalls and sometimes SEPTIC SYSTEMS.

Soap, the oldest known detergent, is composed of carbohydrates rendered from animal fats (see ACIDS, ESTERS). The MAGNESIUM and CALCIUM IONS found in hard water (see DRINKING WATER) react with soap to form a scum composed of mineral salts. Many detergents contain a water-softening agent to prevent scum formation. Artificial detergents avoid this problem by using salts based on inorganic acids (usually sulfuric acid) instead of those based on the fatty acids that are used to make soap. When the salts in artificial detergents react with the minerals in hard water, soluble salts, which can be carried away without losing the oils to which they are attached, are formed. Detergents are sometimes used to assist in the breakup of oil slicks.

Deuterium   a naturally occurring isotope of hydrogen that contains one neutron in addition to the single proton and single electron found in a normal hydrogen atom. About one atom in 5,000 of atmospheric hydrogen is deuterium. Although deuterium is nearly identical to normal hydrogen chemically, the heavier weight of its atomic nucleus changes the physical properties of hydrogen compounds formed with deuterium, and for this reason, the isotope has its own chemical symbol; "D" or sometimes "^{2}H." When two deuterium atoms combine with an oxygen atom, deuterium oxide or HEAVY WATER is formed. Heavy water is used as a radiation shield in nuclear reactors. Lithium deuteride reacts with PLUTONIUM in a fusion bomb to free the neutrons that undergo fusion.

Dibenzofurans   see FURANS

Dibromochloropropane (DBCP)   a toxic chemical once used to control nematodes, microscopic soil organisms that can stunt plant growth. DBCP was widely used as a nematocide on more than 40 crops until it was declared a probable human CARCINOGEN, and was finally banned in 1979. DBCP is very persistent once it infiltrates an AQUIFER, and it has been found in GROUNDWATER in 19 U.S. states, with California especially hard hit. Hawaii's pineapple growers were allowed to continue using the pesticide after the ban because there was no viable alternative at the time, but it is now banned for use on pineapples as well. During congressional debate in 1995 over the renewal of the SAFE DRINKING WATER ACT, House Republicans cited the fact that under the SDWA, community water systems throughout the United States were required to conduct expensive monitoring for DBCP, and argued that this revealed the outrageous nature of the law. But aquifers in many parts of the country still contain DBCP from the days when it was still widely used as a nematocide, and from spills and dumping from plants that manufactured the nematocide in California and elsewhere. PUBLIC WATER SYSTEMS are required under the Safe Drinking Water Act to monitor concentrations of DBCP (MAXIMUM CONTAMINANT LEVEL = 0.0002 milligrams per liter). The BEST AVAILABLE TECHNOLOGY for the removal of DBCP is granular ACTIVATED CARBON or packed-tower AERATION.

Diesel   a PETROLEUM derivative that is used primarily as a FUEL. Diesel is a heavy mineral oil that is used in about 20% of highway vehicles, primarily trucks. Less than 2% of U.S. passenger cars are diesel powered. The ENVIRONMENTAL PROTECTION AGENCY (EPA) released a study in July 1990 linking fumes from diesel engines with cancer in humans. The agency listed the PARTICULATES released by diesel-burning engines collectively as a CARCINOGEN in 1983. Diesel particulates absorb sulfuric acids and volatile organic compounds (VOCs), which can then become lodged deep in the lungs along with the particles to which they are attached. Congress ordered the EPA to impose limits on the emissions from diesel vehicles in the 1977 CLEAN AIR ACT, but the agency took no action until after loosing a federal court suit filed by environmentalists. The agency was close to imposing emissions regulations on trucks and buses in 1988, but the proposed rules were shelved by the Reagan Administration.

In 1994, the CALIFORNIA AIR RESOURCES BOARD imposed tighter controls on diesel fuel that were expected to cut particulate emissions from diesel burning vehicles in the state in half by the year 2000. The board estimated that 58,000 tons of diesel particulates were emitted into California's air in 1990. California already had the nation's toughest controls on diesel exhaust before the board's action. Diesel trucks sold in California in 1994 were about 80% cleaner than those sold before 1988. Further reductions were achieved beginning in October 1994 when a board rule requiring cleaner diesel went into effect statewide over the loud protests of truckers and farmers. The California EPA released a "preliminary draft" study in June 1994 that stopped just short of calling diesel a proven human carcinogen. The study was based on a review of 20 different laboratory tests on the effects of diesel exhaust on animals and 20 occupational studies of people who had worked around diesel exhaust most of their lives. The studies suggested that such high exposure to diesel fumes can increase the risk of lung cancer by 20 to 70 percent. See also AUTOMOTIVE AIR POLLUTION, CALIFORNIA CLEAN AIR ACT.

Diffuse Pollution   POLLUTION that comes from many small sources as opposed to a POINT SOURCE of pollution, which has a larger, more concentrated origin. Sources of diffuse AIR POLLUTION, GROUNDWATER POLLUTION and WATER POLLUTION include nutrients and PESTICIDES in the RUNOFF from farms and forest lands; oil, HYDROCARBONS, ASBESTOS and LEAD in runoff from city streets; erosion; DUST from roads, LANDFILLS and construction projects; and emissions from mobile sources such as motor vehicles and airplanes (see AUTOMOTIVE AIR POLLUTION, MOBILE SOURCE). Diffuse pollution can be much harder or even impossible to control than emissions from a point source because even identifying the source may be difficult, and collecting and removing pollutants is generally expensive. Sources of diffuse pollution such as nitrates from pastures and fertilized fields can be controlled only through the application of new management practices. And while centralized sources of pollution are governed by pollution laws, many diffuse sources are not. The CLEAN WATER ACT and the CLEAN AIR ACT include provisions intended to control diffuse pollution. Compare STATIONARY SOURCE.

Dilution   the introduction of fresh air into an enclosed space or the emission of toxins into the atmosphere or into a body of water to dilute the concentration of the toxins to a safe level. Toxic emissions have long been governed by the maxim "dilution is the solution to POLLUTION." In this age of photochemical smog, ACID RAIN and global air pollution, the limitations of this maxim are becoming obvious. The saying still has validity when it comes to INDOOR AIR POLLUTION—a sufficient supply of fresh air passing through an interior space will dilute airborne toxins to manageable level, thus dilution is one solution to bad indoor air quality. Adequate ventilation will also help to control the buildup of indoor humidity in the less humid climates, helping to control microbial growth (see MICROORGANISM) in the process. But the dilution theory has its limits, just as the atmosphere, lakes, rivers and aquifers can only absorb so many contaminants without experiencing harmful effects. Approximately 19 million tons of SULFUR DIOXIDE are emitted into the atmosphere in the United States each year, making adequate atmospheric dilution next to impossible.

The dilution maxim is taken to its extreme in modern COAL-fired power generation stations—where smokestacks towering a thousand feet over the plant are not unusual. The tall stacks allow the operators of the power plants to meet local air quality standards without buying expensive AIR POLLUTION CONTROL equipment. The 1990 amendments to the CLEAN AIR ACT called for a reduction in U.S. sulfur dioxide emissions to 9 million tons a year. But the net effect of the tall stacks stretches the meaning of the word dilution by stretching the ability of the environment to disperse unwanted toxins. Global problems such as the greenhouse effect and the thinning of the ozone layer may spell the beginning of the end for the practice of using dilution as an easy way to dispose of toxic waste products. Even the CARBON DIOXIDE that is released when coal is burned must be of concern, if the buildup of atmospheric carbon dioxide that may lead to GLOBAL WARMING is to be reversed.

Much of the acid precipitation that falls in the northeastern United States and southeastern Canada has been traced by its chemical signature to coal-burning power plants in the Ohio Valley. The plants burn cheap, high-sulfur coal mined locally and vent combustion byproducts—which includes traces of arsenic and selenium and lots of sulfur dioxide—up stacks that are 400 to more than 1,000 feet high. One plant alone, the General James M. Gavin plant near Gallipolis, Ohio, emits 350,000 tons of sulfur dioxide into the atmosphere from its 1,105-foot stack each year. It is the largest of 17 similar plants dotted along a 150-mile stretch of the Ohio River. It cost about $10 million to build the stack in the mid 1970s. Outfitting it with SCRUBBERS today would cost in the neighborhood of $800 million.

The ENVIRONMENTAL PROTECTION AGENCY (EPA) first approved the tall-stacks concept in 1972. Despite defeats in the courts and incessant opposition by environmental groups, more than 100 such stacks have been erected around the country since 1970, most of them at coal-fired power plants. The Reagan EPA gave the concept of using tall stacks to disperse pollution a boost at a time when it had been losing favor. The measures called for in the 1990 amendments to the Clean Air Act, if they are enforced, will eventually put an end to the release of untreated emissions into the atmosphere. The CLEAN WATER ACT has already reduced the level of contamination found in most rivers and streams to a more acceptable level by putting tighter controls on just what could be released into surface water for dilution. See also AIR POLLUTION, GROUNDWATER POLLUTION, WATER POLLUTION.

Dioxin any of a group of closely related aromatic compounds (see AROMATIC HYDROCARBON) that consist of two BENZENE RINGS attached to one another by a pair of oxygen atoms, with CHLORINE atoms attached to the molecule at two or more of the free corners on the benzene rings. Tetra dioxin or TCDD (2,3,7,8-tetra-chloro-dibenzo-para-dioxin) is the most toxic member of the dioxin family, and is one of the most toxic substances known. TCDD has a 50-percent lethal dose (that is, a dose sufficient to kill 50 percent of a test population), or LD_{50}, of 0.0006 for guinea pigs—which would translate to three one-hundredths of a gram killing 500 humans—if the two species' reactions to the chemical are similar. Most of the 74 other dioxins that are theoretically possible are much less toxic, tens of thousands of times less toxic in many cases. TCDD is produced as an impurity in the manufacture of PESTICIDES such as 2,4,5-T and Silvex. Concern about birth defects caused by dioxin exposure was largely responsible for the suspension of aerial spraying of the HERBICIDE Agent Orange during the Vietnam War. Because of changes in the manufacturing process, the concentration of dioxin in 2,4,5-T is 1,000-fold less today than it was when it was used as a component of Agent Orange during the Vietnam War. Nonetheless, 2,4,5-T has been banned in the United States for all uses except the treatment of rangelands and rice crops, because of the herbicide's dioxin content. Dioxin contamination from the waste oil sprayed on gravel roads and two horse arenas in Times Beach, Missouri, resulted in the

town's permanent evacuation in 1983. Dioxins are found as an impurity in POLYCHLORINATED BIPHENYLS (PCBs) and PENTACHLOROPHENOL (PCP).

Dioxin gets into surface water in the EFFLUENT from industrial operations that use chlorine to bleach PAPER and other products, from surface spills and from old electrical transformers. Levels may be higher in the tissue of fish and other aquatic organisms that live in the dioxin-tainted body of water than in the water itself. Dioxin gets into groundwater from leaking underground storage tanks, from surface PCB spills and from facilities such as post and pole and railroad-tie treatment plants. It gets into air and water from the production, transport and use of pesticides and the disposal of pesticide-production wastes. It gets into outdoor air anywhere aromatic hydrocarbons are burned; from municipal garbage incinerators and trash burners where POLYVINYL CHLORIDE and other materials that contain chlorine are burned and from auto exhaust. Dioxin may be released into indoor air from tobacco smoke and from the use of chlorine-base cleansers. Dioxin has no commercial uses. It cannot be seen or smelled at the levels found in the environment, and degrades after a day or so of exposure to sunlight. It can, however, persist for many years in the soil or in living tissue, and can be hard to flush away from contaminated surfaces because it is not very water soluble. There is a great deal of controversy about the toxicity of dioxins. There is no known antidote for dioxin poisoning.

Dioxin can affect health when it is inhaled or ingested. It is thought to accumulate in fatty tissue. Eating fish contaminated with dioxin is another path of exposure. Bleached paper and CARDBOARD products including food containers, toilet paper, diapers and tampons are potential sources of dioxin. In 1989 the Food and Drug Administration (FDA) ordered that manufacturers alter the bleaching process for milk cartons to drastically reduce their dioxin content. The agency estimates that five out of each million people who drink milk out of bleached white cardboard containers will get cancer. Proposed new regulations governing allowable dioxin residues in food containers would presumably reduce that risk. The symptoms of dioxin exposure—headaches, dizziness, aches and pains and disorders of the digestive tract, nerves (see NEUROTOXIN) and liver—are similar to those associated with PCBs, to which they are closely related. Chloracne (an eruption of the skin that resembles acne) can result from exposure to concentrations as low as one part per billion. The National Academy of Sciences announced in July 1993 that Agent Orange was linked to soft-tissue sarcoma, Hodgkin's disease, non-Hodgkin's lymphoma and *porphyria cutanea tarda* (a rare metabolic disorder). Animal tests have demonstrated that dioxin is a CARCINOGEN and a *teratogen*, that it can be lethal to a developing fetus and that it causes damage to the heart, brain, liver and kidneys.

PUBLIC WATER SYSTEMS are required under the SAFE DRINKING WATER ACT to monitor concentrations of 2,3,7,8-TCDD. The MAXIMUM CONTAMINANT LEVEL for TCDD is 0.00000008 (5×10^{-8}) micrograms per liter. The dioxin content of surface water is regulated under the CLEAN WATER ACT, and its disposal is regulated under the TOXIC SUBSTANCES CONTROL ACT. 2,3,7,8-tetrachlorodibenzo-p-dioxin is on the list of hazardous air pollutants in the 1990 amendments to the CLEAN AIR ACT.

Disinfection the removal of biological contaminants such as viruses, bacteria and parasites from water. Pathogenic bacteria can get into water from RUNOFF contaminated by livestock waste, from tanneries and the EFFLUENT of pharmaceuticals and food-processing manufacturers (see WATER POLLUTION). Improper disposal of pathogenic wastes from health care and medical research facilities can also lead to water pollution. Chlorination is the leading method of DRINKING WATER disinfection (see CHLORINE), but the use of OZONE and other alternatives has increased because of concern about the production of TRIHALOMETHANES (see WATER TREATMENT). The presence in water of FECAL COLIFORM BACTERIA, which are not themselves pathogenic, often indicates the presence of other, more dangerous microbes such as SALMONELLA bacteria. Disinfection was required for public water systems that use surface water, or groundwater that is directly impacted by surface water, in the 1986 amendments to the SAFE DRINKING WATER ACT to guard against pathogenic microorganisms in drinking water. No less than 0.2 milligrams per liter of disinfectant must remain in FINISHED WATER before it is pumped into the distribution system, and detectable levels of disinfectant should be found in at least 95% of the water samples drawn from anywhere in the distribution system. Analysis of the total heterotrophic plate bacteria counts (HPC) can be substituted for measurement of residual disinfectant. Ozone, either by itself or together with reduced levels of chlorine, is seeing increased use because of concern about trihalomethanes (THMs). For giardia lamblia, heterotrophic bacteria and Legionella, treatment techniques rather than MAXIMUM CONTAMINANT LEVELS are specified in the Safe Drinking Water Act. Disinfection can be used in lieu of or in tandem with filtration, as long as the treatment technique removes 99.9% of the giardia cysts and 99.99% of viruses.

An estimated 1.5 billion people, primarily in developing countries, drink water tainted by sewage, natural

bacteria and parasites. The U.S. Centers for Disease Control and Prevention has developed a water purification system for use in remote areas, featuring a small generator that uses electrolysis to turn ordinary salt and water into a disinfectant that kills pathogens in drinking water in about an hour. The generator, which uses about as much electricity as a 120-watt light bulb, produces enough disinfectant in about 12 hours to provide electricity for 500 people for a month. The Pan-American Health Organization is raising funds to help communities buy the generators, which cost about $2,500 apiece. Each generator can produce two different types of disinfectant, depending on the metal used in its electrolysis cell. A stainless steel cell produces a mild chlorine disinfectant, while an iridium or titanium cell produces a stronger mixture of chlorine, ozone, chlorine dioxide and hydrogen peroxide that will kill very strong parasites such as CRYPTOSPORIDIUM. Five-gallons jugs, which were specially designed to minimize microbial contamination, are produced locally as part of the clean drinking water project. Water can be poured into these jugs along with disinfectant, and be drawn out as needed through a large spigot at the bottom. A pilot program in Bolivia was so successful that similar endeavors were launched in Colombia, Nicaragua and the Dominican Republic during 1994, and officials in India had expressed an interest in trying the disinfection system. See also INDUSTRIAL LIQUID WASTE.

Disposable Diapers diapers made primarily of paper and plastic that are designed to be used once, then thrown away. Although "disposable" seems to imply that there is an option involved when it comes to discarding a disposable diaper, the only real option is whether the used diapers will go to a LANDFILL or end up behind a bush alongside some roadside turnout. SOURCE SEPARATION or RECYCLING are really not options. A few individuals who are concerned about the potential for soiled diapers to cause pollution at the local landfill may go so far as to rinse them out in the TOILET, but their next move is to throw them away.

Numerous studies have analyzed the life-cycle cost of traditional cloth diapers vs. disposables, and the results have varied depending on the assumptions made by the researchers. About the only point that almost everybody seems to agree on is that disposables cost more, but the majority of parents with kids in diapers use them anyway because they are convenient. It is a matter of faith among many ENVIRONMENTALISTS to be against disposables, although even most environmentally aware parents will use them when away from home. But a dispassionate analysis of the costs of manufacturing a plastic and paper disposable and throwing it away after a single use as compared to those of producing a cotton diaper and washing it 100 or more times shows energy consumption and pollution of the two practices to be about equal. (see RESOURCE DEPLETION COST, ENVIRONMENTAL COST). The primary difference is that the disposable's environmental liability is run up primarily in its creation, packaging and delivery to the consumer and disposal in a landfill, while much of the cloth diaper's use of energy and contribution to environmental degradation is run up in the many times it is washed. As with paper versus plastic grocery bags (see PAPER OR PLASTIC?), the real issue tends to be based more on moral judgments than science. Epidemiological studies have shown that babies who wear disposable diapers have fewer rashes.

Disposable Products products that are made for a single use followed by disposal. Many disposable products are made of PAPER, CARDBOARD, PLASTIC or METAL foils, and include diapers, dishes, silverware, lighters, beverage containers and even cameras. Some disposable products are recyclable (see RECYCLING), but are composed of materials that are permanently bonded together (see COMPOSITE PACKAGING). Compare REUSE.

Dissolved Oxygen Oxygen carried in SOLUTION in water (see SOLUBILITY, SOLVENTS). If the dissolved oxygen level in a lake or stream becomes too low, severe damage to the aquatic ecosystem can result (see EUTROPHICATION). Waterfalls and rapids in rivers aerate the water (see AERATION) and thereby increase its dissolved oxygen level. Warm water can hold more dissolved oxygen than cooler water, and fresh water can hold more than saline water. Oxygen dissolved in blood is essential for metabolism. Cold, fresh water is capable of holding about 10 parts per million of dissolved oxygen. See also ANAEROBIC STREAM, BIOCHEMICAL OXYGEN DEMAND.

Dissolved Solids solids that are carried in SOLUTION in water (see SOLUBILITY, SOLVENTS). The *volatile fraction* of dissolved solids will evaporate when the solution is heated to 650°C (1,200°F) for 30 minutes, while the *fixed fraction* remains in solution. Sea water contains 35,000 parts per million (ppm) of dissolved solids, the result of eons of erosion and the chemical weathering of rocks. The Great Salt Lake contains about 350,000 ppm dissolved solids. See also TOTAL DISSOLVED SOLIDS.

Distillation the process of heating a liquid to vaporize some or all of its components, which are then captured and recondensed. In FREEZE DISTILLATION, brine is chilled to below the freezing point of water, causing particles of ice that are composed of pure water to form (see DESALINATION).

The product of the distillation process, liquid distillates may then be drawn off for use in their pure form (as is the case with water), or may then be subjected to further processing (as is the case with many PETROLEUM distillates). The residue of unevaporated materials left behind by distillation may be discarded or may go through further processing. Ethyl alcohol (see ETHANOL) can be distilled from most plant materials, but 95% of U.S. ethanol comes from corn. To produce ethanol, corn beer (a pungent yellow brew that resembles curdled milk, made by mixing ground corn, yeast, sugar, water and enzymes) is heated to 173°F, forcing the alcohol it contains to evaporate. When the evaporated gases are recondensed, the resulting liquid is about half alcohol. The mixture must be taken through the distillation process at least two more times to yield pure grain alcohol. Turpentine is distilled from wood waste. Distillation is one of the principal methods of purifying DRINKING WATER for home use (see WATER TREATMENT), and is the process by which distilled alcoholic beverages such as brandy and whiskey are produced. Even METALS can be distilled. MERCURY, for instance, can be distilled from a molten mixture of GOLD and mercury by heating the mixture to 356°C, the boiling point of mercury.

Distillation takes place in an enclosed chamber called a *still* or *retort* in which the liquid to be distilled is heated. The gases that evaporate are collected and passed through a condensation chamber where they are cooled to the point that they reliquify. The liquid condensate then drains into a finished-product container. When the liquid to be distilled contains several components with boiling points that are too close together, fractional distillation must be used. The process involves heating the liquid mixture and routing the vapors released through a fractionating column, a tower that contains baffles or plates. Part of the vapors condense and run back down the column toward the heated chamber, forcing the rising vapors to bubble through the liquid, which is why the column is often called a bubble tower. The heated vapors redistill part of the condensate as they bubble up through the tower, losing energy in the process.

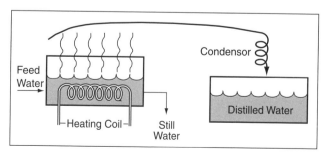

Simple distillation process

Eventually there is too little heat left to revaporize the liquid with the highest boiling point, and it will remain a liquid in the lowest part of the tower. Liquids with lower boiling points will similarly coalesce at progressively higher levels in the fractionating column, where they may be drawn off. More plates are required in the fractionating column as the number of components in the mixture to be distilled increases.

PETROLEUM has so many components that is not practical to distill each one to its pure form. The fractions derived from petroleum distillation, for this reason, consist of mixtures of several compounds with nearly identical boiling points. In the fractional distillation of petroleum, crude OIL is slowly heated to evaporate numerous petroleum products. METHANE and butane are released first, followed by the "petroleum ethers,"—pentanes and hexanes, which evaporate between 20° to 60°C. GASOLINE—a mixture of petroleum distillates that evaporate between 38° and 205°C—consists of pentanes and nonanes. Kerosene, then FUEL oil and lubricating oil evaporate next from the crude oil. Jet fuel, a derivative of kerosene and fuel oil make up about half of every barrel of crude oil, depending on the crude being refined and the refinement process used. The residue of the distillation process is either asphalt or coke. Gasoline produced by the fractional distillation of petroleum is known as *straight run gasoline*, which may represent anywhere from 1 to 50% of the raw crude. Straight-run DISTILLATION supplied most of the nation's gasoline before 1920, but now represents a small fraction of total production.

In *destructive distillation*, a complex substance is heated in the absence of oxygen or other reactive (see REACTIVITY) substances to break the substance down into simpler substances. The simpler substances are usually released as a vapor, which may be separated from the other vapors produced and drawn off as a liquid in a fractionating column. Methanol can be derived from the destructive distillation of wood. The destructive distillation of petroleum, known as cracking, is used to refine gasoline and other fuels. *Pyrolization*, a form of destructive distillation that occurs in the first stages of wood combustion, produces a cleaner burning fuel. Wood stoves and furnaces that pyrolize efficiently are much cleaner burning.

In *flash distillation*, used in the desalinization of seawater, a large quantity of saline water is heated to above its boiling point in a pressurized chamber. The super-heated water is then routed into a low-pressure chamber where it flashes into steam. The super-heated steam thus produced is first used to preheat the next batch of saline water in the pressure chamber, and is then condensed into pure water.

Drainage Liquor   the liquid that escapes from a waste disposal site. Drainage liquor is made up of LEACHATE and liquids leaking from drums, tanks or containers on the site. See also LANDFILL, WATER POLLUTION.

Drilling Mud   a viscous fluid used to reduce friction in the drilling of deep wells. Drilling mud is a colloidal suspension of fine particles of bentonite or barite in water and sometimes oil. It is pumped down the hollow center of pipes that connect the drill bit to the drilling rig, and returns back up the space between the pipe and the sides of the well, carrying with it the mineral waste created by drilling. The particles of ground up rock are filtered out of the drilling mud before it is re-injected into the well. Drilling mud seals cracks to prevent the intrusion of groundwater into the well, and largely eliminates the need for a metal-pipe well casing. See also BRINE, GROUNDWATER POLLUTION.

Drinking Water   potable water intended for human consumption. Community water systems deliver about 25 billion gallons of drinking water each day, and private wells add to the total. An average of 16,000 gallons of fresh water is consumed each day for each resident of the United States. The Delaware River supplies 13% of the people in the United States with drinking water. Although all drinking water is presumably potable, only a small percentage of the water delivered as drinking water is actually swallowed or used for cooking. Most is used to water lawns, flush TOILETS, manufacture products and wash cars, dishes and people. More than 200,000 public water systems supply American homes and businesses. Of these, more than 70 percent serve seasonal facilities such as ski lodges and marinas, and a substantial number are too small to be regulated as PUBLIC WATER SYSTEMS under the SAFE DRINKING WATER ACT.

Several factors can reduce the potability of water: Contamination with toxic substances (see GROUNDWATER POLLUTION, WATER POLLUTION); Tastes and odors caused by decaying organic matter, living MICROORGANISMS, IRON, MANGANESE and other metallic products of corrosion and volatile organic chemicals such as benzene and phenols; excessive COLOR, TURBIDITY; MINERAL content (*hard water* contains lots of minerals, while *soft water* contains relatively few); and excessive ACIDITY (see ACID), which can cause water to corrode plumbing systems and pick up contaminants in the process. Hard water tends to coat the inside of pipes with a hard mineral shell, while soft water is more likely to erode pipe walls and pick up contaminants in the process. AGGRESSIVE WATER is soft *and* acidic, and can cause rapid corrosion, which can lead to contamination with lead, copper and other metals found in the pipes that make up the

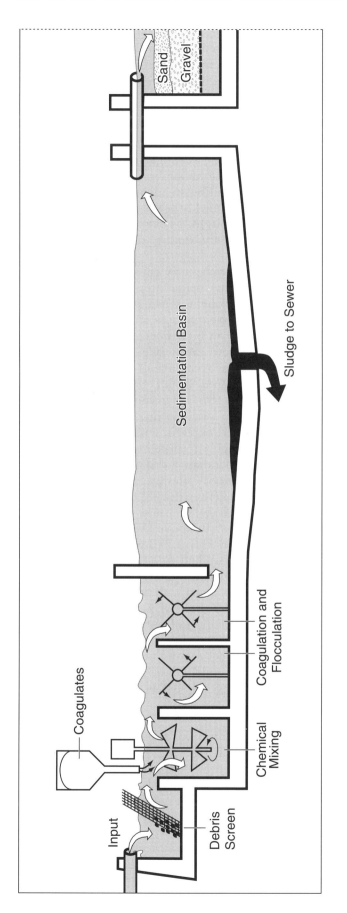

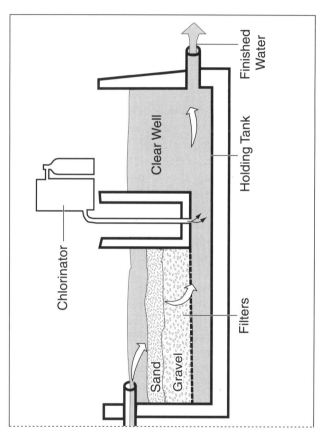

Drinking-water purification process

water distribution system. The ENVIRONMENTAL PROTECTION AGENCY (EPA) estimated in 1994 that lead concentrations in the drinking water delivered by 800 community water systems exceeded federal standards.

Water with more than 500 parts per million (ppm) of DISSOLVED SOLIDS is considered unfit for human consumption in the United States. Livestock can tolerate water containing up to 2500 ppm dissolved solids. Regions with water with low levels of calcium in water have been shown in epidemiological studies to have higher rates of heart disease. Water in the southwestern United States and in the northern Great Plains contain more dissolved solids than that in the rest of the country because the regions' rocks are more vulnerable to the water's solvent effects.

TRIHALOMETHANES are produced when drinking water that contains POLYCYCLIC ORGANIC MATTER is treated with CHLORINE. Finished drinking water can also pick up impurities as it passes through the distribution system (a process that can be accelerated by ACID RAIN). The systems that deliver water to many U.S. cities is a patchwork, with portions of many water systems dating to the nineteenth century. The cracks that leak water when a water system is pressurized can leak contaminants into the distribution system when the water is shut off. A study by the Natural Resources Defense Council, based on data supplied by the water-treatment industry, found that 90% of the nation's large public water systems still use technology developed before World War I.

Drought　the lack of precipitation. "When the well is dry, we know the worth of water," wrote Benjamin Franklin in *Poor Richard's Almanac*. The truth of Franklin's words becomes evident every time a drought threatens a region's supply of WATER. At such times, it becomes obvious that every human endeavor relies on an adequate supply of water. A drought reminds people that DRINKING WATER, PROCESS WATER, IRRIGATION WATER, water to float ships and barges down major rivers, water to wash cars and to spray sweaty children on a hot summer day is actually a precious commodity. Beyond being an extreme inconvenience to people of all walks of life, drought can lead to WATER POLLUTION when the aquatic ecosystem with its natural cleansing ability breaks down. Since a smaller volume of water is available in lakes and streams to dilute pollutants dumped into them, the concentration of all contaminants can become higher. Severe drought such as that experienced during the 1930s in the Dust Bowl—the southern part of the Great Plains region of the United States—can lead to widespread air pollution. Air heavy with the region's soil darkened skies along the Atlantic Coast.

Droughts are a regular occurrence in most parts of the dry western United States, with heavily populated California being one of the most frequently hit. Even wet regions such as the southeastern United States experience occasional severe droughts, however, as residents of the area were reminded during the late 1980s. Municipal water systems, irrigators and industrial water users can all face cutbacks or cutoffs when water levels fall during a drought, and water can get so low—as it did on the Mississippi in 1988—that shipping is curtailed. Getting an adequate supply of clean drinking water can quickly become an emergency situation during a drought, as the available supply of water shrinks and the quality of what is available declines.

Mandatory and voluntary conservation measures can stretch shrinking water supplies to a surprising degree during a drought, when everyone is thinking about water. Nonessential uses such as lawn watering and car washing can be banned until the supply of water increases, and the public and businesses are capable of remarkable reductions in water use when under the (water) gun. Residential water use in San Francisco dropped 65%, for instance, during the 1977 drought. During the two-year drought, ways of saving water were a subject of public enthusiasm and debate, but water use quickly returned to normal when the drought ended. Mandatory and voluntary conservation measures reduced water use in New York City by one-sixth below average during the dry summer of 1985 at a time when water use would normally have been at least one-sixth above normal. City workers installed locks on 30,000 fire hydrants located in neighborhoods where they had traditionally been opened for bathing and car washing.

Tucson, Arizona, faced with an expanding population and a limited supply of water, encourages the reduction or elimination of outdoor watering by its residents. The water utility encourages *xeriscaping*, the use of plants that require no more than seasonal watering to thrive in the area's desert climate. See also DESERTIFICATION, GLOBAL WARMING, WATER CONSERVATION.

Dump Mining　the excavation and processing of materials in landfills with the objective of recovering the resources therein. Some resource specialists predict that the day when the material in twentieth-century dumps is the richest "ore" available for many raw materials is not far away.

Dumpster　a large, usually rectangular receptacle for the disposal of municipal GARBAGE (see MUNICIPAL SOLID WASTE). Mammoth construction-site dumpsters must be delivered and hauled to the dump on the back of a semi truck (see CONSTRUCTION WASTES). Dumpsters are usually emptied into a COMPACTION TRUCK on a regular basis, often weekly, and the trucks haul the garbage to a LANDFILL, incinerator (see INCINERATION), TRANSFER STATION other facility for disposal.

Dust bits of solid matter and ORGANIC MATERIAL small enough to remain airborne for a period of time. Household dust commonly contains dead viruses and bacteria, pollens, skin flakes and dust mite FECES (see INDOOR AIR POLLUTION). Dust does not diffuse after it enters the air, as does a gas, but eventually settles under the influence of gravity. Wind, volcanoes and earthquakes produce large volumes of atmospheric dust. Mechanical processes such as sweeping, drilling, grinding, sanding, crushing, screening, blasting, drilling, excavating and demolition are among the man-made sources of dust. *Road dust* is composed primarily of metals including copper, zinc, lead, iron, manganese, chromium, magnesium and nickel, with the materials of which the road is made determining the composition of the dust. The regular inhalation of *wood dust* can lead to respiratory problems, and airborne wood dust is potentially explosive. A dust explosion can also result when airborne mineral dusts are ignited. A dust explosion often starts with a smaller primary explosion—which can dislodge a large amount of dust that has settled—setting the stage for a larger, secondary explosion. Flour and grain mills are common sites of dust explosions because of the high concentration of airborne dust commonly present. See also NUISANCE PARTICLES, FIBROGENIC DUST, PARTICULATES, AIR POLLUTION, WATER POLLUTION.

Dustfall the descent to Earth of PARTICULATES too heavy to remain suspended in the atmosphere. Particles larger than one micron in diameter tend to settle as the result of the pull of gravity. Dustfall in an area with particularly bad air pollution can be as much as 2,000 tons per square mile per month. A *dustfall jar* is the simplest, least expensive way of measuring dustfall. It is simply an open mouthed container about six inches in diameter and eight inches tall that is made of plastic or GLASS that collects particulates large enough to settle out of the atmosphere. It is left open for a predetermined period of time, usually one month, at a location high enough that it will not be affected by ground-level dust and the particulates that it has collected are then weighed and analyzed (see AIR POLLUTION MEASUREMENT). An *adhesive sheet* or *tacky sheet* can also be used to measure settled particulates. The sheet covered with sticky material is left out for a period of up to a week and is then visually inspected, sometimes with a microscope, to estimate the amount and type of particulates it has collected.

E

Earth First! a radical environmental organization with the motto "no compromise in defense of Mother Earth." Earth First! is more an anarchistic association of small local action groups than a structured traditional organization. The group has no permanent national headquarters, no specific agenda other than the "no compromise" dictum and little or no control over the action of local groups. The term *ecotage* was coined to describe the act of driving bridge spikes into trees to prevent their conversion into lumber, and other disruptive tactics popularized by Earth First! members (although the group now disavows the practice). The most visible activity of Earth First! members in recent years is direct action to stop timber sales in the remaining roadless areas in the federal timberlands of the American West. Members typically chain themselves together to block logging roads or sit in a platform high up a tree to prevent its being cut down. See also ENVIRONMENTALIST. Compare CITIZENS' CLEARINGHOUSE FOR HAZARDOUS WASTE, SIERRA CLUB.

Effluent something that flows out. Effluent is most commonly used in association with a waste product that is in the process of being discharged into the environment, as opposed to *influent*, which is WASTEWATER being introduced into a sewage system or treatment process. Any waste product—whether a gas, a liquid or a solid—can technically be called an effluent, although the term is applied primarily to liquid waste. Most effluent contains some form of contamination that pollutes the receiving body of water. In manufacturing, the discharge of effluents into the environment is an EXTERNALITY that is not usually counted as a part of the cost of a product (see ENVIRONMENTAL COST).

Effluent Standard an air- or water-quality standard based on the concentration of contaminants in the waste stream that is being discharged (see EFFLUENT). Effluent standards set specific limits for such contaminants as SUSPENDED SOLIDS, THERMAL POLLUTION, PH,

HEAVY METALS, ORGANIC COMPOUNDS, NUTRIENTS and PESTICIDES. See also BIOCHEMICAL OXYGEN DEMAND.

Electric Power Research Institute a research organization supported by the annual membership dues of the approximately 700 electric utilities that are members. The Electric Power Research Institute (EPRI) was founded in 1972 by a consortium of electric utilities with the goal of developing new technology that will benefit its members and the public. The more than 350 scientists and engineers employed by EPRI manage more than 1,600 projects located throughout the world. EPRI publishes technical papers and books relating to the generation and transmission of electric power, and a series of pamphlets on subjects such as GLOBAL WARMING, ACID RAIN and emissions control for the general public. Address: 3412 Hillview Ave, Palo Alto, CA 94303. Phone: 415-855-7928.

Electrodialysis a method used to remove dissolved materials from water and blood. An electrodialysis cell is a closed container divided inside into three sections by permeable membranes (see PERMEABILITY). All three sections are filled with water, and electrical current is applied to the cathode and an anode (see corrosion) that are attached to the two end sections, causing negative ions to migrate into the chamber containing the anode and positive ions to migrate into the chamber containing the cathode. Purified water is then drawn out of the central section. Electrodialysis has successfully been used to produce potable water from seawater in small-scale tests but has not been applied on a commercial scale.

Electromagnetic Spectrum the full range of wavelengths of radiant energy. Electromagnetic radiation has no mass or electric charge and moves through a vacuum at the speed of light. The shorter the wavelength of the radiation, the more energy the radiation contains. The electromagnetic spectrum runs from very energetic short-wavelength cosmic rays and GAMMA RAYS to microwaves and radio waves. The *visible spectrum* (0.35 to 0.75 microns), which is made up of light that can be detected by the human eye, accounts for 46% of the sun's energy. *Infrared radiation* has a wavelength longer than 0.75 microns, and makes up 49% of the sun's output of energy. The remaining 5% of the sun's energy is *ultraviolet radiation*, which is shorter than 0.35 microns. The energy content of the sun's radiation is 429.2 Btu per square foot per hour outside the earth's atmosphere, a value called the *solar constant*.

Spectral analysis, which measures a substance's ability to transmit or absorb different kinds of electromagnetic radiation, is used to identify and analyze many contaminants (see POLLUTION MEASUREMENT). Atmospheric pollution plumes are often traced through aerial or space photography in the ultraviolet, infrared, microwave and radio wavelengths.

Electrostatic Precipitator an air-purification device that removes PARTICULATES (see AIR POLLUTION) from air or FLUE GASes by giving them a positive electrical charge, and then capturing them on a collecting surface with a negative electrical charge. Electrostatic precipitators are used to clean indoor air both as free-standing units and to clean air in furnace and air-conditioning ductwork. They are also used to take samples of flue gas (see INDUSTRIAL AIR POLLUTION MEASUREMENT) and to control industrial emissions. A precipitator is capable of removing suspended particulates as small as 0.01 microns. Particulates with a high RESISTIVITY are the most difficult to collect, and condition agents such as sulfur trioxide are sometimes added in industrial precipitators to improve the efficiency with which they are collected.

A typical "two-stage" electrostatic precipitator imparts the electrical charge on airborne particles in the first stage by passing the air to be cleaned through a series of wires and plates with a strong electrical charge. The plates create free electrons, which bump electrons from the air molecules in the air to be processed. These ionized molecules have a positive charge because they have lost their electrical equilibrium when the negatively charged electron was driven away. One or more of the ionized air molecules, in turn, attach themselves to particles in the airstream, giving the particles a positive charge. In general, the larger the particulate, the more positive ions it attracts and the stronger the resulting positive charge. This ionization phase occurs in about one-hundredth of a second. In the second cleaning stage, the process air full of particles with attached positive ions next flows through a series of thin METAL plates with alternate

positive and negative charges induced by a high DC current flowing through them. The positively charged particles are attracted to the negatively charged plates, much as a magnet attracts IRON filings. Once the particles strike the plate's surface, they take on its positive electrical charge. They stay attached via molecular adhesion and cohesion to particles already collected on the plate. Collector plates are sometimes coated with an adhesive to improve the plates' holding power. Most particles, however, contain sufficient oils, tars and sticky residue so that eventually removing them from the plates is more of a problem than having them blown away by the airstream. The efficiency of a two-stage electronic air cleaner falls slowly as its particle-collecting plates get dirty. The plates must be washed every three to six months to assure efficient operation of the air cleaner.

In a "charged-media" or "single-stage" precipitator, polarization and collection of particles occurs in a single step. A typical single-stage electrostatic precipitator consists of a dielectric material that is supported on a frame and which is alternately charged with high-voltage direct-current electricity and grounded. The dielectric material, usually fiberglass or cellulose, produces a strong electrostatic field when the DC current of approximately 12,000 volts is applied, but does not actually conduct electricity. Metal plates broken with perforations or louvers are generally used in a single-stage precipitator to maintain the electrostatic field. Particles enter the alternately charged and grounded gridwork are polarized and drawn to the collecting media. Only particles that can be easily polarized (given a positive and negative pole) are arrested by a single-stage precipitator. See also AIR FILTERS, AIR POLLUTION TREATMENT, INDOOR AIR POLLUTION. Compare NEGATIVE ION GENERATOR.

Element a substance made up of only one kind of atom, which therefore cannot be broken down into other substances through purely chemical means. The characteristics of an element are determined by the number of protons, neutrons and electrons and the arrangement of the electrons around the atom's nucleus. An element consists of atoms with the same number of protons in each atomic nucleus. The number of protons is called the element's atomic number. An ISOTOPE of an element has a different number of neutrons in its nucleus, while an ION of an element has a different number of electrons orbiting its nucleus. There are 87 naturally occurring elements, and another 23 man-made elements, the product of atomic physics. It was once thought that uranium was the heaviest element, but researchers have discovered 18 more since that time, each heavier than the last. The discovery of the as yet nameless 110th element was made in 1994 by an international team at the Heavy Ion Research Center who bombarded lead atoms with nickel atoms with a nuclear accelerator.

Emergency Planning and Community Right to Know Act   a law passed by Congress in 1986 as part of the Superfund Amendments and Reauthorization Act (see COMPREHENSIVE ENVIRONMENTAL RESPONSE, COMPENSATION AND LIABILITY ACT) that requires businesses with more than ten full-time employees to disclose releases of more than 300 toxic chemicals, which are listed in Appendix A of the Chemical Emergency Preparedness Program Interim Guidelines, into the environment. The act also requires that companies disclose which hazardous chemicals they store and use, and that they cooperate with local officials in formulating a plan for emergency response in the event of the release of any of the specified toxic chemicals. The resulting data, known as the Toxics Release Inventory, is available to the public (via computer) at 301-496-6531. Information about the EPCRA is available from the RCRA/UST, Superfund and EPCRA Hotline, telephone 800-424-9346 (382-3000 in Washington, DC).

Emission Standard   the legal limit for emissions of a particular POLLUTANT into outdoor air from a single source. Emission standards are normally expressed as the concentration of a pollutant in parts per billion (ppb) or parts per million (ppm) that is allowed by law in waste gases. If a pollutant is allowed only at concentrations of less than one ppb, this means no more than one molecule of the substance can occur, on average, in each billion molecules of air. For STATIONARY SOURCES of POLLUTION, emission standards are sometimes expressed as the total amount of a pollutant that can be emitted in a given amount of time or for a specified unit of production (kilograms per hour, for instance). See also AMBIENT QUALITY STANDARDS, AIR POLLUTION, CLEAN AIR ACT, GROUNDWATER POLLUTION, WATER POLLUTION.

Engine Oil   mineral oil that is used to lubricate an engine (see INTERNAL COMBUSTION ENGINE, OILS). Oil is a good lubricant because it never wears out, but it is—for the same reason—a threat to the environment once released (see BIODEGRADABILITY, GEODERGRADABILITY, PERSISTENT COMPOUND). Inside an engine, oil additives eventually break down and the oil slowly turns dark as it accumulates a burden of contaminants from the engine, including HEAVY METALS such as LEAD, zinc, ARSENIC, CADMIUM, CHROMIUM and BARIUM. Most of the oil changed in service stations (about half the total) is collected for re-refining or reprocessing (see RERE-

FINED OIL), but much of the rest is sent to the dump, burned, poured on the ground, dumped down storm sewers or otherwise disposed of in a way that can cause the contamination of soil and water. One pint of oil can create a slick on a lake or stream that covers an acre. One quart can ruin the taste of 250,000 gallons of water. It is estimated that far more oil gets into oceans from RUNOFF from land than from oil-tanker spills. Many municipalities have instituted the curbside pickup and recycling of used motor oil, but the capacity to reclaim used oil is limited because only a few refineries have the equipment necessary to process it.

An estimated 1.4 billion gallons of used oil is removed from engines each year in the United States alone, with about 200 million gallons coming out of vehicle crankcases and most of the rest from industrial equipment. Of this, some is recycled, primarily from service stations, and the rest is usually dumped, either down a drain to become part municipal SEWAGE, down a STORM SEWER or on the ground. Used oil picks up metals from engine wear and hydrocarbons from fuels and lubricants, and it can be highly toxic. Oil in sewage can upset the metabolism of microbes at the treatment plant, and oil that is dumped on a road or into the ground can pollute soil or runoff. A significant amount of engine oil and used oil filters also end up in municipal garbage. Oil filters that can be cleaned with soap and water and reused are now on the market.

Putting containers of used oil and oil filters in garbage cans or dumping it into sewers is illegal in most communities and dumping on gravel roads or into the ground is either frowned on or illegal, so the home handyman can be left with few alternatives. Some neighborhood service stations will allow customers to pour their used oil in with that collected by the station for recycling, and an increasing number of communities are offering curbside pickup of oil and oil filters. Oil from service stations is often filtered and sold as a low-grade industrial fuel oil. Arsenic, zinc and lead pass through the filters into finished oil, however, and the large industrial furnaces in which it is burned are generally inefficient, which can lead to the atmospheric emission of heavy metals.

An oil-recycling facility run by Book Cliffs Energy in Green River, Utah, uses a catalytic cracking process to extract GASOLINE, liquid petroleum gas and DIESEL from used oil. The company also plans to convert rubber tires and worn-out plastic products into oil and subsequently into gasoline at the facility. The state of California has a program through which used oil filters are dismantled, the half a cup or so of oil they typically hold removed for recycling and the remaining STEEL melted down and made into steel rebar, a construction material used to reinforce concrete. Under a state law that became effective in 1995, a four-cent tax was added to the sale of quarts of motor oil to fund oil recycling programs coordinated by the state Integrated Waste Management Board. Under state law, motor oil is classified as a hazardous waste. See also HAZARDOUS WASTE, HOUSEHOLD HAZARDOUS WASTE, PYROLYSIS, WASTE-TO-ENERGY PLANT.

English System   a system of measurement originally based on dimensions of the human body. The units of measurement of the English system, which is now used primarily in the United States, are the familiar foot, yard, pound, acre and Fahrenheit degree. In the Fahrenheit temperature scale, the freezing point of water at sea level is 32 degrees Fahrenheit (°F), and the boiling point of water is 212°F (see MEASUREMENTS). The scale was developed by Gabriel Daniel Fahrenheit, who also developed the use of mercury in thermometers. The relationship between degrees Centigrade and degrees Fahrenheit follows the equation F = 1.8(C) + 32. See also METRIC SYSTEM.

Environmental Cost   the cost of damage to the ecosystem caused by the introduction of toxins into air, water, soil or living things. Environmental costs can include death or disease among plants, animals and humans; degradation of the quality of air or water or soil to the point that it is no longer available for normal uses; visual degradation (see VISUAL POLLUTION) as with a clearcut, smoggy air or turbid water (see TURBIDITY); or the loss of the tranquility of a place due to excessive noise (as when a gravel plant or a freeway is built next to a residential area). Arriving at a dollar figure for an environmental cost can be very difficult. This is because many environmental costs have as much to do with the quality of life as with direct damage to living things in the form of disease or death. Determining the cost of a river that is ugly to look at, hazardous to swim in and devoid of game fish, for instance, depends primarily on the values of the individual making the assessment. Some direct economic costs—such as the absence of fishermen, boaters and swimmers and an estimate of the money they would spend locally—can be charged to the POLLUTION, but this is only a small part of the total environmental cost of such pollution. Some would say that the aquatic ecosystem that has been lost is beyond price, while others would assign it little value.

One approach to assessing environmental costs, albeit a partial one, is to compute the cost of cleanup. The expense of installing a water treatment plant capable of improving the quality of water from a polluted river to the point where it is fit to be used as drinking water is one example. Another, much larger, figure would be arrived at if one were to try to compute the cost of restoring a polluted river to pristine condition by cleaning up the water that is entering it and removing pollutants from its bed. In fact, such restoration is normally at best inadvisable and impossible at worst, since a process such as dredging to remove polluted SEDIMENTS would devastate whatever aquatic life remains in the stream. The installation of pollution control equipment as opposed to the unrestricted dumping of effluent is a real bargain when compared to the price tag of complete restoration. Assessing environmental costs is essential, however, because if no estimate of the value of a natural resource has been made, it might be assumed by some that there is no cost associated with a project. See also AIR POLLUTION, AVOIDED COST, EXTERNALITIES, GROUNDWATER POLLUTION, SOCIAL COST, WATER POLLUTION.

Environmental Defense Fund   an organization that focuses on legal action, lobbying and participation in regulatory reform in the defense of the environment. The Environmental Defense Fund (EDF), founded in 1967, works on issues relating to the environmental release of toxins and the use and production of energy. The EDF has 250,000 members in the United States. The EDF was criticized by some environmental groups (especially the CITIZENS' CLEARINGHOUSE FOR HAZARDOUS WASTE) for working with McDonald's rather than opposing them in the MCTOXICS CAMPAIGN. The EDF publishes technical papers relating to ACID RAIN, global climatic change, biotechnology, RECYCLING, INCINERATION, SOLID WASTE management, wetlands and wildlife. Address: 257 Park Ave. S. New York, NY 10010. Phone: 212-505-2100.

Environmental Protection Agency (EPA)   the federal agency responsible for the enforcement of laws relating to environmental quality. The Environmental Protection Agency (EPA) was created as part of the executive branch of the federal government under an amendment to the CLEAN AIR ACT of 1970. The agency is responsible for the enforcement of laws relating to the control and remediation of pollution of indoor and outdoor air, surface water and groundwater (see AIR POLLUTION, INDOOR AIR POLLUTION, GROUNDWATER POLLUTION, LEAKING UNDERGROUND STORAGE TANKS, WATER POLLUTION), the disposal of HAZARDOUS WASTE (see HAZARDOUS WASTE DISPOSAL), and the protection of public health. The CLEAN WATER ACT, SAFE DRINKING WATER ACT, COMPREHENSIVE ENVIRONMENTAL RESPONSE, COMPENSATION AND LIABILITY ACT (Superfund), FEDERAL INSECTICIDE, FUNGICIDE RODENTICIDE ACT (FIFRA) and TOXIC SUBSTANCES CONTROL ACT are among the laws the agency is charged with enforcing. The EPA is required by law to analyze any federal proposal that will have an impact on the environment, and to publish its analysis if it is found that the proposal will have negative impacts. The administrator of the EPA reports directly to the president. There have been several attempts in Congress over the years to elevate the EPA to a cabinet-level agency, but all have been abandoned in the face of stiff opposition by interests that believed this would give the agency too much credibility and power.

There are four program offices within the EPA:

* the Office of Air and Radiation develops national programs, technical policies and regulations relating to air pollution.
* the Office of Water develops programs, policies and regulations related to the pollution of surface water and groundwater, the protection of marine water and estuaries and the development of standards for water quality and effluents.
* the Office of Solid Waste and Emergency Response develops programs, policies and regulations concerning hazardous waste and emergency planning. The Superfund waste-cleanup program, the leaking underground storage tank program and community right-to-know programs are administered by the office.
* the Office of Pesticides and Toxic Substances develops programs, policies and regulations relating to toxic chemicals.

All research performed under the auspices of the EPA is managed by the Office of Research and Development. Litigation is supervised by the Assistant Administrator for Legal and Enforcement Counsel and General Counsel. Standards and regulations are set by the Associate Director for Policy And Resource Management. EPA regional offices are located in each of the ten federal regions.

The EPA was, in its early days, staffed with environmental professionals eager to exercise its regulatory powers to protect the environment to the maximum extent possible. Partially as a result of this early enthusiasm, the agency has become increasingly ineffective. It is often characterized by industrial interests as an intrusive agency with an unjustified and potentially costly interest in the way they do business. As a result, industry lobbyists have taken every opportunity to limit the agency's authority and its budget. Partially as a result of such animosity, the agency has never become the aggressive steward of the environment envisioned by Congress in 1970. Its history of missed deadlines, delays, concessions and exemptions has led to the agency being sued by virtually every national environmental organization that undertakes such litigation. The agency reached a low point early in the Reagan administration when EPA Administrator Anne Gorsuch Burford narrowly escaped prosecution for contempt of Congress for refusing to give Congress sensitive documents on the toxic waste cleanup program. She said she was simply following orders she received from President Reagan and the Justice Department.

The EPA and the OCCUPATIONAL SAFETY AND HEALTH ADMINISTRATION announced an agreement in 1990 to work together to crack down on businesses that violate federal environmental and worker safety regulations. The petrochemical and lead smelting industries were targeted as the agencies' first priority. The agreement ended two decades of noncooperation and animosity between employees of the two agencies. Because the EPA has the authority to seek criminal penalties relating to certain workplace injury, the arrangement was seen as advantageous to OSHA, which can only assess civil penalties for workplace safety violations. The agreement arranges for the exchange of information, data and training to facilitate the prosecution of employers who are threatening their employees' health by exposing them to environmental toxins.

Address: 401 M Street SW, Washington, DC 20460. Public Information Center phone: 202-260-7751. EPA hotlines:

* Asbestos and Small Business Ombudsman Hotline: 800-368-5888. Technical assistance regarding the control of air pollution for small businesses. The 1990 amendments to the Clean Air Act (CAA) required the EPA and state governments to set up the hotlines.
* National Response Center Hotline 800-424-8802. A 24-hour-per-day hotline that accepts calls about potential hazardous waste sites.
* RCRA/UST, Superfund Hotline: 800-424-9346 (382-3000 in Washington, DC)—information for the general public and regulated businesses on the RESOURCE CONSERVATION AND RECOVERY ACT, underground storage tanks, Comprehensive Environmental Response, Compensation and Liability Act and source reduction.
* EPCRA hotline: 800-535-0202. Information on the Toxic Release Inventory and the emergency planning and community right-to-know act.
* Toxic Substances Control Act (TSCA) hotline: 202-554-1404. A source of information relating to the TSCA, the ASBESTOS HAZARD EMERGENCY RESPONSE ACT and the Asbestos School Hazard Abatement Act.
* Acid rain hotline: 617-674-7377. Information about acid rain and sulfur-dioxide pollution.

- Stratospheric ozone hotline: 800-296-1996. Information about depletion of the ozone layer.
- Safe Drinking Water Act Hotline: 800-426-4791. Answers to questions about the Safe Drinking Water Act and access to publications relating to drinking water quality.

Environmental Tobacco Smoke   the combination of two forms of smoke from burning tobacco products: *side stream smoke*, which comes from an idling cigarette, pipe, or cigar; and *mainstream smoke*, which is exhaled by the smoker. Environmental tobacco smoke (ETS) is one of the principal causes of INDOOR AIR POLLUTION. Cigarette smoke is responsible for 39 percent of all exposure to toxins, according to ENVIRONMENTAL PROTECTION AGENCY (EPA) estimates. About half the smoke generated by a cigarette is side stream smoke, which contains an even higher level of toxins than that found in mainstream smoke because it has not been filtered by the smoker's lungs. The majority of the approximately 71,000 tons of tobacco grown in a typical year during the late 1980s and early 1990s by U.S. growers went into the production of more nearly 700 billion cigarettes that were sold in each of those years.

Although some American Indians smoked tubes stuffed with tobacco, it was the cigar that the Spanish Conquistadors took home to Europe. Cigars were smoked primarily by the wealthy, but beggars in Seville, Spain, picked up discarded cigar butts, shredded the tobacco and rolled the first known cigarettes in scraps of PAPER. The primitive cigarettes were called *cigarillos*, or "little cigars." The French started to call them *cigarettes* during the Napoleonic wars. King James I of England issued the first known condemnation of tobacco use as unhealthy in 1604.

Cigarette smoke has been linked to heart disease, cancer and sexual impotence in men. An estimated 1,000 smokers die every day in the United States from smoking-related disease, and cigarette smokers are also more vulnerable to the health effects of environmental toxins in general. Of the more than 4,000 substances found in tobacco smoke, about 60 are associated with cancer or tumor formation. The risk of dying from lung cancer is 20 to 30 times greater for a heavy smoker than for a nonsmoker. Regular exposure to the PARTICULATES found in tobacco smoke damages the lungs and increases the chances of developing such conditions as chronic bronchitis or emphysema. Chewing tobacco and snuff increase the risk of cancer of the mouth. In addition, an estimated 3,000 nonsmokers die of cancer each year as the result of breathing side stream smoke. The American Heart Association estimates that 37,000 people each year die from heart disease caused by side stream smoke. Reports by the U.S. Surgeon General

and the National Academy of Sciences' National Research Council conclude that children of parents who smoke have an increased rate of respiratory disease and reduced lung function. U.S. Centers for Disease Control estimates that the health care costs of smoking were $50 billion in 1993. Former Health Secretary Joseph Califano, addressing a House subcommittee in 1994, estimated that 80 percent of the $20 billion a year in Medicare costs are the result of cigarette smoking. The Office of Technology Assessment estimates that lost productivity caused by smoking cost the U.S. economy $47 billion in 1990. In 1993, the EPA declared secondhand tobacco smoke to be a CARCINOGEN. Younger smokers can reverse some lung damage by quitting, but much of the damage to middle-aged and older smokers seems to be irreversible. The risk of developing a smoking-related cancer depends on how long a person has been smoking, the number of cigarettes smoked per day, the tar content of the cigarette, and how deeply the person inhales the smoke.

The most dangerous components of tobacco smoke are CARBON MONOXIDE and particulate substances including NICOTINE, which are collectively known as tar. Nicotine is a poisonous alkaloid found in tobacco smoke and in other tobacco products, and is the substance to which many smokers are addicted (although the tobacco industry still contests the addictiveness of nicotine). Well-known alkaloids not found in tobacco smoke include cocaine and morphine. AMMONIA, FORMALDEHYDE, hydrogen cyanide, PHENOL, BENZENE and nitrosamine are also found in tobacco smoke. As a result of all the bad news about the health effects of ETS, bans on smoking in public places have been passed in many U.S. cities, and state-wide bans are in process in many states. The first—Vermont's Clean Indoor Air Law, which bans smoking in public places throughout the state—was passed in 1993. Since 1966, cigarette packages sold in the United States have been required to carry a health warning, and cigarette advertising on television has been banned since 1971. Smoking on domestic airline flights was banned in the United States in 1990. The EPA declared ETS a Group A carcinogen in 1993. The International Civil Aviation Organization plans to ban smoking on international flights in 1996. See also ACTION ON SMOKING AND HEALTH.

Environmentalist   a loosely defined term meaning an individual who supports or works on environmental causes such as the preservation of endangered species or the prevention of the pollution of water or air. Timber-sale protesters, U.S. Forest Service employees, professional foresters, neighborhood action group members and mothers protesting pesticide residues in foods have all, at one time or another, been referred to as environmentalists. Animal rights activists and hunters may both be termed environmentalists, even though they are unlikely to be on the same side of many environmental issues.

In general, an environmentalist might be defined as one who believes in the importance of a healthy environment that is free of contaminated air, soil and water, whether or not the individual acts on such ideals in public. There are doubtless many closet environmentalists—people who worry at the grocery store whether paper or plastic is more environmentally correct (see PAPER OR PLASTIC?), and who recycle every recyclable scrap of their household garbage—or people in the bowels of the very corporations that are today lobbying for the relaxation of environmental regulations, who must keep their convictions to themselves if they are to keep their jobs.

The fact is that polls have consistently revealed over the years that most Americans share such "environmentalist" values as a desire for environmental quality, natural diversity and clean air and water. In 1995, while Congress was loudly debating the dismantling or at least the de-horning of key environmental legislation such as the CLEAN WATER ACT and the SAFE DRINKING WATER ACT, a poll paid for by the National Wildlife Federation found that 60% of Americans believe environmental laws should be either strengthened or maintained. The independently conducted poll concluded that 74% of westerners and 76% of all Americans thought Congress should strengthen the Safe Drinking Water Act, and that 39% of westerners and 46% nationally thought that environmental laws do not place a heavy enough burden on businesses to curb the flow of pollutants and to pay for the effects of pollution. Only 23% of westerners and 20% of all Americans thought environmental laws place an unfair burden on business. A total of 54% of westerners and 57% of all Americans polled said Congress should maintain the Endangered Species Act, rather than weakening or eliminating it.

Spectrum

The spectrum of environmentalism can be a good deal more complicated than the standard rendering that has ranchers, miners, loggers, snowmobilers, four-wheel drive vehicle enthusiasts and corporate executives lined up against university professors, people with advanced degrees in botany and wildlife biology and Outward Bound instructors. The controversy over just who is not an environmentalists has been fanned sometimes for personal gain, by such disparate people and organizations as James Watt, the SIERRA CLUB and corporate public relations departments. Even self-pro-

claimed environmentalists are often at odds with one another—as when an Earth Firster calls the Sierra Club part of the establishment that is "ripping off" the earth, and is accused in turn of ruining the credibility of the environmental movement through acts of eco-terrorism. A logger, a miner or a rancher, who might publicly denounce environmentalists as no-growth communists who would take the country back to the Stone Age, if given the chance, and who may be seen by the environmentalist as a despoiler of the land, may himself have a deep-reaching environmental ethic. On the flip side of the coin is the individual who wears a cloak of environmentalism without any true understanding of the workings of the natural world nor deep feeling of connection with such processes. See CITIZENS' CLEARINGHOUSE FOR HAZARDOUS WASTE, EARTH FIRST!, GREENPEACE

EPA   See ENVIRONMENTAL PROTECTION AGENCY

Epichlorohydrin   a colorless liquid with an odor similar to CHLOROFORM. Epichlorohydrin is used primarily to make solvents, glycerol and epoxy resins, and is also used in the production of plasticizers, pharmaceuticals, adhesives, POLYMERS and paper-sizing agents. Epichlorohydrin gets into air and water from industrial effluents and from spills. It is highly water soluble, and is slightly persistent in water, with a HALF-LIFE of two to 20 days.

Epichlorohydrin can affect health when inhaled or absorbed through the skin. ACUTE AFFECTS of exposure to the liquid include burns on skin or eyes. The vapor strongly irritates the eyes, nose and lungs. Higher exposure can cause a fluid build-up in the lungs. Epichlorohydrin is a probable CARCINOGEN in humans. There is some evidence that it causes lung cancer in humans, and it has proven to cause stomach, skin and nasal cavity cancer in animals. Epichlorohydrin may decrease fertility in males. High or repeated exposure may damage the lungs, liver or kidney.

The OCCUPATIONAL SAFETY AND HEALTH ADMINISTRATION has established a permissible exposure limit for airborne epichlorohydrin of 5 parts per million, averaged over an eight-hour workshift. Epichlorohydrin is on the Hazardous Substance List and the Special Health Hazard Substance List. It is also cited in AMERICAN CONFERENCE OF GOVERNMENT INDUSTRIAL HYGIENISTS, Department of Transportation, NATIONAL INSTITUTE OF OCCUPATIONAL SAFETY AND HEALTH and ENVIRONMENTAL PROTECTION AGENCY regulations. Public drinking water systems are required under the SAFE DRINKING WATER ACT to monitor for epichlorohydrin. Treatment technology rather than a MAXIMUM CONTAMINANT LEVEL is specified.

Esters   any organic compound created by the reaction of ALCOHOL and an ACID. Esters give fruits and flowers their taste and fragrance, and are used in drugs such as aspirin. Lipids—a large class of organic compounds that include fats and substances such as OILS and waxes that have properties similar to fats—are esters, as is the explosive nitroglycerin. Polymers formed from esters are called polyesters. Fats and oils are esters of trihydroxy alcohol, and waxes are esters of long-chain monohydroxy alcohols. An ester can be recognized by its name, which is formed by adding "yl" to the root of the alcohol family's name and "ate" to the end of the acid's name. So, for example, oil of wintergreen, which is formed by the reaction of methanol and salicylic acid, is called methyl salicylate. See also GREASE.

Ethanol   ethyl alcohol (C_2H_5OH), also called grain alcohol. Ethanol is added to GASOLINE to create the oxygenated FUEL known as GASOHOL (see AUTOMOTIVE AIR POLLUTION, CLEAN AIR ACT), which produces less CARBON MONOXIDE than gasoline when burned. Ethanol currently costs more than gasoline in most countries, so it is usually sold as a fuel only as a constituent of gasohol. Fuel-grade ethanol is *denatured*—it is mixed with enough gasoline or methyl alcohol to make it unpalatable. Ethanol is relatively easily distilled (see DISTILLATION) from a wide variety of plant materials, but 95% of U.S. ethanol comes from corn. In 1989, about 4% of the nation's corn crop was processed in 39 distilleries to produce 825 million gallons of ethanol. In Brazil, with its heavy dependence on imported oil and an ample sugar crop, ethanol produced from sugar cane is widely used as a fuel. Cars in the Philippines burn ethanol distilled from coconut husks. Ethanol is slightly more poisonous than methanol, but the body is able to metabolize small amounts of ethanol into CARBON DIOXIDE and WATER. Ethanol can also be refined from the petrochemical ethylene.

Most automobile engines can be adjusted to burn ethanol. The ENVIRONMENTAL PROTECTION AGENCY announced in July 1994 that at least 15% of the additives used to make oxygenated fuel must come from renewable sources during 1995, and that by 1996, at least 30% must come from renewable sources. The requirement was a boost to corn farmers. The U.S. demand for corn is expected to increase by 250 million bushels a year as a result of the ruling.

Ethanol is distilled from corn by first grinding corn into a powder, which is then mixed with water and enzymes. The mixture is piped into a settling tank, where corn starch changes into sugar. Yeast is then added, and the sugar ferments into corn beer, which is then heated to 173°F, forcing the alcohol to evaporate.

When the evaporated gases are recondensed, the resulting liquid is about half alcohol. The mixture must be taken through the distillation process at least two more times to yield pure alcohol. Ethanol can be processed to produce ethyl-tertiarybutylether (ETBE), an additive that, unlike ethanol or METHANOL—used to make METHYL-TERTIARYBUTYLETHER (MTBE)—can be blended into gasoline at the refinery and shipped through pipelines to storage terminals, a distinct advantage for petroleum refiners. See also ALTERNATIVE MOTOR FUELS ACT, AIR POLLUTION.

Ethyl Benzene   a colorless FLAMMABLE LIQUID with an aromatic odor. Ethyl benzene is used in the production of STYRENE and synthetic polymers, as a SOLVENT and as a component of automotive and aviation fuels. It gets into air and water primarily from PETROLEUM-based industrial discharges or spills. Ethyl benzene is moderately soluble in water, and is nonpersistent in water, with a HALF-LIFE of less than two days. Ethyl benzene can affect health when it is inhaled or when it passes through the skin. Contact can irritate the skin. Prolonged exposure can cause drying, scaling and blistering of the skin. Exposure to low levels can irritate the eyes, nose and throat. Exposure to higher concentrations can cause dizziness and lightheadedness, and may damage the liver. There is limited evidence that ethyl benzene is a TERATOGEN.

PUBLIC WATER SYSTEMS are required under the SAFE DRINKING WATER ACT to monitor for ethyl benzene (MAXIMUM CONTAMINANT LEVEL = 0.7 milligrams per liter). The OCCUPATIONAL SAFETY AND HEALTH ADMINISTRATION has established a permissible exposure limit for airborne ethyl benzene of 100 parts per million, averaged over an eight-hour workshift. It is also cited in AMERICAN CONFERENCE OF GOVERNMENT INDUSTRIAL HYGIENISTS and Department of Transportation regulations. Ethyl benzene is on the Hazardous Substance List, the Special Health Hazard Substance List and the HAZARDOUS AIR POLLUTANTS LIST. Business handling significant quantities of ethyl benzene must disclose use and releases of the chemical under the provisions of the EMERGENCY PLANNING AND COMMUNITY RIGHT TO KNOW ACT.

Ethylene Dibromide   a heavy, colorless to brown liquid with a sweet, CHLOROFORM-like odor. Ethylene dibromide (EDB) is used as an anti-knock additive in GASOLINE, a grain fumigant, a general SOLVENT and in waterproofing preparations. It gets into air and water from industrial discharges and spills. It gets into groundwater from LEAKING UNDERGROUND STORAGE TANKS. EDB is highly soluble and slightly persistent in water, with a HALF-LIFE of two to twenty days. EDB can affect health when inhaled or passed through the skin. Contact can severely irritate and burn the eyes and skin. Inhaling EDB vapor can damage the nose and throat. Breathing higher concentrations can burn the lungs and cause the build up of fluid, coughing and shortness of breath. Dizziness, drowsiness, vomiting, unconsciousness, and death can follow severe exposure. Repeated or severe exposure can damage the liver and kidneys. EDB is a proven TERATOGEN in animal tests and is a possible human teratogen. There is some evidence that EDB damages human reproductive systems. Ethylene dibromide is a probable human CARCINOGEN. There is evidence that it causes stomach and liver cancer in humans and it has been shown to cause stomach and liver cancer in animals. Most pesticide uses of the chemical were restricted in 1984.

PUBLIC WATER SYSTEMS are required under the SAFE DRINKING WATER ACT to monitor concentrations of ethylene dibromide (MAXIMUM CONTAMINANT LEVEL = 0.00005 milligrams per liter). The BEST AVAILABLE TECHNOLOGY for the removal of EDB is granular ACTIVATED CARBON or packed-tower AERATION. The odor threshold for EDB is 26 parts per million (ppm). The OCCUPATIONAL SAFETY AND HEALTH ADMINISTRATION has established a permissible exposure limit for airborne EDB of 20 parts per million, averaged over an eight-hour workshift, with concentrations not to exceed 50 ppm during any 15 minute work period. It is also cited in AMERICAN CONFERENCE OF GOVERNMENT INDUSTRIAL HYGIENISTS, NATIONAL INSTITUTE OF OCCUPATIONAL SAFETY AND HEALTH, Department of Transportation and National Toxicology Program regulations. Ethylene Dibromide is on the Hazardous Substance List and the Special Health Hazard Substance List.

Ethylene Glycol   an ALCOHOL—also called *dihydric alcohol*—that has two hydroxyl groups replacing hydrogen atoms. Ethylene glycol is the most commonly used automotive antifreeze. It is on the HAZARDOUS AIR POLLUTANTS LIST.

Eutrophication   an increase in the rate of biological productivity of a standing body of water such as a lake or reservoir (see TROPHIC CLASSIFICATION SYSTEM). SEDIMENTATION, increases in temperature (see THERMAL POLLUTION), increases in the concentration of NUTRIENTS and decreases in DISSOLVED OXYGEN (see BIOCHEMICAL OXYGEN DEMAND) are contributing factors to eutrophication. Warm, shallow water that contains many nutrients is capable of supporting a much higher eutrophic state than cooler, deeper water and water containing fewer nutrients.

Eutrophication occurs naturally as part of the process by which lakes slowly fill in to become meadows. As the lake's water gets shallower as the result of sedimentation, the water gets warmer, plant growth and nutrient levels increase, filling the lake entirely over a period of centuries. Eutrophication resulting from WATER POLLUTION, in contrast, can cripple a healthy aquatic ecosystem in a short period. High rates of erosion caused by farming, logging, road building and over-grazing (see RANGE MANAGEMENT) can accelerate sedimentation. Nutrients—primarily PHOSPHATES and, to a lesser degree, NITRATES from household DETERGENTS—are discharged from municipal sewage plant outfalls, in industrial EFFLUENT and are also found in RUNOFF. Algal blooms often occur in water receiving excess nutrients (see RED TIDE). Thermal pollution resulting from the cooling of power plants can further accelerate the process. When water is too warm, many species of fish have trouble reproducing, especially trout and salmon. See also AMMONIA, TURBIDITY.

Evaporation   the conversion of a substance from its liquid to its gaseous state. Evaporation occurs when a MOLECULE gains sufficient energy to break its bond with the surface of a liquid and escape into the ATMOSPHERE. The heat energy required to cause this transformation is called the *heat of vaporization*. A WATER molecule must absorb five times as much heat as that required to bring its temperature from 32°F to 212°F in order to evaporate. This energy is latent heat, because it does not increase the molecule's temperature. A sweating human and his panting dog are both utilizing evaporation to cool off. A swimmer getting out of the water into a stiff breeze will be cool even on a warm day because of the water evaporating from his or her skin.

The evaporation of water molecules is a key part of the *hydrologic cycle*, the process that delivers fresh water to land, and plays a key role in the formation of weather systems. When water vapor condenses (see CONDENSATION) to form precipitation, its latent heat is released. An estimated 1,500 cubic miles of pure water is created each year by evaporation, primarily of seawater. It has been estimated that Lake Mead on the Colorado River loses one cubic mile of water per year to evaporation. Evaporation of VOLATILE ORGANIC COMPOUNDS is a primary cause of AIR POLLUTION and toxic precipitation (see ACID RAIN). When air is at 100% relative humidity, it is saturated with as much water vapor as it can hold, and no further evaporation can occur. See also AEROSOL, GASOLINE, OZONE, SUBLIMATION, VAPOR PRESSURE, VOLATILITY.

Evaporation Pond   an outdoor pond where WASTEWATER is allowed to stand and allow the action of sun, wind and microbial action to neutralize wastes and where volatile contaminants are allowed to evaporate into the atmosphere.

Excrement see FECES

Exfiltration the passage of a gas or fluid out of a porous substance or container through small cracks or holes. Exfiltration is used primarily to denote the passage of indoor air out of a building through cracks and holes in the building envelope. Compare INFILTRATION. See also AIR CHANGES PER HOUR, INDOOR AIR POLLUTION, VENTILATION.

Exhaust Air stale air that passes out of a building as the result of EXFILTRATION or mechanical ventilation. See also INDOOR AIR POLLUTION, VENTILATION.

Expanded Foam a plastic material that has been blown full of tiny bubbles to decrease its density and increase its insulating materials. Expanded POLYSTYRENE is used in rigid building insulation board, egg cartons, meat trays, coffee cups, packing peanuts and to make molded forms for the shipping of sensitive equipment such as electronics. Extruded expanded polystyrene is called blue board or Styrofoam. Only about 10% of the foam plastics manufactured between 1980 and 1989 were used for fast food PACKAGING, and the total volume for all expanded foam products is about 1% of total LANDFILL space, according to estimates by the Garbage Project. See MCTOXICS CAMPAIGN, PLASTICS, PLASTICS RECYCLING, STYRENE.

Explosive Material a material that can easily explode if exposed to flame, air or another chemical. Most explosive materials are classified as HAZARDOUS MATERIALS, regardless of their TOXICITY.

Externalities the costs and benefits of an economic transaction not borne by the producer or the consumer of the goods or services in question. Since externalities have no direct effect on the profitability of an endeavor, they have typically been ignored in the past. But the externalities of a transaction, especially as they apply to the quality of the environment or the economic well-being of a local community, are receiving increased attention.

Externalities may be positive or negative. A logging business may decide, for instance, that the most economically expedient way to remove the commercial timber from their land is to cut everything and then stack and burn all the trees that are too small to haul to the sawmill. The negative externalities that can result from such a practice, such as increased erosion and loss of habitat, are well known, but these are not

expenses that appear in the company's profit-and-loss statements, so they are not considered. But positive externalities may also result. For instance, new growth in the clearcut may provide needed forage for certain animals, skiers may enjoy the open slopes during the winter and future wildfires may be stopped when they reach the area due to a lack of fuel. Unfortunately, negative externalities generally far outweigh the positive effects.

Many of today's social and environmental problems are the result of the indirect effects of development being ignored. The owners of an industrial plant built with no consideration of the POLLUTION it will emit or the raw materials it will consume typically attempt to continue to operate the facility throughout its life on the same economic assumptions on which it was built. When confronted with a drastic change in that economic formula because of the depletion of the natural resource being exploited or because of tougher environmental standards being enacted, the plant's management may decide to abandon the operation. But the externalities that were invisible on the balance sheet can remain with the community indefinitely. Polluted soil and water, a depleted resource base and a depressed economy often become the legacy that is left behind.

Environmental externalities are considered by many state public utility commissions when deciding what mix of new power generation sources to approve. Public utility commissions have the authority to consider externalities because they are charged with protecting the public health, welfare and safety, which includes health and environmental effects of power production. The externalities considered include the air and water pollution (see AIR POLLUTION, GROUNDWATER POLLUTION, WATER POLLUTION) that will be emitted by a thermal power plant and the future cost and availability of fuels used to generate electricity. Because of this practice, the use of renewable energy sources such as geothermal, wind and solar power are becoming more economically viable. But even these technologies come with environmental externalities. The construction of a wind power-generation installation, for instance, may entail tree cutting, road construction and the use of significant resources for the manufacture and installation of the wind generators. Once the installation is in use, birds may be killed by flying into the turbine blades, radio reception may be disrupted and, to many, a permanent visual blight may result. But the long-term environmental impact of such an installation is generally less than it would be for a conventional power plant, because no pollution is emitted and no fuel must be extracted.

There are two basic ways of changing externalities into a part of the economic equation that must be internalized by a company that wants to do business. *Regulation* involves the establishment of standards for factors such as the siting of plants, allowable emissions into air and water and the transportation and disposal of wastes. As long as these standards are met (or as long as the business is not caught breaking them), many of the plant's effects remain externalities. Only when the business is caught breaking the regulations must some of these effects be internalized in the form of fines, plant closures or even jail terms. *Taxation* involves the assessment of fees for siting, emissions waste disposal and the

like. Since these fees are a given part of doing business, the company will be motivated to minimize its taxes by minimizing emissions and waste production and building where siting fees are low. Critics of the taxation approach consider it a license to pollute. The best approach to the externalities problem may be a combination of taxation and regulation, in which the maximum allowable level of environmental impacts is specified, and credit is given for efforts to reduce it further. See also AVOIDED COST, ENVIRONMENTAL COSTS, RESOURCE DEPLETION COST, SOCIAL COST.

F

Fabrics materials that are produced from FIBERS through processes such as weaving, knitting or mechanical or chemical bonding. Natural and manufactured filaments, yarns, threads and the cloth and other materials that are made from them …are referred to collectively as textiles. The textile industry produces fabrics for clothing, draperies, carpets and curtains, upholstery, shoes, backpacks, tote bags and countless other products. Many clothes are resold, handed down to the next smallest child in a family or circle of friends or donated to a charity or second-hand store. Clothing that is unusable (or undesirable) often gets reused as rags. The highest-quality *rag paper* (see PAPER) contains fibers from recycled cotton and linen rags. There was a thriving trade in rags reclaimed from municipal garbage at the turn of the century (see SHODDY), with the fibers in rags being used in the production of paper and cloth. RAG PICKERS were paid the 1990 equivalent of $350 per ton for rags at the time, comparable to the price now paid for aluminum. By the 1920s, virgin wood pulp was replacing recycled paper and rags in the production of paper, and rags were increasingly being made out of virgin materials and disposed of after use. A far smaller percentage of rags is recycled today. *Molded rag shoddy* is produced from scraps created in the manufacture of textiles and is used to make the acoustical padding found in the headliner of automobiles. An estimated 1.25 million tons of clothing is discarded annually by second-hand stores and charities. About half a million tons are exported for sale in third-world countries, some is recycled into asphalt shingles and paper, and some becomes municipal solid waste. Rags make up an estimated 2% by weight of household waste, with the majority coming from low-income households. See also MUNICIPAL SOLID WASTE, POST-CONSUMER WASTE, RECYCLING, WOOL PRODUCTS LABELING ACT.

Faculative Organism an organism that can live in two or more different modes. A *faculative anaerobe* can live with or without free oxygen, although most do better in the presence of oxygen. A *faculative parasite* can live with or without its host, but favors its parasitic phase. See also FACULATIVE POND.

Faculative Pond a SEWAGE TREATMENT lagoon that is used part of the time as an oxygenation tank and part of the time as an ANAEROBIC DIGESTER. A facultative pond is generally used as an oxygenation tank in warmer weather, when ice is not present and breezes can stir the water and keep its level of dissolved oxygen at acceptable levels. In winter, when the pond is covered with ice and snow, the light and oxygen required for the metabolism of aerobic organisms (see AEROBIC BACTERIA) is not available, and anaerobic organisms take over the job of breaking down organic materials in WASTEWATER. Sewage sludge is often left in a faculative pond for a full year so that it can go through a full cycle of aerobic and anaerobic treatment. More FACULATIVE ORGANISMS are generally found in a faculative pond than are found in either an oxygenation tank or an anaerobic digester. See also EVAPORATION POND, WASTEWATER TREATMENT.

Fansteel an Oklahoma company that disposed of the RADIOACTIVE WASTE it had accumulated as a byproduct of the production of the rare metals *tantalum* and *columbium* by selling it to a Thailand company for "recycling." Fansteel had dumped wastes from the production of the rare metals, which are used in electronic equipment and nuclear reactors, into a lagoon near its Muskogee, Oklahoma, plant. In the late 1980s, the market for the metals bottomed out, and Fansteel started to sell off the portion of the company that was involved with their production. By that time, an estimated 25 tons of URANIUM and 65 tons of *thorium* had accumulated in the lagoon in the 40 years they had been in use. The U.S. NUCLEAR REGULATORY COMMISSION listed the sludge ponds in 1991 as being among the nation's 46 most troublesome nuclear-waste disposal sites, and estimated cleanup costs at $100 million. Intimidated by the cleanup estimate, Fansteel officials came up with a scheme to empty the waste ponds by selling it to a Thailand company that would process the sludge to remove the valuable tantalum and columbium it contained. Fansteel applied for and received an export license from the nuclear regulatory commission to export the wastes. The company's plan for disposing of the wastes was short-circuited after the first 14,700 ton shipment was sent when environmental groups got wind of the shipment and alerted officials and the public at both ends.

The ease with which Fansteel arranged for the export of its hazardous waste is not surprising, since there is no international system to monitor toxic exports, and U.S. environmental regulators have little power to stop such deals even if they know the wastes will be inadequately disposed of in a poor country. Similar shipments of hazardous materials including nuclear waste, used automobile batteries and tires, old paints, contaminated soil, chemical solvents, asbestos, incinerator ash and plastics are commonly made by U.S. companies looking to avoid high fees for dumping at domestic hazardous waste disposal facilities. A 1992 international agreement to ban the export of hazardous waste to developing countries has not yet been ratified by the United States, and will be difficult to enforce even if it becomes law.

Fast-Food Packaging   containers and wrappers made of PAPER, CARDBOARD and PLASTIC. The disposable packaging used to deliver hamburgers, french fries, coffee and shakes to customers at fast food restaurants has been the subject of far more public debate (and corporate embarrassment) than far larger sources of MUNICIPAL SOLID WASTE. Fast-food packaging is a major component of LITTER, and its resulting visibility is probably one reason it is held in such low esteem by the public (see VISUAL POLLUTION). See also CONVENIENCE PACKAGING, DISPOSABLE PRODUCTS, MCTOXICS CAMPAIGN.

Fecal Coliform Bacteria   a bacteria associated with mammalian FECES. Fecal coliform bacteria are relatively harmless, but their presence in DRINKING WATER is cause for concern because they are associated with more virulent organisms such as *E. coli*. See also COLIFORM BACTERIA, MICROORGANISMS, PATHOGENIC WASTE, SAFE DRINKING WATER ACT, SEWAGE SLUDGE.

Feces   a waste material excreted by animals and humans—also called excrement, manure, etc. Much of what are termed *solids* in SEWAGE TREATMENT are feces, and PRIMARY TREATMENT is concerned primarily with their removal. Feces from pets is commonly disposed of along with other household garbage (see HOUSE-HOLD HAZARDOUS WASTE, INFECTIOUS WASTE). In an ANAEROBIC DIGESTER, the organic materials in feces are broken down by ANAEROBIC BACTERIA to produce methane gas. In a compost pile, feces is broken down by aerobic bacteria to produce carbon dioxide, water vapor and a nutrient-rich organic material called *compost*. FECAL COLIFORM BACTERIA are associated with mammalian feces, and DRINKING WATER is routinely tested for the bacteria. The bacteria *E. coli*, SALMONELLA and CRYPTOSPORIDIUM (see MICROORGANISM, WATER TREATMENT) are also associated with feces and occasionally show up in finished drinking water.

Federal Insecticide, Fungicide and Rodenticide Act   the federal law governing the registration and labeling of PESTICIDES. The Federal Insecticide, Fungicide and Rodenticide Act (FIFRA) originally became law in 1947, and has been amended many times since. No insecticide, herbicide, fungicide, rodenticide or other poison used to control unwanted pests and plants can be sold legally in the United States unless it first is registered under FIFRA. The manufacturer of the product must prove that the product they want to sell will not compromise the health of people or the environment if it is used as directed. The ENVIRONMENTAL PROTECTION AGENCY can suspend a product's registration at any time, but before the registration can be canceled, the product's manufacturer must be given a chance to offer arguments in response to evidence favoring cancellation at a hearing. All pesticides must display a registration number, an ingredient list and instructions for proper use and disposal. Many pesticides are also regulated under the CLEAN AIR ACT, the CLEAN WATER ACT, the SAFE DRINKING WATER ACT and other federal ordinances.

Federal Water Pollution Control Act   see CLEAN WATER ACT

Feed Materials Production Center   a federal facility at Fernald, Ohio, 20 miles northwest of Cincinnati, that makes URANIUM ingots and rods for use in the fuel rods of NUCLEAR REACTORS that produce material for nuclear weapons. The DEPARTMENT OF ENERGY has acknowledged that the plant has emitted thousands of pounds of uranium dust and that radioactive and toxic wastes were buried on the site in unlined pits. The Department of Energy and the Fernald Environmental Restoration Management Corporation started cleanup of the site in 1995, a process that was expected to take ten years and to cost a little less than $2 billion. See also HANFORD NUCLEAR RESERVATION, ROCKY FLATS, SAVANNAH RIVER, URAVAN URANIUM MILL.

Feedlot a facility for feeding a large number of livestock or poultry in a relatively small space. Feedlots for livestock are generally fenced pastures with rows of covered troughs for feed. Feedlot rearing produces a meat with a higher fat content in a shorter time, because the feed is brought to the animals without their having to expend energy foraging. About half of U.S. beef is raised on wheat and corn at feedlots in the Midwest. In the western United States, feedlots are often used to fatten range-fed cattle before they go to market. An estimated 500 million tons of manure are produced on U.S. farms every year (although nobody has actually gone out and measured it). Because a large quantity of waste is produced in a concentrated area in a feedlot, the ability of the soil to filter out contaminants and of natural organisms ability to break down wastes is overwhelmed. RUNOFF from feedlots is a leading cause of the pollution of surface water and groundwater with AMMONIA and NITRATES (see WATER POLLUTION, GROUNDWATER POLLUTION). Pollution of surface water by runoff has been reduced by the requirement in most states that it be collected by farmers, but groundwater contamination has proven much harder to curtail.

Fertilizer a nutrient-rich substance used to boost plant growth. Most fertilizer is added to the soil, although some is applied directly to leaves. Runoff and irrigation return flow (see IRRIGATION WATER) from fertilized fields can cause excessive NUTRIENT loading of lakes and streams, and groundwater has been contaminated with nitrates in many agricultural regions. Some fertilizers are obtained from natural sources such as manure, vegetable material, PHOSPHATE rock and bone meal, and others are produced synthetically. The atmosphere is about 78% nitrogen by volume, and nitrogen is produced commercially by drawing air through heated COPPER or IRON elements. Of the 16 elements considered essential to plant growth, only oxygen, HYDROGEN and carbon are readily available in air and water. The rest must come from the soil. Nutrients are removed from the soil when crops are produced, but since most of the animal or vegetable fiber produced is removed, the nutrients are not returned to the soil, so some form of fertilizing must be done if the practices are to be continued without seriously depleting the soil. Naturally produced fertilizers include composted manure and vegetable matter. Green manures are crops that are grown to be plowed under to add nutrients to the soil. Nutrients must generally be available in ionic form to be usable by plants. Plants are awash in a sea of atmospheric NITROGEN, for instance, but only a few, primarily legumes, are capable of using it directly as a nutrient.

Nitrogen (N) in the form of urea and related ammonia and ammonium sulfate is readily usable by plants and such compounds are used in fertilizers. Phosphorous (P) is provided in the form of orthophosphoric acid obtained from bone meal or PHOSPHATE ROCK, and potassium (K) as potassium carbonate (from wood ashes) or potassium chloride (from sylvite or sylvinite ores).

Fibers hair-like structures of animal, vegetable, mineral or synthetic origin used in the production of woven and felted FABRICS, rope and twine (called *cordage*), insulation and other products. Fibers can be used as stuffing for furniture and other items; woven, knitted, knotted or felted to produce fabrics; braided to produce twine or rope; or mechanically or chemically bonded to produce nonwoven fabrics. Air and water filters use synthetic and mineral fibers to capture contaminants.

Animal fibers include *silks*, which are spun in continuous filaments by certain species of insects and spiders. *True silk*, which is a cylindrical thread spun only by the silkworm, was woven into fabric as early as the 27th century B.C. according to Chinese legend. *Wild silk*, which is spun by relatives of the silkworm, is more irregular and more rectangular in shape. *Hairs*, *furs* and *wools* from mammals must be spun into threads or turned into felt before they can be used. Wool is the most widely used animal fiber, but alpaca, angora and cashmere goats, camel and horse hair is also woven, and hair from rabbits, cows and cats is used to produce fur felt.

Vegetable fibers consist primarily of cellulose. The soft hair that grows around cotton seeds is spun, and the similar hairs that grow around kapok seeds are used as upholstery stuffing. *Cotton* was cultivated and its fiber woven into cloth in South and Central America in pre-Columbian times. *Basts* are tough fibers that make up the entire stem of some grasses. They are also found in leaves and their stems and between the leaf and stem of certain plants. *Linen* thread is made from flax bast and coarser cloths; and rope and twine are made from the fiber of hemp, jute and other vegetable fibers. The entire stems of some grasses are woven for use as hats and mats, and baskets are woven from reeds and the small branches of willows and other bushes.

Mineral fibers include *fiberglass*, which is made by drawing or blowing molten GLASS into thin, flexible threads. The use of the naturally occurring mineral fiber ASBESTOS, has been drastically reduced due to concern about its toxicity. *Rock wool*, which is used as insulation, is a fibrous material made from limestone, siliceous rock or SLAG from STEEL mills.

Synthetic fibers include *rayons*, which are based on cellulose; *nylon*, which is made from natural proteins;

and several *plastic fibers*, produced from the vinyl and amyl groups (see PLASTICS). Old soft-drink bottles can be pulverized, treated with chemicals and reconstituted into synthetic fibers that look and feel like cotton.

Fibrogenic Dust   dust that causes scar tissue to form in the lungs when inhaled in excessive amounts. See PARTICULATE. Compare NUISANCE PARTICLES.

Filtration   the physical removal of contaminants from air or water by passing it through granular or fibrous filter media (see AIR FILTERS, WATER FILTERS). Contaminants are filtered out of surface water when it passes through streambed sand, and soil filters out contaminants in RUNOFF seeping into the ground (see GROUNDWATER POLLUTION, PERCOLATION). A filter that uses only mechanical processes such as straining and impaction to remove contaminants from air or water is called a *mechanical filter*. *Fabric filters* use natural and synthetic fibers to remove contaminants from air or water. The type of fabric used, the pattern of flow through the filter, the rate at which air or water is passed through the filter, the surface area of the fabric being used and the frequency with which a filter is cleaned or replaced all have a bearing on the amount of contaminants such a filter will remove. Fabric filters are used to remove PARTICULATES from industrial GASEOUS WASTE (see BAGHOUSE, INDUSTRIAL AIR POLLUTION TREATMENT), and to remove suspended solids from household drinking water, often prior to advanced treatment with a granular ACTIVATED CARBON filter, REVERSE-OSMOSIS unit or distiller.

Filtration is used extensively in the treatment of SEWAGE (see SEWAGE TREATMENT, WASTEWATER TREATMENT), DRINKING WATER HAZARDOUS WASTE (see HAZARDOUS WASTE TREATMENT) and industrial emissions. Filters can become clogged quickly in air or water containing many contaminants. If a filter is not replaced soon enough, it will constrict the flow of air or water, and contaminants already trapped by the filter can reenter water or air in a dirty filter. Filters are used in air- and water-pollution measurement instruments to measure airborne particulates (see AIR POLLUTION TREATMENT). The screens used to remove organic material at different stages of the sewage treatment process are a form of very coarse filtration. See also ELECTRODIALYSIS.

Finished Water   water that has gone through all the treatment processes in a drinking-water treatment plant. Finished water is sometimes called product water. See DRINKING WATER, SAFE DRINKING WATER ACT, WATER TREATMENT.

Firch   items of value that are removed from the SOLID WASTE STREAM from garbage cans, dumpsters and dumps. See RECYCLING, REUSE.

Fire   heat and light resulting from the rapid combination of oxygen with other materials. The light released by fire is seen as flame, which consists of glowing particles of the burning material and gases that are luminous at the temperature of the burning material. A substance that is combustible, heat that equals or exceeds the substance's *kindling temperature* and adequate oxygen are the basic ingredients necessary for fire to exist. See also COMBUSTION, COMBUSTION AIR, COMBUSTION BYPRODUCTS.

Fission   a self-driving process in which a free neutron from a split atomic nucleus scores a direct hit on the nucleus of another atom, breaking it into pieces and, in turn, emitting energy, neutrons and other particles. The intense heat released as the result of controlled fission in a NUCLEAR REACTOR is used to generate electricity, and the rapid release of energy in an uncontrolled fission reaction provides the explosive force of the atomic bomb. More than 100 *fission products* are released when the nucleus of an atom disintegrates. These RADIONUCLIDES are radioactive—they emit ALPHA PARTICLES, BETA PARTICLES and GAMMA RAYS (see RADIOACTIVITY). A few fission products—such as the radioisotopes of iodine and phosphorous that are used in medical and biological research, and nuclear weapons components such as PLUTONIUM and TRITIUM—are useful, but the vast majority are waste products that must be discarded. The safe disposal of nuclear waste is one of the chief liabilities of nuclear power (see HIGH-LEVEL WASTE, LOW-LEVEL WASTE, RADIOACTIVE WASTE, REACTOR WASTE). The first controlled nuclear fission reaction was initiated in December 1942 in a secret underground laboratory at the University of Chicago when physicist Enrico Fermi manipulated the control rods of a primitive nuclear reactor loaded with URANIUM. By achieving the controlled reaction, Fermi was validating the theory of nuclear physics that would lead to the production of the world's first atomic bomb.

Uranium-235 (U_{235}) is the only naturally occurring substance capable of supporting fission. The bombardment of atomic nuclei inside a nuclear reactor, however, can produce other *fissile materials*. Uranium-238 (U_{238}), for instance, can gain an extra neutron to become U_{239}, and rapid decay and the emission of a beta particle can change U_{239} into neptunium-239. Emission of another beta particle transforms this element into fissile plutonium-239. Thorium-232 can be transformed inside a reactor into fissionable uranium-

233. Fissionable materials have a large, unstable nucleus.

When a *critical mass* of a fissile material is brought together, a runaway fission reaction consumes the entire mass, producing an explosion of heat and light. The fission of plutonium is used to trigger the fusion reaction in a hydrogen bomb. In a nuclear reactor, the rate of radioactive decay is controlled by the fact that there is space between the uranium atoms, which decreases the likelihood that a free neutron will hit and split another atomic nucleus, and by a *moderator*, which controls the speed of the nuclear particles. Control rods that can be lowered into the reactor core to stop the fission process altogether offer a further degree of control. The fission of one pound of uranium-235 produces as much heat as does the combustion of more than 2 million pounds of coal, although much of this energy is removed from the reactor by the cooling system without having produced electricity.

See also BREEDER REACTOR, HANFORD NUCLEAR RESERVATION, HEAVY WATER, OAKRIDGE NATIONAL LABORATORIES. Compare FUSION.

Flammable Gas a gas that can easily be ignited by static electricity, an electrical spark or an open flame. Flammable gases often evaporate from other PETROLEUM distillates (see METHANE). See also HAZARDOUS MATERIAL. Compare FLAMMABLE LIQUID, FLAMMABLE SOLID.

Flammable Liquid a liquid with a FLASH POINT of less than 100°F. The more volatile fraction of a flammable liquid evaporates readily to become a FLAMMABLE GAS. See also HAZARDOUS MATERIAL, PETROLEUM, METHANE. Compare FLAMMABLE SOLID.

Flammable Solid a solid material that is easily ignited and that will burn rapidly. Flammable solids may also easily give off flammable vapors, have a low ignition point and undergo spontaneous chemical change into flammable forms. See HAZARDOUS MATERIAL. Compare EXPLOSIVE MATERIAL, FLAMMABLE LIQUID, FLAMMABLE GAS.

Flash Point the temperature at which a material begins to give off easily ignitable flammable vapors. See COMBUSTION.

Floater see FLOATING SOLIDS

Floating Liquids liquids such as OILS found in SEWAGE and WASTEWATER that float on the surface because they are lighter than water and relatively insoluble. See SEWAGE TREATMENT, SKIMMING TANK.

Floating Solids lighter-than-water solids found in SEWAGE and other WASTEWATER. Large floating solids are removed from wastewater before it goes into a treatment plant with grates and screens, while smaller solids can be skimmed from the water's surface in a settling tank (see SEWAGE TREATMENT, WATER TREATMENT).

Flocculation a process used in a sewage treatment plant to remove biodegradable organic material from wastewater by mixing it with oxygen and a concentrated mass of MICROORGANISMS for several hours in an AERATION tank. Flocculation produces *activated sludge*—also called *floc* or *zoogloeal*—that is removed from wastewater in a settling tank. The level of dissolved oxygen in wastewater can be increased mechanically—atmospheric air is entrained as a result of the motion of blades, brushes, impellers, propellers or turbines—or compressed air can be introduced through diffuse plates, nozzles or injection jets. The air agitates the mixture to keep organic solids suspended and "activates" microorganisms to consume organics, grow and reproduce. The bacteria feed vigorously on the organic matter (and on each other) to produce the foamy mass of floc. Flocculation removes 60 to 95% of wastewater's BIOCHEMICAL OXYGEN DEMAND. The light, fluffy biofloc produced by flocculation is densified by further microbial action (see SEWAGE SLUDGE). The treatment of industrial wastewater is often finished in large, shallow *oxidation ponds* where natural aeration and photosynthesis break down remaining suspended organic material. See also SEWAGE TREATMENT.

Fluoride a mineral that occurs as a natural byproduct of the decomposition of certain rocks, and that is added to many drinking water systems to improve dental health. The fluoridation process has been rated with immunization as one of the greatest boons to public health of all time. However, a small but dedicated group of critics consider it a public health menace. While science would seem to be on the side of fluoride's supporters, there is little doubt that some sensitive individuals can react adversely to fluoride in their drinking water. Low concentrations of fluoride—on the order of 2 milligrams per liter—protect teeth, especially children's teeth, from decay. Slightly higher concentrations lead to dental fluorosis, a mottling of the surface of the teeth. PUBLIC WATER SYSTEMS are required under the SAFE DRINKING WATER ACT to monitor concentrations of fluoride (MAXIMUM CONTAMINANT LEVEL = 4 milligrams per liter [4 mg/L]). A secondary standard of 2 mg/L has been established by the ENVIRONMENTAL PROTECTION AGENCY (EPA) to control dental fluorosis (a condition not viewed as a health threat by the EPA).

Flue Gas gaseous COMBUSTION BYPRODUCTS (see GAS) that are released into the atmosphere through a chimney. See also ASH, COMBUSTION, ELECTROSTATIC PRECIPITATOR, SCRUBBER.

Fly Ash ASH from 1 to 200 microns in diameter that results from the COMBUSTION of COAL and other fuels. Fly ash is carried up the smokestack with flue gases where it becomes an airborne PARTICULATE. Fly ash that is captured by AIR POLLUTION CONTROL devices often contains toxins such as DIOXIN and MERCURY that attach themselves to the ash particles. Toxic fly ash can be segregated from the much greater volume of BOTTOM ASH, rather than combined as is generally done now, resulting in less expensive ash disposal. Experiments with melting fly ash into a nontoxic glassy building material are under way, which may further simplify disposal. Fly ash is often mixed with bottom ash in a quench tank where it becomes part of a slurry that is hauled in a truck to an ashfill, a landfill or other disposal site. See also FLUE GAS, HAZARDOUS WASTE, HAZARDOUS WASTE DISPOSAL.

Food & Drug Administration see PUBLIC HEALTH SERVICE

Food Scraps scraps from the processing and preparation of food, table scraps and food that is thrown away before it is used. The decomposition of food scraps is responsible for much of the disagreeable odor of GARBAGE and REFUSE. Food scraps serve as breeding places for pathogenic organisms (see INFECTIOUS WASTE, MICROORGANISMS) and as food for rats and other vermin. A piece of cooked meat can become home to potentially lethal pathogens after just a few hours of standing in a warm garbage can (see OFFAL). For all these reasons, the prompt collection of garbage is a matter of utmost political (and practical) importance in a large metropolitan area. COMPOSTING can be used to biologically break down food scraps, either in small-scale, backyard compost piles, or large-scale centralized municipal operations. See also ORGANIC MATERIAL, YARD WASTE.

Formaldehyde a colorless gas that is an essential chemical building block used in the production of textiles, PESTICIDES, cosmetics, glues, disinfectants, preservatives and other products. Formaldehyde is used in the synthesis of new chemicals, and to bind dissimilar materials together to form a new material. Its pungent odor is associated with biology labs where

the closely related compound, formalin, is used to preserve frogs and other potential dissection subjects. Formaldehyde is synthesized from methyl alcohol (see ALDEHYDES), and is sold primarily in SOLUTION with alcohol. Urea formaldehyde (UF) resins are widely used in products found inside the home including plywood, paneling, floor underlayment, and pressed board products used in cabinets and furniture. Curtains and permanent-press fabrics resist mold and fire and maintain their shape because of the formaldehyde they contain. Formaldehyde gets into indoor air by OUTGASSING from such products. Keeping indoor humidity levels below 35% can reduce concentrations. The concentration of formaldehyde inside mobile homes that are less than five years old can be especially high because of the large amount of wood products glued together with UF resin that are usually present. Levels taper off after five years as outgassing declines. The air inside some homes insulated with urea formaldehyde foam insulation can have exceptionally high formaldehyde levels. Other indoor formaldehyde sources include tobacco smoke (see ENVIRONMENTAL TOBACCO SMOKE) and leaky wood stoves and furnaces. Formaldehyde is a normal metabolic product of living cells, so respiration is another indoor source. Formaldehyde gets into outdoor air as the result of incomplete combustion in stoves, furnaces, incinerators and power plants and from industrial emissions. It degrades quickly, however, when exposed to direct sunlight, so outdoor levels are normally much lower than those experienced inside. Formaldehyde is highly water soluble, but little is known about its aquatic effects.

Formaldehyde can affect health when inhaled or by passing through the skin. Airborne concentrations as low as 0.05 parts per million (ppm) can cause cold-like symptoms, bronchitis, insomnia, headache, fatigue, nausea and dermatitis. There is considerable controversy over the effects of long-term exposure to low concentrations. Exposure to very high levels (greater than 100 ppm) can cause formaldehyde poisoning with symptoms including abdominal pain, anxiety, nose and throat irritation, central nervous system depression, coma, convulsions, diarrhea, headache, nausea and respiratory disorders. Animal tests have shown that formaldehyde can cause nasal cancer in rats. It has also been shown to cause mutations in bacteria, yeasts, fruit flies, mammalian and human cells. Epidemiological studies show an increased incidence of brain cancer, leukemia and cirrhosis of the liver among those regularly exposed to formaldehyde on the job. The ENVIRONMENTAL PROTECTION AGENCY classifies formaldehyde as a probable human CARCINOGEN.

Formaldehyde is on the HAZARDOUS AIR POLLUTANTS LIST. The Department of Housing and Urban Development has issued standards governing its use in building products in an attempt to reduce concentrations of the gas in indoor air. Businesses handling significant quantities of formaldehyde must disclose use and releases of the chemical under the provisions of the EMERGENCY PLANNING AND COMMUNITY RIGHT TO KNOW ACT.

Formation Process the lifestyles and patterns of behavior that result in a particular kind of garbage being discarded (see CONSUMERISM, GARBOLOGY, HORIZON MARKER, LANDFILL, MATERIALISM, MIDDEN, PLANNED OBSOLESCENCE).

Fossil Fuel a FUEL that is derived from the organic remains of a fossilized organism. Fossil fuels are recovered from the rock strata in which the fossilized organisms are found. COAL, which is composed primarily of carbon, is made up primarily of fossilized swamp and bog plants. PETROLEUM is composed primarily of the fossilized remains of MICROORGANISMS that were deposited on the beds of ancient seas. NATURAL GAS is believed to be a byproduct of the process through which petroleum is formed.

Freeze Distillation a method of removing contaminants from water and other liquids that involves lowering liquid water's temperature below its freezing point, causing the formation of pure ice. It can be used to desalinate seawater (see DESALINATION). Salt water enters a vacuum chamber, which speeds up the evaporation of water molecules and cools the remaining water, changing it into a mixture of ice and BRINE. The slurry then goes into a separator where brine is drained and ice crystals are washed with part of the product water. The water vapor that was drawn off in the evaporation chamber is then compressed and blown through the ice crystals, where it condenses, causing the crystals to melt. The process produces potable water, uses less energy and causes less corrosion of system components than does DISTILLATION. See also WATER POLLUTION TREATMENT.

French Drain a STORM SEWER that consists only of a storm drain connected to a sump normally located within 10 feet of the ground's surface. French drains are generally practical only in areas where soil infiltration rates are relatively high and storm runoff rates relatively low. Since they are designed to hasten the percolation of storm water into the ground, polluted RUNOFF that runs into a French drain can pollute GROUNDWATER, especially in regions with shallow aquifers and coarse soils (see GROUNDWATER POLLUTION).

Fresh Kills Landfill the world's largest municipal garbage dump. Started in 1949 in a salt marsh at the edge of Staten Island, New York, Fresh Kills now covers 3,000 acres and is more than 155 feet tall. About two-thirds of New York's trash is dumped at Fresh Kills, about 17,000 tons per day. More than 1 million gallons of LEACHATE ooze out of the bottom of the dump each day. The LANDFILL has no comprehensive gas collection system, so the METHANE, CARBON DIOXIDE and volatile organic chemicals emitted by the dump pass directly into the atmosphere. The dump is scheduled to be filled to capacity by 2005, when it will have reached 505 feet above sea level, and will be the highest geographic feature along a 1,500-mile stretch of the Atlantic seaboard running from Maine to southern Florida. The landfill will then be covered with topsoil, planted with grass and trees and made into a park.

Fuel a material that is burned to produce power or heat. The hydrocarbon FOSSIL FUELS (see HYDROCARBON), which are the basis of industrial society, include PETROLEUM, NATURAL GAS and COAL. GASOLINE, DIESEL, kerosene and heating oil are derived from petroleum. Peat, WOOD and dung have been used as fuels since long before the beginning of recorded history. REFUSE DERIVED FUEL is produced from garbage. Trash-to-energy plants use garbage as a fuel to produce electricity, steam and hot water. ALCOHOL and METHANE are fuels that are produced from organic material. HYDROGEN is a fuel that may be produced by the electrolysis of water, or can be derived from various hydrocarbons. Although URANIUM is used to produce power and heat, it is not technically a fuel since it is not burnt. Electricity and renewable energy sources such as sunshine, falling water, blowing wind or geothermal heat are similarly not fuels since they are not burned. The oxygenated fuel and reformulated gasoline programs that have been instituted to reduce air pollution have challenged the U.S. gas pipeline system, with different mixes of fuels to be delivered even within the same urban area, in some instances. The CLEAN AIR ACT requires the use of OXYGENATED FUELS that contain alcohol or METHYL-TERTIARY BUTYLETHER (MTBE). GASOHOL is a mixture of gasoline and alcohol. See also ALTERNATIVE MOTOR FUELS ACT, AUTOMOTIVE AIR POLLUTION, ETHANOL, MOBILE SOURCE, METHANOL.

Fuel Cell a device in which the energy from a chemical reaction is converted directly to electricity. Fuel cells are theoretically capable of converting a fuel into electricity with nearly 100% efficiency. A fuel cell has four essential elements: a fuel, an oxidizer (see OXIDA-TION, OXIDIZING AGENT), positive and negative electrodes and an electrolyte. When the fuel enters the cell, it is decomposed in the electrolyte into negative and positive ions (see ION). The negative ions (electrons) flow to the negative electrode and pass through the electrical circuit, producing an electric current, while the positive ions remain in the electrolyte. When the electrons reach the positive electrode, they react with the oxidizer to form more negative ions, which in turn react with the positive ions to form a waste product, thus completing the conversion process. Fuel cells using HYDROGEN as a fuel and oxygen as an oxidizer show the most commercial promise. Research is also being conducted into fuel cells that use HYDROCARBON fuels and even decomposing GARBAGE as a fuel.

Fuller, R. Buckminster an architect, engineer, inventor, and poet who developed the geodesic dome. Born on July 12, 1985, in Milton, Massachusetts, Richard Buckminster Fuller established a reputation as one of the most original thinkers of the second half of the twentieth century. Fuller held that technological design is the only basis for solving the world's physical and social problems, and the central theme of his thinking was to solve world problems by doing more with less. Fuller became research professor at Southern Illinois University (Carbondale) in 1959. He was awarded the Presidential Medal of Freedom shortly before his death in Los Angeles, California on July 1, 1983. He was intrigued with the concept of *synergy*, a state in which the whole accomplished by a system or design is more than a sum of its parts. Life is synergistic in that the actions of a living being cannot, for instance, be predicted by an examination of its physical structure. Fuller saw the world as "spaceship earth," and likened the world to an egg in which fossil fuels are the nutrient that will allow the embryo, mankind, to develop to the point that it has the strength to break out of the shell and fend for itself, evolving to the point that sustainable sources of energy are harnessed. Fuller invented the geodesic dome, a free-standing structure based on geometric solids and the triangle that encloses a given amount of space with the least possible amount of materials. He also originated the World Game, a computerized effort of researchers and students around the world to better allocate world resources to alleviate shortages and suffering.

Fumes very fine solids formed by the condensation of vapors that result from the COMBUSTION of materials that are normally solid. Metallic fumes are generated from molten metals. They usually occur as oxides (see OXIDE, OXIDATION, OXIDIZING AGENT) because they are chemically reactive (see REACTIVITY) with free oxygen in the atmosphere. Fumes tend to *flocculate* into clusters of larger size with time. See also AIR POLLUTION, INDOOR AIR POLLUTION.

Fungicide a substance capable of killing a fungus. Most fungicides are compounds of zinc, NITROGEN or SULFUR made to control athlete's foot, ringworm and other fungal infections in humans and domestic animals. *Bordeaux mixture*, a mixture of copper sulfate and calcium hydroxide, was once widely used to control fungal growths on agricultural crops and landscape plants, but its use has been largely discontinued due to its toxicity. Fungicides are regulated under the FEDERAL INSECTICIDE, FUNGICIDE AND RODENTICIDE ACT.

Furans a large group of compounds that contain a furan ring (four carbon atoms and one oxygen atom bonded in a cyclic form). Certain sugars and alcohols are furans. *Dibenzofurans*, which consist of a furan ring integrated into a pair of BENZENE RINGS, are a product of incomplete combustion typically found in tobacco smoke, automobile exhaust and incinerator flue gas. Dibenzofurans are on the HAZARDOUS AIR POLLUTANTS LIST. *Polychlorinated dibenzofurans*, a dibenzofuran to which chlorine atoms have attached to one or more of the outer carbon atoms, are closely related to POLY-CHLORINATED BIPHENYLS. Businesses handling significant quantities of furans must disclose use and releases of the chemical under the provisions of the EMERGENCY PLANNING AND COMMUNITY RIGHT TO KNOW ACT.

Furniture beds, chairs, couches, tables, dressers, book cases and other items used to furnish a house. Furniture is made of WOOD, METAL, FIBERS, FABRICS, PLASTICS and other synthetic materials. Investigations by the Garbage Project revealed that furniture and appliances are diverted from the wastestream far more often than is generally indicated in statistics, which are based on the number of new APPLIANCES purchased in a given year and its anticipated life expectancy. When someone buys a new couch or other piece of furniture, they are pretty likely to either sell the old one or put it in the basement. Contributions to needy friends, relatives and strangers take care of most of the rest, and the few sticks of furniture that do make it out to the alley for collection are frequently hauled off by a scavengers for their own use or for resale. Therefore, not as much furniture is hauled to landfills as would be expected, and most of what does is in pretty bad shape. The outgassing of volatile toxins from glues, fabrics and synthetic padding used in furniture is a source of INDOOR AIR POLLUTION. See also MUNICIPAL SOLID WASTE, REUSE.

Fusion a process in which two very small nuclei combine at extremely high temperature to form a third, larger, nucleus. The fusion of HYDROGEN into helium that occurs at a very high temperature and pressure in the sun and other stars is the process that gives them their energy and light. The fusion of deuterium (also known as HEAVY WATER) with TRITIUM to produce helium and free neutrons is the source of the phenomenal release of energy in a hydrogen bomb. The 100 million °C temperature necessary to initiate the fusion process in an H-bomb is provided through the ignition of an atomic bomb. The fusion reaction—in which two small nuclei are merged—is the nuclear opposite of FISSION, in which a large nucleus is split in two. The release of energy that accompanies fission or fusion, however, results from the loss of mass inherent in either process. The mass is converted to energy according to Einstein's famous formula: $E = mc^2$, or energy produced equals the mass lost times the speed of light squared.

Research and development involving the fusion process has occupied physicists for decades, but the commercial production of fusion-generated electricity, if it occurs, is still decades away. The basic problem that must be solved before fusion energy can be sustained and harnessed for peaceful purposes is the enormous amount of energy required to force the particles to be fused to collide in spite of the electrical force trying to hold them apart. A temperature of 50 million °C to 100 million °C must be maintained to force the fusion of the two elements most commonly used in experiments—deuterium and tritium. To achieve the fusion of two deuterium atoms, a 500 million °C temperature is necessary. Such extreme temperatures are very difficult to maintain for longer than the split second necessary to ignite a hydrogen bomb. If the materials to be fused are touching other matter, even a gas, the necessary heat will quickly be drained away. To get the nuclear particles hot enough to be fused, they must be suspended in a vacuum, which can be achieved through two processes: magnetic or inertial confinement. With *magnetic confinement*, a deuterium/tritium mixture is energized to the point that the electrons are stripped from atoms creating a plasma of charged electrons and nuclei that can be suspended in a vacuum by a magnetic field. With *inertial confinement*, a frozen droplet of deuterium/tritium that is less than 1 millimeter in diameter is placed in the center of an evacuated chamber and bombarded from all sides with laser beams for about one-billionth of a second. As a result, the pellet is heated so rapidly that it will reach ignition temperature before it can expand and reduce its density. Deuterium and the light METAL lithium, which is changed to tritium once it is inside the reactor, are the raw ingredients for the fusion reaction. Fusion has the potential of creating 340 million Btu per gram of deuterium consumed. Because no critical mass is required to sustain fusion, there is no risk of core meltdown, as can occur with a fission reactor. Any malfunction would disrupt the flow of plasma and stop the reaction. Fusion has been achieved in experiments by using a particle accelerator to hurl a deuterium particle against a stationary target of deuterium or tritium.

Princeton University's experimental fusion reactor set a record late in 1994 by generating 10.7 million watts of power. The one-second burst of energy, which was enough to momentarily power 2,000 to 3,000 homes, occurred in the Tokamak Fusion Test Reactor, which was designed in 1976 with the goal of producing up to 10 million watts of power. The reactor is operated by the Princeton University Plasma Physics Laboratory, and is scheduled to be dismantled this fall. Some of its auxiliary systems are to be reused in the next planned experimental reactor at the lab, which is considered to be the leading U.S. fusion research facility. Its replacement, the $750 million Tokamak Physics Experiment, is being designed, and is scheduled to be operational by the year 2001. A third experimental reactor at Princeton—the $10 billion International Thermonuclear Experimental Reactor, which is a joint project of Russia, Japan, the United States and a European coalition—is scheduled for completion by 2005. U.S. spending on fusion research has been cut in half since its peak in the early 1980s, and physicists say the reduced funding has slowed progress toward a reactor capable of sustained fusion. Researchers say commercial fusion reactors may be possible by about 2035.

The Lawrence Livermore Laboratory was chosen by the Department of Energy as the site for the $1.1 billion National Ignition Facility late in 1994. In the facility, which is slated for completion in 2001, 192 laser beams will be trained on a tiny deuterium/tritium pellet to produce fusion. The stadium-sized installation will be built within Lawrence Livermore's square mile of property, and will be the largest laser project in the world. The project's total price tag of the project, which has a 15-year life expectancy, will be $1.8 billion including its operating budget.

G

Galvanic Series a ranking of metals and their alloys according to their tendency to lose electrons. A metal is said to be more noble if it is higher in the galvanic series and less noble when lower in the series. Nickel, silver, graphite, GOLD and platinum are near the top of the series, while magnesium, zinc, galvanized STEEL and aluminum are near the bottom. When two metals are connected with a wire and immersed in an electrolyte, an electron flow will be established in the wire moving away from the metal that most easily sacrifices electrons. This electron flow leads to CORROSION of the less-noble metal and to the deposition of a thin layer of atoms from the electrolyte on the more-noble metal. The magnitude of the electron flow and the rates of corrosion and deposition depend on how far apart the two metals are on the galvanic series. A tank built to store highly corrosive materials should be built from materials as far toward the noble end of the galvanic series as possible, and materials used in the tank should be as close together as possible in the series. See also RESISTIVITY.

Gamma Rays high-energy, short-wavelength electromagnetic waves that are generated by the radioactive decay of certain NUCLIDES (see RADIOACTIVITY, RADIONUCLIDE). Gamma rays are the same kind of radiation as visible light, but have a shorter wavelength and a higher frequency. (Gamma rays are defined as electromagnetic radiation with a wavelength shorter than 1×10^{-10} meters, while visible light has wavelengths between 4×10^{-7} to 7×10^{-10} meters.) Although gamma rays may be emitted by a radioactive material along with ALPHA PARTICLES and BETA PARTICLES, they are seldom if ever emitted from the same nucleus at the same time. Gamma rays are capable of deep penetration into many solid materials, and are the most damaging form of IONIZING RADIATION. The presence of gamma rays is the principal reason for the heavy radiation shielding on NUCLEAR REACTORS. Gamma rays can cause cellular damage that can lead to cancer and genetic mutations. Recent evidence indicates that gamma rays are on the borderline between matter and energy, and display some of the characteristics of both. The inspection of castings and welds for imperfections can be achieved at lower cost with equipment that generates gamma rays from COBALT or cesium than with X-RAYS.

Garbage materials that have been discarded or allowed to escape (as byproducts of an industrial process) as useless. In its most specific usage, "garbage" refers to FOOD SCRAPS from homes and restaurants. MUNICIPAL SOLID WASTE, wet REFUSE and dry TRASH from homes, businesses and construction sites (see SOLID WASTE, CONSTRUCTION WASTES) are also commonly called "garbage." It is the willful act of discarding a thing that makes it garbage. Most garbage is worn out, used up, broken, rejected or otherwise worthless, but new and nearly new items also enter the waste stream. The scientific study of garbage, called GARBOLOGY, has made great strides in recent years, with the GARBAGE PROJECT at the University of Arizona in the forefront. By methodically sorting through garbage disposed of in garbage cans and dumpsters, and by excavating landfills and classifying samples as carefully as those unearthed at the excavation of an ancient midden, researchers at the Garbage Project and elsewhere are building a clearer picture than has ever before been available of the true composition of garbage and the physical processes and effects of large masses of it in landfills, as well as providing new insights into American culture. Garbage is regulated under the terms of the RESOURCE CONSERVATION AND RECOVERY ACT (RCRA). There are several components of garbage:

• MUNICIPAL SOLID WASTE consists primarily of PAPER, WOOD, GLASS, METAL, FABRICS, PLASTICS, food scraps and YARD WASTE. The ENVIRONMENTAL PROTECTION AGENCY estimates that 80% of the nation's municipal solid waste is dumped in landfills. Most of the rest is either recycled or burned in incinerators. ASH from incinerators and sewage sludge is also frequently disposed of in municipal landfills, although it is not considered to be a part of municipal solid waste.

- INDUSTRIAL SOLID WASTE is sometimes dumped in a municipal landfill and may contain many of the same materials as municipal solid waste. Most industrial waste, however, is disposed of on site, used as the raw material for another industrial process or shipped to a HAZARDOUS WASTE dump (see HAZARDOUS WASTE DISPOSAL).

- LIQUID WASTE can technically be termed garbage (any material discarded as useless), although it is seldom called that. Kitchen waste is transformed from solid to liquid waste when it is run through the garbage disposal, where it is chopped into particles small enough to be washed down the drain. Solid FECES is handled as liquid waste when it is flushed down the TOILET through the sewer, but once again becomes solid waste after it has been processed at the sewage treatment plant.

- even AIR POLLUTION is a kind of garbage that is thrown away because it is considered worthless—although it is seldom thought of in this way (see AEROSOLS, GASEOUS WASTE). The FLY ASH and other particles (see PARTICULATES) removed from industrial smokestacks become a form of liquid or solid waste in the treatment process. Garbage dumped in landfills pollutes air through the evaporation of toxic substances and the generation of methane and other byproducts of biological metabolism in the landfill. The EVAPORATION of the volatile organic compounds (VOCs) from industrial waste ponds and DUST from plowed croplands and unpaved roads are also major sources of air pollution. The EPA estimates that 200,000 metric tons of VOCs are emitted from U.S. landfills each year.

- INFECTIOUS WASTE from households and some hospitals is discarded at municipal landfills, and food scraps and other wet wastes that were clean when thrown away can quickly become contaminated with pathogens as they get warm and decay sets in. A greasy chunk of cooked meat can become a source of potentially lethal pathogens after just a few hours of standing in a warm garbage can, because it offers an ideal environment for multiplication of environmental spores, bacteria and virus. It has been estimated that as many as 90% of the flies found in urban areas breed in the organic material found in (and around) garbage cans, and flies are known to carry a multitude pathogens. Rats, another vector of disease, similarly thrive in the presence of garbage.

Municipal garbage contains toxins. In fact, more than 200 of the nation's priority hazardous-waste cleanup sites are old municipal dumps with no liner (see COMPREHENSIVE ENVIRONMENTAL RESPONSE, COMPENSATION and LIABILITY ACT). Ordinary household garbage includes a variety of toxic materials including PESTICIDES, PAINTS, SOLVENTS and household cleansers that can become air or water POLLUTANTS if disposed of at a landfill (see AIR POLLUTION, GROUNDWATER POLLUTION, HOUSEHOLD HAZARDOUS WASTE, WATER POLLUTION). Science fiction writers have long visualized a future where dumps are mined because they are the richest remaining source of resources such as metals and plastics. The use of dump mining has become reality, and is likely to become much more common if resources are as scarce in the future as many predictions hold (see POPULATION GROWTH). See also ANIMAL REMAINS, APPLIANCES, DEBRIS, LITTER, MEDICAL WASTE, NEWSPRINT, OFFAL, ORGANIC MATERIAL, PACKAGING, PATHOGENIC WASTE, PESTICIDE MANUFACTURING WASTES, PHOSPHATE, RUBBISH, SPACE JUNK, TIRES.

Garbage Barge barges that haul garbage to sea for disposal (see OCEAN DISPOSAL) or to another area for land disposal. Garbage barges were once common in the northeastern United States until outlawed by the Ocean Dumping Act.

In March 1987, the garbage barge Mobro 4000 embarked from Islip, New York, with more than 3,000 tons of Long Island municipal garbage. The vessel achieved brief international fame by traveling more than 6,000 miles in a fruitless two-month search for a place to dump its load before returning to Islip with its cargo somewhat riper but still intact. The garbage was finally interred in a New York City landfill.

Garbage Can a container made for the disposal of garbage. A garbage can is generally a cylindrical container made of PLASTIC or STEEL with a lid. The right of police and journalists to look through garbage cans for evidence (or tidbits) has been repeatedly affirmed in court, because the contents of the can have been discarded and put in a spot where they are open for inspection or even removal by any member of the public. See also COMPACTION TRUCK, DUMPSTER.

Garbage Compactor see TRASH COMPACTOR

Garbage Crisis a crisis characterized by an increasing volume of garbage and a shrinking number of places to dispose of it. Americans lead the world in the production of garbage. The ENVIRONMENTAL PROTECTION AGENCY (EPA) estimates that nearly 200 million tons of municipal solid waste was produced in the United States in 1990. PLANNED OBSOLESCENCE, inefficient PACKAGING, CONSUMERISM and the popularity of disposable products all contribute to the volume of garbage produced. LANDFILLS have a bad track record for polluting water with contamination bad enough at a couple of hundred sites to qualify them for Superfund status (see COMPREHENSIVE ENVIRONMENTAL RESPONSE, COMPENSATION AND LIABILITY ACT). To correct this problem, the Hazardous and Solid Waste Amendments of 1984 (see RESOURCE CONSERVATION AND RECOVERY ACT) instituted tight new controls on how landfills are constructed and operated, resulting in the closure of many of the nation's dumps. Urban growth has reduced the space available for new waste disposal facilities, but the not-in-my-back-yard syndrome (see NIMBY) is the primary cause of the crisis: everyone wants his garbage picked up, but no one wants a waste disposal facility nearby. As a result of the new regulations and public opposition to proposed new sites, licensing and siting a new landfill can take years. The cost of waste disposal has increased markedly as a result.

One solution to the garbage crisis is to reduce the volume of solid waste hauled to the dump. The volume of garbage can be reduced through a variety of SOURCE REDUCTION techniques:

- REUSE of serviceable items is one way of reducing the amount of garbage that is landfilled. Several municipalities have passed laws allowing or even encouraging businesses that will sort through landfills and remove items with resale value.
- RECYCLING raw materials such as GLASS, ALUMINUM, NEWSPRINT, CARDBOARD and PLASTICS is another way of reducing the volume of garbage that goes to the dump (see PAPER RECYCLING, PLASTICS RECYCLING). The EPA estimates that 13% of the nation's municipal solid waste is recycled. Many states have passed laws setting goals to recycle between 25% and 70% of municipal solid waste in the future.
- COMPOSTING the organic materials found in garbage is another way of removing a valuable product from municipal solid waste while reducing its volume.
- INCINERATION is yet another means of solid-waste reduction that is being adopted in many communities. Usable electricity or steam is produced by many garbage incinerators.
- Reducing wasteful packaging and the production of items designed to be used once and thrown away is another way of reducing solid waste.

See also POPULATION GROWTH.

Garbage Disposal a kitchen appliance that grinds up food scraps so that they can be washed down the sink and carried in suspension to the sewage treatment plant or septic tank for processing. BLACK WATER is the EFFLUENT from TOILETS and garbage disposals. The use of household garbage disposal increases the weight of primary sludge an estimated 25 to 40%. See also SEWAGE SLUDGE.

Garbage Incinerator an incinerator that burns municipal solid waste. A modern garbage incinerator is a high-tech, computerized facility that can cost more than $500 million. INCINERATION is second only to land disposal (see LANDFILL) as a method of the disposal of MUNICIPAL SOLID WASTE (MSW), and is a major source of AIR POLLUTION. Incineration may be the most environmentally benign way and is usually the least expensive method of trash disposal in areas with high water tables or that are otherwise not suitable sites for landfills. The ASH created by garbage incinerators can contain toxins such as LEAD, CADMIUM, MERCURY, DIOXIN and ARSENIC (see BOTTOM ASH, FLY ASH). About 140 incinerators burned more than 30 million tons of MSW in 1992, about 16% of the total. It typically costs $90 to $100 a ton to incinerate garbage in the United States.

The city dump could long be easily traced by the plume of smoke it emitted from the OPEN BURNING of refuse, a practice that continued at many landfills into the 1970s. The first garbage incinerator, called a DESTRUCTOR, was opened in Nottingham, England, in 1874. The first garbage incinerator in the United States, called a *creamator*, was built in New York City by the U.S. Army in 1885. By the 1940s about 700 municipal waste incinerators were in operation. Many were shut down during the 1950s principally because of the foul odors and pollution they generated, with waste going instead to landfills. Fewer than 70 incinerators were still in operation when the CLEAN AIR ACT (CAA) passed in 1970. The energy crisis in the mid-1970s spawned a new wave of garbage incinerators that produced steam to drive electrical turbine generators or to provide energy for some other industrial process (see WASTE TO-ENERGY PLANT). Combustion in the incinerators was more efficient than in their malodorous ancestors, and at least rudimentary pollution control equipment was used. The ENVIRONMENTAL PROTECTION AGENCY (EPA) issued standards in 1989 that required the installation of additional emissions control equipment on existing garbage incinerators and the adoption of other measures that would reduce the release of atmospheric contaminants. There are about 150 waste-to-energy plants in operation in the United States. Operators of incinerators receive a TIPPING FEE for each load of MSW dumped and payment for the electricity or steam produced.

At a typical incinerator, GARBAGE TRUCKS dump their loads on an expansive *tipping floor*, usually inside a building or at least under a roof, where materials such as BATTERIES, twine and potentially EXPLOSIVE MATERIALS are removed. The wastes are then placed on a conveyer belt, which carries them through a series of hammermills where solids are reduced to chunks four inches across and smaller. A magnet then pulls ferrous metals out of the pulverized waste, and vibrating grates shake out heavy solids such as rocks and glass. The processed garbage is then carried into the combustion chamber, where extra air is introduced to improve combustion efficiency. The temperature inside the combustion chamber and the overall efficiency of combustion depend on the exact mix of materials in a particular load of garbage and on the material's water content.

Conflict

The construction of a large garbage incinerator is a large investment, one that can commit a community financially and contractually to incineration to the exclusion of SOURCE REDUCTION techniques such as RECYCLING. Researchers at the Massachusetts Institute of Technology concluded that it would cost $200 per ton less to dispose of waste in the state if more than $1 billion had not been invested in the construction of nine waste incinerators, while little attention was paid to recycling. The companies that operate incinerators often have contracts with municipalities that guarantee delivery of a certain volume of MSW will-be-delivered. Recyclable materials such as PLASTICS and PAPER make some of the best fuel found in MSW, and municipalities with successful recycling programs have been forced to recruit garbage from other cities to fulfill their MSW quota. Expanded polystyrene foam is especially valued at incinerators because of its high heat content and because the polystyrene helps to break down complex molecules found in other GARBAGE. The heat content of various fuels is as follows: coal and coke, 11,000 to 14,000 British Thermal Units/pound (Btu/lb); wood, 8,000 to 10,000 Btu/lb; newsprint, 8,500 Btu/lb; kraft paper, 7670 Btu/lb; corrugated board, 7,400 Btu/lb; food cartons, 7,700 Btu/lb; waxed milk cartons, 11,680 Btu/lb; plastic film, 13,780 Btu/lb; polystyrene, 15,730 Btu/lb; polyethylene, 14,890 Btu/lb; typical municipal solid waste, 5,000 Btu/lb.

Emissions

A large incinerator can produce a large amount of air pollution. Clean Water Action estimates that the incinerator in Seamass, Massachusetts, emits more than 2,000 tons of toxic materials into the atmosphere annually, including 590 tons of mercury. Dioxins, lead, cadmium, arsenic, chromium and FURANS are among the other toxins that can be emitted from an incinerator, and massive ash blowouts have also occurred at many incinerators. The EPA estimates that more than five million tons of ash are generated each year by U.S. MSW incinerators. Incinerator emissions depend on the material being burned:

- Leaves and vegetables can cause nitrogen oxide emissions.
- Dyes and paints can release HYDROGEN CHLORIDE.
- Household batteries can release lead.
- INSECTICIDES and FUNGICIDES release arsenic and mercury.

New compounds can also be formed by the interaction of combustion byproducts in the smokestack and after release into the atmosphere. The dioxins found in municipal solid waste such as paper products are destroyed when they are burned in an incinerator, but evidence suggests that new dioxins may be formed and become attached to particulates as exhaust gases cool while going up the smokestack. Most modern incinerators typically achieve temperatures of 1,800°F, and 1,500°F has been shown to destroy most of the organic compounds found in flue gases, but variations in the makeup and moisture content of garbage can result in reduced temperatures.

New Source Performance Standards

The EPA issued NEW SOURCE PERFORMANCE STANDARDS in late 1989 that required source separation and recycling in cities with waste incinerators as a way of reducing the volume and toxicity of garbage being burned in incinerators and the installation of scrubbers on incinerators to reduce emissions. The new standards will require large expenditures—it can cost well over $100 million to retrofit an existing incinerator to bring it up to snuff—and disputes between companies that run incinerators and the communities they serve over how to distribute the costs of the requirements are widespread. Most incinerators must install scrubbers to remove acidity and particulates and in many cases an electrostatic precipitator to further purify effluent.

Garbage Project a research project at the archaeology department of the University of Arizona in Tucson. Much as archaeologists sift through ancient middens to uncover the secrets of long-dead civilizations, Garbage Project researchers conduct digs at modern landfills (see GARBOLOGY), and carefully sort, weigh and classify the items found, thereby gaining a unique insight into modern civilization.

The work of the Garbage Project was the subject of the 1992 book *Rubbish! The Archaeology of Garbage* by William Rathje, director of the Garbage Project, and Cullen Murphy. It proved quite popular. See also APPLIANCES, ARCHAEOLOGY, AUTOMOTIVE PRODUCTS, BIODEGRADABLE PLASTICS, CONSTRUCTION WASTES, EXPANDED FOAM, FURNITURE, GARBAGE, GLASS, HOUSEHOLD HAZARDOUS WASTE, LIGHT-WEIGHTING, NEWSPRINT, PACKAGING, PAPER, PLASTICS WASTE, POLYSTYRENE, PULL-TAB TYPOLOGY.

Garbage Truck   any of a variety of trucks used in the collection of garbage and recyclable materials. See COMPACTION TRUCK, SOLID WASTE DISPOSAL.

Garbology   the examination of GARBAGE to determine social patterns. Much as archaeologists sift through an ancient MIDDEN and record its contents in an attempt to decipher patterns of behavior of ancient humans, a GARBOLOGIST takes samples from a modern LANDFILL and examines them to learn more about the habits, the way of life and the patterns of consumption of modern people. Studies of the garbage generated by families or businesses are also a part of garbology. The demographic information gained through such studies is invaluable in consumer research and marketing. The analysis of the items discarded by a large number of people can also prove useful to urban planners and government agencies, archaeologists, anthropologists, historians, and others involved with keeping track of human behavior.

The garbage heap is a kind of daily diary, recording people's interests (and disinterests), their likes and dislikes, their successes and failures. The indelible evidence found in the trash is a reflection of what people do, as opposed to what they say they do. The study of material culture—the artifacts left behind by an ancient society or the materials associated with a modern society—can reveal much about the habits and way of life of the people that daily used the materials in question. Material-culture analysis can also reveal ways in which material goods (such as tools) can change a culture, and can find links between the physical evidence available and people's attitudes and patterns of behavior. Such intangibles as the number of infants in a community can be estimated with the aid of exclusive demographic markers found in household garbage. The number of babies in a household or a community can be estimated by counting DISPOSABLE DIAPERS (about two per infant, per day.). This average may seem low, but it includes all infants, including those who wear cloth diapers. The number of children in a population can be estimated by the number of toys and discarded toy packages (2.52 per child, per week).

The number of women can be estimated by the amount of discarded PACKAGING from feminine hygiene products (1.58 per woman, per week). There are no reliable demographic markers for men, and their numbers must be estimated by less direct means. A FORMATION PROCESS is the totality of lifestyles and patterns of behavior that produce the garbage found in an ancient midden—or a modern garbage dump. See also PULL-TAB TYPOLOGY, STRATIGRAPHIC LAYER.

Gas   a formless fluid that expands to occupy the available space uniformly. Gases can be converted to a liquid or solid state by the combined effect of increased pressure and decreased temperature. Gas molecules are less than 0.0001 microns in diameter. See GASEOUS WASTE, VOLATILITY. See also BIOGAS, FLUE GAS, GAS COLLECTION SYSTEM, GAS MIGRATION, NATURAL GAS, OUTGASSING.

Gas Collection System   a network of PERMEABLE PIPES that is installed in a LANDFILL to collect METHANE and other gases. Methane is produced by the decomposition of organic materials in a landfill, and explosions and seepage of gas into the basements of nearby homes has been a problem at many sites (see GAS MIGRATION). For this reason, pipes are laid at many landfills as garbage is dumped, expanding the system as the level of garbage builds up. Another approach is to drill holes into the accumulation of garbage five years or so after it is dumped, and sink gas-collection pipes in the bore holes. Once the gas is collected, it is either burned on site, or is collected, purified and either sold or used by GARBAGE TRUCKS and other dump vehicles. An estimated 2% of the atmospheric methane that is contributing to the greenhouse effect comes from U.S. landfills. The RESOURCE CONSERVATION AND RECOVERY ACT requires that methane gas in landfills be managed, and the CLEAN AIR ACT calls for the collection of volatile organic compounds in landfills.

Gas Migration   the horizontal movement of METHANE gas created in a LANDFILL. Gas migration has led to the movement of methane gas into the crawl spaces and basements of buildings adjoining landfills and to deaths and injuries as the result of explosions, asphyxiation or poisoning. The federal RESOURCE CONSERVATION AND RECOVERY ACT requires that gas-collection systems be installed in new landfills to prevent the dangerous buildup of methane gas.

Gaseous Waste   gases that are released into the ATMOSPHERE or indoor air as a byproduct of a process of some kind. Gaseous waste includes COMBUSTION

BYPRODUCTS such as CARBON DIOXIDE or CARBON MONOXIDE; VAPORS from substances such as GASOLINE or BENZENE; and gases released as the result of chemical reactions used in industrial processes. FORMALDEHYDE from the OUTGASSING of glued building materials is a form of gaseous waste that may not be released into the air for some years after the original manufacturing process in which it was used. METHANE gas created as the result of decomposition of organic materials in a LANDFILL, swamp or compost pile is another example of gaseous waste. The FUMES resulting from the condensation of the vapors of combustion of a solid material are actually very fine solids, and are commonly referred to as a form of gaseous waste. Fumes more properly might be referred to as extremely small solid waste. The situation is similar with SMOKE, which actually consists of minute solids and liquids. See also AIR POLLUTION CONTROL, BIOGAS, FLUE GAS, GAS COLLECTION SYSTEM, GAS MIGRATION, NATURAL GAS.

Gasohol a mixture consisting of approximately 90% GASOLINE and 10% ETHANOL (ethyl or grain alcohol). Gasohol is an OXYGENATED FUEL (see CLEAN AIR ACT, AUTOMOTIVE AIR POLLUTION) that reduces the emission of CARBON MONOXIDE from vehicles by up to 30%. The ENVIRONMENTAL PROTECTION AGENCY announced in July 1994 that at least 15% of the additives used to make oxygenated fuels must come from renewable sources during 1995, and that by 1996 at least 30% must come from renewable sources. The move was a big boost to corn farmers since 95% of ethanol is distilled (see DISTILLATION) from corn. Gasohol typically costs 3¢ to 5¢ a gallon more than gasoline. Ethanol for gasohol must be moved by truck or by tank car because it tends to bond with water in pipelines. It must be blended with fuel at the last moment before gasoline is delivered to a service station. In addition to helping corn growers, ethanol has further advantages. It contains more oxygen than MTBE, so refiners need to add less of it to gasoline to boost the oxygen content. See also METHANOL, METHYL-TERTIARYBUTYLETHER, FUEL.

Gasoline a mixture of some of the lighter liquid derivatives of PETROLEUM, used primarily as fuel for INTERNAL COMBUSTION ENGINES. Approximately 100 billion gallons of gasoline are produced in the United States each year. Gasoline is a blend of some extremely toxic and extremely volatile (see VOLATILITY) ORGANIC COMPOUNDS. Gasoline's components evaporate rapidly when exposed to air. Once released, these VOLATILE ORGANIC COMPOUNDS (VOCs) figure in the production of PHOTOCHEMICAL SMOG and OZONE.

Numerous small releases of gasoline vapors from filling stations and vehicles with fuel leaks in a large urban area can cause a major air quality problem. Gasoline vapors are also released from auto accidents and spills. Gasoline fumes can get into indoor air from storage containers, or from a leak or a spill in an attached garage. While gasoline spilled into a lake or stream evaporates fairly quickly, in groundwater, where it floats to the surface of the AQUIFER, gasoline's components can be long-lived. When water from such a gas-tainted aquifer is aerated, as in a shower, gasoline fumes are released into indoor air. Many components are also capable of passing directly through the skin into the bloodstream. LEAKING UNDERGROUND STORAGE TANKS are a leading source of GROUNDWATER POLLUTION.

Gasoline produced by the *fractional distillation* of petroleum is known as *straight run gasoline*, which may represent anywhere from 1% to 50% of the raw crude. Fractional DISTILLATION supplied most of the nation's gasoline before 1920, but now comprises only a small fraction of total production. Gasoline can also be produced by:

- condensation or ADSORPTION from NATURAL GAS containing *natural gasoline*;
- catalytic or thermal *cracking* of petroleum;
- *hydrogenation* of refined petroleum oils or COAL at high temperature and pressure in the presence of a catalyst;
- polymerization (see POLYMER) of HYDROCARBONS of lower molecular weight.

Gasoline is classified according to its *octane rating*. Increased branching within the hydrocarbon chain generally yields a higher octane rating. A gasoline-fueled engine knocks when its fuel has an inadequate octane rating. The petroleum distillate, *n-heptane*, is assigned the value zero on the octane rating scale. Engines knock furiously when fueled with pure heptane. Another distillate, *isooctane*, causes very little knocking, even when used in a high-compression engine, and it is arbitrarily assigned an octane rating of 100. A variety of anti-knock compounds including BENZENE and other AROMATIC HYDROCARBONS are added to gasoline to boost its octane rating. Tetraethyl lead, a COMPOUND that contains LEAD, was the principal anti-knock additive until the phase-out of lead additives begun in 1975 as mandated by the 1970 CLEAN AIR ACT.

The exhaust from gasoline-burning engines is the principal source of AIR POLLUTION in most urban areas (see AUTOMOTIVE AIR POLLUTION, MOBILE SOURCES). Cleaner-burning blends of gasoline and alternative fuels (see ETHANOL, GASOHOL, HYDROGEN, METHANE,

METHANOL, NATURAL GAS) are a crucial part of the effort to reduce urban air pollution. *Reformulated gasoline* is produced by using an extra step in the refining process, resulting in a cleaner-burning gas. Reformulated gasoline is required in the 1990 Clean Air Act for areas with OZONE and VOC problems. Reformulated gasoline is one way for fleet vehicles to meet federal low-emission vehicle standards.

Reducing the EVAPORATION of gasoline and other fuels is also part of the effort to reduce ozone and VOC levels in urban areas. Commuters are advised during alerts to wait until evening to buy gasoline, as less ozone is created when it is cooler, and to replace their gas cap as tightly as possible. Vapor hoods or sleeves on gasoline pumps are required in some areas, but are unpopular because they make pumping gas awkward. Gasoline sold in targeted areas is also reformulated to reduce its volatility. Even tougher standards are required in California, where 13 billion gallons of gasoline are consumed each year.

Reformulated Gasoline Programs

Cities with dangerously high ozone levels were required by the ENVIRONMENTAL PROTECTION AGENCY (EPA) to use reformulated gasoline (RFG) during the ozone season. Los Angeles, Baltimore, Chicago, Houston, Milwaukee, New York, Philadelphia, San Diego and Hartford were required to use the fuels. Other areas that exceed the ozone standard could voluntarily participate in the RFG program, and 40 cities elected to do so rather than opting for other methods of pollution control such as stricter car inspections or tighter smokestack controls. States that chose reformulated gasoline did not have to write the complicated new rules required for industrial emissions. Using the federal gasoline standard also makes enforcement the responsibility of federal, not state officials.

EPA representatives said that in addition to reducing the emission of nitrogen oxide, the principal precursor of ozone, the reformulated fuels would result in a 25% reduction in the emission of other toxins such as benzene, sulfur and particulate matter. Reformulated gasoline emits about 15% fewer volatile organic compounds. It also discharges 15% less poisonous carbon monoxide into the atmosphere—about 83 pounds less per year for each of the more than 50 million vehicles that use the fuel.

Reformulated gasoline was first proposed by the petroleum and automotive industries. When the Clean Air Act was up for renewal in 1990, the industries convinced Congress to go for reformulated gas instead of pushing for alternate fuel vehicles such as battery-powered cars, with the aim of protecting the market for conventional gasoline vehicles. The petroleum industry has spent billions of dollars over the past several years upgrading refineries and producing or purchasing necessary process chemicals to produce RFG. The pipeline system that delivers gasoline around the nation from refining centers such as Houston has been hard-pressed to deliver the various types of gasoline required under the Clean Air Act to different parts of the country. Consumers are likely to pay an extra 5 to 15 cents per gallon for the new gasolines.

Geiger Counter   an instrument that measures radiation. Inside the Geiger counter is a glass tube with a small metal cylinder at its center and a thin metal wire passing through the center of the cylinder, but insulated from it. When the counter is in use, a high voltage is applied across the cylinder and wire. A small amount of vapor, often a mixture of alcohol and argon, is present inside the glass tube. When radiation passes through the tube, it ionizes the gas for an instant, making it possible for electricity to flow from the metal tube to the wire. The passage of electricity across the tube registers as a click on a set of earphones attached to the Geiger counter. The intensity of radiation may be indicated by the number of clicks and a reading on a meter, or may be recorded graphically on a sheet of paper. See also RADIOACTIVITY.

Geodegradability   the ability of a substance to be broken down by sunlight, wind, water or other natural agents other than the action of living things. Compare BIODEGRADABILITY.

Giardia   a parasitic protozoan—*giardia (lamblia) intestinalis*—that causes the flu-like gastrointestinal ailment *giardiasis*. The FECES of aquatic animals such as beaver is the most common source of the giardia cyst in water. "Pure" mountain streams can be infested with the cysts, hence the nickname "backpacker's lament" for giardiasis. The giardia parasite can live in the intestinal tract of any warm-blooded animal, and once established, the freeloader becomes a permanent resident—unless evicted by a potent antibacterial agent such as quinacrine hydrochloride or furizolidone. The giardia cyst has an incubation time of from one to three weeks after ingestion. Although the effects of giardiasis can be relatively mild or even unnoticeable in some individuals, others contract a flu that will not go away. The side effects of the drugs used to treat the malady are also quite severe. Giardia cysts can be killed by boiling water twenty minutes, and can be removed from water by a reverse osmosis filter. The protozoan became established in North America only in 1975, and is of European origin.

PUBLIC WATER SYSTEMS are required under the SAFE DRINKING WATER ACT to monitor for giardia. The MAXIMUM CONTAMINANT LEVEL GOAL (MCLG) for giardia is 0. A treatment technology is specified rather than a MAXIMUM CONTAMINANT LEVEL (MCL) for giardia. The treatment technique used—generally either disinfection or filtration—must remove 99.9% of giardia cysts in the water supply. The cysts survive chlorination at normal levels, but can be eliminated by exposing them to higher concentrations of chlorine compounds for longer time periods. To avoid the requirement that water systems drawing from surface sources install filtration equipment, a utility must practice disinfection and demonstrate that giardia concentrations in its source of drinking water meets ENVIRONMENTAL PROTECTION AGENCY guidelines. See also DISINFECTION, MICROORGANISMS.

Glass an amorphous material that is transparent or translucent. Most glass is a mixture of silicates, borates or phosphates. Glass is cooled from a liquid state to a solid state without crystallization. Glass is brittle and easily broken. Glass is used to make windows for automobiles and buildings, electric light bulbs and light fixtures, mirrors, glass jars for food and glass bottles for beverages. The Glass Packaging Institute estimates that 35% of all glass containers were recycled in 1993. ARSENIC, SALT and SELENIUM are among the materials sometimes used in glass production.

Tempered glass—which is three to five times stronger than regular glass of the same thickness—shatters into small cubes when broken rather than into larger shards. Tempered glass is created by heating regular glass, and then slowly cooling it. The surface of tempered glass is slightly distorted because the edges tend to get hotter in the tempering process, and as a result shrink slightly more than the middle. *Heat strengthened glass* is about twice as strong as regular glass. It is made through the same basic process as tempered glass, but at lower temperatures. *Patterned glass* is embossed with a pattern made by a roller while it is still semi-molten. Sheets of patterned glass are often tempered after imprinting for use as shower doors. *Wired glass* is reinforced with a wire mesh that is rolled into the glass sheet while it is still molten. The glass is then tempered for use in skylights and fire-rated entry doors. *Safety glass*, which is used in automobiles, consists of two sheets of glass with a layer of plastic laminated between them. *Low-iron glass* transmits 90% of the solar spectrum, and is used in passive solar homes and is tempered for use as a cover for solar heat collectors. *Low-e glass* is made by applying a thin coating on the surface of window glass that has a low emissivity—the surface does not emit infrared

energy very well, so the flow of heat through the window is slowed.

A process called LIGHT-WEIGHTING—using the minimum amount of necessary for the item in question to maintain its necessary functional characteristics—has kept the volume of landfilled glass from growing as much as it would have otherwise in recent decades. As a result, researchers at the Garbage Project have found that glass bottles from a shallower strata in a landfill are more likely to be broken when found because they contain less glass and are therefore easier to crush.

Glass Recycling the collection primarily of glass bottles for reprocessing and use in new GLASS products. Glass is relatively inexpensive to produce from virgin materials, and the silica sand it is made from is abundant, so RECYCLING centers have little margin in selling it. About 35% of glass bottles and jars are recycled, but the glass used in windows, mirrors, light bulbs and the like are seldom recycled. The ENVIRONMENTAL PROTECTION AGENCY (EPA) estimates that recycling glass containers saves about 13% of the energy that would be required to make new glass from virgin silica.

The removal of glass is a problem in COMPOSTING and the production of REFUSE DERIVED FUEL. The practice of LIGHT WEIGHTING has greatly reduced the amount of glass used to make a disposable glass bottle.

In many communities, only clear or brown glass can be recycled because that is the only glass scrap the recycling center can sell. Green glass is more difficult to recycle and resell as scrap, and multicolored glass harder yet. An increasing amount of colored glass is being ground down into *cullet*, and recast into dark, single-color glass that is good for such things as wine bottles. Ceramics are generally not recyclable, and programs to replace old water-wasting flush TOILETS with newer, more efficient models has increased the flow of old porcelain toilet bowls to dumps. Some are reused as flower planters, chairs and other novelty items, or crushed and used as fill material or aggregate for concrete, but the majority go to a landfill. See BUILDING MATERIALS, DEMOLITION WASTE.

Bottle Bills

A deposit was placed on most glass bottles as recently as the 1960s. If the bottle was returned to a store, the deposit was refunded and the bottle was returned to the bottling plant for washing and refilling. The advent of the NO DEPOSIT, NO RETURN bottle has replaced the reusable glass bottle, but the practice of putting a deposit on beverage containers that is refunded when they are returned has been revived in states that have passed a BOTTLE BILL.

Global Pollution   AIR POLLUTION that circles the globe in the upper atmosphere. Many pesticides and volatile organic compounds have very small molecules and are persistent once released into the environment. Such toxins will remain airborne until they bond with atmospheric moisture and fall to the earth as precipitation. Rainwater often contains traces of PESTICIDES picked up from contaminated air often from half-way around the world (see AERIAL DEPOSITION). The emission of sulfur dioxide and other precursors of ACID RAIN has caused pollution problems that reach across national boundaries over a distance of hundreds of miles, and the release of greenhouse gases (see GLOBAL WARMING) and chemicals that attack the ozone layer (see STRATOSPHERIC OZONE) have caused pollution problems that circle the globe. Global pollution can also come from natural sources such as forest fires and volcanoes. SULFUR DIOXIDE and PARTICULATES emitted into the upper atmosphere by the eruption of Mt. Pinatubo in 1992 prevented about 2% of incoming solar radiation from reaching the Earth's surface for two to four years, reducing worldwide average temperatures by about 1°F over the period.

Global pollution also comes in the form of HAZARDOUS WASTE. Manufacturing processes that produce lots of toxic waste as a BYPRODUCT are increasingly being located in countries with minimal environmental laws and less enforcement, cheap labor and little or no rules protecting the safety and health of workers. Hazardous byproducts of industrial processes that contain toxic chemicals, HEAVY METALS, SOLVENTS and radioactive materials are routinely exported from industrialized nations to developing countries for processing and/or disposal. Participants at a 1992 United Nations conference in Geneva agreed to ban the export of hazardous waste to developing countries, but the treaty has not yet been ratified by the United States and New Zealand, and will be difficult to enforce if it does become law.

Global Warming   a gradual warming of the Earth's average temperature predicted by many scientists as the result of growing concentrations of CARBON DIOXIDE (CO_2) and other gases in the ATMOSPHERE. Global warming would alter climate and rainfall patterns, and cause flooding of coastal areas because ocean water would become warmer and less dense, and would raise the sea level by as much as six feet. In the most extreme scenario, which includes partial melting of the ice caps, the sea level could rise ten times beyond the increase that would be caused by warmer sea water alone. In 1990 an international panel of 300 atmospheric scientists convened by the United Nations predicted an increase in average temperature of 2°F by 2025 and 6°F by the end of the twenty-first century if nothing is done to combat global warming.

Global warming is called the *greenhouse effect* because CO_2 and the other *greenhouse gases* work much the same as the glazing in a greenhouse, allowing shortwave solar radiation to enter, but intercepting long wave infrared radiation before it can pass back outside. This happens because the gases are transparent to incoming shortwave radiation, but are opaque to the longer, infrared, wavelengths. Carbon dioxide, METHANE and water VAPOR are the principal greenhouse gases, although nitrous oxide and two kinds of CHLOROFLUOROCARBONS (CFCs) also play a limited role. Burning carbon compounds in the presence of oxygen produces CO_2 while breaking it down in a ANAEROBIC DIGESTER (or an animal's or human's digestive system)—where there is no oxygen—produces methane (see ANAEROBIC BACTERIA). Methane is about 20 times as effective as CO_2 at trapping heat inside the atmosphere. It is estimated that as much as 15% of atmospheric CO_2 is produced by the flatulence of livestock.

The concentration of carbon dioxide in the atmosphere is only about 350 parts per million, but this CO_2 along with other greenhouse gases causes the earth to be an average of about 60°F warmer than it would otherwise be, making life possible. After about 150,000 years of relative stability, the concentration of atmospheric CO_2 began to increase in about 1850, primarily the result at first of the combustion of COAL and later of PETROLEUM-based FUELS. The concentration of atmospheric CO_2 has increased from about 280 parts per million (ppm) in 1800 to 356 ppm today. CO_2 emissions are expected to increase another 15% by the year 2000. The global average temperature has increased by about nine-tenths of a degree Fahrenheit in the last 100 years, according to a British study released in March 1990, and only about three-fifths of this increase can be attributed to natural causes. Ocean temperatures in some areas have also increased over the last several decades. The temperature of waters off southern California, for instance, are 1.5°F warmer than in 1950, enough warming to cause about half of the roughly three inches that sea level has increased in the area since 1950. Deforestation contributes to global warming because living trees remove some of the CO_2 from the atmosphere through photosynthesis, and when they are cut down, CO_2 is released to the atmosphere if part or all of the tree is burnt or left on the ground to rot. On a worldwide basis, only one tree is planted for each ten that are cut down, and this deforestation is a major contributor to the greenhouse effect.

Global Cooling

The global warming theory—like any climatic model—is not without detractors. Although most atmospheric scientists agree that the buildup of greenhouse gases will make the biosphere warmer than it would otherwise be, many point to other factors that may offset or even reverse this warming trend. One theory is that heavier cloud cover will be one result of atmospheric warming, and that the resulting cloudier skies will limit incoming solar radiation and thereby reduce average temperatures. Another theory holds that natural and manmade AIR POLLUTION will limit global warming. Climatologists estimate, for instance, that SULFUR DIOXIDE (SO_2) emitted into the upper atmosphere by the eruption of Mt. Pinatubo in 1992 prevented about 2% of incoming solar radiation from reaching the Earth's surface for two to four years, resulting in global average temperatures that were about 1°F cooler. The eruption of Mt. Tambaro in Indonesia in 1815 caused so much cooling that 1816 became known as the "year without a summer," and snow fell several times during the summer in New England. Sulfur dioxide from industrial emissions does not get into the upper atmosphere, so its effects are not as long-lasting, but since it is continually being emitted, its cooling effect has been estimated to be even greater. One study concluded that industrial SO_2 reduces the amount of sunshine falling on the northeastern United States by 7.5%. It is possible, therefore, that the reductions of SO_2 emissions resulting from the effort to reduce ACID RAIN could accelerate global warming.

Solutions

Slowing down or reversing the greenhouse effect is difficult because of the fundamental role of FOSSIL FUEL in modern industrial society. Any measure that could make a big difference in the resulting CO_2 release inevitably has drastic effects on all parts of the economy. A carbon tax on coal, for instance, would lead to higher electricity prices in areas like the American midwest, where coal-fired power plants are the rule, and this increase could in turn lead to a higher cost of living and increased cost of doing business across the entire region. Similarly, improving the efficiency of automobile engines in an effort to reduce CO_2 emissions is a time-consuming and expensive proposition. A high tax on gasoline and other fuels to encourage increased engine efficiency, discourage unnecessary travel and fund restoration attempts has been proposed in the United States and elsewhere but the idea is politically unpopular. (Gasoline taxes and prices are quite low in the United States by comparison with other countries, particularly in Western Europe and Japan.)

Because the greenhouse effect is a worldwide problem, any solution will require worldwide cooperation. More than 2,200 delegates representing 56 countries met in March 1990 at the Globe '90 conference in Vancouver, British Columbia, to discuss worldwide climatic change. Only 15 of the 160 nations that signed the climate accord at the U.N. Conference on the Environment in Rio de Janeiro in 1992 had ratified the treaty by March 1993. The treaty will take effect only after 50 countries have ratified it.

The U.S. Forest Service sponsors research into and monitoring of global climatic change. For information, write: Global Change Research Program, USDA Forest Service, PO Box 96090, Washington, DC 20090, or telephone 202-205-1561. A Global Change Information Packet is available from the NATIONAL AGRICULTURAL LIBRARY, 10301 Baltimore Blvd, Beltsville, MD 20705.

Gold   a bright-yellow metallic element with a high luster that is found in quartz veins and alluvial deposits, either as a free metal or in combination with other elements. Gold is rare but widely distributed, and is usually associated with silver. The rarest form is a nugget of solid gold. The largest nugget ever found, weighing 156 pounds, was turned up accidentally by a wagon wheel in Victoria, Australia, in 1869. Gold is found in seawater at concentrations between 5 and 20 parts per 100,000,000. Gold is extremely soft and very dense, and is practically chemically inert. It is a good conductor of electricity and heat, and is the most malleable and ductile of all metals. Pure gold can be hammered to a thickness of 1/200,000th of an inch, and an ounce of the metal can be drawn into a wire 60 miles long. It is used primarily in the production of jewelry, and smaller amounts are used in dentistry and art work. Prospecting for and mining gold played a central role in the early development of the western United States.

The mining of gold during the nineteenth and early twentieth centuries was responsible for some of the most severe pollution of air, water and land experienced in the United States. Placer mining (see PLACER MINE) and dredging to recover gold from the alluvial deposits that made up the bed and banks of streams decimated countless miles of western creeks and rivers, leaving behind a devastated landscape that is still clearly evident today. HEAP LEACHING of gold ore removed from HARD-ROCK MINES using CYANIDE or MERCURY dissolved in water compounded the damage. Water played an important role in the early gold mines, and the rules established in the early gold camps are the basis of western WATER LAW. The mining and milling of gold is still a source of air and water pollution, although the pollution caused is much less than that caused in the early days. Tons of rock are typically mined and processed to obtain an ounce of gold in a modern mine. See also MINE DRAINAGE, MINERAL WASTE, TAILINGS.

Graveyard a place in which dead animal or human bodies are interred. Graveyards sometimes contaminate underlying GROUNDWATER. Lack of space in and around long-settled urban areas such as those found in Europe has greatly reduced the practice of land burial. In Britain, almost 70% of the dead are now cremated. The City of London Cemetery and Crematorium—Britain's biggest cemetery, with 750,000 graves—will run out of virgin ground for new graves in the next couple years. Cemetery officials have proposed retiring a law passed in Victorian times that prohibits the disturbance of human remains so that existing graves can be dug up, and the remains reburied deeper to make room for one or more new graves closer to the ground's surface. The proposal from the Institute of Burial and Cremation Administration calls for leaving graves that are less than 75 years old alone, and no reinterment would occur if there were objections from relatives. An area where old equipment, automobiles, ships or planes are stored for possible future use is also called a graveyard.

Gray Water domestic WASTEWATER other than that from toilets and garbage disposals (see BLACK WATER). Gray water—also called *sullage*—contains too many impurities to be used as DRINKING WATER, but is generally fine for watering plants, although it can damage species that are sensitive to DETERGENTS. Gray water use is prohibited by sanitation laws in some areas, although regions that experience regular water shortages have repealed prohibitions against its use. Bans on the use of drinking water for lawn and garden watering during droughts have encouraged the use of gray water, which comes from domestic sinks, bathtubs and showers. See also IRRIGATION WATER, SEWAGE.

Grease a wide variety of organic substances that have an unctuous, greasy texture. HYDROCARBONS, ESTERS, OILS, waxes, fats and high-molecular-weight fatty acids are the principal forms of grease found as a contaminant in water. Grease can be extracted from aqueous solution or suspension with hexane. Animal fat that has been released by cooking or other processes and the thickened petroleum oil to which a metallic soap has been added are the most commonly encountered forms of grease. Oils and greases of animal origin are generally biodegradable and can normally be processed at a sewage treatment plant with no problem, while petroleum-based oils generally can-

not be removed from the wastestream through biological processing. Grease can build up in waste handling equipment and clog filters in a SEWAGE TREATMENT plant. See also GREASE TRAP.

Grease Trap a basin where GREASE is allowed to settle out of WASTEWATER before it is discharged into a SEWER or SEPTIC TANK. Grease traps are used primarily at restaurants and industrial facilities that produce lots of grease. See also SEWAGE TREATMENT, WATER TREATMENT.

Great Lakes Water Quality Agreement a 1978 agreement between the governments of Canada and the United States to restore and protect the quality of water in the Great Lakes. The agreement is unusual in that the lakes are recognized as a single ecosystem, undivided by national boundaries, and in that the goal of the cleanup effort for persistent contaminants (see PERSISTENT COMPOUND) is complete removal with no additional discharges (see ZERO DISCHARGE). The INTERNATIONAL JOINT COMMISSION was assigned the task of overseeing the implementation of the agreement. See also WATER POLLUTION.

Greenhouse Effect see GLOBAL WARMING

Greenpeace a radical international environmental organization founded in 1979 whose members are dedicated to stopping environmental degradation through nonviolent direct action. Some of the most publicized Greenpeace actions have focused on the international trade in HAZARDOUS WASTE, whaling, fur-seal harvest and the POLLUTION of the world's oceans. The organization is also involved in domestic, land-based issues such as the INCINERATION of GARBAGE, AIR POLLUTION and WATER POLLUTION. Greenpeace activists have put their boats between harpooners and the whale they were attempting to shoot, and have blockaded freighters trying to enter foreign ports to dump a load of hazardous waste, but the organization also takes part in tamer activities such as lobbying and the publication and dissemination of information relating to the environment. The export of hazardous waste from industrialized nations to Third World countries is a current focus of the organization. Greenpeace works toward this goal by informing government officials and the public about the dangers associated with hazardous waste, and through direct action that publicizes the movement of such materials. Greenpeace has taken actions since its founding to stop France from testing nuclear weapons on an atoll in the South Pacific. In 1985, French commandos sank the Greenpeace flagship, the Rainbow Warrior, while it was moored in Auckland, New Zealand, killing one person. Address: 1436 U Street NW, Washington DC 20009. Phone: 202-319-2444.

Groundwater water located below the WATER TABLE, the top surface of the saturated zone in an AQUIFER. Water found in the soil between the top of the water table and the surface of the ground is called SOIL WATER. Groundwater is precipitation that has percolated down through overlying soils and rocks into an aquifer (see HYDROLOGIC CYCLE, PERCOLATION). It comes to the surface in springs and seeps, and feeds rivers during periods of low flow. Groundwater is being overdrafted in many parts of the world because the water use of an increasing human population is exceeding the capacity of the hydrologic cycle to supply fresh water (see POPULATION GROWTH). An estimated 96% of the fresh water in the United States is groundwater, and aquifers supply about 25% of total U.S. water use. Wells provide DRINKING WATER to most rural residences in the United States and provide more than one-third of the water used by community water systems (see PUBLIC WATER SYSTEM). About one-fifth of U.S. IRRIGATION WATER is pumped from aquifers. Falling water tables in areas such as the

Texas high plains (see OGALLALA AQUIFER) have forced cutbacks in irrigated agriculture. Rights to an increasing amount of groundwater formerly used for irrigation are being purchased by urban areas to supply drinking water systems. GROUNDWATER POLLUTION is also reducing the amount of groundwater available. See also CONE OF DEPRESSION, SUBSIDENCE, WATER POLLUTION.

States are required under the SAFE DRINKING WATER ACT to develop well-head protection programs for the areas around public water wells, with the ENVIRONMENTAL PROTECTION AGENCY (EPA) funding up to 90% of the cost. The EPA is required to develop a Ground-Water Protection Strategy to protect aquifers that are used as a source by a community water system. Water systems with less than 500 connections can get two year extensions indefinitely, as long as the system is making progress toward compliance.

Groundwater Dam a barrier to the flow of groundwater that causes the water table to be higher on one side than on the other. A groundwater dam may be natural—as when an impervious ridge of rock cuts across an AQUIFER—or it may be artificial—as when a thin wall of concrete is poured to intercept contami-

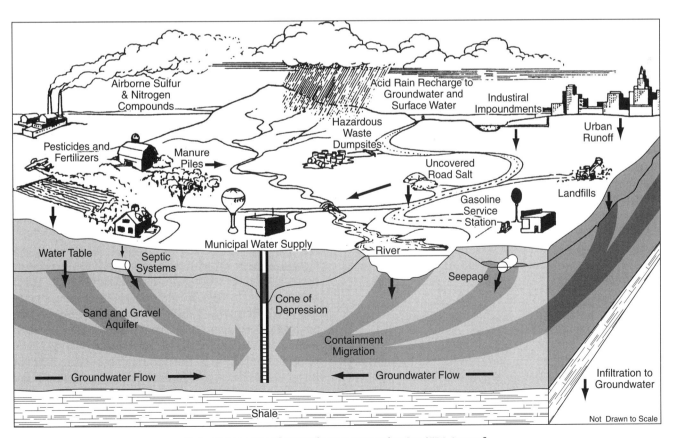

Potential sources of groundwater contamination (*EPA Journal*)

nated water so that it can be pumped out of the ground for cleaning (see GROUNDWATER TREATMENT).

Groundwater Pollution the contamination of GROUNDWATER. It was assumed as recently as the 1970s that groundwater pollution was not a significant concern because any contaminants borne by polluted surface water would be purged as the water percolated down through the soil to the AQUIFER. Although some soils can filter contaminants out of water as it seeps down toward an aquifer (see PERCOLATION), the widespread contamination of groundwater in U.S. agricultural regions with pesticides, nitrates and ammonia has demonstrated that this assumption was incorrect. It turns out that groundwater can actually be more vulnerable to pollution than surface water. VOLATILE ORGANIC COMPOUNDS such as TRICHLOROETHYLENE (TCE) and BENZENE can build up in groundwater because they are unable to evaporate, as they can from surface water (see WATER POLLUTION). TCE has, in fact, been found in groundwater at 2,000 times the highest level recorded in surface water. Groundwater contamination can be very long lived. In 1950, the aquifer under the English fishing village of Norwich was found to be contaminated with whale oil that had soaked into the ground in 1815.

The fact that groundwater is not visible means that contamination can go unnoticed for a long period. Serious contamination of groundwater is normally discovered only after toxic substances make their way into wells and contamination in the wells gets serious enough that it is noticed and the water is tested. The water drawn from public water wells (see PUBLIC WATER SYSTEM) is routinely tested for certain contaminants. And even after the fact of contamination has been established, groundwater pollution is hard to measure, and the shape of a pollution PLUME and its movement can be difficult to trace. Once the contamination is mapped and a cleanup strategy is devised, the cleanup itself can be expensive, and in many cases complete restoration of water quality is not possible (see INTERCEPT WELL, MONITORING WELL). OILS, GREASES and other contaminants stick to the materials of which the aquifer is composed, and it can be next to impossible to remove them.

Sources of Pollution

The routes by which contaminants can enter an aquifer are many and include municipal LANDFILLS, INDUSTRIAL LIQUID WASTE lagoons, LEAKING UNDERGROUND STORAGE TANKS, INJECTION WELLS, leaking sewer lines (see SEWER SYSTEM), SEPTIC TANKS, RUNOFF from city streets (see FRENCH DRAIN, STORM RUNOFF, STORM SEWER) and fuel spills and pipeline ruptures.

Toxic substances often seep into aquifers from HAZARDOUS WASTE DISPOSAL sites. Groundwater has been contaminated in the vast majority of Superfund toxic waste cleanup sites in the United States (see COMPREHENSIVE ENVIRONMENTAL RESPONSE COMPENSATION AND LIABILITY ACT). Polluted water can also infiltrate an aquifer through poorly sealed well casings and abandoned well bore holes. Even polluted rivers can carry contaminants into an aquifer. But leaking underground storage tanks holding PETROLEUM products are the largest single source of groundwater pollution. The ENVIRONMENTAL PROTECTION AGENCY (EPA) has estimated that GASOLINE leaking from such tanks may account for as much as 40% of the total pollution load. The EPA and the American Petroleum Institute estimate that about 5% of the 2 million underground gasoline tanks in the country leak. Industrial process chemicals and toxic waste are also often stored in underground tanks. Groundwater has become contaminated with radioactive materials at most of the military bases involved in the production of nuclear weapons (see ROCKY FLATS, SAVANNAH RIVER, OAKRIDGE NATIONAL LABORATORIES, HANFORD NUCLEAR RESERVATION). Groundwater contaminated with beta particles can be found in areas where nuclear bombs were detonated underground in Rio Blanco County, Colorado during the 1970s in an effort to stimulate natural gas production.

Effects of Pollution

A 1983 Office of Technology Assessment study found that 29% of U.S. cities with a population of more than 10,000 that use groundwater were drawing from a source that exceeded federal drinking-water standards for one or more contaminants. A 1989 survey by Freshwater Foundation concluded that the 35 cities and companies studied lost a combined $67 million as the result of groundwater contamination. The value of clean groundwater is certain to increase as water supplies become more scarce. See also LOVE CANAL, POPULATION GROWTH, WOBURN.

Groundwater Treatment the removal of contaminants from GROUNDWATER or the isolation of contaminated groundwater. Cleaning up polluted groundwater can be a much slower and more expensive process than is normally the case with surface water. Even determining what contaminants are present and mapping the plume of pollution in the aquifer can be time consuming and expensive, and stopping the spread of pollution and removing contaminants can prove even more challenging. Some of the principal techniques used to stop the spread of contaminated groundwater and clean it up include:

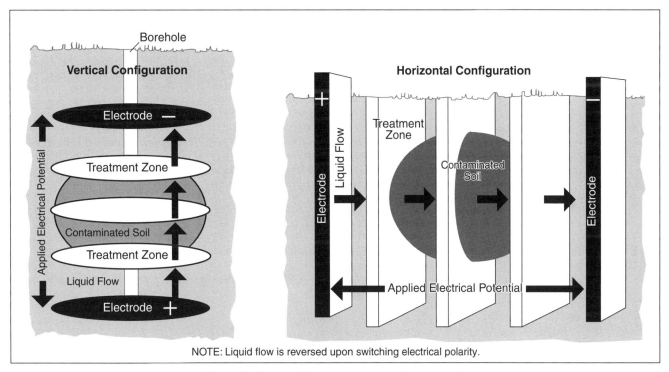

Schematic diagram of "Lasagna Process" (*EPA Journal*)

- the installation of a GROUNDWATER DAM that can be used to intercept the spread of contaminated water or to totally isolate the polluted portion of an aquifer;
- the removal, treatment (see WATER TREATMENT) and re-injection of groundwater. This method is very expensive because treatment must be continued for years, sometimes decades. Its effectiveness is limited by the stubbornness with which some pollutants, especially petrochemicals, stick to the materials of which the aquifer is composed.
- the *in situ* treatment of groundwater contaminants with toxin-eating microbes. This is much less expensive than removal and filtration, and is a quickly developing technology.
- the use of microbes that degrade waste and attack toxins to purify soil or water. Toxin-attacking microbes are injected into an aquifer for *in situ* purification of groundwater. Such bioremediation is often done in conjunction with groundwater pumping and a system that reinjects filtered water to which oxygen and nutrients have been added.

In the *lasagna* process, which is used to remove chlorinated SOLVENTS (see CHLORINATED HYDROCARBONS) from groundwater located in dense, clay-base soils, a low-voltage electric current is passed through the soil. The dense soil to be treated is layered—either vertically or horizontally—between permeable zones created by injecting a mixture of sand and water into the contami-

nated soil at intervals. Chemicals and/or MICROORGANISMS that help with the detoxification process are often injected along with the sand/water mixture. The outer layers of the lasagna thus created contain enough graphite or carbon particles that they are capable of conducting electricity, and act as electrodes. An electric field is created in the area between energized electrode's, and mobile materials in the soil and groundwater are attracted to either the positive or negative electrode. The resulting movement of materials in the area to be treated carries contaminants into one of the permeable treatment zones where they are neutralized.

Gypsum   a common, very soft mineral composed of hydrous calcium sulfate ($CaSO_4 \bullet 2H_2O$). Gypsum is used in the production of building materials and as a hydration retardant in cement (see CONCRETE). Gypsum mixed with lime is used as a plaster finish coat, and plate glass is polished on a bed of gypsum. *Gypsum wallboard*—also called *sheetrock* or *drywall*—consists of processed gypsum rock sandwiched between two sheets of heavy paper. Ease of installation and relatively low cost have caused sheetrock to almost completely replace plaster as an interior wall finish. Drywall scraps from new construction and old wallboard removed during renovation are a substantial component of MUNICIPAL SOLID WASTE (MSW—see CONSTRUCTION WASTES, DEMOLITION WASTES). Almost

17 million short tons of gypsum were produced in the United States in 1993. *Gypsum rock*, a sedimentary rock composed primarily of gypsum, is formed by the PRE-CIPITATION of calcium sulfate from seawater, and often found in association with limestone and shale. Gypsum can be produced when sulfuric acid generated by a volcano makes contact with calcium-bearing minerals. The mineral is found in clays as the result of the action of sulfuric acid on limestone. Gypsum is created by stack SCRUBBERS that use lime or a solution of powdered limestone to remove particulates and acidity from industrial smokestacks. Sulfur dioxide reacts with the limestone in the presence of water to form gypsum. The mining and refinement of gypsum is a source of air and water pollution, and the tailings of gypsum mines and mills can contain URANIUM and other RADIONUCLIDES as well as ARSENIC or ASBESTOS, which are sometimes found in gypsum rock. Gypsum wallboard that contains radium can emit radon.

Slowly heating gypsum causes it to lose part of its water and become a white powder called *plaster of paris*. When plaster of paris is mixed with water, it forms a white paste that quickly sets up into gypsum, the reverse of the dehydration process. The gypsum expands slightly when it rehydrates, filling all the voids if it is in an enclosed space. For this reason, it is used in the production of casts for statues, ceramic objects, dental plates and fine metal parts.

Half-life the amount of time it takes for half of an element undergoing radioactive decay to change into the new element that is the end product of the decay process. The term "half-life" also denotes the amount of time required for half of a water pollutant to degrade to another substance. Over a period of time equivalent to 10 half-lives, the radiation emitted from a RADIONUCLIDE (or the toxicity of a water-borne contaminant) would be approximately one-thousandth the original value. See also RADIOACTIVE WASTE DISPOSAL, RADIOACTIVITY, TOXIC WASTE.

Halide a binary COMPOUND in which one of the two constituents is a member of the halogen family, with the other element usually being a METAL. Halides are usually crystalline solids that separate easily into negative and positive IONS when in water, and are good electrical conductors.

Halogen any of the five elements—fluorine, CHLORINE, bromine, iodine and astatine—that form the halogen series in the periodic table. The halogens are all chemically active, and are similar in their chemical properties. They exist in their free state as diatomic molecules (molecules made up of two atoms).

Halons a family of gases—also called *halocarbons*—that are commonly used in fire extinguishers. Halons' role in the destruction of STRATOSPHERIC OZONE led to an agreement at the Montreal Protocol to ban further production after January 1, 1994. The compounds are extremely stable once released into the atmosphere, and eventually migrate to the ozone layer. Once there, ultraviolet radiation breaks the halons down, releasing bromine gas in the process. The bromine, in turn, can react chemically with ozone to destroy it. Bromochlorodifluoromethane and bromotrifluoromethane are the principal halons. Representatives of the fire-protection industry say there is no easy substitute for halons in firefighting because they are noncon-ducting, have low toxicity and put out fires quickly. The industry sought an exemption to the ban and, barring that, plans to recycle the halons now in place in fire extinguishers for as long as possible in the hope that a viable alternative will be developed in the next several years. Although several businesses that specialize in the production of firefighting equipment have been working on replacements for halons that are based on fluorinated hydrocarbons, none has yet proven nearly as effective.

Hanford Nuclear Reservation a 560-square-mile federal facility located along the Columbia River north of Richland in northeastern Washington state. Hanford was built in 1943 as part of the wartime Manhattan Project that led to the production of the first atomic bombs. The world's first full-scale NUCLEAR REACTOR, the B Reactor, which was constructed in a frenzied 15 months at Hanford, produced the PLUTONIUM for the Nagasaki bomb. The B-reactor and seven other plutonium-producing reactors that were built on the reservation in the 1940s and 1950s are now permanently shut down. A ninth reactor, the N-Reactor, was built in 1963 to produce plutonium and to generate commercial electric power as a byproduct. The N-reactor is similar in design to the reactor that melted down at CHERNOBYL in the Soviet Union in 1986. It was shut down in January, 1987.

In 1986, the U.S. government released 19,000 pages of previously classified documents relating to the operation of Hanford during and immediately after World War II, which revealed a series of intentional radiation releases during the war. In one event in 1949, which has come to be known as the Green Run, an especially large amount of radiation was released into the atmosphere so that its course could be tracked. It was called the "Green Run" because especially radioactive plutonium was released (it had been allowed to cool just 16 days rather than the normal 40). Although many of the details of the Green Run are still classified, scientists have disclosed off the record that the experiment was intended to develop ways of monitoring the production of nuclear weapons in Russia. It has been estimated that 8,000 curies of radiation were released during the Green Run, twice what researchers had planned. Federal experts estimate that a total of 435,000 curies of radiation were released from Hanford from 1945 to 1947. The partial core meltdown at the THREE MILE ISLAND plant, by comparison, released only an estimated 15 to 24 curies of radiation into the atmosphere.

An estimated 61 million gallons of radioactive liquids are stored in 177 underground tanks at Hanford, and about 90,000 fuel rods are stored in concrete basins that were designed in the 1940s to last 20 years, with many of the deteriorating storage basins located near the Columbia. A 1989 Government Accounting Office report on wastes at U.S. nuclear weapons facilities said that 50 of 149 single-shell storage tanks at Hanford had been leaking high-level waste into groundwater for 20 years. A 1990 Government Accounting Office report concluded that there was a risk of an explosion in 22 of the single-shell tanks that contained *ferrocyanide*, which was added in the 1950s to separate nuclear materials from liquids so the liquids could be pumped out. It was later found that a similar report in 1984 had been suppressed by the DEPARTMENT OF ENERGY (DOE). The 1990 report concluded that people outside the Hanford reservation could be exposed to as much as 7.3 rems of radiation by such an explosion (see RADIOACTIVITY, MEASUREMENTS). A class action lawsuit was filed against 13 contractors at the plant claiming that their negligence exposed workers and people that lived near the plant to health damage from toxic chemicals and radioactive materials, and that death, disease, disfigurement, miscarriages, birth defects, sterility and infertility to the 21 plaintiffs or their family members had resulted. The DOE and its predecessor, the ATOMIC ENERGY COMMISSION, were not named in the suit because government agencies' immunity from tort claims would have made it difficult to win. The DOE concluded in 1990 that groundwater contamination at Hanford was spreading more quickly than it was being contained. A report released by Batelle Pacific Northwest Laboratory in Richland in 1992 said that radioactive materials in cooling water and from accidental releases from Hanford, primarily during the '50s and '60s, had reached the Pacific Ocean, and that traces of radiation could be detected on islands and along the banks of the Columbia. The report concluded that people that regularly ate fish caught near the mouth of the Columbia could be exposed to an additional 100 to 200 millirems of radiation per year as a result. Critics said the report greatly underestimated the seriousness of radioactive WATER POLLUTION from Hanford (see GROUNDWATER POLLUTION). The federal government now expects to spend nearly $60 billion to cleanup nuclear wastes and toxic chemicals that have polluted soils and groundwater at Hanford.

Hanford was one of the finalists for a national high-level nuclear waste repository called for in the Nuclear Waste Policy Act of 1982 (see HIGH-LEVEL WASTE). The DOE listed the Hanford Reservation as one of the finalists despite findings by government scientists that a repository located there could soon leak. Washington voters approved a referendum opposing the move in 1987 by a margin of 84% to 16%. Construction of three of the five commercial nuclear power plants being built for the Washington Public Power Supply System (WPPSS) commenced at Hanford during the 1970s. The $2.25 billion default that occurred in 1983 when construction of one of the Hanford WPPSS plants and another in Satsop, Washington, was canceled was the largest municipal bond default ever.

After six years and $7.5 billion, the Hanford cleanup has produced many studies, but very little cleanup. The tanks holding fuel rods and radioactive liquid waste must be maintained until a billion-dollar processing center is completed. Waste will be converted in the center into radioactive glass logs for burial elsewhere. It is expected that the cleanup will take 50 years. See also OAKRIDGE NATIONAL LABORATORY, ROCKY FLATS, URAVAN URANIUM MILL.

Hard-Rock Mine   a mine consisting primarily of tunnels bored and blasted through solid rock. A hard-rock mine causes less surface disruption than a STRIP MINE or a PLACER MINE. Surface and underground drainage from hard-rock mines can pollute aquifers and surface streams. The ore of many of the commercial minerals mined in hard-rock mines contains pyrite, which oxidizes readily when exposed to air in a mine or TAILINGS pile, and can release sulfuric acid into water or SULFUR DIOXIDE into the air, depending on the temperature and humidity at which it oxidizes. Bacteria that metabolize IRON and sulfur in the presence of oxygen also break down pyrite, and contribute to acid RUNOFF from mines and tailings piles. See also HEAP LEACHING, MINE DRAINAGE, MINING LAW, MINERAL WASTE, SPOIL.

Hazardous and Solid Waste Amendments of 1984   see RESOURCE CONSERVATION AND RECOVERY ACT

Hazardous Air Pollutants List   a list of 189 toxins published in the 1990 amendments to the CLEAN AIR ACT. The amendments allow interested parties to submit evidence to the ENVIRONMENTAL PROTECTION

AGENCY to add or delete materials from the list. The agency must accept or deny the petitioner within 18 months, basing its decision on the toxin's ability to harm human health or to have adverse environmental effects, which are defined as any threat of significant adverse effects to wildlife, aquatic life or other natural resources. The disruption of local ecosystems, impacts on threatened or endangered species and significant degradation of environmental quality over broad areas are also to be considered in granting or denying the petitioners request for addition or deletion.

Hazardous Material a material that is dangerous because it is ignitable, explosive (see EXPLOSIVE MATERIAL), reactive (see REACTIVE MATERIAL), corrosive (see CORROSIVE MATERIAL) or toxic. Also, one of the approximately 500 substances regulated by the ENVIRONMENTAL PROTECTION AGENCY under the terms of the RESOURCE CONSERVATION AND RECOVERY ACT (RCRA) of 1976. The solubility of a contaminant in water and fat also plays a key role in the hazardous qualities of a substance.

Although TOXICITY is one important consideration when assessing the threat to environmental and human health posed by a substance, a high toxicity rating does not necessarily mean the substance poses an extraordinary threat to the health of humans and the environment. While a high toxicity rating (see LD_{50}) indicates that a small amount of the material in question can cause damage or death to a living organism, if other characteristics make it unlikely that the substance will be able to affect health by being ingested or inhaled, or by passing through the skin. A hazardous material represents a threat to health, and if a highly toxic material can be transported, handled and stored with little risk that it will escape into the environment to injure humans or other life forms, it may not be considered as hazardous as a less toxic material that is extremely hard to contain.

The sludges that are created as a waste product of treatment sometimes contain sufficiently high levels of toxic materials that must be disposed of as hazardous waste. The TOXICITY CHARACTERISTIC LEACHING PROCEDURE is used to measure the concentration of regulated toxic materials in the sludge. The low-level RADIOACTIVE WASTE that is produced when RADIUM, RADON and URANIUM are removed from drinking water is subject to especially strict disposal requirements. See also CROSS MEDIA POLLUTION, HAZARDOUS WASTE DISPOSAL.

Hazardous Materials Information Exchange a computer bulletin board at the Argonne National Laboratory designed for the distribution and exchange of information about HAZARDOUS MATERIALS. Infor-

mation on the exchange, which is sponsored by the Federal Energy Management Agency and the Department of Transportation, includes data on response to emergencies involving hazardous materials, training in their handling and regulations on their use. The system can be accessed at 708-972-3275 using 8 data bits, no parity and 1 stop bit at modem speeds of up to 14,400 baud. For more information, contact Argonne National Laboratory, DIS900, 9700 S Cass Ave, Argonne, IL 60439, or call 800-752-6367.

Hazardous Materials Transportation Act federal legislation signed into law in 1975 that is intended to reduce the threat to public health posed by the transportation of HAZARDOUS MATERIALS. The act calls for the regulation of the shipping, handling, packaging, labeling and storage of hazardous materials that are shipped interstate or that are transported within a state if that movement is related to interstate shipment. Hazardous material carriers are required to register with the federal government, keep complete records and make periodic reports on their activities. The act also calls for continuing research into safer methods of moving HAZARDOUS WASTE, and forbids the air shipment of nonmedical radioactive materials by passenger aircraft. See also HAZARDOUS WASTE DISPOSAL.

Hazardous Substances Act federal legislation signed into law in 1960 and frequently amended since then that has the intent of limiting public exposure to hazardous substances. Although the law remains in effect, it has been largely superseded by more recent legislation (see COMPREHENSIVE ENVIRONMENTAL RESPONSE COMPENSATION AND LIABILITY ACT, RESOURCE CONSERVATION AND RECOVERY ACT, TOXIC SUBSTANCES CONTROL ACT).

Hazardous Waste waste material that poses a risk to human or environmental health because of its ignitability, explosivity (see EXPLOSIVE MATERIAL), reactivity (see REACTIVE MATERIAL), corrosivity (see CORROSIVE MATERIAL) or TOXICITY. The ENVIRONMENTAL PROTECTION AGENCY (EPA) estimates that 240 million metric tons of reported hazardous waste is generated each year in the United States, 90% of which comes from large facilities that generate more than 500 pounds per month (see INDUSTRIAL HAZARDOUS WASTE). Nearly two-thirds of U.S. hazardous waste is generated in the ten most heavily industrialized states: New Jersey, Illinois, Ohio, California, Pennsylvania, Texas, New York, Michigan, Tennessee and Indiana. The volume of hazardous wastes dumped by applications that are exempt from regulation under the RESOURCE CONSERVATION AND RECOVERY ACT (RCRA), and hazardous materials that are dumped illegally (see MIDNIGHT DUMPING) make it impossible to determine exactly how much hazardous waste is generated each year.

The largest part of the hazardous wastes produced annually in the United States is the corrosive materials, spent acids and alkaline wastes from the metal-finishing, chemical-production and petroleum-refining industries (see INDUSTRIAL HAZARDOUS WASTE). Hazardous waste is also part of residential and commercial GARBAGE (see HOUSEHOLD HAZARDOUS WASTE), and waste from hospitals and research institutions; small businesses including beauty parlors, gas stations, photo developers and laundries; government agencies, especially those involved with the maintenance of roads and equipment and water and sewer utilities; military installations, with those involved with the production of nuclear weapons and the maintenance of equipment producing the most waste; and pesticides and nutrients from farms (see AGRICULTURAL WASTE). Hazardous wastes were discarded regularly in surface and ocean water and into the atmosphere without treatment until concern about resulting environmental damage began to slow the practice in the 1960s. The cleanup of hazardous waste sites dating from the era has become a semipermanent government burden (see COMPREHENSIVE ENVIRONMENTAL RESPONSE COMPENSATION AND LIABILITY ACT).

RCRA Hazardous Waste

Only waste classified as hazardous under the terms of the RCRA must be managed as hazardous waste throughout the United States, but additional substances are classified as hazardous in some states. Classifications of waste regulated under the RCRA include *contaminated cleanup materials and soil* from spills at manufacturing sites, accidents, pipeline leaks, tanker spills or leaking containers; *paint and organic residuals* from the solvents, pigments, dyes and other materials used in the production of and cleanup of paint products; *organic and oily residuals* of petroleum products and machinery manufacturing processes, which come in the for of thick sludges and solids that often contain HEAVY METALS; *organic sludges* generated in the treatment of industrial wastewater (see INDUSTRIAL LIQUID WASTE) and in the degreasing of metal parts with solvents; *solvents and organic solutions* produced in the manufacture of pharmaceuticals, rubber, plastics and organic chemicals; *electroplating wastes* that contain heavy metals, salts, cyanide and sulfides that are often reactive or corrosive; *still bottom*, which is a thick, tar-like substance created in the production of organic chemicals; *solutions and sludges* from metal finishing processes; the production of metal-based inks;

PESTICIDES; *oils and greases* (see ENGINE OIL); *solid inorganic residuals* produced by air-treatment equipment in the metals industry; ASH from incinerators in which certain toxic substances have been burned (see BOTTOM ASH, FLY ASH, INCINERATION, TOXICITY CHARACTERISTIC LEACHING PROCEDURE); and *spent catalysts* from petroleum refining.

Transit of Hazardous Waste

The storage and movement of hazardous materials is regulated by the Department of Transportation (DOT) under its mandate to ensure the safety of highways, railroads and other transportation corridors. The DOT lists approximately 1,700 materials grouped into seven categories: explosives, flammables, corrosives, compressed gases, poisons, etiologic agents (an agent capable of causing disease) and radioactive materials. The DOT specifies how hazardous materials are to be designed, packaged, handled and routed; what substances a given hazardous material should *not* be shipped with; and what paperwork must be completed relating to hazardous waste. The COMMUNITY RIGHT TO KNOW ACT requires that hazardous industrial materials be reported to local authorities, and that businesses handling hazardous materials help develop local emergency plans to be followed in the event of an accidental release.

Hazardous Waste Source Reduction

Cutting the volume of hazardous waste through source reduction, recycling and treatment can create useful process chemicals and energy while reducing the volume of materials bound for disposal. SOURCE REDUCTION involves changes in the manufacturing process that reduce the production of hazardous waste. The purchase of more efficient equipment, changes in plant housekeeping, changes in management philosophy and manufacturing processes and improvement in the purity of the feedstock and in the type of feedstock used are all ways of reducing the volume of waste produced in a given industrial process. Some of the hazardous materials that are removed from the wastestream through chemical and physical treatment (see HAZARDOUS WASTE TREATMENT) can be purified for REUSE in the same way in which they were originally used, or can go through further processing for RECYCLING as feedstock, fuel or industrial chemical. As the disposal of hazardous waste becomes more costly, an increasing number of innovative ways of reducing its volume are being tried. A nylon manufacturer that recycles an acidic byproduct into a raw ingredient used in the production of paints and solvents, for instance, or a manufacturer that recycles caustic waste produced in the production of videotapes into a raw material used in fertilizer manufacturing both reduce the cost and liability associated with the company's production of hazardous wastes, and provides instead a new source of income. See also CENTER FOR WASTE REDUCTION TECHNOLOGIES.

Hazardous Waste Cleanup see COMPREHENSIVE ENVIRONMENTAL RESPONSE COMPENSATION AND LIABILITY ACT

Hazardous Waste Disposal the disposal of HAZARDOUS WASTE, both legally under the terms of RESOURCE CONSERVATION AND RECOVERY ACT (RCRA) and CLEAN WATER ACT (CWA) or CLEAN AIR ACT (CAA) permits, and illegally through MIDNIGHT DUMPING and incomplete reporting of regulated emissions. The improper disposal of hazardous waste has caused severely polluted air, water and soil in virtually every part of the world, and most of the world's governments have been forced to undertake expensive and lengthy cleanups to control the flow of contaminants into the environment from hazardous waste disposal sites. As a result, most governments at least attempt to control the disposal of hazardous materials. The United States is the world's leading source of hazardous waste, creating more than all other industrialized countries combined. Of the 238 million tons of hazardous waste generated in the United States each year, an estimated 13 million tons—much of it scrap metal—is exported legally, and investigators believe an even greater volume is exported illegally, much of it to Mexico. Participants at a 1992 United Nations conference in Geneva agreed to ban the export of hazardous waste from rich to poor countries.

Secure Landfill

The high cost of treating and disposing of hazardous waste and the long-term legal responsibility for any environmental damage caused by the disposal of such wastes has led to a decrease in the volume of wastes produced. Reduction, REUSE and RECYCLING of some of the hazardous components of the wastestream is largely responsible for this decrease (see HAZARDOUS WASTE TREATMENT). But even after every possible step has been taken to reduce the volume, TOXICITY, reactivity, solubility and other objectionable qualities of the waste and to make further use of valuable elements in the wastestream, a sludge or a solid that must be disposed of as hazardous waste usually remains. Environmental cleanup operations also generally produce dangerous materials such as incinerator ash or contaminated soil for storage or disposal (see COMPREHENSIVE ENVIRONMENTAL RESPONSE, COMPENSATION AND LIABILITY ACT). Many hazardous wastes will

remain dangerous for centuries, and the chance that they will eventually find their way into their environment are, as a result, pretty good. The disposal, or, more accurately, long-term storage of treated, solidified, stabilized wastes in SECURE LANDFILLS is one way of storing hazardous materials in a way that at least minimizes the chances of leakage. Even if an earthquake or other natural disaster were to rupture the impermeable layers sealing off the wastes in such a facility, the blocks of insoluble material in which they are stored will not quickly release their load of hazardous materials, allowing time to repair the damage. Treated wastes are prepared for disposal through *solidification*—they are cast into monolithic blocks with a high degree of structural integrity—and *stabilization*—they are chemically treated to reduce their solubility.

Deepwell Disposal

DEEPWELL DISPOSAL was the principal way in which many of the nation's most hazardous wastes were dispatched until the 1980s (see INJECTION WELL). Because of the depth of the formations into which waste is pumped, and the fact that leaks can not easily be detected, traced or remedied at such a depth has led to tight restrictions on the use of injection wells in most states.

Heat Treatment

INCINERATION and PYROLYSIS break down hazardous wastes by exposing them to temperatures between 800° and 3000°F (430° TO 1700°C). Incineration occurs in the presence of oxygen (see COMBUSTION, HAZARDOUS WASTE INCINERATOR), and pyrolysis occurs in an oxygen-starved chamber.

Hazardous Waste Incinerator a thermal-treatment facility at which HAZARDOUS WASTE is burned (see COMBUSTION, INCINERATION) or broken down by PYROLYSIS. The ENVIRONMENTAL PROTECTION AGENCY (EPA) requires that thermal treatment remove 99.99% of the hazardous qualities of the waste being neutralized. Scrubbers must be installed to limit the emission of air pollutants from combustion facilities, and ashes produced by thermal treatment must usually be handled as hazardous waste.

Incineration and pyrolysis break down hazardous wastes by exposing them to temperatures between 800° and 3000°F (430° TO 1700°C). Materials heated in the presence of oxygen will be changed by the rapid oxidation that characterizes combustion, which produces heat and usually light, toxic PARTICULATES (see BENZO-A-PYRENE). Toxic ASH and toxic gases (see AIR POLLUTION, CARBON MONOXIDE) may also be created by combustion, along with CARBON DIOXIDE and

WATER. Pyrolysis is the chemical decomposition of a compound that results when it is slowly heated in the absence of oxygen to a temperature high enough to break the chemical bonds that hold it together. Pyrolysis, which is also used to reclaim METALS from SOLID WASTE (see RECYCLING), neutralizes hazardous materials without the release of toxins associated with combustion. Pyrolysis produces simpler organic compounds than does incineration, as well as a char or ash.

Single-chamber liquid injection systems are the most widely used hazardous-waste incinerators, and they are used to dispose of PHENOLS, POLYCHLORINATED BIPHENYLS (PCBs), still bottoms, reactor bottoms, SOLVENTS, polymer wastes and PESTICIDES. A *rotary kiln* is a cylindrical combustion chamber lined with refractory material in which liquid or solid hazardous waste can be burned. The combustion chamber can be rotated while in use to improve turbulence in the combustion zone. A *fluidized bed incinerator* burns liquid and solid hazardous wastes on a heated bed of inert granular materials that are agitated to promote the rapid exchange of heat, which drives off gases.

The refinement of PETROLEUM creates an oily residue, and the solids that remain after this residue is processed to remove the remaining OILS must be disposed of as hazardous waste. One method of disposal is their conversion into fuels designed to be burned in high-temperature kilns that produce cement. The fuels consist of as much as 40% hazardous waste solids suspended in waste oil. The savings in fuel costs to the company involved can be substantial, and the destruction of the hazardous wastes is complete, with few emissions from a kiln being operated within normal guidelines. The legal and political hurdles of licensing a cement kiln as a HAZARDOUS WASTE INCINERATOR can, however, be substantial (see NIMBY). Hazardous solids are also mixed into a water-based slurry for injection as quench water in COKE refineries, where the solids are converted to fuel grade coke. It costs refineries roughly half as much to dispose of wastes by converting them into coke as it does to make conventional waste-derived fuel or to dispose of wastes in an incinerator. See also INCINERATING TOILET, MEDICAL WASTE INCINERATOR.

Hazardous Waste Treatment   the treatment of hazardous wastes to reduce their volume and toxicity and to separate component chemicals for further use or disposal. Hazardous waste treatment through physical, biological and chemical processes can reduce the wastes' volume and toxicity and make them easier to handle and store and less of an environmental threat.

Physical Treatment

Physical treatment processes similar to those used in a SEWAGE TREATMENT plant or a DRINKING WATER treatment facility (see WATER TREATMENT) are often used to separate components of the wastestream or change its physical form without changing its chemical characteristics, often in preparation for chemical or biological treatment or INCINERATION of the more toxic elements. All but the final steps in the treatment of hazardous waste resemble those used in sewage treatment. Physical processes include: *screening*—the removal of larger suspended and floating solids with racks and screens; *sedimentation*—suspended solids settle to the bottom for removal in quiet water, often with the assistance of a chemical coagulant; *flotation*—tiny air bubbles pass through wastewater picking up suspended solids to a small to be removed by sedimentation and float them to the surface for removal; *centrifugation*—the removal of water from the wastestream in a spinning vessel; *dialysis*—the separation of components of a liquid stream through the use of a selective membrane; *electrodialysis*—used to separate the components of an ionic solution to which an electrical current has been applied through the use of a selective membrane; *reverse osmosis*—effectively removes HEAVY METALS, RADIONUCLIDES, NITRATES, NITRITES and synthetic organic chemicals by forcing water through micropores in a synthetic membrane; *ultrafiltration*—similar to reverse osmosis, but membrane can withstand corrosive fluids better than reverse osmosis and it separates materials of a higher molecular weight from the wastestream; *distillation*—separates liquids with different boiling points through slow heating and evaporation and is used in the refinement of petrochemicals and the recovery of solvents used in industrial processes; *solvent extraction*—a nonsoluble liquid that acts as a solvent to the elements to be removed is added to the wastestream. The solvent is later separated and processed to separate the dissolved material for reuse or disposal; *evaporation*—liquids are boiled off to concentrate non-volatile solids in the wastestream; *ADSORPTION*—traps organic materials on a porous surface, usually ACTIVATED CARBON.

Biological Treatment

BIOLOGICAL TREATMENT—also called *biodegradation*—uses the hungry digestive tracts of MICROORGANISMS to break down hazardous materials (see BIODEGRADABILITY). Microbes break down organic wastes to produce carbon dioxide, water and simple inorganic compounds or less-complex organic substances such as ALDEHYDES and ACIDS. Biological treatment methods include: *activated sludge*—subjects organic materials to rapid oxidation by exposure to oxygen and microbe-rich floc (see AERATION, FLOCCULATION, OXIDATION, SECONDARY TREATMENT); *aerated lagoon*—wastewater is aerated in a large enclosure to increase biological oxidation; *trickling filter*—wastewater trickles through a filter bed composed of aggregate covered with microbial growth (see ROTATING BIOLOGICAL CONTRACTOR); *stabilization pond*—wastes are exposed to biological action in a shallow pond over a long period of time with only natural air movement providing aeration (see EVAPORATION POND, FACULATIVE POND, PONDING); *anaerobic digestion*—wastes are metabolized by ANAEROBIC BACTERIA in a vessel sealed from the atmosphere (see ANAEROBIC DIGESTER). Wastewater was once commonly spread on the surface of the ground at a carefully regulated rate in a process called *land treatment* or *land farming*, and naturally occurring microbes broke down organic wastes as they filtered through the soil, but the process is seldom used today.

Chemical Treatment

Chemical treatment changes the chemical structure of hazardous substances to produce compounds that are less toxic or safer to handle. Chemical processes are one of the principal means of disposing of hazardous materials that have been isolated by the physical processes outlined above (see HAZARDOUS WASTE DISPOSAL). Chemical processing produces few air pollutants, and often can be carried out in a portable laboratory at the site where the hazardous material is stored, eliminating extra handling and transport of the waste. The principal techniques used are: *neutralization*—mixing acids and bases into the waste as needed to neutralize acidity or alkalinity, process used to neutralize acid sludges and wastes from processes such as STEEL production, metal plating, leather tanning and the production of pickles (see ACIDITY, ALKALINITY); *precipitation*—varying the PH and temperature of a SOLUTION to decrease the solubility of heavy metals and other hazardous dissolved materials—forms a solid PRECIPITATE that can be further coagulated, settled out in a sedimentation tank or removed with a filter; *ION EXCHANGE*—removes inorganic ions from solution by passing wastewater over natural or synthetic resin beds that exchange IONS for the dissolved contaminants, process used to recover precious metals such as GOLD and SILVER, heavy metals such as CHROMIUM and LEAD and SALTS from wastewater; *oxidation/reduction (redox)*—breaks chemical bonds by passing electrons from one reactant to another, process used to convert both organic and inorganic hazardous materials such as CYANIDE wastes from electroplating into less toxic, easier to handle form; *dechlorination*—stripping chlorine atoms from chlorinated compounds such as chlorinated hydrocarbons by initiating chemical reactions.

Thermal treatment (see INCINERATION, HAZARDOUS WASTE INCINERATOR, PYROLYSIS), *deep well injection* (see DEEPWELL DISPOSAL, INJECTION WELL) and *land disposal* (see SECURE LANDFILL) are the principal processes used to dispose of treated hazardous waste.

Heap Leaching   a method used to extract metals from crushed ores that relies on the creation of soluble metallic compounds through OXIDATION—also called LIXIFICATION. The process was initially used in the production of copper. Layers of coarse copper ore are interspersed with layers of finer ore in a heap about 20 feet high. Water is poured over the pile periodically, facilitating further oxidation and absorbing and washing away existing oxidation products. The resulting leachate is rich in copper oxides, which can be precipitated from the solution chemically (see PRECIPITATION). Heap leaching is also used to extract GOLD from its ore, but potassium cyanide is first added to the water to cause the gold to dissolve. The resulting leachate is treated with zinc to precipitate the gold from solution. The leachate used in the heap leaching of gold is heavily contaminated with zinc and CYANIDE, and can cause severe environmental damage and WATER POLLUTION if it escapes from the area in which the heap leaching is taking place, as is often the case, even at modern mines. Little effort was taken to confine leachate for longer than was necessary to remove the gold from solution at mines of a century ago, and the resulting damage is still obvious. See also HARD-ROCK MINE, MINE DRAINAGE, MINERAL WASTE, STRIP MINE, TAILINGS.

Heat Recovery Ventilator   a device that blows stale indoor air out of a building while transferring much of its heat to incoming fresh air. A heat-recovery ventilator (HRV), also called an air-to-air heat exchanger, can also be used to pre-cool fresh air being drawn into a building during the cooling season, although this application is relatively rare. Because high infiltration rates cut the device's efficiency (and the need for additional ventilation), HRVs are used primarily in tightly sealed homes. Heat-recovery ventilators are available either as small, wall-mounted units about the size of a room air conditioner, or as larger systems that resemble a furnace with ductwork and sophisticated controls. The smaller units are capable of providing up to 100 cubic feet per minute (CFM) of fresh air to a room or two or to a small apartment, while the larger, centralized, system delivers up to 500 CFM and is capable of providing ventilation for an entire house. A whole-house HRV is much easier, and cheaper, to install during construction of a new building than it is to retrofit in an existing structure. Inside the housing of an HRV are a heat-exchanger core, fans for incoming and out-going air, controls, a defrosting mechanism and a con-densate drain. When the unit is on, outgoing air is pushed through the CARDBOARD, PLASTIC or METAL core, which transfers about 70% of the heat in exhaust air into incoming fresh air. As the outgoing air is cooled, its capacity to hold water decreases, and con-densate forms inside the heat-exchanger core and runs out the drain. The defroster is used to melt ice that can build up inside the equipment during winter, hinder-ing the transfer of heat and eventually blocking the passage of air.

The performance of an air-to-air heat exchanger is dependent on:

- the amount of heat exchange surface inside the core (the greater the surface area, the more heat can be transferred);
- the speed of airflow through the core (the slower the airflow, the more heat transferred);
- the configuration of airflow through the core. *Counter-flow* heat exchangers, where in-going and outgoing air move in opposite directions, are gener-ally more efficient than *cross-flow* systems, where air currents pass one another perpendicularly, while *parallel-flow* patterns are generally less efficient than the former two.

Correct installation is crucial for the efficient perfor-mance of an HRV. If intake and exhaust vents are located too close together, part of the fresh air entering the building can be sucked into the exhaust vent before it provides any ventilation. For this reason, intake and exhaust vents should be well separated. Although the fresh air supplied by an HRV is preheated on its way indoors, it is only warmed to a level somewhat below the exiting, room-temperature air. While a cool human in the line of fire of the 140°F air being blown into a home by a forced-air furnace will be warmed by the air current, even a warm body will be cooled by an airstream that may be between 60° and 70°F (a typical temperature in an HRV). For this reason, fresh-air vents should be located in areas where they will not blow cool air on building occupants. Although the fresh air provided by an HRV may be cool, it is far more comfortable and energy efficient than the normal practice of bringing untempered outdoor air into a building for VENTILATION.

Regular maintenance is crucial if a heat-recovery ventilator is to operate efficiently. Dust from both indoor and outdoor air can accumulate in the systems core, cutting the flow of heat from one airstream to the other. For this reason, the core must be cleaned regu-larly. Many systems incorporate air filters to reduce the introduction of foreign material into the core, and some feature a core that is easily removable for clean-ing. To ensure proper performance, filters should be replaced and/or cores should be cleaned in keeping with the manufacturer's recommendations—more often in installations with lots of dust in the air. See also INDOOR AIR POLLUTION.

Heavy Metal   any of a loosely defined group of metallic elements with high atomic weights and simi-lar health effects. The elements are called "heavy" partly because of their comparatively high density, and partly due to their comparatively low level of chemical activity. Heavy metals are generally defined to include elements as light as CHROMIUM, with an atomic weight of 24, and to include at least two non-metals, ARSENIC and SELENIUM. THALLIUM is one of the most toxic of the heavy metals. Heavy metals such as IRON, COBALT and zinc are essential dietary trace ele-ments. Others can interfere with metabolism. The majority are toxic in relatively small doses. Heavy met-als tend to have the ability to interfere with one another's metabolic actions. LEAD, for instance, may interfere with the body's production of hemoglobin, thereby inhibiting the transfer of oxygen into the bloodstream. A high-protein diet generally inhibits the absorption of heavy metals into the body. Mine and mill TAILINGS frequently contain high concentrations of heavy metals derived from crushed rocks. Industrial wastes also frequently contain high levels of heavy metals. Most of the heavy metals in HOUSEHOLD HAZ-ARDOUS WASTE comes from discarded batteries. See also GROUNDWATER POLLUTION, WATER POLLUTION.

Heavy Water   water in which all the hydrogen atoms consist of the isotope DEUTERIUM ($_1D^2$). Heavy water (D_2O) boils at 101.4°C, freezes at 3.8°C, and is notice-ably more viscous than ordinary water. It is frequently used in scientific research. In natural water, only one water molecule in 6,500 is deuterium. Heavy water is used as a moderator and heat transfer fluid in NUCLEAR REACTORS, and is widely used in scientific research. About one in each 6,000 molecules of natu-rally occurring water is heavy water. See also FUSION, HYDROGEN, TRITIUM.

HEPA Filter   see AIR FILTERS

Heptachlor   a CHLORINATED HYDROCARBON INSECTI-CIDE. Heptachlor was used to control a variety of insect pests on food crops in the past, but is now banned for uses except termite control and the dipping of roots or tops of nonfood plants. It gets into air and water from toxic waste dumps, industrial discharges, agricultural RUNOFF and spills. Heptachlor is moderately soluble in water. It is highly persistent in water, with a HALF-LIFE of more than 200 days.

Heptachlor can affect health after it is inhaled or passes through the skin. ACUTE EFFECTS include a feeling of anxiety and irritability, a headache, dizziness and muscle twitching. Exposure to higher concentrations can cause seizures, unconsciousness and death. Repeated or extreme exposure may damage the liver or kidneys. Heptachlor is a possible human carcinogen, and has been shown to cause liver cancer in animals. There is limited evidence that heptachlor may be a TERATOGEN.

PUBLIC WATER SYSTEMS are required under the SAFE DRINKING WATER ACT to monitor concentrations of heptachlor (MAXIMUM CONTAMINANT LEVEL (MCL) = 0.0004 milligrams per liter), and for heptachlor epoxide (MCL = 0.0002 mg/L). The BEST AVAILABLE TECHNOLOGY for the removal of heptachlor and heptachlor epoxide from water is granular ACTIVATED CARBON. The OCCUPATIONAL SAFETY AND HEALTH ADMINISTRATION has established a permissible exposure limit for airborne heptachlor of 0.5 milligrams per cubic meter, averaged over an eight-hour workshift. It is also cited in AMERICAN CONFERENCE OF GOVERNMENT INDUSTRIAL HYGIENISTS, Department of Transportation and ENVIRONMENTAL PROTECTION AGENCY regulations. Heptachlor is on the Hazardous Substance List, the Special Health Hazard Substance List and the HAZARDOUS AIR POLLUTANTS LIST. Businesses handling significant quantities of heptachlor must disclose use and releases of the chemical under the provisions of the EMERGENCY PLANNING AND COMMUNITY RIGHT TO KNOW ACT.

Herbicide a substance used to kill unwanted vegetation. Herbicides are used to control plants that compete with commercial farm crops, to kill broad-leaf plants such as dandelions in lawns and to kill grass and weeds in gardens. Defoliants such as Agent Orange are used to strip vegetation of its leaves to expose enemy troop movement in wartime. Herbicides are frequently applied to remove deciduous trees and brush to promote the growth of coniferous trees in a clearcut. They are even used to aid in the harvest of crops, as when a chemical such as calcium cyanide is applied to cotton crops before a mechanical harvester goes into the field. The compound causes the cotton plant to loose its leaves, which would stain the cotton green if left in place.

There are several types of herbicides. *Defoliants* cause a plant to drop its leaves, usually by forcing the plant to produce so many growth hormones that they accumulate in leaf nodes, causing the nodal area to grow so quickly that veins that run out to the leaf are pinched off interrupting the flow of nutrients to the leaf. 2,4,5-T (2,4,5-TRICHLOROPHENOXYACETIC ACID) and 2,4-D (2,4-DICHLOROPHENOXYACETIC ACID) are both defoliants. *General herbicides* simply poison the plant. Since they are usually toxic to a wide range of plants and animals, extreme care must be exercised in handling and applying such herbicides. Sodium arsenate and sodium chlorate are widely used general herbicides. *Contact poisons* such as paraquat kill only those parts of a plant with which they come in contact. *Nematocides* kill nematodes, microscopic parasites found in soil that slow a plant's growth by attaching themselves to its roots. DIBROMCHLOROPROPANE is a nematocide. Alachlor is the mostly widely used herbicide in the United States. It is applied primarily to corn and soybean crops. See also PESTICIDE.

Heterotrophic Bacteria bacteria sometimes found in drinking water that can cause gastrointestinal infections. PUBLIC WATER SYSTEMS are required under the SAFE DRINKING WATER ACT to monitor concentrations of *heterotrophic bacteria*. The MAXIMUM CONTAMINANT LEVEL GOAL is a *heterotrophic plate count* of 0. Rather than a MAXIMUM CONTAMINANT LEVEL (MCL), a treatment technology is specified to reduce the heterotrophic plate count. See also GIARDIA, LEGIONELLA BACTERIA, MICROORGANISMS, TURBIDITY, VIRUS.

Hexachlorobenzene an industrial chemical used in the production of other chemicals, as a wood preservative and as a FUNGICIDE. Solid hexachlorobenzene consists of white needles. The compound is often used dissolved in a liquid SOLUTION. It gets into air and water primarily from agricultural RUNOFF and industrial emissions. Residues of hexachlorobenzene are often found in PESTICIDES that are made from it. It is slightly soluble and highly persistent in water, with a HALF-LIFE of more than 200 days. It is highly toxic to aquatic life.

Hexachlorobenzene can affect health when inhaled or passed through the skin. ACUTE EFFECTS include irritation of eyes, skin, nose, throat and lungs. Exposure to high concentrations or repeated exposure to lower levels may damage the liver, immune system, thyroid, kidneys and nervous system. Hexachlorobenzene is a possible CARCINOGEN in humans, and has been shown to cause liver and thyroid gland cancers in animals. Hexachlorobenzene may be a TERATOGEN.

PUBLIC WATER SYSTEMS are required under the SAFE DRINKING WATER ACT to monitor for hexachlorobenzene (MAXIMUM CONTAMINANT LEVEL = 0.001 milligrams per liter). No occupational exposure limits have been set for hexachlorobenzene. It is cited in Department of Transportation, National Toxicology

Program and ENVIRONMENTAL PROTECTION AGENCY regulations. It is on the Hazardous Substance List, the Special Health Hazard Substance List and the HAZARDOUS AIR POLLUTANTS LIST. Businesses handling significant quantities of hexachlorbenzene must disclose use and releases of the chemical under the provisions of the EMERGENCY PLANNING AND COMMUNITY RIGHT TO KNOW ACT.

Hexachlorocyclopentadiene (C-56)　a CHLORINATED HYDROCARBON that is the basis of the cyclodiene PESTICIDES (ALDRIN, dieldrin, CHLORDANE, HEPTACHLOR), and is also used in the production of flame-retardant materials. Hexachlorocyclopentadiene is a yellow to amber colored liquid with a pungent odor. It gets into air and water from industrial discharges, leaching from landfills and from agricultural RUNOFF. Hooker Chemical dumped an estimated 20,000 tons of C-56 at toxic waste dumps in Niagara County, New York, including the dump at LOVE CANAL. C-56 is moderately soluble and nonpersistent in water with a HALF-LIFE of less than two days.

Hexachlorocyclopentadiene can affect health when inhaled or passed through the skin. It is a highly corrosive chemical, and contact can burn the eyes and skin. ACUTE EFFECTS include irritation of eyes, nose, throat and lungs, which may cause tearing, sneezing or headaches. Exposure to high concentrations can irritate and burn the lungs, and can lead to a fatal buildup of fluid. Repeated exposure can damage the kidneys, liver and nervous system.

Hexachlorocyclopentadiene is on the RTK Hazardous Substance List, the Special Health Hazard Substance List and the HAZARDOUS AIR POLLUTANTS LIST. PUBLIC WATER SYSTEMS are required under the SAFE DRINKING WATER ACT to monitor for C-56 (MAXIMUM CONTAMINANT LEVEL = 0.05 milligrams per liter). Businesses handling significant quantities of hexachlorocyclopentadiene must disclose use and releases of the chemical under the provisions of the EMERGENCY PLANNING AND COMMUNITY RIGHT TO KNOW ACT.

High-Density Polyethylene (HDPE)　a polymer used to make PLASTIC milk jugs, yogurt containers, hard plastic toys, kitchenware and office supplies. Of the nearly 9 billion pounds of HDPE that was produced in the United States during 1992, only about 5% was recycled. Public demand for the RECYCLING of HDPE milk and juice jugs is high because they are a highly visible product that is discarded in most households. But recycling HDPE products is difficult because the price of the virgin resin (see VIRGIN MATERIAL) is generally lower than the price recycling centers must charge to break even for the plastics, and because it is technically difficult to guarantee that a given batch of used HDPE products is chemically pure enough to avoid damage to processing equipment. Plastic milk jugs constituted 0.6% of the municipal solid waste stream in 1988, according to the ENVIRONMENTAL PROTECTION AGENCY. See also PLASTICS RECYCLING.

High-Level Waste　the highly radioactive byproducts of nuclear decay (see NUCLEAR REACTOR, RADIOACTIVITY). The vast majority of high-level waste now being produced is irradiated URANIUM produced at commercial nuclear power plants. The U.N. International Atomic Energy Agency (IAEA) estimates that 190,000 tons of irradiated uranium will be in storage at nuclear power plants in operation around the world by the year 2000, and that the 413 commercial nuclear power plants in operation worldwide in 1990 produced about 9,500 tons of irradiated uranium that year. The IAEA estimates that nuclear plants now operating or under construction will have produced 450,000 tons of irradiated fuel by the time they are all closed by the middle of the twenty-first century. Another vast stockpile of high-level waste has been accumulated around the world as a byproduct of the production of nuclear weapons over the last 50 years. All of the high-level wastes produced were either dumped in lakes, streams or unlined pits or discharged into the atmosphere (see RADIOACTIVE WASTE, RADIOACTIVE WASTE DISPOSAL), or were stored "temporarily" in holding tanks, many of which have leaked, and await final deposition. See also HIGH-LEVEL WASTE DISPOSAL.

High-Level Waste Disposal　the permanent disposal of high-level RADIOACTIVE WASTE in such a way that it is no longer an appreciable environmental threat. The permanent, safe disposal of high-level nuclear waste is a problem that is confronting all of the world's nuclear nations. The fact that such waste must be confined for a minimum of 10,000 years—estimated as the amount of time it will take for wastes to decay to the point that they are no more dangerous than unmined URANIUM— is the primary difficulty. Many changes can come in 10,000 years: Volcanoes were erupting in central France 10,000 years ago, and the English Channel did not exist as recently as 7,000 years ago. And some experts contend that wastes would have to be stored even more tens of thousands of year before they are safe. The preferred option among countries looking for a way of disposing of high-level waste is to entomb the wastes in specially built repositories located deep underground in a geologically stable formation (see RADIOACTIVE WASTE DISPOSAL). But given that few civilizations survive for even a few thousand years, just being sure that the facility is clearly marked to warn away potential intruders over the millennia becomes a significant problem. We can have no conception of what language they might speak, or even whether written and spoken language as we now know it will exist 30,000 years from now. But plans for underground repositories carved out of granite, clay, volcanic tuff or salt are underway in many nations. In a typical installation, spent fuel rods (see NUCLEAR REACTOR) packed in stainless-steel/nickel containers would be placed by remote control in cylindrical holes bored into tunnel floors. Once the cylinders are in place, an impermeable material such as clay would be placed around the canisters to retard the movement of groundwater, and the hole would be sealed with a concrete plug. When full, the site would be sealed off from the surface and some form of permanent marking would be installed to warn away the future's curious. A field of massive concrete "thorns," an international symbol representing bodily harm, has been proposed. The canisters holding the waste are expected to disintegrate within 300 to 1,000 years, and the bore-hole lining will decay, exposing the waste. Groundwater penetration is deemed the most likely way for the waste to migrate from the storage site at that time.

The Nuclear Waste Policy Act of 1982 required the Department of Energy (DOE) to develop one high-level repository in the east and another in the west. An unrealistic timetable and the agency's tendency to consider sites that were clearly unacceptable hampered the siting process. The DOE, for instance, ignored findings by government scientists that a repository located at the HANFORD NUCLEAR RESERVATION in Washington would be prone to leakage, and designated the site as one of the western finalists. The attempt to locate the repositories fell apart in 1986, when the eastern states forced the cancellation of an eastern repository. Congress then ordered the DOE to concentrate on Yucca Mountain, a ridge on the west side of the federal nuclear testing grounds in southern Nevada, as the sole high-level repository. The Nevada legislature passed a bill in 1989 that prohibited the storage of high-level wastes within the state, although the law may violate the federal power of preemption on nuclear matters. Congress had the power to override the Nevada law if it acted within 90 days, but no action was taken. The DOE has threatened to sue the state if it is prevented from pursuing its studies of the site, and plans for the repository at Yucca Mountain are proceeding nevertheless. The DOE is spending $6 billion to assess the hydrology, geology and geochemistry of site. Yucca Mountain is composed primarily of volcanic tuff (compacted ash), and the area gets only a few inches of rain per year so has minimal surface RUNOFF

and resulting erosion. The planned repository would feature 117 miles of tunnels located 1,000 feet below surface (and 800 feet above the WATER TABLE) cut through 1,600 acres of unsaturated tuff. The tuff at Yucca Mountain contains zeolite—a mineral that absorbs certain RADIONUCLIDES. If the repository is built, it will not be operational until at least 2010. Critics of the project contend that the more than 30 seismic faults that crisscross the site could be the source of an earthquake that could raise the water table enough so that it would flood the site. Criticism also focuses on a volcano located 20 kilometers from the site. Original estimates were that it had not erupted for 270,000 years, but further investigation revealed that it last erupted 20,000 year ago, raising concerns that it might erupt again before the materials stored there could become inert.

The federal Waste Isolation Pilot Project near Carlsbad, New Mexico, was designed by the DOE to demonstrate the safe, permanent disposal of radioactive waste from the nation's weapons program. The WIPP is a geologic repository with storage vaults located 2,150 feet underground in ancient salt deposits. The DOE expects disposal activities to begin in 1998, and has selected Westinghouse Electric Corporation to manage the installation. In addition to the WIPP, the corporation manages DOE's Hanford Nuclear Reservation, the SAVANNAH RIVER reservation and the Bettis Atomic Power Laboratory, near Pittsburgh. Potential problems with the WIPP site include the fact that disposal areas are sandwiched between two bodies of water and a reservoir of brine is located underneath the storage area. An aquifer that feeds the Pecos River is above the site. About 60% of the radioactive waste to be disposed of at WIPP also contains hazardous chemicals including flammable solvents. The mixed waste gives off gases including explosive hydrogen, increasing the possibility that an explosion at the installation could inject a plume of radioactive slurry into the AQUIFER above and thereby seriously contaminate both the Pecos and the Rio Grande River into which it flows.

High Efficiency Particle Arresting (HEPA) Filters air filters that are designed to remove extremely small particles from the air. High-performance, particle-arresting filters, also called absolute filters, have a minimum of 99.97% efficiency, and collect particles as small as 0.03 microns. They are made of densely packed fibers that are pleated to increase their surface area. The filters, developed during World War II to remove radioactive dust from the exhaust of nuclear plants, are used in laboratory clean rooms and hospital operating rooms. They have a high resistance to air flow and are quite expensive. Vacuum cleaners with HEPA filters must be used in the cleanup of buildings contaminated lead and other fine toxic dust. See also INDOOR AIR POLLUTION, VOLATILE ORGANIC CHEMICALS.

Horizon Marker   an object found in archaeological digs that was in use over a relatively short period of time. The presence of a horizon marker helps an archaeologist to determine when the materials in a particular level of an archaeological dig were deposited. See also DEMOGRAPHIC MARKER, GARBOLOGY, MIDDEN, PULL-TAB TYPOLOGY.

Household Hazardous Waste   HAZARDOUS MATERIALS found in domestic garbage. The average American household generates about 15 pounds of hazardous waste per year. A large percentage of household hazardous waste is dumped at LANDFILLS along with MUNICIPAL SOLID WASTE, although most landfill rules and many local ordinances prohibit such dumping. Household hazardous waste constitutes about 1% of municipal solid waste. Hazardous materials in the waste stream can injure sanitation workers that handle solid waste, and can cause air and water pollution at a landfill. Hazardous materials can also raise havoc at a sewage plant, if they are dumped down storm sewers or household drains. Toxins in hazardous household products also contribute to INDOOR AIR POLLUTION. An increasing number of municipalities either have a permanent hazardous waste disposal facility or offer curbside pickup.

Principal sources of household hazardous waste include:

- paint products including oil-base and latex paints, varnishes, lacquers, wood preservatives, stains, thinners and strippers (see paint);
- AUTOMOTIVE PRODUCTS including motor oil, BATTERIES, antifreeze, FUELS and fuel additives, SOLVENTS and waxes;
- PESTICIDES including insecticides in plant-care products and mothballs;
- household cleaning products including oven cleaners, drain openers, disinfectants, solvents in cleansers and polishes for floors and furniture;
- building and demolition materials including glues, solvents, ASBESTOS and LEAD (see BUILDING MATERIALS, DEMOLITION WASTES);
- personal care products such as deodorant, nail polish (typically contains XYLENE and TOLUENE) and polish remover and hair spray;
- a variety of other hazardous materials including photographic chemicals, medicines, swimming-pool chemicals and DISPOSABLE DIAPERS. Even used

kitty litter contains parasites that can be dangerous to humans, and should not be used in the garden.

Although household products are not required to be labeled as "hazardous waste," law requires that labels contain "signal words" that warn the consumer of the product's potential toxicity. For instance, pesticides and other products that contain toxic chemicals must be identified as "poison" on their label, and other hazardous products must bear the warning "danger" if they are highly flammable, corrosive or toxic. "Warning" or "caution" must appear on the label of other products that contain hazardous materials.

A 1993 ruling in federal district court held that 14 southern California cities were responsible for toxic materials in commercial and household garbage that was hauled to the Operating Industries landfill in Monterey Park, now a 190-acre Superfund site (see COMPREHENSIVE ENVIRONMENTAL RESPONSE COMPENSATION AND LIABILITY ACT). American Communities for Cleanup Equity is a collaboration of about 100 cities that have banded together to lobby for a limit to municipal liability for pollution caused by municipal solid waste or sewage sludge.

The Garbage Project found that car-care products were the most common hazardous waste coming from low-income households. Middle-income homes produced more house-care products such as paints and strippers. Lawn-care products such as pesticides and fertilizers predominated in affluent neighborhoods. Garbage Project excavations have also found that drives to collect household hazardous waste can backfire. Garbage deposited for a month or two after a toxic-trash collection drive typically contains a higher concentration of hazardous waste, apparently because the toxic materials were saved up for the drive in many households. Local garbage companies and health departments can provide details on how and where to dispose of household and commercial hazardous wastes.

Human Waste   the byproducts of human METABOLISM. The CARBON DIOXIDE and water vapor that are created by metabolism are passed from the body through the lungs along with a variety of toxins (see INDOOR AIR POLLUTION). Heat, perspiration, hair and fingernails and toenails are other waste products that are constantly being generated by the body. The roughly three grams of dead skin that flakes off a typical human body each day is a primary source of microbial nutrition (see BACTERIA, DUST, MICROORGANISMS). Disposal of the feces and urine that are byproducts of the digestive process is the most serious aspect of human waste disposal because of the pathogenic substances involved and the increasing importance of water quality (see DISPOSABLE DIAPERS, SEWAGE, SEWAGE TREATMENT, WATER POLLUTION). Human waste products play a significant role in indoor air pollution. A human sneeze is a veritable fountain of microbial contaminants, and the lighter pathogens can remain airborne long enough to carry an infection from human to human. The roughly three pints of water vapor released via respiration and perspiration contributes to indoor humidity problems, which can, in turn, lead to microbial contamination and resulting damage to building materials and release of airborne contaminants. CREMATION is a method of reducing the ultimate piece of human waste—a dead body—to ashes and water vapor, while a GRAVEYARD relies on land disposal for the final deposition of human remains. Hospitals, research facilities and food processors produce INFECTIOUS WASTE that is sometime incinerated (see MEDICAL WASTE INCINERATOR), and is sometimes disposed of in a LANDFILL (see HAZARDOUS WASTE DISPOSAL).

Hydrocarbon a chemical COMPOUND that contains only HYDROGEN and carbon in its molecular structure. Hydrocarbons range from simple molecules such as METHANE (CH_4) to the mixtures of extremely complex molecules found in PETROLEUM. There are two classes of hydrocarbons. AROMATIC HYDROCARBONS contain one or more BENZENE RING. ALIPATHIC HYDROCARBONS have linear, branched or nonalternating cyclic molecules. See also DISTILLATION, POLYAROMATIC HYDROCARBON.

Hydrogen a colorless, odorless gas that is the simplest and lightest of the elements. Hydrogen is extremely light—one liter weighs less than one-tenth of a gram. Hydrogen makes up 90% of the known universe. It occurs sparsely in its free state in the terrestrial environment, but free hydrogen is common within stars. On earth, hydrogen is found as a component of countless compounds, with the largest proportion bonded with oxygen atoms to form water (H_2O). Scientists believe that all other elements are probably derived from hydrogen, primarily as the result of nuclear fusion within stars. Hydrogen burns explosively in the presence of oxygen, and liquid hydrogen is used as a rocket FUEL. Hydrogen's use as a fuel is limited primarily to stationary applications because of the heavy tanks and tight seals that must be used to contain it. Hydrogen is one of the easiest fuels to convert to electricity in a FUEL CELL. Pure hydrogen can be derived from the *electrolysis* of water, or by reaction with steam or by PYROLYSIS from NATURAL GAS and other HYDROCARBONS. There are three NUCLIDES of

hydrogen: Hydrogen-1 is called *protium*. It consists of a single proton with an electron orbiting it, and it makes up 99.985% of atmospheric hydrogen. Hydrogen-2 is called *deuterium*, which has a nucleus containing one proton and one neutron with one electron orbiting the nucleus. Deuterium makes up about 0.015% of atmospheric hydrogen. Hydrogen-3, the radioisotope *tritium* (see ISOTOPE, RADIOACTIVE MATERIAL, TRITIUM), occurs naturally only rarely, but is a product of radioactive decay that is commonly produced by nuclear FISSION. Tritium consists of one proton, two neutrons and one electron. Hydrogen is nontoxic, but it is classified as a hazardous substance because it is flammable.

Hydrogen Chloride a colorless gas with a strong odor, most commonly found in a SOLUTION called hydrochloric acid. Hydrogen chloride is used to make and clean metals; to make PHOSPHATE fertilizers, HYDROGEN and other chemicals, in the treatment of oil and gas wells, and in the removal of scale deposits from boilers and heat-exchange equipment. It gets into air and water from industrial discharges, laboratories, and spills. Hydrogen chloride is a corrosive chemical (see CORROSIVE MATERIAL), and contact can severely burn the skin. Eye contact can lead to permanent damage and loss of sight. Continued contact with dilute solutions can cause a skin rash. Long-term exposure may cause erosion of the teeth. Breathing hydrogen chloride vapor can irritate the mouth, nose, throat, and lungs and can cause bronchitis. Exposure to higher concentrations can cause a build up of fluid in the lungs. There is limited evidence that workers who are regularly exposed to hydrogen chloride have an increased rate of respiratory cancers.

The OCCUPATIONAL SAFETY AND HEALTH ADMINISTRATION has established a permissible exposure limit for airborne hydrogen chloride of 5 parts per million, not to be exceeded at any time. It is also cited AMERICAN CONFERENCE OF GOVERNMENT INDUSTRIAL HYGIENISTS, Department of Transportation and ENVIRONMENTAL PROTECTION AGENCY regulations. Hydrogen chloride is on the Hazardous Substance List and the Special Health Hazard Substance List.

Hydrogen Sulfide a colorless gas produced by the METABOLISM of ANAEROBIC BACTERIA. Hydrogen sulfide (H_2S) can be easily identified by its disagreeable rotten-egg odor, and is moderately water soluble. The gas is highly toxic, and many sewer workers have died from exposure to it. It forms hydrosulfuric acid when dissolved in water. Hydrogen sulfide is corrosive—eggs and mustard contain enough to tarnish silverware—and the concentrations found in sewage can corrode METAL components of the sewage system (see

SEWAGE COLLECTION SYSTEM, SEWAGE TREATMENT). It is used in industry as a reducing agent. Exposure to hydrogen sulfide gas causes nausea, headaches and dizziness at low concentrations, and unconsciousness and death can quickly result with high concentrations. It is especially dangerous because its rotten-egg odor can be detected only for a short while.

Hydrogen sulfide is produced in swamps and hot springs, and bathing in water from a hot spring in an unventilated room has caused numerous deaths. Traces of sulfur in FUEL can cause a CATALYTIC CONVERTER to emit hydrogen sulfide—the source of the rotten-egg smell in an area where lots of cars are idling. Anaerobic bacteria thrive under eutrophic conditions (see EUTROPHICATION), and the AMMONIA and hydrogen sulfide emitted by their metabolism can permeate the air in the area of a eutrophic lake or stream. Large amounts of the gas are produced in

petroleum refinement. Most of the H_2S found in sewage results from anaerobic action, although industrial effluent and the intrusion of groundwater, which sometimes contains the gas, can increase sulfide levels in sewage.

The concentration of inorganic sulfides such as hydrogen in wastewater is generally expressed as the amount of sulfur they contain. If it is reported, for instance, that wastewater contains 4 milligrams per liter of sulfur in the form of un-ionized H_2S, it is understood to mean that the water contains 4 mg/l of sulfur or about 4.25 mg/l of hydrogen sulfide. Hydrogen sulfide can be oxidized into sulfur dioxide in an afterburner (see COMBUSTION, INDUSTRIAL AIR POLLUTION TREATMENT). Sulfide levels in sewage can be controlled by avoiding unnecessary losses of dissolved oxygen, such as may occur if sewage is backed up behind a pumping station, and by injecting compressed air into

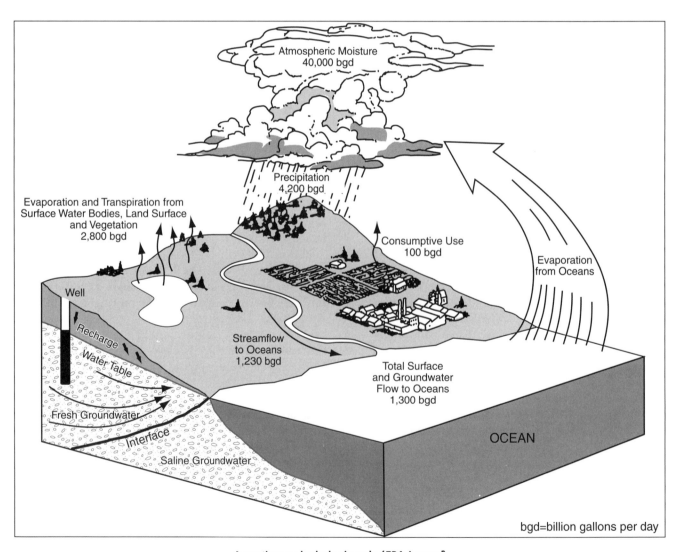

A continuous hydrologic cycle (*EPA Journal*)

wastewater to increase its oxygen content. Wastewater is also aerated by passing over a series of falls, which also increases dissolved oxygen (see AERATION). Gas ABSORPTION is used to remove hydrogen sulfide and *mercaptans* from natural gas. Mercaptans comprise a class of compounds analogous to ALCOHOLS and PHENOLS but with SULPHUR replacing oxygen in their molecular structure.

Hydrologic Cycle   the cycle through which WATER evaporates, primarily from the oceans, but also from bodies of water, land surfaces and plant leaves (see EVAPORATION)—moves through the atmosphere as water vapor, then recondenses into a liquid (see CONDENSATION) and falls to the earth's surface as precipitation—and ultimately returns to the ocean as RUNOFF to complete the cycle. When rainfall hits the ground, all or part may be absorbed, depending on the PERMEABILITY and POROSITY of the soil and the amount of water (or ice) near the surface. Part of the absorbed moisture will be absorbed by the soil, and the rest will pass down through the soil to the *zone of saturation* to become GROUNDWATER. Precipitation that is not absorbed into the ground becomes *surface runoff*, which flows first in trickles and rivulets and then in brooks and streams until it reaches the ocean (see surface water). *Transpiration* is the loss of water vapor from leaves, needles and other parts of plants and trees. An estimated one million gallons of water passes into the atmosphere from an acre of hardwood forest in a year. Approximately one-three thousandth of the water in the hydrosphere evaporates each year to form 1,500 cubic miles of precipitation. The evaporation and condensation of water moderate the climate, carrying heat from equatorial regions toward the polar regions. A large lake can moderate the temperature of an entire region while the oceans moderate the climate of the entire globe. Water molecules are practically indestructible. Most of the same water molecules have been circulating through the hydrologic cycle of evaporation, condensation and freezing since the dawn of time.

I

Ignitable Materials materials that can readily catch on fire. Ignitability is one of the characteristics of a HAZARDOUS WASTE. Liquid SOLVENTS and FUELS that can readily be ignited by a spark, and oily rags that can ignite by spontaneous combustion are examples of ignitable wastes. See also HAZARDOUS MATERIAL.

Impermeable Layer a geologic layer or an artificially constructed layer in a LANDFILL that will not allow the passage of water. Layers containing lots of clay and rock formations that contain no cracks (except for the more porous rocks such as coarse-grained sandstones) can both form an impermeable layer. An AQUIFER that is overlaid by an impermeable layer will be protected from pollution, but will also be prevented from being recharged by water precipitation that seeps into the ground (see OGALLALA AQUIFER). INJECTION WELLS are used to pump HAZARDOUS WASTE into deep aquifers protected by impermeable geologic structures in DEEP-WELL DISPOSAL. An impermeable layer made up of clay, plastic or a combination of the two is used to seal the bottoms of landfills and hazardous waste dumps (see HAZARDOUS WASTE DISPOSAL) to prevent the escape of LEACHATE.

Impoundment a barrier that is used to dam up water or other fluids, either permanently, as in the case of a dam, or temporarily, as with a dike used to prevent polluted water or toxic liquids from draining from the scene of an accidental spill. Impoundments can be used to control the movement of polluted groundwater. See also AQUIFER, GROUNDWATER DAM, GROUNDWATER POLLUTION, INTERCEPT WELL, LEAKING UNDERGROUND STORAGE TANKS.

In-place Pollutant a POLLUTANT that has become part of the bed of a lake or stream. Suspended contaminants that are heavier than water can become in-place pollutants simply by settling out of the water in an area of slow water velocity. Pollutants that are lighter than water can also settle out as sediments if they have the ability to *sorb* with other suspended particles (see ABSORPTION, ADSORPTION). Contaminants in in-place pollutants can make their way into the food chain when picked up by aquatic plants and burrowing organisms, with their effects being magnified in higher life forms, and can contaminate water if resuspended, which is especially likely in times of high flow or mechanical dredging (see SEDIMENTATION).

Incinerating Toilet a TOILET that incinerates human waste, usually using NATURAL GAS as a FUEL. Incinerating toilets are usually used in rural applications. See OUT-HOUSE.

Incineration the COMBUSTION of waste materials in an incinerator. MUNICIPAL SOLID WASTE is burned in an incinerator in about 140 American cities (see GARBAGE INCINERATOR, WASTE-TO-ENERGY PLANT), and hazardous and infectious waste is often neutralized in an incinerator (see HAZARDOUS WASTE INCINERATOR, INCINERATING TOILET, MEDICAL WASTE INCINERATOR). About 80% of the garbage's volume is eliminated in the incinerator, and the rest is ASH, which is usually dumped in a LANDFILL. A crematory is an incinerator designed to reduce human or animal remains to ashes.

The heat produced in an incinerator can be used to produce steam to drive electrical generators or other industrial purposes. The first incinerator to produce electricity from the heat produced by burning garbage was in Great Britain in the mid-1890s, and the practice was common in Europe by 1910. By the 1930s, more than 600 American cities were reducing the volume of municipal solid waste in incinerators, but the heat produced was not used to produce electricity in the United States until the 1960s.

Incinerators are a major POINT SOURCE of AIR POLLUTION, and the ash they produce can be toxic. The kind and quantity of toxins emitted by a incinerator are a function of the makeup of the waste material being burned, its moisture content, the combustion temperature and configuration of the combustion chamber and the emission control equipment used (see INDUSTRIAL AIR POLLUTION TREATMENT). The incomplete combustion of HYDROCARBONS in an incinerator can release ALDEHYDES and oxygenated organic compounds. Incinerators emissions can include POLYAROMATIC HYDROCARBONS including NAPTHALENE and BENZO-A-PYRENE (BAP), which is a product of incomplete combustion that is released especially when the temperature drops too low to consume FUEL completely. DIOXIN, FORMALDEHYDE, MERCURY and FURANS are also sometimes emitted by incinerators. The Clean Water Action Project estimates that the incinerator in Seamass, Massachusetts, emits more than 2,000 tons of toxic materials into the atmosphere annually including 590 tons of mercury.

The ENVIRONMENTAL PROTECTION AGENCY issued NEW SOURCE PERFORMANCE STANDARDS in late 1989 that required source separation and recycling in cities with waste incinerators as a way of reducing the volume and TOXICITY of garbage being burned in incinerators. Before incineration, solid waste is often broken up in a hammer mill and run under either a permanent magnet or an electromagnet that removes ferrous metals from the waste stream. In *catalytic combustion* (see CATALYTIC CONVERTER), temperatures that are as much as 500°F lower than what would be required for incineration can be used to break down wastes. Plastics' high energy content and high combustion temperature make them welcome at trash-to-energy incinerators, although dioxin, vinyl chlorides, formaldehyde and a variety of other toxins can be emitted when plastic products are burned. Explosions have been caused at incinerators that were being charged with too large a volume of plastics, however. PETROLEUM refining creates oily hazardous materials, which can be used as an alternative fuel for the high-temperature kilns that produce cement. Gaseous industrial emissions of solvents, liquid aerosols and hydrocarbons can be flamed off in an afterburner.

According to an Office of Technology Assessment study, medical waste incinerators emit up to 100 times more AIR POLLUTION per pound of waste material burned than a typical incinerator burning municipal solid waste. In 1994 the Environmental Protection Agency imposed penalties totaling $4.2 million on 10 firms for breaking hazardous waste burning regulations. The agency settled out of court with 22 other companies, which paid $3.3 million in civil penalties and carried out environmentally beneficial projects under the Resource Conservation and Recovery Act.

All types of incineration facilities are vigorously opposed in nearly every community where they are proposed (see NIMBY) on the grounds that the air pollution they cause and the transit of wastes through their community will be a threat to public health.

Indoor Air Pollution   the contamination of air inside buildings. The quality of indoor air is of special concern because humans spend much of their time inside buildings. In fact, the average American is indoors 90% of the time, and the very young, the elderly and people with health problems are inside even more than that. Concentrations of contaminants can get much higher indoors than outside because of the comparatively small amount of air inside a building. An ENVIRONMENTAL PROTECTION AGENCY (EPA) survey of eleven toxic air pollutants sampled in Los Angeles and New Jersey found concentrations of the toxins measured were two to five times higher indoors than outside. The EPA estimates that poor indoor air quality costs American businesses $60 billion per year, primarily as the result of absenteeism and lost productivity. According to a study by the National Institute of Safety and Health (NIOSH), 25 million office and factory workers are regularly exposed to polluted indoor air. The World Health Organization estimates that the air inside 30% of the world's buildings is bad enough to constitute a health hazard. The quality of indoor air is especially critical for those whose immune system has been taxed by diseases or chemical sensitivities.

IAP at Home

There are countless sources of the contaminants that commonly show up in indoor air. BUILDING MATERIALS are one major source. Most of the glued-wood products that are commonly used in floors, furniture and cabinets are held together with glues based on urea-formaldehyde resin, and as this resin cures, FORMALDEHYDE and other toxins can escape into indoor air through a process called OUTGASSING. The phenomenon is most pronounced when temperature and humidity are high and the glued materials are new. Building materials can also release toxic fibers including asbestos as they wear or when they are cut, drilled or sanded. ENVIRONMENTAL TOBACCO SMOKE is perhaps the most common and certainly the most talked about of the indoor contaminants. *Combustion sources* are another important contamination source. A wood stove or furnace that is improperly sealed can release toxins such as CARBON MONOXIDE and BENZO-A-PYRENE. Portable kerosene heaters are especially dangerous sources of carbon monoxide, and should always be used with adequate ventilation. Carbon monoxide can also find its way into the house from an automobile or mower engine running in an attached garage. Negative pressure can be established inside a tight house through the operation of exhaust blowers leading to a BACKDRAFT, which can carry toxic COMBUSTION BYPRODUCTS into the home. RADON gas that can seep into the house through cracks in the foundation is a problem in many areas. *Household cleansers*, especially those that contain CHLORINE, can react with other compounds to release chlorine gas, DIOXIN and CHLOROFORM. Many of the SOLVENTS found in all cleansers can also release toxic components into indoor air. *Synthetic materials*, including those used in carpets, drapes and the fabrics used in furnishings, can emit formaldehyde and other toxic VOLATILE ORGANIC COMPOUNDS and particles as they age. *Microbial agents* that may release toxins into indoor air including mold spores, dust-mite feces, BACTERIA, VIRUSes and POLLENS. *Hypersensitivity pneumonitis*, also called *humidifier fever*, is a flu-like disorder caused by toxins released by microbial growths in damp places, such as a humidifier or refrigerator condensate pan. *Polluted outdoor air* is yet another source of toxins indoors. *Remodeling and redecorating* can set LEAD and ASBESTOS loose into indoor air and stir up toxic dust that has collected in cracks and inaccessible corners for years. The solvents in PAINT contain a variety of toxic volatile chemicals that find their way into indoor air.

IAP Away from Home

Most of the same phenomenon that pollute indoor air at home also are operative in schools, public buildings and at work, and a variety of additional POLLUTION sources may also be present. The air in places where people congregate can be full airborne viruses and bacteria. The death of 29 members of the American Legion who had attended a conference at a Philadelphia hotel in 1976 gave their organization the dubious distinction of having a disease and a bacteria named after it—Legionnaires' disease and LEGIONELLA BACTERIA. This bacteria usually is found in water systems, and it seems to become more active after heavy rains and during unusually hot weather. Several outbreaks have occurred since that time, but none with the severity of the first. Other diseases that can be spread by airborne infectious agents include respiratory ailments such as the common cold, tuberculosis, smallpox, measles, chicken pox and staph infections.

A *building-related illness* is a disease whose origins can be traced to a specific source within a school, public building or workplace. Legionnaire's disease is one such affliction, and the upper-respiratory disease and skin irritation that can be caused by the outgassing of glues and solvents is another. NIOSH is promoting the concept of indoor environmental quality, saying multiple factors such as ergonomics, workplace stress, workplace lighting all play a role in building-related

illnesses. *Sick building syndrome* differs from building-related illness in that its source can not be identified. The symptoms of sick-building syndrome—which may include irritation of the eyes, nose, throat and mucous membranes, congestion, respiratory problems, lethargy, fatigue, irritability, dizziness and difficulty with concentration—usually do not fit the pattern of any one disease. The common thread connecting the numerous cases of sick building syndrome that have been reported are that it usually occurs in a tightly sealed building with inadequate or faulty ventilation, and that the symptoms usually clear up when the individual suffering the syndrome gets away from the building for a couple of weeks or more. In those cases where the reasons for sick building syndrome have eventually been at least partially traced, inadequate ventilation has almost always played a role. Microbial contamination, dirty ductwork and emissions from office equipment also frequently have been implicated. Some researchers believe that manmade fibers in ducts and building materials are a principal source of the syndrome.

In industry, the causes of indoor pollution are usually not so subtle. Industrial workers have been the subject of most of the epidemiological studies of the health effects of PESTICIDES and volatile organic chemicals simply because the concentrations of such toxins that workers must endure on a daily basis are many times the maximum seen in other settings. Much of what is known about strictly human reactions (as opposed to animal testing) to toxic chemicals comes from such studies. Because conditions have been so bad in industrial settings, almost no attention has been paid to the quality of air inside office buildings until fairly recently.

Cleaning Up Indoor Air

Inadequate VENTILATION is the basic cause of almost all indoor air quality problems, so improving the flow of outdoor air into indoor spaces to reduce contaminant concentrations is the most important step that can be taken to improve indoor air quality. Eliminating the sources of pollution and removing airborne particles with an AIR FILTER can also improve air quality. See also OCCUPATIONAL SAFETY AND HEALTH ADMINISTRATION. Answers to specific questions concerning indoor air quality are available (during business hours) from the EPA Indoor Air Quality Information Clearinghouse, phone 800-438-4318.

Indoor Radon Abatement Act   federal legislation passed in 1988 as an amendment to the TOXIC SUBSTANCES CONTROL ACT. The law authorized $45 million in expenditures over a three-year period for research and information dissemination regarding residential RADON contamination. A long-term goal was established in the legislation to reduce indoor radon levels to a level consistent with those found outdoors. Funding for state governments to assist in radon measurement and mitigation programs was also included.

Industrial Air Pollution AIR POLLUTION generated in the acquisition and processing of natural resources and their fabrication into finished products. The gaseous wastestream from such processes contains toxic gases, liquids and solids (see GASEOUS WASTE, LIQUID WASTE, SOLID WASTE) that pollute the atmosphere (see AIR POLLUTION), land (see DUSTFALL) and water (see AERIAL DEPOSITION). The amount of SULFUR DIOXIDE and other gaseous pollutants emitted by even a single power plant is staggering. The Navajo Generating Station near Page, Arizona, for example, produces 12 to 13 tons of sulfur dioxide *per hour*. Some of the contaminants in waste are removed (see INDUSTRIAL AIR POLLUTION TREATMENT), and the rest are discharged into the ATMOSPHERE where they are diluted (see DILUTION). Waste gases typically move up through an industrial stack at more than 30 feet per second (about 20 mph), and the hot gases may rise several hundred feet above the stack before beginning to level off. The warm, humid, polluted plume from a major industrial installation can reach miles into the atmosphere under certain weather conditions. Cold plumes and those made up of gases denser than air tend to sink toward the ground on exiting a stack, and then rise and disperse when sufficient mixing has occurred. An industrial smokestack (or a cluster of them) is a POINT SOURCE and a STATIONARY SOURCE of pollution, but more diffuse air pollution (see DIFFUSE POLLUTION) may also be generated in forms such as DUST from roads, dry process materials scattered by wind and fumes from EVAPORATION PONDS.

Industrial air pollution is regulated by the states under the terms of the CLEAN AIR ACT (CAA). The 1990 amendments to the CAA define a *major source* of air pollution as a stationary source that emits more than 10 tons of any one hazardous air pollutant annually, or more than 25 tons per year of a combination of toxins. An *area source* is defined as a group of pollution sources in the same geographical area that together emit more than the 10 tons of one contaminant or 25 tons of two or more contaminants per year that it takes to be classified as a major source (motor vehicles are not counted as an area source). Regulated industrial air pollutants are listed on the HAZARDOUS AIR POLLUTANTS LIST, which is maintained by the ENVIRONMENTAL PROTECTION AGENCY (EPA). The EPA'S administrator is authorized in the CAA to set a lower annual total pollutant limit, or to establish a different criteria for RADIONUCLIDES (see RADIOACTIVITY) on the basis of the material's potency (see LD$_{50}$), persistence (see PERSISTENT COMPOUND), potential for BIOACCUMULATION and other factors. The first national survey of the cancer risk associated with industrial emissions, which took place in 1990, found that 200 plants emissions from 1985 to 1990 posed a risk 1,000 times higher than federal standards.

Industrial Air Pollution Measurement the sampling and analysis of industrial air pollution. Testing methods range from the laboratory analysis of manual *grab samples*, to sophisticated electronic monitoring systems that feed data into an on-site or remote computer. With *manual sampling*, the waste gases to be tested are generally collected through FILTRATION or ABSORPTION, and samples are analyzed in a laboratory, which may be on site for a large installation. A variety of hand-held instruments can be employed to get an instantaneous reading of contaminant levels inside smokestacks and in pollution plumes, and pollution badges and DETECTOR TUBES are often used where people are working to monitor their exposure (see AIR POLLUTION MEASUREMENT). When a computer is interfaced with electronic monitoring instruments, averages, cumulative totals, peak values and other data of interest are available at any time, and the emissions reporting required under the CLEAN AIR ACT can be greatly simplified. Routine equipment checks and maintenance scheduling can also be a part of the system.

Stack sampling is performed to determine what contaminants are present in the stream of GASes and PARTICULATES emitted by a smokestack. Such measurements play an essential role in the enforcement of air pollution regulations. To gauge the efficiency of manufacturing processes and emission-control equipment, and to aid in the development of new pollution control equipment, samples are also taken at locations downwind from industrial smokestacks. A *grab sample* is a sample removed from an exhaust stream over a period of ten to sixty seconds, while a *continuous sample* is obtained over a longer test period. Determining exactly what is going up an industrial smokestack can be more difficult than it sounds. More than 300,000 cubic feet per minute of exhaust gases are typically moving up the stack, so only a minute quantity of total effluent can be thoroughly analyzed. To obtain an accurate representation of the makeup of the exhaust flow, the pattern of airflow inside the stack must be determined, and samples must be taken in several locations representative of typical airflow patterns to get a good average reading. The placement of an air-sampling nozzle and its orientation can distort its readings, showing a higher proportion of either gases and smaller particulates or larger particulates than is present in the exhaust. Sampling nozzles are aerodynamically shaped to minimize the disturbance they cause in the exhaust stream. To obtain an *isokinetic sample*, the speed of airflow across the tip of the sampling nozzle must be matched to the speed of the exhaust gases.

Several factors must be considered to obtain an accurate sample of the exhaust stream:

- Chemical reactions are often still occurring as exhaust gases go up a stack, and if this is happening where the sample is taken, they may continue inside the probe, tubing and sample-collection vessel, skewing results.
- If gases are allowed to condense inside the test apparatus, test results will similarly not reflect actual conditions, so the sample storage vessel must be kept warmer than the condensation point of the gases involved.
- If the materials of which the probe, tubing and sample collector are made are reactant with any of the contaminants in the exhaust stream (see REACTIVITY), chemical processes that will change the makeup of the sample will occur.

Measurement of Gases

A typical setup for measuring the concentration of gases going up a smokestack includes a filter to remove particulates, and a probe containing the nozzle through which the sample is taken, connected with tubing to a gaseous pollutant collector. A flow regulator controls a pump that keeps gases moving at the desired rate, and a meter measures either the rate at which gases are moved through the system or the total volume measured. Materials that intercept gases via absorption or ADSORPTION are commonly used to store samples for later sampling. An *orsat analyzer* can be used to measure the concentration of CARBON DIOXIDE, CARBON MONOXIDE and oxygen in exhaust gases.

Monitoring of air pollution plumes based on point samples becomes increasingly inaccurate and expensive as the area over which air pollution is being measured grows. Aerial or space photography in the ultraviolet, infrared, microwave and radio wavelengths of the ELECTROMAGNETIC SPECTRUM can be used to more-accurately trace broad patterns of air pollution. See also PARTICULATE MEASUREMENT.

Industrial Air-Pollution Treatment   the removal of contaminants from effluent being discharged into the atmosphere from industrial sources (see INDUSTRIAL AIR POLLUTION). FILTRATION (see AIR FILTER, ADSORP-

TION, ABSORPTION, INCINERATION) and electrostatic precipitation (see ELECTROSTATIC PRECIPITATOR) are some of the principal techniques used to purify waste gases.

Filtration is used to remove PARTICULATES from industrial air-streams, generally through some form of aerodynamic capture such as impaction, direct interception or diffusion (see BAGHOUSE, FLY ASH). The load of suspended particles in the wastestream being processed is often heavy, and an efficient method of removing the buildup of contaminants in the filter is essential. Early fabric filters were cleaned by shaking, but reverse airflow or pulse jet techniques are more frequently employed today.

Absorption into water or other liquid sorbent removes contaminants in a SCRUBBER, and absorption is also used to purify industrial products, as when it is used to remove HYDROGEN SULFIDE and mercaptans from NATURAL GAS. Sulfur dioxide can be removed by contact with alkaline water in a scrubber or by bubbling through an alkaline solution. Disposable solutions containing magnesium carbonate or calcium carbonate and regenerable media containing magnesium oxide are used for this purpose.

Adsorption is used in the drying of gases, the recovery of SOLVENTS and the purification of WASTEWATER and GASEOUS WASTE. Temperature and pressure are sometimes used to enhance the adsorption process for certain targeted substances. Mixtures of gases are separated commercially through adsorption, and it is used to remove CARBON MONOXIDE during PETROLEUM processing, to remove sulfur oxides from flue gases and to recover silicon tetrafluoride and hydrogen fluoride from fertilizer production stack gases.

Electrostatic precipitators employ high-voltage, direct current corona discharge to charge particles in flue gas for removal. Electricity passing from a high-voltage discharge electrode to a grounded electrode ionizes the gases between the electrodes. The ionized gas molecules bond to particulates, which then stick to collector plates with the opposite charge. In a *dry precipitator*, reentrainment of captured materials can be a problem, and periodic shutdown for cleaning is necessary, while in a *wet precipitator*, a film of liquid is continually passing over the collection surface, flushing away contaminants. Precipitators do not normally operate with the efficiency of which they are capable because they are sensitive to changes in gas volume and density, the concentration of dust, the distribution of particle sizes in the airstream, and the resistivity of contaminants. A precipitator typically removes 75 to 90% of the targeted contaminants.

Incineration initiates the rapid oxidation of liquid aerosols, HYDROCARBONS and other contaminants by injecting a FUEL, usually oil, into exhaust gases as they enter a flame *afterburner*. Toxic and explosive materials are removed by incineration, and odors and opacity are reduced. Adequate temperature, a turbulent air flow to thoroughly mix oxygen and sufficient time for OXIDATION to occur are the elements of efficient combustion in an afterburner. Hydrogen sulfide can be oxidized into sulfur dioxide in an afterburner. Coffee roasters, rendering plants and asphalt plants are among the industries that use incineration to reduce the emission of air contaminants. Solvents are often removed from flue gas in an incinerator. See also AIR POLLUTION TREATMENT.

Industrial Coatings paints and sealers used by industry to protect materials that are subject to deterioration as the result of rust, CORROSION and other destructive processes. Many traditional industrial coatings are no longer available because of requirements in the 1990 amendments to the CLEAN AIR ACT (CAA), which drastically restricted the type and amount of SOLVENTS coatings could contain. The release of VOLATILE ORGANIC COMPOUNDS (VOCs) from such solvents into the ATMOSPHERE is linked to OZONE pollution (see PAINT). The CAA requirements had an even greater effect on industrial coatings than it did on house paint. The vinyl and chlorinated rubber coatings traditionally used to seal bridges and industrial plants because of their resistance to corrosion can not be reformulated to meet the new standards, but new products are under development. One new process announced by Union Carbide in 1989 uses recycled CARBON DIOXIDE to replace up to 70% of the VOCs normally used in industrial spray painting. It is now being used in the automobile, appliance and furniture industries. The United States signed an agreement in Geneva in 1992 to limit VOC emissions from automobiles, glues, inks, paints and other products. See also CALIFORNIA CLEAN AIR ACT.

Industrial Hazardous Waste wastes generated by industry that contain substances that are reactive (see REACTIVE MATERIAL), corrosive (see CORROSIVE MATERIAL) or toxic (see TOXIC WASTE), or that pose a threat of fire or explosion. About 500 hazardous substances are regulated under the terms of the RESOURCE CONSERVATION AND RECOVERY ACT (RCRA) by the ENVIRONMENTAL PROTECTION AGENCY (EPA). The disposal of more than 200 million metric tons of hazardous waste is reported each year in the United States, primarily from large industrial facilities, and an unknown amount is disposed of illegally (see MIDNIGHT DUMPING). The largest contributor to the flow of hazardous waste in the United States is corrosive materials, spent acids

and alkaline materials from the METAL finishing, chemical and petroleum refining industries. Potentially hazardous waste materials generated by the metals industry include metal sludges, pigments, cyanides, acids, alkaline cleaners and solvents. Chlorinated organic compounds are hazardous byproducts produced in plastics manufacturing, and heavy metals, dyes and solvents are produced by the textile industry. Industrial hazardous wastes are generally treated to reduce volume and hazardous qualities, with nonrecoverable wastes disposed of via INCINERATION or dumped in a landfill (see HAZARDOUS WASTE DISPOSAL, HAZARDOUS WASTE TREATMENT).

Industrial Liquid Waste EFFLUENT that is discharged from an industrial facility as a liquid. The liquid wastestream from an industrial facility carries contaminants in the form of dissolved gases, liquids and solids (see SOLUTION) and suspended liquid and solid particles (see SUSPENSION). An estimated 7 billion gallons of industrial wastewater is processed each day in the United States. Industrial liquid waste is usually treated on site (see SEWAGE TREATMENT, WATER TREATMENT for methods) with effluent discharged into a surface stream under the terms of the CLEAN AIR ACT, but is sometimes dumped into a municipal sewer after PRETREATMENT to neutralize excess ACIDITY or ALKALINITY. As the cost of water and of polluting water supplies gets higher, the reuse of industrial process water and the reclamation of process chemicals from wastewater is increasing (see WASTEWATER RECYCLING). The heat and nutrients in liquid industrial effluent that contribute to biochemical oxygen demand can also fuel the increased growth of algae and other MICROORGANISMS in water that is associated with EUTROPHICATION. See also CHEMICAL OXYGEN DEMAND, INJECTION WELL, MINERAL WASTE, SOLVENTS.

Industrial Solid Waste industrial waste that is a solid material, as opposed to a GAS or a LIQUID (see AIR POLLUTION, GASEOUS WASTE, LIQUID WASTE). Waste materials suspended or dissolved in WATER (see SOLUBILITY, SUSPENDED SOLIDS, WATER POLLUTION) or carried in the air as a PARTICULATE are also technically defined as industrial solid waste in the RESOURCE CONSERVATION AND RECOVERY ACT. Chemicals, WOOD, PLASTICS, ORGANIC MATERIAL, PAPER, METAL, sediment (see SEDIMENTATION, SILTATION) and rocks (see MINE DRAINAGE, MINERAL WASTE, TAILINGS). See also INDUSTRIAL HAZARDOUS WASTE.

Inert Material a substance that will not react with other substances under normal temperatures and pressures. Compare REACTIVE MATERIAL.

Infectious Waste waste that contains BACTERIA, VIRUSes, protozoans, fungi or other materials capable of causing disease. Infectious waste is generated by hospitals, laboratories, research institutions and food processing facilities. Medical wastes have been disposed of in landfills in the past, and studies have shown that the viruses and bacteria found in infectious waste tend to die off in a landfill. The TOXIC SUBSTANCE CONTROL ACT requires, however, that medical wastes be treated as infectious waste, which must be incinerated (see INCINERATION). The quantity of infectious waste generated by a single hospital is surprisingly large. Johns Hopkins Hospital in Baltimore, for instance, produces 16 to 18 *tons* of infectious waste daily. The hospital's waste is hauled to a medical waste incinerator owned by Medical Waste Associates, which charges the hospital about $300 per ton to burn it.

The disease-producing organisms in infectious waste are not necessarily present when the material is discarded. A piece of cooked meat, for instance, can become a source of potentially lethal pathogens after just a few hours in a warm garbage can, which is an ideal environment for the multiplication of spores, bacteria and viruses in the environment. It has been estimated that as many as 90% of the flies found in urban areas breed in the organic material found in (and around) garbage cans, and flies are known to carry a multitude of pathogens. Rats, another vector of disease organisms, similarly thrive in the presence of garbage.

Infiltration the passage of a fluid into a porous substance or container through small cracks or holes. The infiltration of air into a building through cracks and holes in the building envelope, and the infiltration of surface water into an AQUIFER through overlying soils and cracks in bedrock are common uses of the term. GROUNDWATER infiltrating into sewer lines poses problems for sewage plants, whose capacity must allow for the extra volume. Compare EXFILTRATION. See also GROUNDWATER POLLUTION, INDOOR AIR POLLUTION, LEAKING UNDERGROUND STORAGE TANKS, VENTILATION.

Infiltration Capacity the maximum rate at which a soil can absorb rainfall. Bare soils generally have a lower INFILTRATION capacity because rain falling directly on the soil's surface tend to flatten it out and pulverize it, making it more difficult for rainwater to be absorbed. Soils that are dry generally have a higher infiltration capacity than wet soils because dry soil pores are still capable of taking up water. Clay soils generally have a high infiltration capacity when dry because they tend to crack, opening up avenues for

water to seep in. Clay soils, however, have almost no infiltration capacity once wet because the cracks swell shut. Infiltration capacity determines how much precipitation will soak into soils and underlying aquifers, how quickly RUNOFF and EROSION will occur and how easily surface contaminants can make their way into GROUNDWATER. See also GROUNDWATER POLLUTION, PERCOLATION, PERMEABILITY, POROSITY.

Injection Well a well through which HAZARDOUS WASTE is injected into deep AQUIFERS for disposal. The cap of an injection well is fitted with sensors that will record any sudden drop of pressure inside the well, since such a pressure drop may indicate water is leaking out of the disposal area. The casing on an injection well must be tightly sealed down to the impermeable zone over the aquifer into which wastes are being pumped to prevent leaks into overlaying fresh water aquifers (see DEEPWELL DISPOSAL). The well casing and attached fittings must be highly corrosion resistant to avoid deterioration from the often caustic wastes. See also PETROLEUM.

Inorganic Compound any COMPOUND that does not contain carbon. Simple carbon compounds such as CARBON MONOXIDE, CARBON DIOXIDE and CYANIDE are also generally termed inorganic. The number of ORGANIC COMPOUNDS is about 10 times the number of inorganic compounds. The ATMOSPHERE and the lithosphere (the solid portion of the earth) are made up almost entirely of inorganic compounds.

Insecticide a substance used to kill insects. The TOXICITY of many insecticides to life forms other than the insects they were designed to kill has led to partial or complete bans on their use in the United States and other countries. Unfortunately, chemical suppliers have generally resorted to increasing exports of these products when confronted with a domestic ban. One of the first PESTICIDES to be banned for use in this country was the CHLORINATED HYDROCARBON, DDT. Although it was banned in 1972, it is still found in water and in animal tissue. The comeback of the bald eagle—which was having a hard time reproducing because the chemical caused its eggshells to be too thin—is credited largely to the decrease in environmental concentrations of DDT resulting from the ban. LINDANE, HEPTACHLOR and CHLORDANE are other insecticides that were once widely used and have been partially or completely banned. METHYLENE CHLORIDE, NAPTHALENE, XYLENE, PARATHION, METHOXYCHLOR, TOXAPHENE and ALDRIN are all used as insecticides. Because the life-cycle of insects is extremely short, it does not take long for enough generations to develop and to pass on an immunity to the insecticides that are used to control them. Because of such immunities, the increasing cost of large-scale agricultural insecticide use and concern about WATER POLLUTION (see GROUNDWATER POLLUTION), INTEGRATED PEST MANAGEMENT is seeing increased use on the nation's farms.

Institute of Scrap Recycling Industries an association of RECYCLING processors, brokers and industrial consumers of recycled products that was formed in 1987 through the merger of the Institute of Scrap Iron and Steel and the National Association of Recycling Industries. The association's goal is "to discourage regulatory hindrance and create expanded business opportunities" for its members. The typical ISRI member is a small, often third- or fourth-generation family business, although larger-scale concerns are also members. Family scrap yards have been established in the recycling business long before recycling became the green thing to do, and the modern recycling movement has more often than not hindered rather than helped their business by upsetting markets and establishing mandatory recycling programs. An astounding volume of materials are recycled by ISRI members. In 1990, ISRI members recycled 9 million automobiles, 60 million tons of ferrous metals (see IRON, STEEL), 7 million tons of nonferrous metals and 30 million tons of paper, glass and plastic—a total of 100 million tons of material. Members are key players in the international trade of metals (see METAL RECYCLING). Address: 1325 G Street, NW, Suite 1000, Washington, DC 20005. Phone: 202-737-1770. See also JUNKYARD, NATIONAL RECYCLING COALITION.

Integrated Pest Management the control of agricultural pests through a mix of biological, mechanical and chemical means. Natural processes are the first line of defense against unwanted plants, animals, insects and fungal growths on a farm relying on integrated pest management (IPM). Insect populations, for instance, are kept under control by introducing natural predators such as ladybugs or birds. An invasion of weeds might be managed by introducing parasites or pathogens that will kill or weaken the species in question. Chemicals may be used on such a farm, but only as a last resort and in the smallest quantity and lowest overall TOXICITY possible. No attempt is made to eradicate a given pest; rather, populations are managed to keep overall crop damage below a given threshold (often 5%). The philosophy behind integrated pest management is that by maintaining a healthy ecosystem, total losses to pests can be minimized without resorting to expensive and disruptive PESTICIDE use and heavy-handed techniques. The

United States Department of Agriculture funds research and collaborative programs relating to integrated pest management. For information, write: USDA Cooperative State Research Service, Plant and Animal Sciences, Aerospace Building, Washington, DC 20250. The USDA also sponsors clinics, workshops, conferences and tours relating to IPM. For more information, contact: USDA Extension Service, Ag Box 0909, Washington, DC 20250. Phone: 202-720-4395. See also BIOLOGICAL TREATMENT, INSECTICIDES, MICROORGANISMS, SUSTAINABLE AGRICULTURE.

Integrated Waste Management a waste management program that typically includes SOURCE REDUCTION, RECYCLING, COMPOSTING, INCINERATION and a LANDFILL.

Intercept Well a well designed to intercept a pollution PLUME so that contaminated water can be pumped out of the aquifer for treatment. Intercept wells are generally drilled in a group that curves around the downslope edge of a plume. They are also used to intercept salt water that is intruding into freshwater AQUIFERS. See also GROUNDWATER POLLUTION.

Interflow RUNOFF that flows within the upper layers of the soil rather than on top of the land's surface. Some interflow remains in the ground for a time as SOIL WATER, but most either reemerges on the surface or flows into a stream. Interflow occurs during periods of heavy runoff anywhere soil is deep enough and sufficiently permeable.

Intermittent Sand Filter a small-scale sewage treatment system used to upgrade EFFLUENT from a septic tank (see SEPTIC SYSTEM) installed in an area with inadequate room for a *drain field* or a high likelihood of pollution if a conventional septic system were installed. AEROBIC BACTERIA become established among the grains of land, and contribute to the filter's ability to remove organic materials from WASTEWATER. An intermittent sand filter uses pumps to periodically dump effluent into the sand filter in short bursts. The filter rests between bursts, allowing it to become oxygenated. A sand filter costs anywhere from $5,000 to more than $20,000 to install, depending on the application.

Internal Loading the contamination of a lake or stream from sediments on its bed. Internal loading can cause severe pollution problems long after the source of pollution that contaminated the sediments is gone. Heavy rainstorms regularly stir up sediments from stream beds polluted with HEAVY METALS that originated in mining operations, causing fish kills and degradation of aquatic ecosystems. The mining and milling that caused the pollution may have ceased more than a century earlier (see CLARK FORK RIVER).

Internal-Combustion Engine   an engine that converts chemical energy by burning a FUEL in a combustion chamber that is an integral part of the engine. The internal-combustion engine and the mobility it implies has changed the form of modern cities, and is the leading source of AIR POLLUTION in most urban areas. Similar sweeping changes in agriculture and transportation followed the introduction and development of the internal-combustion engine. There are four basic type of internal combustion engines: the Otto-cycle engine, the diesel engine, the rotary engine and the gas turbine.

The *Otto-cycle engine*, named after its inventor, Nikolaus August Otto, is the ubiquitous gasoline engine used in automobiles and airplanes. The combustion chamber in an Otto-cycle engine consists of a cylinder that is closed at one end and that is fitted with a movable piston connected to a crankshaft by a connecting rod. The crankshaft translates the reciprocating motion of the piston in the cylinder into rotary motion. Otto-cycle engines have been built with anywhere from one to 28 cylinders. The engine is driven by liquid fuel, which is sprayed into the cylinder through an intake valve, compressed by the upward motion of the cylinder and ignited with an electrical spark, causing an explosion that drives the cylinder down, turning the crankshaft in the process. Hot exhaust gases leave the cylinder through an exhaust valve, are pushed through an exhaust header and exit through the vehicle's tailpipe. The efficiency of such engines (their ability to convert the thermal energy in fuel into mechanical energy) is limited to about 25% by the energy lost in the form of heat, friction and other factors.

The *diesel engine*, which was also named after its inventor, Rudolf Christian Karl Diesel, differs from the Otto-cycle engine in that air rather than fuel is first drawn into the cylinder. It is compressed by the upward stroke of the cylinder, and heated in the process to approximately 800°F. Vaporized fuel is sprayed into the cylinder through the intake valve at the end of the compression stroke where it instantaneously ignites because of the superheated air already in the cylinder. Some diesel engines have an auxiliary electrical ignition system that ignites the fuel when the engine is first started until it has had a chance to warm up. The combustion of the fuel drives the cylinder back down and turns the crankshaft. Exhaust gases are pushed out through the tailpipe and the process starts over. Diesel engines generally run more slowly than Otto-cycle engines operating in the range of 100 to 750 revolutions per minute (RPM) as opposed to 2500 to 5000 RPM. The diesel engine's efficiency is a little more than 40%, and diesels have the advantage of being able to run on inexpensive fuel oils.

In a *rotary engine*, the piston is replaced by a three-cornered rotor that turns inside a combustion chamber that is roughly oval. An air-fuel mixture sprayed into the combustion chamber from an intake port is trapped between one of the faces of the rotor and the cylinder wall, the mixture is compressed as the rotor turns and then ignited by a spark, and exhaust gases are pushed out an exhaust valve. The process occurs alternately at each face of the rotor so that there are three power strokes for each revolution of the rotor. The rotary engine can be run on low-grade fuel, and is nearly vibration free.

In a *gas turbine*, compressed air and fuel in gaseous, fine-powder or liquid-spray form is injected into a combustion chamber and ignited, and the exhaust gas created in the resulting explosion is used to spin a turbine. The spinning turbine drives the compressor that supplies compressed air to the combustion chamber. A simple gas turbine has an efficiency of 10% to 15%, but this can be boosted to 20% to 30% through addition of a variety of auxiliary equipment. Gas turbines have been used to power locomotives and ships, and a modified version, the *turbo jet* (see JET ENGINE), is used in airplanes.

International Joint Commission   a six-member commission formed under the provisions of the Boundary Waters Treaty of 1909 between the United States and Canada. The Commission's principal responsibility is the restoration and maintenance of water quality in the Great Lakes. To achieve this end, the commission works with Canadian and American federal and local agencies to limit the flow of pollutants into the Great Lakes, and approves applications for dams, canals and water diversions. The commission also investigates water pollution and water use problems that span the two countries' border in areas other than the Great Lakes. The commission is currently studying air pollution as a source of pollution of boundary waters.

Ion   an atom or molecule that has either lost or gained one or more electrons acquiring a positive or negative charge in the process. Metallic elements tend to lose electrons to form positive ions called *cations* (see OXIDATION, OXIDIZING AGENT), while nonmetallic elements tend to gain electrons to form negative ions called *anions*. When metals and nonmetals react chemically to form *ionic compounds*, the metals tend to lose electrons to the nonmetals, with the compound being bound together by the opposite charges of the negative and positive ions created in the process. When an ionic compound is introduced into a POLAR LIQUID such as water, the negative and positive ions dissociate from one another, or ionize. Each ion becomes surrounded by molecules of the liquid, with the pole of each liquid molecule matched to an ion with an opposite charge. The resulting liquid is a conductor of electricity, and is called an *electrolyte*. The properties of an ion are often totally different from the properties of the element from which it is created. Sodium and chlorine, for instance, are both extremely poisonous, but the sodium and chlorine ions found in table salt are essential for human health.

Ions are created in the laboratory through *corona discharge*, a release of electricity similar to lightning, and by passing an electrical current through a radioactive isotope of HYDROGEN. Ions are collected for measurement on a device containing charged aluminum plates. Ions are created in the natural world by the action of sunlight, wind, radioactive materials, fast-moving water and lightning (see ION EXCHANGE, NEGATIVE ION GENERATOR, RADIOACTIVITY, X-RAYS). Compare ISOTOPE.

Ion Exchange

a method of removing contaminants from water by initiating a chemical reaction that exchanges the pollutant's IONS for less harmful and/or more controllable ions of the same charge. It is widely used to remove minerals and RADIONUCLIDES from water. The most cost-effective use of ion exchange is for polishing water in which TOTAL DISSOLVED SOLIDS have been reduced to less than 500 parts per million through filtration or other treatment techniques (see WATER TREATMENT). The process can produce finished water with virtually no dissolved solids. An ion exchange system produces a toxic brine that often must be handled as HAZARDOUS WASTE.

Water softeners rely on a simple ion-exchange process to replace the calcium and magnesium ions that make water hard with sodium ions. This is usually accomplished by bubbling the water through an artificial *zeolite* (*sodium illuminosilicate*), causing the sodium in the zeolite to replace the undesirable ions in the water. The ions that accumulate in the zeolite are periodically washed out of the ion-exchange matrix by flushing it with a salt solution (see SALT, SOLUTION), thereby restoring the sodium. Survival kits sometimes include a purification kit that relies on a complex series of ion-exchange reactions to render seawater potable, and ion exchange is used to isolate certain rare-earth elements (such as transuranium and halfnium) and to remove radioactive materials from the water in holding tanks for nuclear fuel rods (see NUCLEAR REACTOR)

to separate URANIUM and PLUTONIUM from FISSION products. Filters and conditioners for automobile radiators and other cooling systems rely on synthetic ion-exchange resins to clean water. Such resins are also used in antacids and intestinal absorbents; in the treatment of peptic ulcers, migraine headache, heart disease and edema; in the preparation of blood plasma; and for diarrhea control. See also DESALINATION, WATER TREATMENT.

Ionizing Radiation

radiation capable of knocking one or more electrons out of their orbits around an atom, thereby creating an ION. See RADIOACTIVITY.

Irrigation Water

surface water or groundwater that is diverted and used to water crops. Crops that could not be reliably produced with locally available precipitation can be introduced through the use of irrigation, and the yield of other crops can be increased. The supply of irrigation water is usually a cooperative venture, and the nucleus of many early local governments was an association of irrigators. Floodwater catchments and irrigation ditches are a prominent feature at prehistoric ruins around the world. Irrigation is by far the largest user of water in the United States, siphoning off approximately nine-tenths of total U.S. water consumption. The roughly 50 million acres of irrigated farmland in the nation represents about 10% of total farmland, but this land produces 25% of the cash value of U.S. crops. Most of the nation's large-scale irrigation projects would not have been possible without the expenditure of vast amounts of federal money on "reclamation" projects. Irrigators still pay rates as low as $1 per acre foot for water from federal irrigation projects, while cities today may pay as much as $10,000 an acre foot for water delivered from distant sources. (An *acre foot* is the amount of water required to cover one acre one foot deep.)

World Water Use

As world population grows, demands on natural resources including water dramatically increase (see POPULATION GROWTH). Drought, DESERTIFICATION and the drawdown of lakes, streams and aquifers in one part of the world affect worldwide patterns of agricultural production and resultant water use. A shortage of water and growing population in developing nations such as India and China led to the quick disappearance of surplus farm products in the United States and other countries in the mid-1990s. Future water shortages, the decline of world fisheries and the shrinkage of the amount of productive farmland available promise to compound future problems with food production and distribution.

Salinity

When irrigation water is spread out on a field's surface, much of it evaporates, increasing the concentration of natural SALTS in the water that soaks into the soil (see EVAPORATION, RUNOFF, SALINITY). Salts tend to build up in the soil with repeated applications of irrigation water, and can build to the point that soils lose their productivity or even their ability to grow native plants. The return flow from irrigated fields can be heavily laden with salts that add to the salinity of the watercourse.

Conservation of Irrigation Water

The *consumptive use* of irrigation water is about 95% (as compared to about 25% for domestic water use). That is, only about five gallons of return water makes its way back into streams or aquifers for every 100 gallons withdrawn. Most of the water that is consumed evaporates, either from the surface of irrigation ditches and wet soil, or through transpiration from the leaves of plants. Leaky ditches and inefficient irrigation techniques compound the problem. Only about 25% of the water withdrawn for irrigation benefits plants. The storage of irrigation water in reservoirs can also lead to large losses via evaporation. It has been estimated that one cubic mile of water evaporates from Lake Mead on the Colorado River each year. In *flood irrigation*, water spills out into fields from irrigation ditches. *Sprinkler irrigation*, which may consist of a series of sprinkler heads that shoot jets of water out in a circular pattern or of large, perforated plastic pipes, uses considerably less water to irrigate a given area. *Drip irrigation*, in which a measured drip of water is applied directly to a plant's roots under the surface of the soil, is even more efficient because water losses due to evaporation are largely avoided. Improvements such as the lining of ditches and conversion to more efficient irrigation methods, although expensive, can free up water for use in a PUBLIC WATER SYSTEM without a reduction in irrigated acreage or the retirement of irrigated acreage (see WATER FARMING). Switching from crops such as rice, which requires lots of water, to crops such as corn and wheat that can get by with much less, can help save water, as can the development of plants that are more tolerant to drought and to salty irrigation water.

Irritant a toxic material that damages the mucous membrane causing gastrointestinal irritation or inflammation often accompanied by stomach pain and vomiting. ARSENIC, MERCURY, iodine and laxatives are commonly occurring irritants. Irritants include cumulative poisons that can be taken up by the body gradually with no apparent ill effects, and then can cause systemic damage when a threshold level is reached

(see BIOACCUMULATION, PERSISTENT COMPOUND). Compare INERT MATERIAL.

Iron a soft, malleable METAL that is the fourth most abundant element in the crust of the earth. Iron (chemical symbol Fe) is highly reactive, so rarely appears in pure form. Western Greenland is one of the few places in which deposits of relatively pure iron have been found. Iron has a bright, silvery-white surface, which quickly becomes covered with dark iron oxides when it is exposed to air and water vapor. Iron is easily magnetized at ordinary temperatures. Iron beads that were made as early as 4000 B.C. have been found in Egypt, but the Iron Age—the archaeological era when iron was used extensively—did not occur until much later. Iron, usually alloyed with nickel, is found in meteors. Iron is an essential trace element for humans and also an irreplaceable part of metabolism in animals and many other lifeforms. The metal is an essential building block of the hemoglobin that carries oxygen in the bloodstream. Iron is found in water, plants and blood, and is used in the treatment of anemia. Meat, molasses and spinach are good dietary sources of iron.

The production of iron and iron alloys is fourteen times that of all other metals combined. About 53 million metric tons of pig iron and almost 98 million tons of steel were produced by U.S. industry in 1993. The mining and milling of iron is a major source of air pollution. A principal byproduct of iron and steel production is BOILER SLAG—the impurities that float to the top and are removed during smelting. Boiler slag can contain a variety of toxins, depending on the ore used and the nature of the process that created the slag. Much of it must be disposed of as hazardous waste. Bacteria that metabolize iron and SULFUR in the presence of oxygen break down the PYRITE found in the tailings from mines, and contribute to acid runoff from mines and TAILINGS piles (see ACIDIC WASTE).

Products

Pure iron is used in the production of galvanized sheet metal and electromagnets, but alloys of one sort or another are used for most other purposes. Most commercially used iron contains a carefully controlled level of carbon and other impurities, which give it the qualities necessary for the application. Pure iron is very brittle, and impurities increase its strength. Pig iron produced in a blast furnace and the cast iron products made from pig iron generally are made up of about 92% iron, from 2% to 4% carbon, varying amounts of manganese, phosphorous, silicone and a trace of sulfur. Open-hearth iron and wrought iron contain only a few hundredths of a percent carbon, and steel contains anywhere from about 0.04% to

2.25%. Cast iron is used in the production of pipe, engine blocks, manhole covers, machine parts and countless other products. Cast iron, ductile iron and welded steel pipe is used in sewers in places such as stream crossings, where strength and tight-fitting joints are critical. Tannins react with iron to form iron-tannate, which is commonly used to make blue-black ink. Iron sulfate is used in water purification and in the production of ink.

Ores

Hematite—the mineral form of *iron oxide* (Fe_2O_3), and also known as *rust*—is the principal ore from which iron is extracted. The reddish-brown rust that forms on the surface of iron exposed to water and oxygen, and much of the reds and browns in rocks, soils and mine tailings is the result of staining by hematite. The iron ores *limonite* and *taconite* are also iron-oxide compounds. Pyrite, which is also called *iron sulfide* (FeS_2), is not processed for iron because it is difficult to remove the sulfur. Many iron ores are thought to have been deposited by iron-fixing bacteria, whose metabolism causes the precipitation of iron from seawater. The iron then became concentrated in seabed sediments that later ended up on dry land. Most U.S. iron ore is mined in Minnesota, Wisconsin and Michigan. Iron forms *ferrous compounds*, in which it has a valence of two, and *ferric compounds* in which it has a valence of three. Ferrous compounds oxidize into ferric compounds. Iron and ferrous metals are removed from the wastestream flowing into a waste-to-energy facility with a magnet. The formation of rust is accelerated by the presence of salts and lots of moisture—just the conditions found on in the slushy mixture of ice and salt that covers roads in many areas in the winter. Cars that have spent much time around such areas can be identified by the distinctive brown trim of rust on fenders and rusted-through spots. Rusting iron is a common example of chemical CORROSION—a reaction of oxygen and iron in the presence of water that yields hydrated iron oxides.

Mining/Smelting

Iron has been refined from ores since at least 3000 B.C. Iron ore is crushed, and then mixed with limestone and heated by burning COKE in a blast furnace to more than 1,536°C (2,797°F), the melting point of iron. A blast furnace is a tall, cylindrical retort with a hole in the bottom out of which molten iron can be tapped, and a higher outlet that can be opened to drain off slag. Gases are vented out the top of the furnace, which operates continuously with new loads being introduced every 10 to 15 minutes. The air that provides the blast furnace's blast is preheated to 1,000° to 1,600°F. The slag that floats to the top of the molten iron is drawn off, and relatively pure iron is left behind. The CARBON MONOXIDE created by the combustion of the coke provides carbon to the molten mixture inside the converter, which strengthens the resulting iron by facilitating the production of nickel and iron carbonyls. Carbon monoxide emissions from the conversion process can sometimes be a problem, but the gas is usually captured as a fuel for furnaces used in the steel-making process. The limestone in the retort acts as a flux, and combines with impurities in the iron ore so that they will not recombine with the iron when it cools. After the purified pig iron is drawn from the furnace, it can undergo further processing to become steel, which is an alloy of iron and carbon that can be poured into a cast to make cast iron products.

Isomer a compound that has the same chemical formula as another compound, but which varies from the other compound in structure. A compound and its isomer contain the same elements in the same proportions, but the elements are connected to each other in different ways. The two compounds may have completely different chemical characteristics. For instance, acetic acid, which gives vinegar its bite, is a strongly caustic acid with a pungent odor. Methyl formate, acetic acid's isomer, is pleasant-smelling, less caustic and more flammable. Most complex compounds have at least one isomer. The large molecules characteristic of ORGANIC COMPOUNDS may have several.

Isotope a modified form of an element that has a different number of neutrons in the nucleus of its atom. A single element may have several isotopes. Uranium, for example, occurs naturally in three forms: U^{234} with 142 neutrons, U^{235} with 143 neutrons and U^{238} with 148 neutrons. All three isotopes have 92 protons. A radioactive isotope (or radioisotope) emits charged particles or energy waves (see ALPHA PARTICLES, BETA PARTICLES, GAMMA RAYS). Compare ION.

J

Jackson Turbidity Unit a unit used in the measurement of the TURBIDITY of a water sample. The Jackson Turbidity Unit is named after hydrologist and chemical engineer Daniel Dana Jackson, who developed it around 1900 by measuring how much water had to be poured into a tube with a flat transparent bottom before the outline of the flame of what came to be called the *Jackson candle* that was burning under the vessel could no longer be made out. This depth was recorded on a scale based on water in which one milligram of diatomaceous earth per liter had been suspended. One centimeter depth of the standard solution in the standard tube is a JTU. Hydrologists today use a spectrophotometer, calibrated in JTUs, to measure turbidity.

Jet Engine an *atmospheric engine* that provides thrust for forward motion through the rearward expulsion of a high-velocity gaseous stream. Jet engines compress a large volume of air to the point that it ignites fuel in a combustion chamber, and expands the resulting hot combustion gases to low pressures through a nozzle in order to in order to get a high jet exit velocity. *Turbojets* are the most widely used atmospheric engine. In a turbojet, air is drawn into the engine, pressurized by a compressor and then routed into the combustion chamber. A turbine between the combustion chamber and the nozzle drives the axial-flow compressor that is used to pressurize incoming air. Turbojet engines would be more efficient if combustion temperatures were allowed to be the maximum obtainable from the complete combustion of the oxygen/fuel mixture, but since this would exceed the 2,000 °F limitations of the materials of which the engine is made, some compromises must be made. The problem is solved by separating the flow of compressed air, and routing only part of it into the combustion chamber to be mixed with fuel and ignited. The rest is introduced into the combustion chamber as needed to keep the temperature of the engine within design limits. The result is more emissions and a higher cost per mile of travel. After the air/fuel mixture is ignited in the combustion chamber, the exhaust gases are expanded through a nozzle pointed opposite the desired direction of flight (hopefully). An external starting engine gives the engine its initial spin, and the fuel is then ignited with a heated plug. Once the turbo jet is lit, however, it requires no sparkplug to keep it running. A turbojet can draw more air (and oxygen) in when the temperature of intake air is lower because the cooler air is denser. As a result, the thrust of a turbo jet increases as the temperature of outside air decreases. Extra thrust for a takeoff on a hot day can be provided by injecting water at the compressor inlet so that the resulting EVAPORATION will cool the airstream. Additional thrust is obtained in military jets through the addition of a second burner, called an *afterburner*, just in front of the nozzle. Enough fuel can be injected into the afterburner to consume the oxygen left in the exhaust stream from the first combustion chamber, increasing air volume and jet velocity in the process.

A *turbofan*, also called a *bypass engine*, is an improvement on the basic turbojet design. In a turbofan, only part of the incoming air passes through the high-pressure compressor upstream of the combustion chamber. The rest bypasses the combustion chamber in a shell on the outside of the engine, and rejoins the hot combustion gases before they pass through the nozzle. By cooling the exhaust stream, greater thrust, increased efficiency and quieter operation are achieved. With a *turboprop* engine, a propeller mounted in front of the air intake provides about 90% of the thrust, with the rest coming from the exhaust jet. The propeller is driven either by a second turbine or by a second stage in the turbine that drives the compressor. Turboprops work well on small- to medium-sized planes. A *ramjet* engine takes advantage of the *ram effect*, a buildup of pressure in front of a fast-moving jet engine. When airspeed is more than 200 miles per hour, the ram effect provides enough pressure that neither a compressor nor a turbine is necessary to compress air sufficiently to support fuel combustion. Ramjet engines have been called a flying stovepipe because they are nothing more than a (carefully shaped) tube that is open at both ends with fuel nozzles mounted in the middle. A ramjet will not work until it is brought up to its designed operating speed—which is supersonic in military planes—by other means.

Junk Mail mail consisting of unwanted advertisements, catalogs and solicitations. Exactly what constitutes junk mail is a highly personal matter. The mail-order catalog that is devoured in one household will quickly make its way into the GARBAGE CAN in another. See also CONSUMERISM, MATERIALISM.

Junkyard a facility where wrecked and worn-out machinery, mostly automobiles and other vehicles, is dismantled, with most of the parts and materials sold for reuse or recycled as scrap metal. Junkyards have been centers of reuse and recycling since decades before the terms became environmentally correct. More than one third of the ferrous scrap recycled in the United States comes from junkyards. Most modern junkyards are high-turnover operations. More than 11 million automobiles, trucks, buses and motorcycles pass through junkyards every year. Auto recycling is the 16th largest industry in the United States, with $5 billion in annual sales.

Most junk automobiles are purchased from insurance companies after an accident that was fatal, at least to the vehicle. Most areas have some kind of junk-vehicle program through which derelict cars are hauled away from the curbside or field where they were abandoned and taken to junkyards. When a junked car comes in, fluids such as oil, gas and antifreeze solution are drained, and the heat-transfer gas in the air conditioner is drained and saved for reprocessing. All the parts that still have value are stripped from the chassis and, along with engine and drive-train parts and serviceable body parts, are sorted out and stored for sale. Modern junkyards are hooked up to an international network of scrap dealers electronically. If a yard does not have a part on its computerized inventory, the inventories of junkyards across town or across the nation can be consulted and the part ordered on line. Scrap exporters can access a similar network of international dealers in recyclable METALS, PAPER, PLASTIC, BATTERIES and other commodities via satellite. Some parts, including body and engine parts, tires and other parts recycled from junked vehicles are resold as is, while others, such as radiators, starter motors and alternators, are returned to manufacturers for rebuilding. After everything with resale value has been removed from a junked vehicle, the frame, engine block, body and any other remaining metal parts are collected, crushed into blocks and sold as scrap.

Three-fourths of the material in a junked car is typically reused or sent to a scrap dealer. The leftovers—a conglomeration of GLASS, rubber, rust and fluid—is run through a shredder to produce a substance commonly referred to as *car fluff* which is usually hauled to a landfill for disposal. General Motors researchers are experimenting with recycling car fluff by heating it in the absence of oxygen, a process called PYROLYSIS. The end products of the pyrolysis process are gases (about 26% by weight), oily fuels (about 21%) and water (10%), with the remaining 43% being a black powder residue than can be used to make car parts, concrete, roofing shingles and other products. See also INSTITUTE OF SCRAP RECYCLING INDUSTRIES, METAL RECYCLING.

K

Kenilworth Dump a municipal waste dump in Washington, DC, used from the mid-1940s through the late 1960s. The dump, which was located near the Capitol, adjacent to a black neighborhood, was a breeding ground for rats, and smoke from OPEN BURNING and odors from decaying refuse was unpleasant, at best. In response to public pressure to close down the dump, the city had reduced its size and converted some of the land to a park. The bulk of Washington's MUNICIPAL SOLID WASTE was burned in one of four ancient incinerators located around the city. The incinerators could not handle the city's entire flow of solid waste, so city officials were forced to keep the dump open until an alternative site could be found. Wood, paper, plastic and other fuels in the cities garbage produced lively fires on a windy day, and they were poorly monitored. On a breezy day in February 1968, three boys were playing in the dump when the winds shifted, and one of them was overtaken and killed by the flames. Coming as it did at a time when not only Washington, DC, but the entire nation was confronting problems relating to solid waste disposal, the Kenilworth blaze helped speed the passage of tighter controls of the disposal of municipal solid waste (see SOLID WASTE DISPOSAL ACT, RESOURCE CONSERVATION AND RECOVERY ACT).

L

Lagoon any relatively small, shallow body of water. Lagoons used in the treatment of sewage are called polishing ponds.

Landfill a place where garbage is dumped, mechanically compacted and covered with dirt or other material. The modern landfill started as a solution to the unsanitary conditions common at the OPEN DUMPS in which most garbage was disposed of until after the Second World War. Sanitary landfills, in which wet waste materials are layered with dry rubbish or soil, were introduced in the 1940s in the United States. It was the adoption of the sanitary landfill as the preferred method of disposing of the troops' trash during the war by the U.S. Army Corps of Engineers that helped popularize the approach to garbage disposal. A handful of early American dumps—including dumps in Champaign, Illinois (1904), Dayton, Ohio (1906), and Davenport, Iowa (1916)—were early adopters of the landfill approach to trash disposal, but it wasn't until the 1930s and 1940s that the practice began to catch on. Before the advent of the landfill, smoke-belching incinerators were used in many cities to burn the more combustible trash (see GARBAGE INCINERATOR), and the more edible garbage was fed to pigs (until the practice was widely banned by local health departments because of the well-documented association of garbage-fed pigs and trichinosis). The landfill was seen as a vast improvement over the open dumps that ringed most cities and that were the source of foul odors and disease. The first known use of the term "sanitary landfill" is credited to Jean Vincenz, who coined the term while he was the commissioner of public works in Fresno, California, in the early 1930s.

Today, the local landfill is a source of concern to many members of the public, especially if a new one is proposed in any kind of proximity to one's home (see NIMBY), and there is good basis for such fanaticism. Nearly all older landfills have involved at least some degree of environmental degradation, especially polluted groundwater (see GROUNDWATER POLLUTION), and the pollution associated with at least 200 old dumps is so severe that they are now Superfund toxic-waste cleanup sites. This is partly due to the presence of a host of HAZARDOUS MATERIALS that should never have been disposed of in a landfill in the first place, and the fact that most early landfills were situated in wetlands and other low-lying areas made matters worse. It was thought that by locating a landfill in low spots, that unusable land could be made available for other purposes when filled in. In fact, much of the dry land in and around many cities was "reclaimed" in this way. But with our modern understanding of the enviromental importance of wetlands and the mechanics of water pollution, it is easy to see the downside of this approach. Few older landfills had liners, and many were located on permeable soils over vulnerable AQUIFERS and near surface streams.

The hazardous and solid waste amendments of the RESOURCE CONSERVATION AND RECOVERY ACT (RCRA) were passed in 1984 with the intention of clearing up these kinds of problems at municipal landfills and at hazardous waste disposal sites. Companion legislation had previously been passed in 1980, addressing the cleanup of abandoned and inactive hazardous waste disposal sites (see COMPREHENSIVE ENVIRONMENTAL RESPONSE, COMPENSATION AND LIABILITY ACT), and conditions at landfills have improved as a result. New landfills are subjected to rigorous technical and public scrutiny before they can be built—a process that can and usually does take years. Under the terms of the RCRA, new landfills must be situated in an area where soils are relatively impermeable (see PERMEABILITY, POROSITY) so that the chance that liquid toxins will percolate (see PERCOLATION) into underlying groundwater or into nearby surface water are minimized. To further reduce the chance that toxins will find their way out of the landfill, the pit where new landfills are to be situated must be lined with several feet of dense clay, and a thick plastic liner composed of strips that

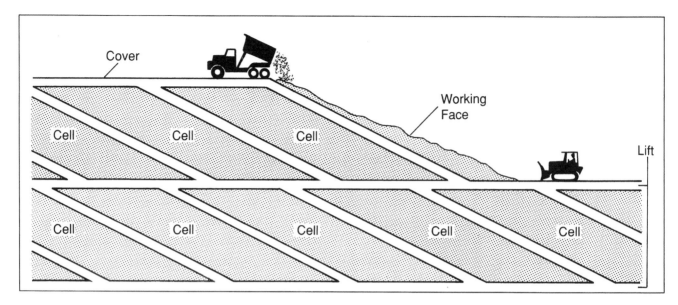

Sanitary landfill (*Encyclopedia of Environmental Studies*)

have been hot-sealed together is installed over the clay. The liner is then covered with several feet of gravel or sand to protect it from accidental puncture. New landfills are generally sculptured to facilitate the collection of liquid LEACHATE that gathers under the garbage (see LEACHATE COLLECTION SYSTEM). Many landfills have an on-site treatment plant that remove contaminants (see LEACHATE TREATMENT SYSTEM). The SLUDGE that is the end product of the treatment process is either dumped in the landfill or is incinerated. Leachate from other landfills goes into the sewer system for treatment at the sewage plant. A new landfill is required to install monitoring wells to detect any contaminants that escape containment.

Many existing dumps have also been required to install monitoring wells, and the ENVIRONMENTAL PROTECTION AGENCY (EPA) estimates that 25% of the nation's landfills now have such pollution-detection wells. METHANE gas is produced by the decomposition of organic materials in a landfill, and explosions and seepage of gas into the basements of nearby homes have been a problem at many sites (see GAS COLLECTION SYSTEM, GAS MIGRATION). Garbage that is dumped at a landfill is compressed by bulldozers and compactors (see COMPACTION), massive machines with studded rollers that are about five feet wide. At the end of a day of dumping, the cell of compacted garbage is covered, usually with a clay-rich dirt that discourages the penetration of water. When the site is full, a thick CAP of clay and a plastic liner are installed, topsoil is added over that and vegetation is planted. A landfill that has been so retired often becomes a golf course or a park. The Dubuque Greyhound Park in

Dubuque, Iowa, Mount Trashmore Park in Virginia Beach, Virginia, Mile High Stadium in Denver, Colorado, and the John F. Kennedy Presidential Library in Boston, Massachusetts, are all built over retired landfills.

All the new requirements have drastically increased the cost of building and operating a landfill. In 1992 the EPA announced new regulations, to be phased in from 1993 to 1996, requiring all landfills to have liners and monitoring wells, although many states already required this. The EPA estimates the cost of complying with the regulations at $330 million, and that landfills that handle less than 100 tons of garbage a day generally will not be able to afford compliance and will be forced to close or get an exemption, which will be issued only to a few small landfills in remote or arid regions. The Solid Waste Association of North America estimates that new landfills must be at least 50 acres to collect enough trash to cover the costs of the new regulations. There were 16,000 landfills in the United States in 1982, and the number is expected to shrink to about 3,000 by 1997. It is estimated that approximately three-fourths of the nation's garbage goes to landfills. The rest is incinerated, recycled or dumped illegally (see RECYCLING, MIDNIGHT DUMPING). See also ASHFILL, DUMP MINING, GARBAGE CRISIS, SECURE LANDFILL, SOLID WASTE DISPOSAL, TIPPING FEE, WASTE STREAM.

LD$_{50}$ the amount of a toxin required to kill 50% of the test population. The toxicity of a substance is expressed as its LD$_{50}$, which stands for *l*ethal *d*ose for *50%* of test animals exposed. TCDD (see DIOXIN), one of the most toxic substances known, has an LD_{50} of 0.0006 for guinea pigs. An LC_{50} denotes the lethal concentration for 50% of exposed test animals for a toxic substance found in water.

Leachate the liquid created in a LANDFILL (or HEAP-LEACHING operation in mining) as the result of the PERCOLATION of RUNOFF and the squeezing of liquid from materials by the weight of overlaying materials. Leachate, which tends to collect at the bottom of the landfill, is normally laced with toxins, and the migration of untreated leachate into aquifers and surface water is a major source of WATER POLLUTION. VOLATILE ORGANIC COMPOUNDS, HEAVY METALS, AMMONIA and NITRATES are among the toxins commonly found in leachate, which typically has a BIOCHEMICAL OXYGEN DEMAND of more than 20,000 milligrams per liter, about 100 times that of sewage. The RESOURCE CONSERVATION AND RECOVERY ACT (RCRA) requires low permeability CAP on landfills to keep rainwater and surface runoff out to minimize the production of leachate, and a low permeability liner to prevent the escape into the environment of whatever leachate is produced. Monitoring wells to detect the migration of leachate into groundwater and a LEACHATE COLLECTION SYSTEM are also required. Once collected, the leachate is sometimes recirculated through the landfill, but is usually treated, either on site or at a sewage treatment plant (see LEACHATE TREATMENT SYSTEM, SEWAGE TREATMENT).

MUNICIPAL SOLID WASTE contains liquids. Although potentially toxic waste products such as spent SOLVENTS and old PAINT are not supposed to be dumped in the trash, containers holding such liquids routinely go to the landfill. Absorbent solid materials also may be soaked with liquids. Liquids contained in solid waste join rainwater and percolate down through the garbage, picking up toxins along the way, and collect at the bottom of the pile. In a modern landfill, this leachate is stopped by a liner, and is collected by a drainage system. The leachate is usually collected in a storage tank, which is periodically hauled to a sewage treatment plant in a tank truck. Many older dumps have no bottom liners or have leaks that allow leachate to percolate through the ground and contaminate groundwater. See also BATTERY, FRESH KILLS LANDFILL, NAPTHALENE.

Leachate Collection System a drainage system required in new landfills by the RESOURCE CONSERVATION AND RECOVERY ACT that collects LEACHATE for treatment. A typical leachate collection system consists of an impermeable liner, permeable drain pipes (see PERMEABLE PIPE), pumps and one or more storage

tanks. Since leachate can be highly corrosive, the leachate collection and storage system must be made of corrosion-resistant materials. Many new dumps have an on-site leachate treatment, while others either dump the leachate into the local sewer system or haul the leachate to the SEWAGE TREATMENT plant in tank trucks for treatment. See also GROUNDWATER POLLUTION, LEACHATE TREATMENT SYSTEM.

Leachate Treatment System a water treatment system that employs many of the same techniques used at a SEWAGE TREATMENT plant to process leachate. AEROBIC BACTERIA and ANAEROBIC BACTERIA are used to remove the ORGANIC MATERIAL and NITROGEN, either in a ROTATING BIOLOGICAL CONTACTOR or an ACTIVATED SLUDGE process. FILTRATION, SEDIMENTATION and aeration may also be used. See also HAZARDOUS WASTE TREATMENT, LIXIFICATION.

Lead a soft gray METAL widely used in industry because of its malleability, high density, low melting point, resistance to CORROSION and ability to stop GAMMA RAYS and X-RAYS. Lead is used to make metal alloys, lead-lined pipe and containers for the storage of corrosive gases and liquids, and electric storage batteries, and is an additive in certain PLASTICS. Almost 400,000 metric tons of lead was produced by U.S. manufacturers in 1993. Lead gets into outdoor air from industrial and automotive emissions (see AUTOMOTIVE AIR POLLUTION) and from the removal of old lead-based PAINT. Atmospheric lead levels have fallen dramatically in the United States during the last two decades because the CLEAN AIR ACT of 1970 initiated a phase-out of leaded GASOLINE. It gets into indoor air primarily from lead based paint. Mine and mill TAILINGS, LANDFILLS and AERIAL DEPOSITION all contribute to the lead found in lakes and streams. Lead enters groundwater from industrial emissions and from LEAKING UNDERGROUND STORAGE TANKS and spills involving leaded gasoline. Lead occurs widely in the Earth's crust, so water that is naturally tainted with lead exists in some areas. Lead and its compounds range in water SOLUBILITY from highly soluble to practically insoluble. Lead compounds are highly persistent in water, with a HALF-LIFE of more than two-hundred days. Lead and its compounds have high acute TOXICITY to aquatic life (see ACUTE EFFECTS). Lead can get into DRINKING WATER from the lead solder used to connect copper water-supply systems installed before it was banned in 1987. Older homes with lead pipe and community water distribution systems with lead connecting pipes also may have lead-tainted water. Mineral-rich hard water is less likely to pick up lead from plumbing and distribution systems than is soft water, because the lead is soon covered with a coating of minerals. The ENVIRONMENTAL PROTECTION AGENCY (EPA) estimated in 1994 that lead concentrations in the drinking water delivered by 800 community water systems exceeded federal standards. Some pottery glazes contain lead. It has been theorized that the lead in the wine glasses used by Roman nobility played a role in the decline of that empire. Lead is found in wine, especially wines sold in Europe.

Health Effects

Lead can affect health when inhaled or when ingested with food or drink. It can cause learning disabilities, hyperactivity, high blood pressure, kidney disease and brain damage. Children under the age of six and pregnant women run the greatest risk. The symptoms of lead poisoning generally manifest a month or more after exposure. Lead accumulates in body tissue, so repeated exposure to low levels can eventually result in levels that cause damage. Symptoms of low-level lead poisoning include tiredness, mood changes, headaches, stomach problems, constipation and trouble sleeping. The Centers for Disease Control in 1991 reported "pervasive evidence" that 10 micrograms per deciliter in a child's blood—a level found in one-sixth of American children—can cause mental impairment.

Exposure to higher levels of lead may cause aching, weakness, anemia, problems with memory and concentration, and permanent kidney or brain damage. The symptoms of *pediatric plumbism*, the affliction resulting when a child eats sufficient lead paint, include a numbing or deadening of nerves (see NEUROTOXIN) and subsequent loss of bodily control. Such severe lead poisoning also damages developing brains and nervous systems and causes associated learning and behavioral problems. Lead exposure increases the risk of high blood pressure. Lead is a probable TERATOGEN in humans. It may decrease fertility in males and females.

Before 1940, lead carbonate was the primary white pigment in paints, and many products in use at the time were more than half lead. Paint based on red lead was also universally used as a metal primer at the time. Although lead paint was especially popular on window trim, where it proved particularly resistant to weathering, it is often found on other trim and in wall paint. One reason for lead paint's durability is that the surface "chalks"—that is, it oxidizes (see OXIDATION). The chalked surface wipes off to reveal clean paint underneath. Although this was once seen as a desirable quality, the DUST given off by such oxidation contains lead, so even simply leaving old lead paint alone can be hazardous.

Although the lead content of paints dropped steadily over the next three decades, it wasn't until the Lead Based Paint Poisoning Prevention Act was passed in 1971 that the metal was outlawed altogether for use in paints. The act initially capped the lead content of paints at one percent with the limit reduced to half that amount in 1973 and to .06 percent in 1977. So although there is likely to be more lead in older homes, at least some lead-based paint can be found in the majority of houses and apartments built before 1970. According to the Environmental Protection Agency, 42-million American homes have a serious lead-based paint problem, and 25 to 40 million of these homes have lead paint accessible to children. Surveys have indicated a lead problem in 60% to 75% of New Orleans homes and 60% to 80% of the homes in Boston, New York, Philadelphia and Chicago. Title X of the Housing and Community Development Act of 1992 outlines procedures to control lead poisoning in children from lead-based paint in older housing. It is administered by the Department of Housing and Urban Development.

PUBLIC WATER SYSTEMS are required under the SAFE DRINKING WATER ACT to monitor concentrations of lead. Treatment technology rather than a MAXIMUM CONTAMINANT LEVEL is specified for the metal. The concentration of lead in drinking water is measured at the user's tap, since the metal can be picked by aggressive finished water (see AGGRESSIVE WATER) as it passes through the distribution system and household plumbing (action level = 0.015 milligrams per liter). Acidic DRINKING WATER must be neutralized to make it less corrosive, and the public must be educated about ways of reducing exposure. Water utilities are required to replace lead service lines in the water distribution system (but not privately owned lead pipes that connect individual customers to the water system). The OCCUPATIONAL SAFETY AND HEALTH ADMINISTRATION has established a permissible exposure limit for airborne lead of 0.05 milligrams per cubic meter averaged over an eight-hour workshift. Lead is also cited in AMERICAN CONFERENCE OF GOVERNMENT INDUSTRIAL HYGIENISTS and NATIONAL INSTITUTE FOR OCCUPATIONAL SAFETY AND HEALTH regulations. It is on the RTK Hazardous Substance List and the Special Health Hazard Substance List. Lead compounds are on the HAZARDOUS AIR POLLUTANTS LIST. Businesses handling significant quantities of lead or lead compounds must disclose use and releases of the chemical under the provisions of the EMERGENCY PLANNING AND COMMUNITY RIGHT TO KNOW ACT. See also TETRAETHYL LEAD.

Leaking Underground Storage Tanks

storage tanks made of METAL, PLASTICS and CONCRETE that have developed leaks. Leaking underground storage tanks (LUST) are the largest single source of groundwater pollution in the United States. The ENVIRONMENTAL PROTECTION AGENCY (EPA) has estimated that tanks leaking GASOLINE alone may account for as much as 40% of all U.S. groundwater contamination. The EPA and the American Petroleum Institute estimate that about 5% of the 2 million underground gasoline tanks in the country leak. Even a slow leak of gasoline and other petroleum products can put several water wells out of commission, and just a couple of tanks with relatively serious leaks can contaminate an entire AQUIFER if the leaks go undetected.

The EPA has estimated it will cost service station owners $50 billion to install tanks that will comply with the requirements passed in amendments to the 1984 Resource and Conservation Act and that one-third of the nation's small stations would go out of business as a result of the new standards. The act requires that station owners have leak-proof tanks installed and that they buy liability insurance that will cover the cost of any cleanup that may be necessary. There are many who criticize the new laws for going too far, but others contend that it will save more than it costs in the long run by preventing at least some of the gasoline contamination. In addition to gasoline, the underground tanks that hold other petroleum products, toxic chemicals and hazardous waste also commonly leak part of their contents into underlying aquifers. See also AQUIFER SENSITIVITY, GROUNDWATER POLLUTION.

Legionella Bacteria the bacteria that causes Legionnaire's Disease (see INDOOR AIR POLLUTION). PUBLIC WATER SYSTEMS are required under the SAFE DRINKING WATER ACT to monitor for Legionella Bacteria. The MAXIMUM CONTAMINANT LEVEL GOAL for Legionella Bacteria is 0. A treatment technology rather than a MAXIMUM CONTAMINANT LEVEL (MCL) is specified for enteric viruses. Treatment, whether through disinfection or filtration, must remove 99.99% of the Legionella Bacteria in the water supply. See also CRYPTOSPORIDIUM, MICROORGANISMS, VIRUS.

LeGrand Rating System a method developed in the mid-1970s by the U.S. Geological Survey to assess the risk posed to groundwater by hazardous substances. A *hydrogeologic rating* based on the distance to the nearest well or spring, the depth and slope of the WATER TABLE, the PERMEABILITY of AQUIFER materials and their ability to absorb contaminants (see ADSORPTION, ABSORPTION); an *aquifer sensitivity rating* (see AQUIFER SENSITIVITY); and a *hazard potential rating* based on the characteristics of the materials present at the site (see hazardous materials). The three ratings are combined, and a rating of the site's suitability as a repository for haz-

ardous waste is given a rating from an excellent "A" to a failing "F."

Light-weighting a reduction in the amount of materials put into a manufactured item gained without sacrificing the product's strength or durability. Since it allows a manufacturer to turn out more finished items while keeping the investment in materials the same, it is generally adopted as a way of maximizing profit. The effects of light-weighting can be seen at a LANDFILL, where the process works as a kind of SOURCE REDUCTION. Newer plastic bottles compress into a smaller space since they contain less plastic. POLYETHYLENE TEREPHTALATE (PET) soda bottles that weighed an average of 68 grams in 1977, for example, now average just 51 grams. Similarly, a one gallon HIGH DENSITY POLYETHYLENE milk jug, which weighed about 120 grams in the 1960s was down to 65 grams by 1992, and plastic grocery bags that averaged 30 microns in thickness in 1976 were down to an average of 18 microns by 1992. Glass beverage bottles have also gone on a crash diet, with a 16-ounce bottle weighing about 44% less today than in 1974, and half-pint paper milk cartons have shed 23% of their weight over the same period. Researchers at the Garbage Project have concluded that although the number of plastic items found in landfills has increased in recent decades, the volume they occupy has remained constant because of lightweighting. See also PLASTICS RECYCLING, SOLID WASTE.

Light Year see MEASUREMENTS

Lindane an organochlorine COMPOUND that was a widely used household INSECTICIDE until concerns about its health effects surfaced in the 1970s. Lindane, also called *cyclohexane* and *hexachlorocyclohexane*, is used to treat farm animals and pets for external parasites, to protect plants and seeds from bugs and for mosquito control. It is a component of aerosol-spray insecticides, floor wax and mothballs. It gets into air and water from industrial discharges, the application of insecticides and spills. It gets into indoor air from its use as an insecticide and FUNGICIDE. Lindane is moderately soluble and moderately persistent in water, with a HALF-LIFE of 20 to 200 days. Lindane has high acute and chronic TOXICITY (see ACUTE EFFECTS, CHRONIC EFFECTS) to aquatic life, and is highly persistent once it has entered the environment. Some uses of lindane were banned in 1983.

Lindane can affect health when inhaled or passed through the skin. Symptoms of exposure include restlessness, insomnia, anxiety, irritability, poor appetite and headache. Exposure to higher levels can cause muscle twitching, seizures, convulsions and even death. Lindane is a possible human CARCINOGEN and TERATOGEN. It has been shown to cause liver, lung and endocrine cancer in animals. There is limited evidence linking lindane to leukemia in humans, and it may decrease fertility in females. Repeated exposure may cause liver damage, nervous system damage and poor coordination. A drastic reduction in the total number of blood cells (*aplastic anemia*) or in the white blood cell count (*agranulocytopenia*) may result after exposure to lindane.

PUBLIC WATER SYSTEMS are required under the SAFE DRINKING WATER ACT to monitor concentrations of lindane (MAXIMUM CONTAMINANT LEVEL = 0.0002 milligrams per liter). The BEST AVAILABLE TECHNOLOGY for the removal of lindane from DRINKING WATER is granular ACTIVATED CARBON. The OCCUPATIONAL SAFETY AND HEALTH ADMINISTRATION has established a permissible exposure limit for airborne lindane of 0.5 milligrams per cubic meter, averaged over an eight-hour workshift. It is also cited in AMERICAN CONFERENCE OF GOVERNMENT INDUSTRIAL HYGIENISTS, Department of Transportation and National Toxicology Program regulations. It is on the Hazardous Substance List, the Special Health Hazard Substance List and the HAZARDOUS AIR POLLUTANTS LIST. Businesses handling significant quantities of lindane must disclose use and releases of the chemical under the provisions of the EMERGENCY PLANNING AND COMMUNITY RIGHT TO KNOW ACT.

Liquid   one of the three fundamental states of matter. Liquids demonstrate some of the properties of a gas and some of the qualities of a solid. See INDUSTRIAL LIQUID WASTE, LIQUID WASTE, SEWAGE.

Liquid Waste   waste that is in liquid form. Water carrying solid, liquid and gaseous waste materials in SUSPENSION and SOLUTION is the most common form of liquid waste (see GASEOUS WASTE, SOLID WASTE, WASTEWATER, WATER, WATER POLLUTION). Some of the most toxic municipal liquid waste is collected and treated (see SEWAGE, SEWAGE TREATMENT), but much enters watercourses untreated (see COMBINED SEWER SYSTEM, MIDNIGHT DUMPING, RUNOFF, STORM RUNOFF, STORM SEWER). Most industrial liquid waste is treated and disposed of on-site, with wastewater disposed of in a lake or stream (see CLEAN WATER ACT) or an INJECTION WELL (see DEEPWELL DISPOSAL). Significant quantities of liquid RADIOACTIVE WASTE is in storage in underground tanks at commercial and military facilities associated with the production of electricity and weapons from nuclear energy (see RADIOACTIVE WASTE DISPOSAL).

Litter   dry solid waste materials that are scattered around. Litter wastestreams include the type deposited by a *litterbug*—an individual who throws away unwanted items such as pop cans or waste paper along paths or roadways or in a public place— or it may be the species that is scattered by natural forces such as gravity, wind or WATER, as is the case with rocks and gravel broken down by mechanical and chemical weathering, or the *organic litter* of partially decayed leaves, needles, bark, branches and other debris on the floor of a forest. Decaying food in litter attracts rats, cockroaches and pathogenic MICROORGANISMS, and wildlife can be injured when they ingest litter in the belief that it is food, or when they get tangled up in litter such as plastic six-pack holders or bailing twine. Some state laws impose taxes on products such as beverage containers (see BOTTLE BILL), DISPOSABLE DIAPERS and FAST-FOOD PACKAGING (see MCTOXICS CAMPAIGN) that often end up as litter to fund cleanup efforts. It has been estimated that about 1,500 visible pieces of litter can be found along the average mile of American highway. Although litter makes up a relatively small percentage of MUNICIPAL SOLID WASTE, the fact that it is readily visible makes it especially objectionable to many (see VISUAL POLLUTION). The placement of plentiful GARBAGE CANS that are emptied frequently has been found to limit the littering of public places. Fast-food packaging is a major component of litter, and its resulting visibility is probably one reason it is held in such low esteem by the public. See also DEBRIS, TRASH. Compare GARBAGE, REFUSE.

Lixification   a technique for removing metals from ore by washing the ore with a solvent and recovering the dissolved METAL from LEACHATE. See HEAP LEACHING.

Long-Term Turbidity   TURBIDITY that does not clear up when a WATER sample is allowed to sit for a specified period, usually seven days. Long-term turbidity is caused by clays that are held in colloidal suspension (see COLLOIDAL SOLIDS), and is an important consideration in the treatment of DRINKING WATER and in the effects of water quality on aquatic life. Compare SHORT-TERM TURBIDITY.

Love Canal   a neighborhood near Niagara Falls in upstate New York that suffered lethal groundwater contamination from a nearby abandoned toxic-waste dump. The Love Canal neighborhood was declared a disaster area by President Jimmy Carter in 1978, and evacuation of those living closest to the 40-acre waste dump commenced. More than 1,000 families were eventually forced to abandon their homes, which were purchased with funds from the state and federal governments. The houses nearest to the waste dump have been demolished; the rest are now boarded up and abandoned. Drainage trenches have been dug to intercept contaminated water flowing from the site and carry it to a treatment plant, where the toxins are removed. A high-density polyethylene barrier and a clay mound intended to keep toxics in and keep rainwater out now covers the site. Love Canal is one of the best-publicized HAZARDOUS WASTE sites, although it actually was not one of the most dangerous or most contaminated.

The clay-lined canal was built by a man named Love in 1892. It was the first phase of a project intended to generate electricity by routing water around Niagara Falls. The scheme was abandoned when Love ran out of funds. Hooker Chemical Corporation dumped 55-gallon drums holding an estimated 20,000 tons of process slurries, waste solvents and PESTICIDE residues into the 60-foot-wide by 3,000-foot-long excavation from 1942 until it was full in 1953. The site was then covered with a layer of clay and soil and deeded to the local board of education in exchange for a token payment of one dollar. The deed included a warning that the site had been used as a toxic waste dump and that the company would take no responsibility for any future effects of the dangerous chemicals stored there. A school and playground were eventually built on the site.

In 1976, after several unusually wet years, water laced with toxins began to kill vegetation and seep into basements. Most of the drums holding the toxic waste had rusted through and water from rain and melted snow was seeping into the pit, causing the contaminants to rise to the surface. A chemical stench enveloped the neighborhood most of the time. By 1978, it had become apparent that the air and water pollution (see AIR POLLUTION, GROUNDWATER POLLUTION, WATER POLLUTION) emanating from the dump was the cause of a high rate of miscarriages, low birth weight and birth defects, and area residents were convinced that the pollutants were also responsible for a rash of more serious afflictions such as cancer. As a result, the school and playground were closed, and 238 homes adjacent to the dump were evacuated and demolished. In 1980, another 792 homes were evacuated, although health officials have since questioned whether the second evacuation was necessary. In 1988 a federal judge ruled that Occidental Chemical was liable for the cleanup of the site, a cost estimated at $250 million at the time. (Hooker had changed its name to Occidental Chemical because of all the bad publicity that resulted from the Love Canal debacle.) Suits and countersuits over what entities should share these costs continued for more than two decades.

Low-Level Waste   materials that are moderately radioactive as the result of exposure to or contamination with RADIONUCLIDES. Low-level waste is often described as radioactive material (see RADIOACTIVITY) with a HALF-LIFE of 30 years or less. In practice, however, low-level waste often contains traces of materials such as PLUTONIUM (with a half-life of 24,000 years) and iodine-129 (with a half-life of 15.8 million years). For this reason, most countries classify many materials that are handled as low-level waste in the United States as medium-level waste. Low level waste typically includes used resins, filter sludge, worn parts and cleaning materials. Very low level liquid waste from nuclear plants—which can include materials such as cooling water, final wash water, effluents from laundry and decontamination centers, floor drainage from cleanup operations, personnel shower water and effluent from the final stages of liquid waste purification—can actually pose a more difficult disposal problem than more highly contaminated wastes simply because of the huge volume normally involved. In 1989, 76,000 cubic meters of low-level waste associated with weapons production and another 46,000 cubic meters from civilian sources were buried in shallow trenches in the United States. The federal classification system for low-level waste designates Class A wastes as those that are hazardous for 100 years, Class B wastes as hazardous for 300 years and Class C wastes as hazardous for 500 years. See also LOW-LEVEL WASTE DISPOSAL, RADIOACTIVE WASTE, RADIOACTIVE WASTE DISPOSAL.

Low-Level Waste Disposal   the disposal of moderately RADIOACTIVE WASTES for as long as they are hazardous (up to 500 years). Two of the six shallow federal burial sites for commercial LOW-LEVEL WASTE have leaked, and three have been closed. Starting in 1993, the three national nuclear waste repositories in Nevada, Washington and South Carolina have been free to refuse shipments of low-level waste. The states were charged with either building their own low-level waste facilities, or getting together with neighboring states to build a regional center. Public fear and loathing of any waste with the word "nuclear" attached to it (or any waste at all, for that matter—see NIMBY) has made even the process of coming up with a site for such an installation, let alone building it, difficult at best. The furor surrounding the siting of a low-level radioactive waste-disposal facility in Allegheny County, New York, demonstrates the public's fear of nuclear wastes. New York's siting commission had been looking at potential sites for three years, facing stiff local opposition every step of the way, and had narrowed its list of potential sites down to five. Opposition to all five sites was intense, and a few of the people who lived near one of the five potential sites actually chained themselves to a bridge off-ramp to keep officials from coming to survey the potential site. The Ward Valley low-level nuclear waste dump being proposed in California's eastern Mojave Desert is facing similar opposition, even though it is in a sparsely populated area. Obviously the states are facing a long uphill battle before even one low-level repository is built.

Lowest Achievable Emission Rate   the lowest emission rate achievable for a given STATIONARY SOURCE of industrial AIR POLLUTION. The lowest achievable emission rate is defined in the CLEAN AIR ACT as the most stringent standard established in any STATE IMPLEMENTATION PLAN, or the lowest emission rate that has been achieved for the pollutant in the United States, whichever is lower. This emission rate becomes the national standard. To receive a license for a new industrial facility, a company must either meet the standard or prove to the ENVIRONMENTAL PROTECTION AGENCY that it is impossible to meet. See also BEST AVAILABLE CONTROL TECHNOLOGY.

M

Magnetic Separator a permanent magnet or an electromagnet used to remove ferrous metals from a wastestream passing into a WASTE-TO-ENERGY PLANT. See INCINERATION, REFUSE DERIVED FUEL.

Mandatory Recycling recycling that is mandated by law. An increasing number of states and communities are passing legislation that requires a certain percentage of the municipal waste stream to be recycled. The ENVIRONMENTAL PROTECTION AGENCY issued NEW SOURCE PERFORMANCE STANDARDS in late 1989 that required source separation and recycling in cities with waste INCINERATORS as a way of reducing the volume and TOXICITY of garbage being burned in incinerators. The standards were issued to resolve a controversy over HEAVY METALS and other toxic materials being emitted by an incinerator in Spokane, Washington.

A 1989 law—the California Integrated Waste Management Act—requires cities and counties to eliminate half of the garbage headed for landfills by the year 2000. RECYCLING is key to meeting that goal. Communities that fail to meet the goal face fines of $10,000 per day. In Fremont, California, BROWING FERRO INDUSTRIES will begin new programs to recycle mixed PAPER, such as JUNK MAIL, and so-called "green waste," such as lawn clippings and tree branches. Under a mandatory program, all city residents would be required to receive and pay for garbage and recycling services unless they can demonstrate that they recycle everything on their own or do not generate any waste or recyclables. In order to gain an exemption from the $4.08-a-month recycling program, residents will have to demonstrate they recycle everything recycled by BFI, including newspaper (see NEWSPRINT), mixed paper, GLASS, PLASTIC, cans, drink boxes, milk cartons, used motor oil (see ENGINE OIL) and corrugated CARDBOARD. A growing number of companies are finding profitable business opportunities in recycling, waste reduction and COMPOSTING in California as a result of the act. A study by the Californians Against Waste Foundation concluded that the passage of the Beverage Container Recycling and Litter Reduction Act in 1986 led to the establishment of nearly 1,800 companies operating more than 3,000 facilities across the state. According to the study, more than 400 of the 495 curbside recycling programs that now reach 11 million Californians were established as a result of the BOTTLE BILL. See also SOURCE REDUCTION.

Manganese a reddish-gray or silvery hard brittle METAL used primarily as an alloying ELEMENT and cleansing agent in the production of STEEL, cast IRON and nonferrous metals. Manganese is a component of dry cell batteries and as a substitute for LEAD in GASOLINE, and is a part of many commonly occurring minerals. Manganese compounds get into air and water from natural and industrial sources. Manganese and its compounds vary in their solubility in water from very soluble to insoluble, and are highly persistent in water, with a HALF-LIFE of more than 200 days.

Although manganese is an essential trace element of plants and animals, larger doses have toxic effects. Manganese can affect health when inhaled. Heated manganese can release fumes causing a flu-like illness with chills, fever and aches. A pneumonia-like chest congestion can also occur. Repeated exposure can damage the kidneys and liver, and cause a lung allergy. Permanent brain damage can also result. Early symptoms of manganese poisoning include poor appetite, weakness and sleepiness. Changes in speech, balance and personality similar to the effects of Parkinson's Disease can follow.

The OCCUPATIONAL SAFETY AND HEALTH ADMINISTRATION has established a permissible exposure limit for airborne manganese of 5 milligrams per cubic meter, not to be exceeded at any time. It is also cited in AMERICAN CONFERENCE OF GOVERNMENT INDUSTRIAL HYGIENISTS and ENVIRONMENTAL PROTECTION AGENCY regulations. Manganese is on the Hazardous Substance List. Manganese compounds are on the HAZARDOUS AIR

POLLUTANTS LIST. Businesses handling significant quantities of manganese or manganese compounds must disclose use and releases of the chemical under the provisions of the EMERGENCY PLANNING AND COMMUNITY RIGHT TO KNOW ACT.

Martin Grate   see GARBAGE INCINERATOR

Materialism   a world view in which personal worth is equated with wealth and material possessions. The *conspicuous consumption* that is characteristic of materialistic cultures is linked to the depletion of natural resources (see RESOURCE DEPLETION) and the excessive production of wastes (see AIR POLLUTION, LIQUID WASTE, SOLID WASTE, WATER POLLUTION). In a society in which status is usually linked to possessions, the acquisition of *things* tends to become an obsession. Once the basics have been acquired, replacing the original possessions with new things that are of higher quality or newer design can become a self-perpetuating condition (see CONSUMERISM). While the robust economy that is associated with such behavior is desirable, concern about the corresponding boom in the production of waste products tends to lag behind interest in the production and marketing of new consumer products. To break the cycle of production and environmental degradation, the full cost of a product—including its ENVIRONMENTAL COST and energy investment—must be reflected in the selling price. See also ARCHAEOLOGY, GARBOLOGY, PLANNED OBSOLESCENCE. Compare SUSTAINABLE AGRICULTURE, SUSTAINABLE SOCIETY.

Materials Recovery Facility   a facility in which SOLID WASTE is sorted and recyclable materials are separated from the waste stream. Most recycling operations collect materials that have already been sorted (see SOURCE SEPARATION), but unsorted trash is processed at a materials-recovery facility. Typically, the operator of such a facility contracts with newspaper publishers, business offices, government agencies, schools, construction companies and other large-volume producers of waste PAPER and other materials to remove all the waste they produce, generally for considerably less than would be charged by a normal garbage-collection service. Once it reaches the facility, the TRASH is sorted both by hand and machine, and recyclable materials are removed from the waste stream. Recyclable materials are sold to paper mills, and to companies that process recycled GLASS, PLASTICS and METALS, and the remaining trash is sent to a landfill. Northwest Resource Recycling Inc. spent nearly $2 million on a new 40,000-square-foot materials-recovery facility in Eugene, Oregon, that opened in 1995—the third such installation in Oregon. Company officials are limiting their pick-up service to commercial businesses, schools and government agencies that generate enough waste to regularly fill huge dumpsters with 15 to 30 cubic yards of capacity. All sorting at the installation will be carried out indoors using $600,000 in sorting and baling equipment.

Maximum Contaminant Level (MCL)   the maximum level of a CONTAMINANT allowable in drinking water under the SAFE DRINKING WATER ACT standards. The maximum contaminant level is defined in the act as the highest permissible level of a water contaminant as measured at the user's tap—except for TURBIDITY, which is measured at the point of entry to the water utility's distribution system. The Safe Drinking Water Act requires the ENVIRONMENTAL PROTECTION AGENCY (EPA) to develop standards for regulated contaminants. A survey of available scientific literature on the contaminant or family of contaminants is first undertaken by the agency. In instances where the scientific evidence on a substance's TOXICITY is scant or inconclusive, the EPA may arrange for additional analysis. The agency next publishes the proposed MCL and MAXIMUM CONTAMINANT LEVEL GOAL (MCLG) for the substance in question, and takes comments on their validity from scientists, water-industry representatives and the general public. If a challenge to the validity of the proposed guidelines is found to be legitimate, the proposed MCL and MCLG are revised accordingly, proposed guidelines are once again published, and comments once again accepted.

MCLs are to be set as close as is practicable to MCLG, but the cost and availability of treatment technology must be considered in establishing the MCL. MCLGs are target levels that are considered along with new data on toxicity and water-treatment technology any time an MCL is adjusted. The EPA set MCLs for 26 contaminants between passage of the SAFE DRINKING WATER ACT (SDWA) in 1974 and passage of the 1986 SDWA amendments, which required that MCLs be established for an additional 57 specified contaminants by 1989 and for yet another 25 by 1991.

Water suppliers are not responsible for contaminants that get into water under circumstances controlled by the user (with the exception of CORROSION of piping and plumbing caused by AGGRESSIVE WATER—see DRINKING WATER, GROUNDWATER POLLUTION, WATER POLLUTION). The standards apply to all PUBLIC WATER SYSTEMS (any water system with at least 15 year-round service connections of 25 seasonal connections). Violations of maximum contaminant levels are common, and fines for violations of the standards are very unusual. MCLGs for noncarcinogens are based on a *no-effect level*—the level at which even regular exposure over a lifetime will have no negative health—and it is assumed that the body can neutralize a low level of a toxic agent without ill effects. MCLGs for carcinogens, in contrast, are based on the assumption that even a minute quantity of the carcinogenic agent is capable of causing cancer, and is set, at as matter of policy, at zero (see ZERO DISCHARGE).

Maximum Contaminant Level Goal (MCLG)

the concentration of a DRINKING WATER contaminant that can be present without causing adverse health effects. Maximum Contaminant Level Goal—called *recommended maximum contaminant levels* before the 1986 amendments to the SAFE DRINKING WATER ACT—are set without regard to the cost of reaching the goal.

Maximum Disinfectant Residual Level (MDRL)

the maximum concentration of disinfectant allowed in the finished water from a PUBLIC WATER SYSTEM under the terms of the SAFE DRINKING WATER ACT. See also MAXIMUM CONTAMINANT LEVEL, MAXIMUM CONTAMINANT LEVEL GOAL.

MCL   see MAXIMUM CONTAMINANT LEVEL

McToxics Campaign   a boycott of McDonald's restaurants organized by the CITIZENS' CLEARINGHOUSE FOR HAZARDOUS WASTE (CCHW) in 1989 and 1990 with the goal of convincing McDonald's to discontinue the use of POLYSTYRENE foam "clamshells" used as a container for hamburgers and other hot sandwiches. The campaign was successful—McDonald's switched to sandwich wrappers made of a composite paper/plastic (see PAPER, PLASTICS) material in November 1990. McDonald Corporation president Edward Rensi said in announcing the move that "although some scientific studies indicate that foam packaging is environmentally sound, our customers just don't feel good about it." The victory was ironic in that the restaurant chain abandoned one of the most aggressive PLASTICS RECYCLING efforts in the United States to make the switch, and the new composite wrappers are not recyclable. The Pro Environment Packaging Council, which has ties to the paper industry, was one of the citizens' groups taking part in the campaign. McDonald's buys more than $200 million worth of recycled products annually.

McDonald's switched to the expanded-polystyrene containers in 1976, partially to satisfy customers who were complaining about the number of trees cut down to make paper hamburger wrappers, and partially to keep the burgers warmer and make the package less droopy and greasy. The company announced in 1987—again as the result of public complaints—that its package supplier would no longer use CFC-11 and CFC-12, which had been implicated in the depletion of the OZONE layer. McDonald's had recycling bins installed in many of its restaurants, and clamshells and polystyrene foam cups deposited in the bins by customers were shipped to companies such as Rubbermaid that used the resins in the recycled clamshells to make durable plastic products. Problems were encountered with the program because foreign material thrown into the recycling bins meant that the quality of the recycled material was low, and because about half the plastic foam used left the restaurants in carryout bags. McDonalds established a joint task force with the ENVIRONMENTAL DEFENSE FUND (EDF) in 1990 to study how best to minimize the SOLID WASTE generated by the company's 8,500 U.S. franchises. EDF suggested that the company recycle polyethylene shrink-wrap, compost egg shells and coffee grounds, reuse of shipping pallets and purchase recycled materials including tables and chairs made from recycled plastic. McDonald's bowed to the public pressure created by the CCHW campaign in the end, however, due in no small part to corporate horror over school children across the country dressed up in their "Ronald McToxic" outfits and sending their used hamburger clamshells in to corporate headquarters. McDonald's is now looking into the possibility of composting its new paper and plastic sandwich wrappers, and is, in the meantime, coping with customers who complain that their burgers are cold.

Measurements   systems to gauge the length, volume, weight, heat, energy content, velocity, emission of radiation and other characteristics of matter and energy. Measurements are expressed in many different ways in different countries and within different disciplines. Scientific work is based on the *metric system*, and most of the world's nations have officially adopted the metric system as their official system of weights and measures. The *English System* is still used for nonscientific purposes in the United States, and a variety of other systems are used for local commerce in other parts of the world. Governments specify standard measurements for everything from the weight of half a cup of chopped meat (1/2 pound is the United States standard) to the pounds of pickled fish in a barrel (the U.S. standard is 200 pounds). Early measurements were based on parts the dimensions of parts of the human body such as the palm of the human hand and the foot. The *yard* was originally defined as the length of an arm, and the distance from the elbow to the tip of the middle finger was called a *cubit*. The properties of water are used in the development of

Commonly Used Measurements

Length and Distance

1 **micrometer** or **micron** (**m**) = 0.001 millimeter (mm) = .0000393 inch

1 **millimeter (mm)** = 0.04 inch

1 **mill** = 0.001 of an inch

1 **centimeter (cm)** = 0.032808 feet = 0.01 meter

1 **inch (in.)** = 2.54 cm

1 **foot (ft.)** = 30.48 centimeters = 12 in.

1 **meter** = 3.2808 ft

1 **yard** = 0.9144 meters

1 **mile** = 1.6093 kilometers (km)

1 **kilometer** = 0.62137 miles

1 **micron** = 0.0001 centimeters

1 **fathom** = 6 feet

1 **nautical mile** = 1,852 meters

1 **knot** = 1 nautical mile per hour

1 **hand** = 4 inches (measuring horses)

1 **rod (rd.)** = 5.5 yards

1 **light year** = 5.88×10^{12} miles

Area

1 **square centimeter** = 0.0010764 square feet

1 **square meter** = 1.196 square feet

1 **acre** = 43,560 square feet = 0.004047 square km = 4,047 square meters = 160 square rods

1 **square kilometer** = 247.1 acres = 0.3861 square miles

1 **square mile** = 640 acres

1 **hectare** = 1 square hectometer (hm^2) = 10,000 square meters

Volume

1 **board foot** = 144 cubic inches

1 **cord** = 8 cord feet

1 **cord foot** = 16 cubic feet

Weight

1 **microgram (μg)** = one millionth of a gram = 0.00001 gm

1 **milligram (mg)** = one thousandth of a gram = 0.001 gm

1 **gram** = 0.035274 ounces

1 **pound** = 0.45359 kilograms

1 **kilogram** = 2.2 pounds

1 **hundredweight (cwt.)** = 100 pounds

1 **short ton (S.T.)** = 2,000 pounds = 907.2 kilograms

1 **metric ton** = 2,204.6 pounds = 1,000 kilograms = 1.102 short tons = 0.9842 long tons

1 **long ton (L.T.)** = 2,240 pounds = 1,016 kilograms = 1.016 metric tons

1 **carat** = 200 milligrams (gem weight)

Volume and Liquid Measure

1 **cubic centimeter (cc)** = 1 **milliliter (ml)** = 3.5314667 cubic feet = 0.001 liters = 0.0338 U.S. fluid ounces

1 **cubic foot** = 7.4805 U.S. liquid gallons = 28.31685 liters

1 **cubic inch** = 16.387 cubic centimeters = 0.0005787 cubic feet = 0.004329 U.S. liquid gallons

1 **cubic meter** = 35.314667 cubic feet = 1,000 liters;

1 **liter** = 0.001 cubic meter = 1,000 milliliters = 0.0353 cubic feet = 0.2642 U.S. liquid gallons

1 **U.S. liquid gallon** = 231 cubic inches

1 **cubic yard** = 27 cubic feet = 0.76455 cubic meters

1 **cubic inch** = 16 cc

1 **U.S. liquid ounce** = 29.57 cc

1 **pint** = 28.875 cubic inches

1 **quart** = 0.9463 liters = 3,785.4 cc = 3.7854 liters

1 **acre foot** = 43,560 cubic feet = 1,233.5 cubic meters

1 **barrel** = 5.6146 cubic feet = 42 U.S. gallons = 158.98 liters

1 **U.S. gallon** = 231 cubic inches

1 **imperial gallon** = 277.77 cubic inches

Concentration

1 **part per million (ppm)** = 0.000001 = 1×10^{-6}

1 **part per billion (ppb)** = 0.000000001 = 1×10^{-9} = 1,000 ppm

1 **milligram per liter (mg/L)** = 1,000 micrograms

Temperature and Heat

1 **degree Centigrade** = (5/9)(F – 32)

1 **degree Fahrenheit** = (9/5 C) + 32

absolute zero = –273.16°C or –459.69°F

1 **calorie** = the amount of heat required to raise the temperature of one pound of water 1°C = 3.08596 foot pounds = 1.1622×10^{-6} kilowatt hours

Commonly Used Measurements (*Continued*)

1 **British Thermal Unit (Btu)** = 251.99 calories, gram = 777.649 foot-pounds = 1,054.35 joules = 0.000292875 kilowatt-hours

1 **Btu/hr** = 0.000292875 kilowatts

1 **watt** = 3.4144 Btu/hr = 14.34 cal/min = 0.001341 horsepower

1 **joule** = 0.0009485 Btu

1 **kilowatt (kw)** = 56.90 Btu/min = 1.34102 horsepower

1 **megaton** = 1,000,000 tons

Radiation

1 **roentgen (pronounced "renkin")** = amount of x-ray or gamma radiation required to create 2 billion ions in one cubic centimeter of air

1 **rad** = amount of radiation that will cause the absorption of 0.01 joules of energy per kilogram of mass

1 **rem** (roentgen equivalent in man) = the damage caused by one rad of gamma radiation

1 **millirem (mR)** = 0.001 rem

1 **curie** = radiation emitted by one gram of pure radium

1 **picocurie (pCi)** = one-trillionth of a curie = 1×10^{-12}

Misc.

1 **bolt of cloth** = 40 yards

1 **ream** = 500 sheets of paper

1 **bundle** = 2 reams = 1/4 case

1 **candle power** = 12.566 lumens

---

weights and measures. The freezing point of water (32°F) is 0 on the Celsius scale, and its boiling point (212°F) is 100 degrees Celsius. The *calorie* and the *gram* are also based on water's properties. For specifics on the development of the two systems of measurement, see ENGLISH SYSTEM and METRIC SYSTEM.

Mechanical Composting   see COMPOSTING

Mechanical Ventilation   see VENTILATION

Medical Waste   waste generated by hospitals and other medical-care facilities. Medical waste contains INFECTIOUS WASTE such as blood and other bodily fluids, disposable syringes, body parts and DISPOSABLE DIAPERS (see SALMONELLA), and may contain HAZARDOUS WASTE such as SOLVENTS, RADIOACTIVE WASTE produced by x-ray equipment and other medical devices, discarded medicine and PESTICIDES. But PLASTICS make up as much as 60% of medical waste. It is estimated that between half a million and three million tons of medical waste are generated in the United States each year; but because its disposal is only loosely regulated, exact figures are not available. Some medical waste is dumped in landfills, and infectious liquids are sometimes dumped into sewer systems, although many sewage treatment plants are incapable of breaking it down. Most medical waste, however, is burned in one of 6,000 incinerators in the United States (see MEDICAL WASTE INCINERATOR). Syringes and other medical waste washing up on beaches are an economic and esthetic problem, not to mention a public health

threat. An estimated $2 to $3 billion was lost by coastal businesses in New York, New Jersey, Connecticut Rhode Island and Puerto Rico in 1988 and 1989 when medical waste washed up on beaches and closed them during the tourist season. Medical waste can contain high concentrations of pathogenic bacteria, with intensive care, pediatric and psychiatric wards typically producing the most. About half the outbreaks of salmonellosis in the United States occur at hospitals. Hospital refuse chutes have frequently been found to be a source of microbial contamination of INDOOR AIR POLLUTION.

The volume of medical waste is increasing because more disposable items are being used in hospitals. Sterility is essential with equipment such as a syringe, and the single use of a sterile needle followed by disposal is an easy way to assure this. But practically everything used in a hospital now is disposable: Dishes and silverware, bedpans, linens, clothing and countless other items that were once washed and reused are now used once and discarded. The need for tighter regulation of the disposal of medical waste has been frequently cited by experts. Especially needed are rules to prevent the same refrigerated trucks from hauling medical waste and then food and rules governing the shipment of medical waste through the mail—both of which are now legal. Needles and other sharp objects in garbage bags filled with medical waste are a threat to all who handle the waste, and there is a danger of exposure to pathogens such as hepatitis or the HIV virus. Rules requiring nonpuncturable packaging of medical waste are also needed.

An Illinois company, Stericycle, runs four medical waste disposal facilities in different parts of the country. Part of the plastics are removed and some are ground up for recycling. Pathogens in the waste are destroyed by using high-energy radio waves to cook the material, a process called *electrical thermal deactivation*, not unlike what would happen in a very large microwave oven. After disinfection, the waste is ground into a fine powder for disposal at a landfill. Plans are underway to use the powder as cement-kiln fuel.

Medical Waste Incinerator   incinerators used to burn medical waste. There are about 6,000 medical waste incinerators in the United States, most of them small and designed to handle only the waste generated by the facility where they are located. Most are in cities, and New York City alone has more than 50. Few such incinerators have more than the most rudimentary pollution-control equipment, and most are operated on a sporadic basis by an unskilled employee such as a janitor. According to an Office of Technology Assessment study, such incinerators emit 10 to 100 times as much air pollution to burn a given weight of waste as does a typical MUNICIPAL SOLID WASTE (MSW) incinerator.

More than half of the waste burned at a typical incinerator consists of PLASTICS, which contain chlorides and HEAVY METALS that become toxic air contaminants when the plastics are burned and create toxic ash. Medical incinerators often emit DIOXINS, acidic gases and VOLATILE ORGANIC COMPOUNDS. The ENVIRONMENTAL PROTECTION AGENCY issued rules under the Medical Waste Tracking Act of 1988 that encourage the use of on-site incinerators. The act requires that any wastes that are removed from a medical facility's premises must be reported, while materials burned on-site need not be reported. The standards for the disposal of the ASH produced by on-site medical waste incinerators are also relatively lax. The rules were intended to prevent future contamination of beaches such as occurred in the northeastern United States in 1988. About two-thirds of U.S. hospitals have their own incinerator.

Pollution control equipment that can clean up the emission from hospital incinerators is readily available. An *acid-gas scrubber*, for instance, sprays a mixture of lime and water into acidic flue gas to form calcium salts and thereby reduce the acidity of exhaust gases, and an ELECTROSTATIC PRECIPITATOR can remove most of the PARTICULATES from emissions. But such equipment is seldom used because of the lax enforcement of AIR POLLUTION standards, the relatively low volume of incinerator use and the relatively high cost of cleaning up EFFLUENT. Such pollution control equip-ment can cost millions of dollars, and is not usually practical for a single hospital to finance the upgrade by themselves. Medical wastes in West Germany go to regional incinerators, where the installation of state-of-the-art combustion and pollution-control equipment is more cost effective. In Switzerland, the most toxic and infectious medical waste is burned at hazardous waste incinerators, and the rest is burned in MSW incinerators. See also INCINERATION.

Merchant Plant   a waste-to-energy facility that is sited at or adjacent to a landfill. A merchant plant converts energy found in solid waste to refuse derived fuel or produces steam in an incinerator, with the ashes being disposed of in the landfill. Merchant plants are often run through cooperative agreements between companies that are primarily in the landfill business and incinerator operators. Such an agreement between WHEELABRATOR and WMX TECHNOLOGIES in 1988 is expected to have a major impact on the future shape of the garbage-disposal industry, although thus far changes have been slow.

Mercury   an ELEMENT that is a heavy, silvery liquid METAL at room temperature. Mercury is one of the least abundant elements in the Earth's crust. It is a good electrical conductor, and is used in electrical switches, thermocouples, fluorescent lamps, thermometers and barometers. It is an ingredient in mirror coatings, fumigants and mildew-proofing PAINTS, and is used in the production of CHLORINE, caustic soda and PAPER. Mercury is used as a catalyst in the production of ORGANIC COMPOUNDS. Approximately 60 metric tons of mercury was produced by U.S. manufacturers in 1993. Mercury is a corrosive chemical (see CORROSIVE MATERIAL). Elemental mercury easily oxidizes (see OXIDATION) into inorganic mercury, which gets into water from industrial or municipal waste-treatment plants, streambed SEDIMENTS and the natural weathering of rocks. Bacteria can convert inorganic mercury to methylmercury. The concentration of water-borne inorganic mercury may increase as the result of ACID RAIN, because the acidic precipitation absorbs atmospheric mercury and increases mercury's release from sediments. Mercury is highly persistent in water, with a HALF-LIFE of more than 200 days. Inorganic mercury and methylmercury have high chronic TOXICITY to aquatic life (see ACUTE EFFECTS, CHRONIC EFFECTS). Eating mercury-tainted fish has caused mercury poisoning in humans, birds and land animals. On-site disposal of mercury used in the production of nuclear weapons at the Department of Energy's Oakridge, Tennessee, nuclear facility between 1950 and 1966 resulted in the release of an estimated 2.4 million pounds of the metal into the surrounding environment, polluting groundwater and nearby streams.

Mercury can affect health when inhaled or passed through the skin. Exposure to high concentrations can cause chest pain, shortness of breath, and a buildup of fluid in the lungs which can lead to death. Repeated exposure can lead to tremors, kidney disease and gum disease, problems with memory and concentration, hallucinations and changes in mood and personality. Prolonged exposure can cause clouding of the eyes. The term "mad as a hatter" refers to the mercury poisoning that was common among hatters, who were routinely exposed to fumes from the mercury they used to cure felt. There is limited evidence that mercury may cause spontaneous abortions in humans. Mercury can cause a skin allergy. Once the allergy develops, very small future exposures can cause itching and rash. Exposure to mercury may lower sex drive.

PUBLIC WATER SYSTEMS are required under the SAFE DRINKING WATER ACT to monitor concentrations of mercury (MAXIMUM CONTAMINANT LEVEL = 0.002 milligrams per liter). The BEST AVAILABLE TECHNOLOGIES for the removal of mercury from DRINKING WATER are granular ACTIVATED CARBON, coagulation-filtration, lime softening or REVERSE OSMOSIS (see WATER TREATMENT). The OCCUPATIONAL SAFETY AND HEALTH ADMINISTRATION has established a permissible exposure limit for airborne mercury of 0.1 milligrams per cubic meter, not to be exceeded at any time. It is also cited in AMERICAN CONFERENCE OF GOVERNMENT INDUSTRIAL HYGIENISTS, Department of Transportation and NATIONAL INSTITUTE FOR OCCUPATIONAL SAFETY AND HEALTH regulations. Mercury is on the RTK Hazardous Substance List and the Special Health Hazard Substance List. Mercury compounds are on the HAZARDOUS AIR POLLUTANTS LIST. Businesses handling significant quantities of mercury or mercury compounds must disclose use and releases of the chemical under the provisions of the EMERGENCY PLANNING AND COMMUNITY RIGHT TO KNOW ACT.

Metabolism   a set of reactions that takes place within a living cell that allow the conversion of energy and chemicals necessary to support life and the construction and maintenance of tissue. *Anabolic metabolism* or *anabolism* is the process by which energy is consumed in the construction of the complex molecules that make up living tissue, and *catabolic metabolism* or *catabolism* is the conversion of complex molecules into simpler ones and an associated release of energy. The *metabolic rate* is the speed at which these reactions take place, and *basal metabolism* is the rate at which metabolism takes place when an organism is at rest.

Metabolism is essentially the same in plants, animals, fungus, bacterium and protista. *Adenosine triphosphate* (ATP) contains three phosphate molecules, one of which is weakly bonded to the rest of the molecule. As a result, it can be easily removed, releasing energy in the process—and then replaced, consuming energy in the process. This characteristic of ATP is used to move energy within the metabolic cycle, with the oxidation of sugar providing the energy produced by photosynthesis in plants and digested from external sources by animals. The metabolic process is extremely complex. A molecule of sugar, which can undergo oxidation in a single step in the atmosphere, may go through more than 100 steps to be oxidized inside a cell.

Metal   an element that gives up an electron easily enough to form a *metallic bond*, in which a single electron may be shared by three or more atoms. There are no whole atoms in pure metals, but they consist rather of a sea of electrons swarming around metallic IONS, bonded first to one, then to another group of atoms. It is this characteristic that makes most metals excellent conductors of electricity (see RESISTIVITY) and of heat (see CONDUCTIVITY), and is what makes it possible to fashion many metals into sheets and wires. Pure metals are a silvery gray, ranging from bright silver to dark lead, but reddish-yellow copper and yellow gold are unique. Metals are used in the production of building materials—from structural STEEL to ALUMINUM rain gutters—and metal foils, sheets, straps, rods, castings, pipes, nails and screws are an integral part of countless products. Pure metal is used to produce machine parts, auto frames and bodies, boat hulls, aircraft shells and other parts where strength and durability are important. Products such as furniture, appliances, automotive accessories (see AUTOMOTIVE PRODUCTS) and batteries all contain metals. Metallic fabric can be used to produce fireproof clothing and other products where resistance to heat is essential. The average American accounts for the use of 16 tons of metals in his or her lifetime, according to estimates by the WORLDWATCH INSTITUTE. PLASTICS have partially replaced metals in many applications such as BUILDING MATERIALS, beverage containers and auto parts. COPPER was undoubtedly the first metal used by humans. Naturally occurring native copper is fairly widely distributed, and the soft metal is easily worked, even without heating. The evolution of man is expressed as a function of the kind of metallic artifacts left behind by the cultures, as in the Bronze Age and the Iron Age.

Metals are recycled at a higher rate than other materials because of their value (see JUNKYARD, METAL RECYCLING). The BIODEGRADABILITY of most metals is low, and many metals and metallic compounds are toxic. The largest part of the HAZARDOUS WASTE produced annually in the United States is the corrosive materials, spent acids and alkaline wastes from the metal finishing, chemical production and petroleum refining industries. Reactive industrial waste comes primarily from the chemical production and metal finishing industries. The BOTTOM ASH from incinerators contains pure metals and metallic compounds. SLAG is a byproduct of metal smelting that consists primarily of silicates.

The Properties of Metals

It is the unique properties shared by most metals that has led to their universal use in every conceivable kind of application. Among these qualities are:

- *ductility*—the ability to be drawn into a wire;
- *malleability*—the ability to be pounded into a thin sheet;
- *metallic luster*—which give metals their shiny surface;
- the ability to form *alloys*;
- *conductivity*—the ability to conduct heat and electricity.

Most metals are subject to CORROSION, and most uncoated metals are unsuitable for use as storage tanks for this reason. Paints and platings of noncorrosive metals are used to combat corrosion. The *noble metals* are more resistant to corrosion. A metal is said to be more noble if it is higher in the galvanic series and less noble when lower in the series. The galvanic series is a ranking of metals and their alloys according to their tendency to lose electrons. Noble metals such as NICKEL, SILVER, graphite, GOLD and platinum are near the top of the galvanic series, while metals such as magnesium, zinc, galvanized steel and aluminum are near the bottom. Noble metals are less affected by processes such as weathering and corrosion. Most metals will crystallize in two or more forms, and are said to be *polymorphic* or *allotropic*. Metals in solid waste can cause problems in compost piles (see COMPOSTING), and metals in SEWAGE can cripple a SEWAGE TREATMENT plant. Metal fragments are allowed to settle out of sewage in a *grit* chamber. Vaporized metals can lead to fast deterioration of a CATALYTIC CONVERTER in an automobile. Many salts are formed by the union of a metal with a nonmetal, as in sodium chloride. SELENIUM is a *metalloid*, a nonmetal with some of the characteristics of a metal. A HEAVY METAL is one of a loosely defined group of metallic elements with a high density and a comparatively low level of chemical activity. See also BARIUM, BERYLLIUM, CADMIUM, CHROMIUM, IRON, LEAD, MANGANESE, MERCURY, OXIDIZING AGENT, RADIUM, THALLIUM, URANIUM.

Metal Recycling   the collection and processing of METAL items for REUSE in new products. Most of the precious metals, and the majority of the COPPER, brass, ALUMINUM, STEEL and IRON discarded is recycled. As with other materials, a reliable market for recycled metals is essential to a successful recycling effort. In 1993, the proportion of aluminum beverage cans recycled was down to 63% from 68% the year before. The decline was caused by a weak metals market created by the sale of large amounts of metals by former Soviet-bloc countries. The price paid for recycled aluminum and the demand for used beverage cans decreased as a result.

In contrast to the volatile aluminum market, the market for recycled steel is relatively stable because of the large number of customers worldwide, for steel scrap and the immense volume of steel they buy, and because of the necessity of using recycled materials in the steel production process. The U.S. scrap recycling industry has been doing a lively international business in discarded U.S. metals, primarily iron and steel, for decades (see STEEL RECYCLING, STEEL RECYCLING INSTITUTE). About 63 million tons of steel scrap was recycled in 1993.

Methane   a HYDROCARBON, chemical formula (CH_4), that is produced by the decomposition of organic matter. Methane is lighter than air, colorless and odorless. It is highly flammable (see FLAMMABLE GAS), and is widely used as a FUEL. Methane is the major constituent of the atmosphere of the outer planets (Jupiter, Saturn, Uranus and Neptune). It is produced in the laboratory by the hydrogentation of carbon, by the action of water on aluminum carbide and by heating sodium acetate with alkali. Methane is used in the production of HYDROGEN, hydrogen cyanide, AMMONIA, acetylene and FORMALDEHYDE. Methane is the first member of the *paraffin* (see ALIPHATIC HYDROCARBON) or *alkane* series of hydrocarbons. It occurs naturally in pockets under the Earth's surface, often in association with PETROLEUM, where it is called NATURAL GAS. Almost 19 quadrillion Btus of dry natural gas and another 2.4 quadrillion Btus of liquid natural gas products were produced in 1993. Natural gas is produced in such great quantities by decaying organic material in swamps that an airborne bubble of *swamp gas* is occasionally mistaken for a UFO. Sources of atmospheric methane gas include wetlands (20.2%), rice fields (19.4%), cud-chewing animals (14.0%), the burning of forests (9.7%), leaks from pipelines carrying natural gas and other petroleum distillates (7.9%), termites (7.0%), coal mining (6.2%), landfills (6.2%), animal waste (5.0%) and sewage (4.4%). *Firedamp*, a gas found in coal mines, is mostly methane.

Methane gas passing into the atmosphere constitutes an estimated 15% of all emissions of greenhouse gases (see GLOBAL WARMING). It is emitted from LANDFILLS as the result of the partial decomposition of the ORGANIC MATERIAL found in garbage. It is has been estimated that methane emissions from U.S. landfills could represent as much as 2% to the worldwide buildup of greenhouse gases, a number equal to the emissions of 10 million automobiles (see AUTOMOTIVE AIR POLLUTION). If not vented to the atmosphere, explosions in landfills (and nearby basements) can occur when the buildup of unvented methane is ignited (see gas migration). The RESOURCE CONSERVATION AND RECOVERY ACT requires that methane gas in landfills be managed, and the CLEAN AIR ACT calls for the collection of VOLATILE ORGANIC COMPOUNDS in landfills (see GAS COLLECTION SYSTEM).

Methane gas emitted by rotting organic material in garbage must be vented to prevent dangerous buildups of the gas. The gas may be released directly into the atmosphere, burned with COMBUSTION BYPRODUCTS (BYPRODUCT) entering the open air, used to fuel an electrical generator, sold to a nearby business that needs an inexpensive heat source or purified and sold to a natural gas utility. About 1,500 of the nation's landfills either vent or burn the methane gas that has been collected. At present, gas sufficient to produce two megawatts of electricity must be present before a commercial power generation venture is economically feasible. It is estimated that approximately 2% of worldwide atmospheric methane emissions come from U.S. landfills. Garbage incinerators emit gases and particulates including toxic DIOXINS and FURANS. About 80% of the garbage's volume is eliminated in the incinerator, and the rest is ASH, which is usually dumped in a landfill.

Higher prices for FOSSIL FUELS and the fact that the gas-collection systems must be installed anyway have led to increased use of methane for the generation of electricity and to power vehicles. Acrion Technologies and Consolidated Natural Gas of Cleveland have developed technology for refining the gas that is produced in landfills by removing CARBON DIOXIDE and trace contaminants to make it usable as a fuel. The companies plan to have modified trucks filling up with liquefied natural gas derived from landfills by 1996. The companies estimate that a landfill with five million tons of trash produces enough methane to fuel 200 GARBAGE TRUCKS or 3,600 cars. Methane is one of the first gases to be released in the fractional DISTILLATION of petroleum. Most commercially traded methane comes from natural gas wells and from the distillation of petroleum and COAL. An ANAEROBIC DIGESTER produces methane from organic material with the help of

ANAEROBIC BACTERIA. The use of natural gas for the generation of electricity has increased markedly over the last decade because of its low price and the relatively low cost of building a gas-fired generation. The fact that such power plants can be added to a utility's power grid in small increments is also appealing to utilities, many of which have suffered massive financial setbacks as the result of unwise investments in mega-plants powered by coal or nuclear power. The buildup of methane gas in landfills can lead to explosions and injuries (see GAS MIGRATION).

The Los Angeles County Transportation Authority plans to put the nation's largest fleet of natural-gas-fired buses on the streets. Plans for the new buses, which will be fueled with compressed natural gas, are going ahead even though prototypes of the buses were plagued by mechanical trouble during a test program in which the engines of all ten of the test vehicles melted down after traveling an average of 2,600 miles. The plan to build the 196 new buses at an estimate cost of $61.5 million came after a $102 million methanol bus project was abandoned because the buses broke down after traveling an average of 4,000 miles. If the fleet is completed it will be the largest of its kind in the United States, according to statistics kept by the American Public Transit Association. There are currently 297 buses in the country that run on compressed natural gas. In Richmond, Iowa, the first of four planned fueling stations that will provide compressed natural gas for city vehicles was dedicated in 1994. Nearly 300 city vehicles are slated to use the fuel by 2001. A car converted to use the fuel has an onboard tank capable of holding about 600 cubic feet of natural gas (the equivalent of about five gallons of GASOLINE), which gives the cars a range of about 110 miles before refueling. When the natural gas runs out, the car can switch to gasoline.

Methanol   methyl alcohol (CH$_3$OH), also called *wood alcohol*. Methanol is derived primarily from COAL or NATURAL GAS, although it can also be obtained from the destructive DISTILLATION of WOOD. ETHANOL is slightly more poisonous than methanol, but the body is able to metabolize small amounts into CARBON DIOXIDE and water, whereas methanol is metabolized in the body into formic acid.

Methanol fuel is a mixture of 85% methanol and 15% unleaded regular GASOLINE and is called *M85*. An M85-fueled car will have more power than one fueled by gasoline because the fuel has a lower density than gasoline, which allows more air to be forced into the combustion chamber. This improved air-to-fuel ratio means a better burn and more power, but also leads to poorer fuel economy—about half the mileage achieved by pure gasoline. The American Petroleum Institute and Ford, Chrysler and General Motors have opposed the widespread introduction of methanol. Unlike gasoline, methanol conducts electricity, which could necessitate extensive redesign of fuel injection systems. Methanol is also corrosive, which could necessitate new materials for the entire fuel-delivery system. Refinery owners are not anxious to produce methanol because they do not have the facilities to switch to a fuel that is derived primarily from coal and natural gas. A study funded by the three auto makers and 14 oil companies in 1992 concluded that methanol was no less polluting than gasoline. The CALIFORNIA AIR RESOURCES BOARD, on the other hand, has sponsored research that concluded that methanol powered vehicles produce only half as much smog-forming hydrocarbons as gasoline-powered vehicles. Businesses handling significant quantities of methanol must disclose use and releases of the chemical under the provisions of the EMERGENCY PLANNING AND COMMUNITY RIGHT TO KNOW ACT. Methanol is on the HAZARDOUS AIR POLLUTANTS LIST.

Methoxychlor a CHLORINATED HYDROCARBON INSECTICIDE used to control insects on fruit and shade trees, vegetables, dairy and beef cattle and home gardens. Methoxychlor gets into air and water from agricultural RUNOFF, from industrial discharges and from spills. It is slightly soluble and moderately persistent in water, with a HALF-LIFE of 20 to 200 days. Methoxychlor has high acute and chronic TOXICITY to aquatic life (see ACUTE EFFECTS, CHRONIC EFFECTS).

Methoxychlor can affect health when inhaled or passed through the skin. Dizziness, headache, disorientation, weakness, a sensation of "pins and needles," muscle twitching and tremor can result from overexposure (see NEUROTOXIN). Repeated overexposure may damage the kidneys. Methoxychlor is a possible human CARCINOGEN that has been shown to cause liver, testes, ovary and spleen cancer in animals, and may damage human testes. There is limited evidence that methoxychlor is a human TERATOGEN.

PUBLIC WATER SYSTEMS are required under the SAFE DRINKING WATER ACT to monitor concentrations of methoxychlor (MAXIMUM CONTAMINANT LEVEL = 0.04 milligrams per liter). The BEST AVAILABLE TECHNOLOGY for the removal of methoxychor from drinking water is granular ACTIVATED CARBON. The OCCUPATIONAL SAFETY AND HEALTH ADMINISTRATION has established a permissible exposure limit for airborne methoxychlor of 10 milligrams per cubic meter total DUST and to 5 mg/m³ respirable fraction averaged over an eight-hour workshift. It is also cited in AMERICAN CONFERENCE OF GOVERNMENT INDUSTRIAL HYGIENISTS and

Department of Transportation regulations. Methoxychlor is on the Hazardous Substance List and the HAZARDOUS AIR POLLUTANTS LIST. Businesses handling significant quantities of methoxychlor must disclose use and releases of the chemical under the provisions of the EMERGENCY PLANNING AND COMMUNITY RIGHT TO KNOW ACT.

Methyl Alcohol see METHANOL

Methyl Chloroform a colorless liquid with a CHLOROFORM-like odor that is used primarily as a cleaning SOLVENT and degreasing agent. Methyl chloroform, an isomer of trichloroethane, gets into air and water from industrial and municipal waste treatment plant discharges and from spills. It is moderately soluble and nonpersistent in water, with a HALF-LIFE of less than two days.

Methyl chloroform can affect health when inhaled or by passing through skin. Symptoms of exposure include dizziness and lightheadedness. Contact can irritate the skin and eyes. Exposure to higher concentrations can lead to unconsciousness, irregular heartbeat and death. Liver and kidneys damage may also result from such exposure. Prolonged contact can cause thickening and cracking of the skin. Methyl chloroform may cause genetic changes in living cells, but it is not clear whether it is associated with cancer and reproductive damage. The odor threshold for methyl chloroform is 120 parts per million (ppm).

The OCCUPATIONAL SAFETY AND HEALTH ADMINISTRATION has established a permissible exposure limit for airborne methyl chloroform of 350 ppm, averaged over an eight-hour workshift. It is also cited in AMERICAN CONFERENCE OF GOVERNMENT INDUSTRIAL HYGIENISTS and NATIONAL INSTITUTE FOR OCCUPATIONAL SAFETY AND HEALTH regulations. Methyl chloroform is on the Hazardous Substance List and the HAZARDOUS AIR POLLUTANTS LIST.

Methyl Ethyl Ketone (MEK) a colorless, volatile liquid with a sharp, mint-like odor similar to that of acetone. Methyl ethyl ketone is popular as an industrial SOLVENT, and is used to make lacquers and varnishes, PLASTICS, lubricating oils and artificial leather. It is also employed in the manufacture of drugs and cosmetics. Certain PAINT strippers and specialty adhesive products (including products made to glue fishing rods and china) contain MEK. The human nose can generally detect MEK before it builds up to toxic levels. Although acetone is not particularly toxic, many related ketones with a similar odor that may be present in MEK are very toxic. When spilled, it quickly evaporates into the atmosphere, where it is easily broken down. MEK mixes well with water and oils, and can enter into solution with surface water when it is spilled before it has a chance to evaporate. It evaporates easily when surface water is agitated as in a waterfall, but it can accumulate to toxic levels in groundwater. Small amounts of MEK can get into DRINKING WATER that stands in plastic water supply pipes for more than a few hours. MEK can explode when exposed to heat or flame. Approximately 300,000 tons of MEK are produced annually. An estimated three million workers routinely exposed to MEK vapors.

MEK can affect health when inhaled or passed through the skin. Symptoms of exposure include irritation of the eyes, mucous membrane and skin. Headache and throat irritation are common at levels near the maximum allowed in the workplace. At higher levels, numbness of the fingers and arms are experienced. Long-term exposure to MEK vapors can result in dermatitis, and long-term exposure to very low levels can lead to decreased memory and slowed reflexes. MEK is expelled from the body by exhalation and through the kidneys. The presence of MEK fumes can increase the toxicity of other solvents including CARBON TETRACHLORIDE and CHLOROFORM. MEK is on the HAZARDOUS AIR POLLUTANTS LIST. Businesses handling significant quantities of napthalene must disclose use and releases of the chemical under the provisions of the EMERGENCY PLANNING AND COMMUNITY RIGHT TO KNOW ACT.

Methyl-tertiarybutylether a GASOLINE additive used in the OXYGENATED FUELS designed to reduce AIR POLLUTION in areas that do not meet federal standards for CARBON MONOXIDE and OZONE. Methyl-tertiarybutylether (MTBE) is blended with fuel at the refinery and the resulting oxygenated gasoline can be delivered to local distributors through pipelines. Refiners add between 11% and 15% MTBE to gasoline to boost its oxygen content from zero to between 2% and 2.7% MTBE is obtained by processing methanol.

Reports of headaches and nausea in New Jersey, Montana and Alaska have been widely associated with the presence of MTBE in gasoline, symptoms that have confounded public health officials. Although MTBE is a possible CARCINOGEN, and engines burning the fuel emit slightly more formaldehyde than those burning conventional fuels, health researchers have not found it to be associated with such symptoms. Jack Snyder, M.D., Ph.D., a professor of Toxicology at Thomas Jefferson University in Philadelphia, says that the adverse health effects being attributed to

MTBE are actually more consistent with exposure to gasoline. He said that MTBE's distinctive odor may cause some people to be more aware of gasoline fumes than they normally would be, resulting in the symptoms. The U.S. Geological Survey found MTBE in tests of shallow aquifers, precipitation and snowpack in the Denver area in tests conducted in 1993 and 1994, apparently the result of the city's switching to oxyfuels in 1988.

Methylene Chloride a colorless, volatile SOLVENT with a pleasant odor. Methylene chloride is used in PAINT removers, degreasers, cleaning fluids, INSECTICIDES and in fire extinguishers. It is moderately soluble and slightly persistent in water, with a HALF-LIFE of two to 200 days. Methylene chloride gets into air and water primarily from industrial discharges. The use of paint strippers with inadequate ventilation can produce toxic levels of methylene chloride in indoor air.

Methylene chloride can affect health when inhaled or passed through the skin. Contact with the liquid can cause skin irritation and can irritate eyes. Mild exposure can cause headaches, unsteadiness and "drunken" behavior. The chemical decreases the blood's ability to carry oxygen, which can cause fatigue, shortness of breath and heart pain. Inhaling the vapor can irritate the nose, throat and lungs and cause coughing. Exposure to higher concentrations can cause a life-threatening buildup of fluid in the lungs, severe shortness of breath and liver damage. Methylene chloride is a possible human CARCINOGEN that has caused liver and lung cancer in animal tests. Long term exposure may cause brain damage and memory loss, reduced coordination and impaired thinking ability. Repeated contact can cause thickening and cracking of the skin. The odor threshold for methylene chloride is 250 parts per million (ppm). Methylene chloride must be stored and disposed of as a HAZARDOUS WASTE in some states.

The OCCUPATIONAL SAFETY AND HEALTH ADMINISTRATION has established a permissible exposure limit for airborne methylene chloride of 500 ppm, averaged over an eight-hour workshift, with a maximum allowable concentration of 2,000 ppm for 5 minutes in any two-hour period. It is also cited in AMERICAN CONFERENCE OF GOVERNMENT INDUSTRIAL HYGIENISTS, National Toxicology Program, NATIONAL INSTITUTE OF OCCUPATIONAL SAFETY AND HEALTH, Department of Transportation and ENVIRONMENTAL PROTECTION AGENCY regulations. Methylene chloride is on the Hazardous Substance List and the HAZARDOUS AIR POLLUTANTS LIST.

SI Units of Weights and Measures

Base Units

length—meter (m)
mass—kilogram (kg)
time—second (s)
electric current—ampere (A)
temperature—kelvin (K)
amount of substance—mole (mol)
luminous intensity—candela (cd)

Metric Equivalents

1 avoirdupois ounce = 28.3495 grams
1 apothecaries' or troy ounce = 31.1035 grams

1 apothecaries' or troy pound = 373.242 grams = 0.37324 kilogram
1 avoirdupois pound = 0.45359 kilogram
1 short ton = 0.9072 metric tons
1 long ton = 1.016 metric tons
1 metric ton = 1.102 short tons
1 metric ton = 0.9842 long ton
1 imperial gallon = 1.2 United States gallons

The English and American mile is 1,760 yards (statute mile). Other nations measure differently. A nautical or geographical mile is one minute of the arc on the Earth's surface at the equator. A knot is a speed sufficient to travel one nautical mile in one hour.

Metric System a system of weights and measures in which all units are related to one another by powers of ten. The metric system has been adopted by the majority of the world's governments, and is the standard measurement system for all scientific work. The system was first adopted as France's official system of weights and measures in 1790. The *meter* was defined as one ten-millionth of the distance from the equator to the north pole on a line running through Paris. By 1900, the metric system was based on the meter-kilogram-second (mks) system in which the unit of mass was the kilogram and the unit of time, the second. The metric system later was based on the meter-kilogram-second-ampere (mksa) system. In 1821, President John Quincy Adams advocated the adoption of the metric system in a report to Congress, which did not pass legislation legalizing the use of metric weights and measures until 1866. In 1893, the National Bureau of Standards adopted the metric system as the standard against which the conventional yard and pound were measured.

Earlier systems were replaced in 1960 with the International Standard System or *Systeme International* (SI). The SI system has now been adopted worldwide. The system has six basic units: The *meter* and the *second* are defined in terms of nuclear properties; *Celsius degrees* are defined in terms of the *triple point* of water, which is the point just above freezing where the solid, liquid and gaseous forms of water can exist together in equilibrium. The Celsius, or Centigrade, temperature scale defines the freezing point of water as zero and its boiling point as 100. The *kilogram* is based on a cylinder made of platinum and iridium that is kept in a vault in Paris. The *candela* is defined as the light emitted from one-sixtieth of a square meter of surface area when a perfect radiator of heat called a *black body* is heated to the melting point of platinum. The *ampere* is defined in terms of the force generated by electric current flowing through parallel wires running through a vacuum exactly one meter apart. In 1965, Great Britain started to urge other world governments to switch to the metric system. The Metric Conversion Act of 1975 had the complete conversion of the United States to the metric system as its goal, although no timetable was specified. So far it has met massive public resistance.

The length of the meter was redefined in 1960 in terms of the wavelengths of light. The meter, for example, is defined as 1,650,763.73 wavelengths in a vacuum of the orange-red line of the spectrum of krypton-86.

The following prefixes modify the number with which they are associated by the following amounts: *giga*—multiply by one billion; *mega*—multiply by one million; *kilo*—multiply by one thousand; *hecto*—multiply by one hundred; *deka*—multiply by ten; *centi*—divide by ten; *milli*—divide by one thousand; *micro*—divide by one million; *nano*—divide by one billion. So a *giga*watt is in one billion watts (or one million *kilo*watts), and a microgram is one-millionth of a gram (or one thousand *nano*grams).

Centigrade degree is the unit measurement of a temperature scale that divides the difference between the freezing point (32°F) and the boiling point of water (212°F) into 100 equal units.

Microbe see MICROORGANISM

Micronutrient see TRACE ELEMENT

Microorganisms living organisms—also called *microbes*—that are too small to see without magnification. Microorganisms are found in virtually all air, soil and water. Among the million-plus known species of microorganisms are *monera* (which include BACTERIA and blue-green ALGAE), most *Protista* (primarily one-celled organisms including protozoans, algae and slime molds), *fungi* (such as yeasts and smuts) and some animals (including rotifers, hydra and the larvae of many marine invertebrates). VIRUSES are often referred to as particles rather than microbes because they are too small to see with an optical microscope and because it is not firmly established whether or not they are living. Botulism, which is caused by toxins emitted from a bacterium called *clostridium botulimum*, is generally regarded as the most life-threatening of the diseases that can be carried by food. Microorganisms are responsible for vital processes such as the digestion of food and the putrefaction and decay of dead organic material. Humans have been using microbes for more than 9,000 years for fermentation processes such as the brewing of alcoholic beverages and the production of cheese and bread.

Microbes, their waste and dead bodies are a major source of INDOOR AIR POLLUTION. Even in clean environments such as hospitals, where HIGH-PERFORMANCE PARTICLE-ARRESTING FILTERS are used to clean air, some microbes are found. *Hypersensitivity pneumonitis*, or *humidifier fever*, is a flu-like disorder caused by toxins released by microbial growths in damp places, such as a humidifier or refrigerator condensate pan. At least some microorganisms are found in even the purest water and ice, and pathogenic organisms often spread through water systems (see DISINFECTION, DRINKING WATER, WATER POLLUTION). Water with unusually high levels of microbes can quickly contaminate WATER FILTERS (see WATER TREATMENT).

Pathogenic bacteria such as e-coli and SALMONELLA get into water from mammalian FECES. Drinking water is routinely tested for fecal coliform bacteria, which are not toxic but are easier to detect than some of the more pathogenic organisms that are usually found in company with them. TURBIDITY and heterotrophic plate bacteria counts are also measured in drinking water, and the ENVIRONMENTAL PROTECTION AGENCY (EPA) recommends that both be kept as close as possible to zero, although no MAXIMUM CONTAMINANT LEVEL GOAL has yet been proposed. Specific treatment techniques can be recommended for substances such as GIARDIA lamblia and other microorganisms where a MAXIMUM CONTAMINANT LEVEL is not appropriate. The EPA allows PUBLIC WATER SYSTEMS that use surface water to avoid the requirement for filtration in the 1986 amendments to the SAFE DRINKING WATER ACT if they can show that disin-fection has removed 99.9% of the giardia cysts and 99.9% of viruses from the water being processed.

The metabolism of microorganisms is used to break down contaminants, a process called *biotransformation* or BIOREMEDIATION. AEROBIC BACTERIA and other microorganisms have been used to neutralize sewage since the first treatment plants (see SEWAGE TREATMENT). Naturally occurring microbes with the ability to break down toxic compounds may be sought out, or new ones may be produced through genetic engineering. The metabolism of microbes is also being harnessed to produce basic chemicals from biomass that are now largely derived from petroleum. Microorganisms are also seeing increased use to leach valuable minerals from their ores. About 10% of the COPPER now produced in the United States is now removed from copper ores by naturally occurring microbes. The use of microbes to control insects and other agricultural pests is also increasing (see PESTICIDES, SUSTAINABLE AGRICULTURE). Concern about the release of genetically engineered microorganisms into the environment has so far kept the practice to a minimum. See also ACTIVATED SLUDGE, AEROALLERGENS, AEROSOL, ALLERGEN, BIOLOGICAL AEROSOLS, CRYPTOSPORIDIUM, DUST, POLLEN.

Midden an accumulation of refuse. *Midden* is a Scandinavian word that originally referred to a dunghill. Much of what we know about ancient cultures is the result of archaeologist's excavation and analysis of ancient middens. Shell middens, the remnants of Indian feasts on clams and oysters, are common along Atlantic and Gulf coasts of the United States. One at Pope's Creek, Maryland, along the Potomac River, covers 30 acres to a depth of 10 feet. See also ARCHAEOLOGY, ARTIFACT, GARBOLOGY, STRATIGRAPHIC LAYER.

Midnight Dumping the practice of illegally disposing of HAZARDOUS WASTES, often by hauling it to a secluded spot and dumping it on the ground or in a watercourse at night. An unknown but significant portion of the hazardous waste produced in the country is disposed of through illegal dumping. Since the materials disposed of cannot be traced to a given source, it can be very difficult to apprehend midnight dumpers. Organized crime is playing an increased role in the illegal disposal of hazardous waste, a trend that is deeply disturbing to regulators. As the costs associated with hazardous waste treatment and disposal (see HAZARDOUS WASTE TREATMENT, HAZARDOUS WASTE DISPOSAL) increase, the potential for making money by illegally disposing of waste materials increases.

Milligram see MEASUREMENTS, METRIC SYSTEM

Mine Drainage GROUNDWATER and RUNOFF that seeps through tunnels and bore holes created by mines. Mine drainage may run out of a mine mouth to contaminate surface water, or may percolate through underlying soils and cracked bedrock to contaminate groundwater. See also HARD-ROCK MINE, HEAP LEACHING, MINERAL WASTE, SEDIMENTATION, TAILINGS.

Mineral a naturally occurring element or compound that is found as a constituent of rock. Most but not all minerals are solids, the exception being MERCURY and WATER. Minerals are generally inorganic (see INORGANIC COMPOUND), although pure carbon is usually classed as a mineral. Groundwater with a high mineral content is called hard water.

There are approximately 2,000 known minerals; feldspar, which makes up an estimated 30 to 50% of the earth's crust, and quartz, which constitutes 28 to 30%, are the most common. The calcite group, which includes limestone; the HALIDE group; the micas; and the IRON-oxide compounds, which include the iron ores limonite, taconite and hematite, are other widely occurring minerals. Minerals generally occur in the form of crystals too small to be seen with the naked eye (*microcrystals*). Larger, easily visible crystals form in areas where a large body of magma has come close to the surface and cooled very slowly, and where mineral-rich groundwater has infiltrated through a rock body containing cavities. Crystals that are particularly rare, well formed or attractive often have an economic value. All the rocks that form the Earth's crust are minerals. Most rocks are made up of several minerals, although pure limestone (calcium carbonate) and pure quartzite (silicon dioxide) are exceptions. A mineral is classified according to its chemical composition, crystal class, hardness and appearance (color, luster and opacity). Metallic minerals that are mined because of their value are termed ores. Mineralogy is the study of the properties, identification, origin and classification of minerals, and geology is the study of the earth's largely mineral surface. In mineralogy and geology, only the chemical elements and inorganic compounds are termed minerals. Petroleum and COAL, which formed from the decomposition of organic material, are not minerals, although they are sometimes called minerals. See also MINERAL FIBERS, MINERAL WASTE.

Mineral Fibers pieces of MINERAL crystals from rocks that have been rendered small enough to become airborne by erosion or by mechanical crushing or abrasion. Mineral fibers get into outdoor air primarily from mines, quarries and gravel roads, where the constant pounding provided by passing vehicles crushes rocks.

A variety of industrial sources also can emit mineral fibers. Household DUST contains mineral fibers such as ASBESTOS and fiberglass that come from BUILDING MATERIALS and other fibers that originated as outdoor dust that was blown into the house through cracks, windows or doors or as soil particles that were tracked into the house. All can become airborne if they are disturbed. Proposed mines are often challenged in court on the grounds that mineral fibers released by the operation will be a local source of AIR POLLUTION. Mineral fibers are on the HAZARDOUS AIR POLLUTANTS LIST. See also PARTICULATE.

Mineral Waste waste consisting primarily of rocks that have been broken down into smaller sizes or pulverized, and the MINERAL DUST released by such operations. The TAILINGS of mines, mills and quarries are a form of mineral waste. Such tailings frequently contaminate water with heavy metals or radioactive materials. Structures built on tailings from URANIUM mines and mills have frequently been found to have unacceptably high radiation levels. The majority of mineral waste is nontoxic, and can be safely used as fill.

Mining the removal of mineral deposits and mineral ores from the earth (see HARD-ROCK MINE, PLACER MINE, STRIP MINE). Mining is a major source of diffuse pollution, and the milling of mineral ores is a major point source of both AIR POLLUTION and WATER POLLUTION. TAILINGS piles at mines and mills can leach HEAVY METALS and ACIDS into RUNOFF for centuries (see ACIDIC WASTE, PYRITE, TAILINGS DAM, TAILINGS POND). The extraction of GROUNDWATER at a rate faster than that at which an AQUIFER is being recharged with runoff is sometimes called *water mining*, and logging at a rate faster than natural forest regeneration is similarly called *timber mining*. *Hydraulic mining* was accomplished by washing away stream banks with a high-pressure hose and running the resulting slurry through a *sluice box*—an especially devastating mining technique. See also MINING LAW.

Mining Law laws that apply to mining. Mining law applies to the extraction of minerals including PETROLEUM and NATURAL GAS. Mining policy in the United States is governed primarily by ordinances that went into effect in 1866, 1870 and 1872, at a time when the development of the West—not the protection of its natural resources—was considered paramount. The laws were based on rules that had been worked out by miners in Colorado and California. All public lands except those specifically withdrawn from mineral development were open to mineral exploration and development, and miners were granted the right to develop water sources for use on their claims and to build roads to access mining properties located on public land. The water rights outlined in the regulations are the basis of the first-in-time, first-in-right doctrine used to appropriate water in the western United States (see WATER LAW). Miners are allowed to tap natural water sources on federal land and to transport it to their claim in a ditch or flume, with the rights to use the water going to the first miner who went to the trouble and expense of developing the source. When the claim played out and the water source was no longer used, the right was to end. The mining laws also spelled out the allowable dimensions for mining claims, and allowed the public land in a mining claim to be patented—that is, to become the property of the individual or company that owned the claim. Federal properties in which mining is not allowed include national parks, national monuments, national wildlife refuges, wilderness areas, military reservations and federal land in the states of Wisconsin, Minnesota, Michigan, Alabama, Missouri and Kansas. The *Aspinall Amendment* to the Wilderness Act of 1964 allowed mineral exploration within wilderness areas to continue for 20 years after passage of the law, and for mining to continue on both patented and unpatented mining claims thereafter.

One of the few changes to federal mining law is that the definitions of *locatable* and *leasable* minerals have been somewhat restricted. A *locatable mineral* is one for which a mining claim can be filed under the provisions of federal mining law, and consist primarily of metals and gems. Common minerals such as gravel, stones for building, phosphate rock, sulfur or conglomerate (sedimentary rock containing chunks of parent rock) are not considered to be locatable minerals nor are petroleum or natural gas. The rights to mine a *leasable material* can be leased from the federal government, but claims cannot be patented on the mineral lodes, and management of the area reverts to the government when the mining is complete. Energy resources such as oil shale, petroleum, natural gas and coal and deposits of phosphate rock, sulfur, potassium and sodium are considered leasable materials. Several attempts to pass a new federal mining law have been made during the 1990s, and mining companies, regulators and environmentalists agree that an overhaul of antiquated federal mining laws is long overdue. But the mining industry, fearful that any change will make it more difficult and expensive for miners to remove minerals from federal lands, have so far effectively blocked passage of any new legislation. See also HEAP LEACHING, MINE DRAINAGE, MINERAL WASTE, SEDIMENTATION.

Mist a liquid dispersed into very small droplets (see AEROSOL, SUSPENSION). Mists can be produced by atomizing or spraying, and through chemical reaction, bubbling a gas through a liquid or allowing a gas to escape from a liquid under pressure. Sneezing produces a mist that contains microbial contaminants (see MICROORGANISM, INFECTIOUS WASTE).

Mixed Liquor the mix of zoogloeal and WASTEWATER in a FLOCCULATION tank used in secondary SEWAGE TREATMENT (see SECONDARY TREATMENT). The mixed liquors consist of the stream of wastewater flowing into the tank for treatment (the influent) and a much smaller stream of ACTIVATED SLUDGE that is added to the influent to initiate the microbial action that occurs in the tank. Mixed liquor suspended solids are solid particles suspended in wastewater that is going through activated-sludge treatment. See SEWAGE SLUDGE.

Mixing Zone a stretch of river downstream from a POINT SOURCE of pollution where concentrations of certain pollutants are allowed to be higher than federal guidelines. WASTEWATER mixes with river water in the mixing zone reducing the concentration of contaminants. Some of the more volatile contaminants will evaporate in the mixing zone. See also CLEAN WATER ACT, EFFLUENT, WATER POLLUTION.

Mixture a substance containing two or more compounds or elements that are not chemically bonded to one another, and that retain their unique chemical properties. A mixture is heterogeneous—different parts are composed of varying proportions—in contrast to a homogenous SOLUTION, in which all parts are the same. Since no chemical bonding is involved, a mixture can be separated by mechanical methods such as FILTRATION, SEDIMENTATION or exposure to heat. A mixture has no set of distinguishing characteristics of its own, but instead has a mix of the qualities of the various homogenous materials it contains. A SUSPENSION and a colloidal suspension (see COLLOIDAL SOLIDS) are mixtures.

Mobile Source a source of AIR POLLUTION that is movable such as an automobile, train, ship or aircraft. Although each individual mobile source is not fixed, mobile sources are often treated as a kind of POINT SOURCE of air pollution, as is the case with the effort to reduce tailpipe emissions in a given AIRSHED. Mobile sources may also be treated as an *area source* of pollution, as when the amount of air pollution that typically can be expected from a highway, a parking facility or an airport is addressed as part of an effort to curb local air pollution problems. Mobile sources are the single largest source of air pollution in most urban areas, and are a principal source of HYDROCARBONS, nitrous oxides and CARBON MONOXIDE. See CLEAN AIR ACT. Compare STATIONARY SOURCE, DIFFUSE POLLUTION.

Molecule the smallest unit that a COMPOUND can be divided into while maintaining its physical and chemical characteristics. A molecule's composition is reflected by its chemical formula. Carbon dioxide (CO_2), for instance, is composed of two oxygen (O) atoms attached to one carbon (C) atom. Most elements are *monatomic*, that is, they exist as individual atoms in their pure state; but a few, such as hydrogen (H_2), chlorine (Cl_2) and oxygen (O_2), are diatomic and some, such as ozone (O_3) are *triatomic*. Molecules are bound together by shared electrons. In a *covalent bond*, two atoms share a pair of electrons that orbit around both atomic nuclei.

Monitoring Wells a well drilled to study the quality and flow patterns of GROUNDWATER. Monitoring wells are the only means of determining the size and shape of a PLUME of pollution (see GROUNDWATER POLLUTION) in an AQUIFER and of measuring the severity of the contamination. Because of the complexity and unpredictability of groundwater flow, numerous monitoring wells must be used. The RESOURCE CONSERVATION AND RECOVERY ACT requires that landfills have monitoring wells that will detect contamination leaking from a landfill. A *vapor well*, also called a *sniff well*, is drilled only as deep as the top of the water table, and is fitted with an instrument capable of detecting volatile organic chemicals. A *wet well* extends below the surface of the water table, and can be used to take samples for analysis of groundwater contamination and to determine water flow characteristics within an aquifer. A *single-interval well* is cased with impermeable pipe above the water table and permeable pipe below the groundwater's surface. A *well cluster* is a close-together group of wells drilled to different depths to measure water quality and movement. A well cluster is sometimes made up of several well casings inserted in a large-diameter bore hole. A *nested well* employs several sampling screens deployed at different depths and connected to the surface with separate pipes within the same well casing.

Monofill see ASHFILL

Monomer a chemical COMPOUND, or a portion of a compound that is composed of nonrepeating parts. Elements of the compound may appear more than once, but the relationship of the repeating elements to the surrounding constituents of the molecule will be different in each place they occur. See also PLASTICS. Compare POLYMER.

Motor Oil see ENGINE OIL

Municipal Solid Waste household waste and certain commercial and industrial waste that is disposed of in an area served by a municipal LANDFILL. MUNICIPAL SOLID WASTE (MSW) consists primarily of solid materials discarded by householders, with a smaller proportion coming from commercial and industrial installations. About 200 million tons of MSW are generated annually in the United States. PAPER and paperboard account by weight for 73 million tons, or 37.5% percent of the nationwide total in 1990, according to the ENVIRONMENTAL PROTECTION AGENCY (EPA). The agency estimates that 80% of the nation's municipal solid waste is dumped in landfills, with most of the rest being either recycled or burned in incinerators. ASH from incinerators and SEWAGE SLUDGE are also frequently disposed of in municipal landfills.

Municipal solid waste contains liquids. Although potentially toxic waste products such as spent SOLVENTS, used ENGINE OIL and old PAINT are not supposed to be dumped in the trash, containers holding such liquids routinely go to the landfill. Absorbent solid materials also may be soaked with liquids. Liquids contained in solid waste join rainwater and percolate down through the garbage, picking up toxins along the way, and collect at the bottom of the pile (see LEACHATE).

Mutagen a substance capable of damaging an organism's reproductive DNA, causing birth defects and other changes in offspring. Such changes are *heritable*, birth defects that will also be passed on to the offspring's progeny. Mutagens are capable of penetrating a cell's wall and reaching the nucleus and reacting chemically with the DNA molecule to alter its structure. Radiation that passes through the cell and damages DNA is also mutagenic. Many mutagens are also CARCINOGENS, which also cause cellular changes. Compare TERATOGEN.

N

Napthalene the active ingredient in mothballs and moth flakes. Napthalene is used in home carpet cleaners, typewriter correction fluid, adhesives, INSECTICIDES, veterinary medicines and as a deodorant in toilet-bowl cleansers. It is used in the production of dye compounds, SOLVENTS, explosives, PLASTICS, lubricants and motor fuels. It is a white crystalline solid at room temperature. Napthalene gets into water from industrial and municipal wastewater treatment plants and spills. It gets into indoor air from mothballs, cleansers, adhesives and other products of which it is a component. It is commonly found in LEACHATE. Naphthalene is the most abundant single constituent of coal tar, from which it is derived. It is moderately soluble and slightly persistent in water, with a HALF-LIFE of two to 20 days.

Naphthalene can affect human health when inhaled or passed through the skin. Exposure can irritate the eyes, nose and throat. Higher concentrations can cause headaches and nausea; can damage the red blood cells causing hemolytic anemia; can damage the nerves of the eye (see NEUROTOXIN) and the liver and kidneys; and, in extreme cases, can cause death. Repeated exposure can lead to a clouding of the eye lens that can damage vision. Naphthalene is a possible human TERATOGEN.

The OCCUPATIONAL SAFETY AND HEALTH ADMINISTRATION has established a permissible exposure limit for airborne napthalene of 10 parts per million (ppm), averaged over an eight-hour work-shift. It is also cited in AMERICAN CONFERENCE OF GOVERNMENT INDUSTRIAL HYGIENISTS and Department of Transportation regulations. Naphthalene is on the Hazardous Substance List and the HAZARDOUS AIR POLLUTANTS LIST. Its odor threshold is 0.084 ppm. It may be necessary to store and dispose of naphthalene as a HAZARDOUS WASTE, depending on state regulations. Businesses handling significant quantities of napthalene must disclose use and releases of the chemical under the provisions of the EMERGENCY PLANNING AND COMMUNITY RIGHT TO KNOW ACT.

National Agricultural Library the library of the U.S. Department of Agriculture. Information regarding farming, forestry and climate are available from the library. The library's Global Change Information Packet is one of the best sources on the climatic change available from the federal government. The NAL publishes directories, bibliographies, reference briefs and factsheets on subjects such as sustainable agriculture, pesticides, erosion, irrigation water, biotechnology, food and nutrition and climate. Address: United States Department of Agriculture, National Agricultural Library, 10301 Baltimore Blvd, Beltsville, MD 20705. Phone: 301-504-5755.

National Ambient Air Quality Standards federal standards governing the level of POLLUTION allowable in AMBIENT AIR. The National Ambient Air Quality Standards (NAAQS) were established in the 1970 amendments to the CLEAN AIR ACT. Maximum allowable levels are set for individual pollutants and classes of pollutants by the ENVIRONMENTAL PROTECTION AGENCY. Classes of contaminants are called *criteria pollutants* because they are regulated on the basis of criteria published by the agency. The seven criteria pollutants are: TOTAL SUSPENDED PARTICULATES, SULFUR DIOXIDE, CARBON MONOXIDE, nitrogen dioxide (see NITROGEN OXIDES), OZONE, HYDROCARBONS and LEAD. *Primary standards* are set to protect human health and are theoretically low enough to prevent a significant number of individuals exposed to airborne toxins from suffering adverse health effects as a result. *Secondary standards*, which are stricter, are designed to protect the health of wildlife, plants and other living organisms and to prevent damage to property, clothing and intangibles such as visibility that have only an indirect effect on human life. The law does not allow the economic impacts of compliance with NAAQS to be considered, but individual states may address economic impacts when they establish control strategies to meet the standards. See also AIR POLLUTION, AIR QUALITY

CLASSES, AMBIENT QUALITY STANDARD, NATIONAL EMISSION STANDARDS FOR HAZARDOUS AIR POLLUTANTS, POLLUTANT STANDARD INDEX, STATE IMPLEMENTATION PLAN.

National Association for Plastic Container Recovery an association of manufacturers, suppliers and businesses involved with the production of PET plastic products (see POLYETHYLENE TEREPHTHALATE PLASTICS) founded in 1987. The association offers technical assistance, help with market development and educational materials to U.S. communities and organizations interested in recycling PET plastics. Address: 3770 Nations Bank Corporate Center, 100 N Tryon Street, Charlotte, NC 28202. Phone 704-358-8882.

National Drinking Water Priority List a list of contaminants found in public drinking-water systems that could have an adverse health affect on water users. The ENVIRONMENTAL PROTECTION AGENCY was required by the 1986 amendments to the SAFE DRINKING WATER ACT to compile the National Drinking Water Priority List (NDWPL), which was initially published in 1988, with updated lists issued every three years. The EPA must propose MAXIMUM CONTAMINANT LEVEL GOALS and MAXIMUM CONTAMINANT LEVELS for 25 contaminants on the list within two years of publication. The EPA is required to assemble an advisory group to assist in compiling the list, and to consider chemicals listed as toxic under COMPREHENSIVE ENVIRONMENTAL RESPONSE, COMPENSATION AND LIABILITY ACT and Federal Insecticide Fungicide and Rodenticide Act for inclusion. PUBLIC WATER SYSTEMS must perform monitoring for unregulated contaminants on the list at least every five years. See also NATIONAL PRIMARY DRINKING WATER REGULATIONS, NATIONAL SECONDARY DRINKING WATER REGULATIONS.

National Emission Standards for Hazardous Air Pollutants federal standards governing the emission of AIR POLLUTION from STATIONARY SOURCES. National Emission Standards for Hazardous Air Pollutants (NESHAP) were established as part of the CLEAN AIR ACT. Pollutants governed by NESHAP must not already be covered by NATIONAL AMBIENT AIR QUALITY STANDARDS, and must be proven to be dangerous to human health. The ENVIRONMENTAL PROTECTION AGENCY (EPA) is granted authority to set the standards in section 112 of the act, but this authority may be turned over to individual states if the STATE IMPLEMENTATION PLAN is deemed by the EPA to be adequate to control the pollutant in question. The EPA has established standards for ASBESTOS, MERCURY, BERYLLIUM and VINYL CHLORIDE under the NESHAP

rules, and has prepared draft standards for BENZENE and ARSENIC.

National Environmental Policy Act a federal law that establishes a framework for the review of actions that may have environmental impacts. Under the National Environmental Policy Act (NEPA), signed into law in 1970, federal agencies are required to file an environmental impact statement before taking significant actions "affecting the quality of the human environment." In the impact statement, factors that might pollute air, water or soil or that might negatively affect plants or animals—especially that are threatened or endangered with extinction—must be identified and analyzed. The creation of the Council on Environmental Policy was mandated in the act. NEPA establishes a framework for cooperation on environmental matters between state, local and federal government agencies and between federal agencies and foreign governments. There have been continuous legal challenges to NEPA and its enforcement, primarily surrounding when and how environmental impact statements should be prepared. The question of which federal actions are adequately significant in terms of their environmental impact to require a statement is especially contentious. Many impact statements have been challenged in court for not meeting the criteria for impact assessment set down in the act. The act has generally been interpreted strictly by the courts, and numerous projects have been slowed down for redesign and a few have been stopped altogether as a result.

National Institutes of Health see PUBLIC HEALTH SERVICE

National Institute of Occupational Safety and Health a subdivision of the OCCUPATIONAL SAFETY AND HEALTH ADMINISTRATION that conducts research and testing to identify workplace hazards and to develop safety standards. NIOSH is not to be confused with the National Institute *for* Occupational Safety and Health, which is a subdivision of the Centers for Disease Control in Atlanta, Georgia.

National Pollutant Discharge Elimination System a national system of permits governing the release of EFFLUENT into surface water that was established under the provisions of the CLEAN WATER ACT. NPDES is administered by the ENVIRONMENTAL PROTECTION AGENCY (EPA). State governments that are deemed by the EPA to have an adequate water-quality management plan are given the authority to manage (with EPA supervision) NPDES permits within their boundaries, and roughly 70% of the states now do so. A permit is required for any POINT SOURCE that discharges contaminants into a waterway. Sewage plants and industrial outfalls are required to have NPDES permits. Discharges from power-plant cooling systems, SPOIL from dredging, bilge water and other wastes from ships and boats, drainage ditches and RUNOFF from feedlots are also governed under the system. Under NPDES provisions, permits are to be granted only when applicants have proven the TOXICITY of the effluent they plan to emit will be treated using the BEST AVAILABLE TECHNOLOGY, and that the effluent discharge will not reduce the water quality of the receiving body of water below the AMBIENT QUALITY STANDARD that has been established for the body of water in question. If granted, the permit specifies the maximum amount of certain pollutants that can be emitted. Before the permit can be renewed, which is generally after five years, the holder must demonstrate that its terms have been met during the period covered, and that the water-quality standards specified in the Clean Water Act are being met in the body of water into which emissions are being dumped. See also STATE POLLUTANT DISCHARGE ELIMINATION SYSTEM, WATER POLLUTION.

National Polystyrene Recycling Council a joint venture of eight large plastics manufacturers formed in the mid-1980s to facilitate and promote the recycling of POLYSTYRENE products. The council announced a program in 1993 through which users of disposable polystyrene food trays could return boxes of used trays to the company for RECYCLING via United Parcel Service. The charge of $7.50 per box for the service covers the UPS fee plus NPRC's handling and reprocessing costs. The program has thus far received a lukewarm response, at best, from hospitals and industrial cafeterias, but has been welcomed enthusiastically by schools, with 300 institutions signing up for the program in its first six weeks. The fact that many schools teach students about recycling and the environment and that many PTAs support recycling made it natural for schools to participate in the program. Critics of the NPRC program say that it is nothing more than a public subsidy paid by those least able to afford it to support the plastic industry's polystyrene recycling costs.

National Primary Drinking Water Regulations regulations that specify the maximum allowable concentration of toxic substances found in DRINKING WATER. The National Primary Drinking Water Regulations (NPDWR) are enforced by the ENVIRONMENTAL PROTECTION AGENCY under the authority of the SAFE DRINKING WATER ACT. The EPA is required under the act to either assign a listed pollutant a MAXIMUM CON-TAMINANT LEVEL—the highest acceptable concentration that can legally be found in drinking water coming out of a PUBLIC WATER SYSTEM's treatment plant—or to specify one or more suitable techniques for removing the contaminants from water. See also NATIONAL DRINKING-WATER PRIORITY LIST. Compare NATIONAL SECONDARY DRINKING WATER REGULATIONS.

National Priorities List a list of hazardous waste sites eligible for cleanup under the COMPREHENSIVE ENVIRONMENTAL RESPONSE, COMPENSATION AND LIABILITY ACT. A site that has been recommended is first investigated by the ENVIRONMENTAL PROTECTION AGENCY (EPA), which uses a Hazard Ranking System to quantify the risk to public health posed by a site. If the site scores high enough, it is then proposed for listing by the EPA. A site may also be proposed for inclusion on the National Priorities List if the Agency for Toxic Substances and Diseases Registry issues a health advisory for the site, or if the site is chosen as a state's highest priority cleanup site. The proposal for listing is published in the Federal Register, and the public is given an opportunity to make comments in support or opposition of the listing. It typically takes about five years from the time a site is identified as contaminated until it is placed on the priorities list, although the EPA is attempting to reduce the red tape involved to expedite the listing process.

National Recycling Coalition a nonprofit coalition of groups and individuals interested in recycling with 4,000 members in the United States including recyclers, waste haulers, government agencies and environmental organizations. The coalition promotes recycling as a way of reducing SOLID WASTE, conserving resources and energy and promoting social and economic development. The NRC sponsors conferences, provides information and educational packages and works to develop markets for recycled materials and products. The coalition's *Recycling Advisory Council*—an 18-member panel including representatives of the recycling industry, community recycling operations, environmental organizations, public interest groups, business and industry—worked with the Chicago Board of Trade during 1993 to establish a demonstration trading system for recycled materials. The *Buy Recycled Business Alliance* is an alliance of the NRC and business leaders that encourages the use of products containing recycled materials. Among NRC's publications are a bimonthly newsletter, the "NRC Connection", two quarterly newsletters, "Market Development NewsLink" and "Buy Recycled Newsline"; and an annual "State of Recycling" report that features news and recycling statistics for the previous year.

National Research Council   a private, nonprofit organization that organizes and performs research at cost for the federal government. The National Research Council (NRC) was founded in 1906 as an offshoot of the National Academy of Sciences (NAS), and has since become the principal administrative arm of the NAS and the related National Academy of Engineering and the Institute of Medicine. All four organizations bring together committees and study groups composed of scientists from the public and private sectors, government officials and representatives of the public to produce studies, reports and surveys of issues of pressing public importance. Controversial issues such as automotive fuel efficiency standards or global warming are addressed by the bipartisan groups under the auspices of the NRC. A list of publications is available from the National Academy Press, 2101 Constitution Ave, NW, Washington, DC 20418; phone: 800-624-6242.

National Secondary Drinking Water Regulations guidelines that specify the maximum allowable concentration of substances that can give DRINKING WATER undesirable qualities such as disagreeable odor or taste. The National Secondary Drinking Water Regulations (NSDWR) are enforced by the ENVIRONMENTAL PROTECTION AGENCY under the authority of the SAFE DRINKING WATER ACT. Secondary regulations are not enforceable, and are intended to protect public welfare, as opposed to public health. Secondary maximum contaminant levels (SMCLs) are specified in the regulations. They include:

- color: 15 cu
- odor: 3 threshold odor number
- corrosivity: non corrosive
- foaming agents: 0.5 milligrams per liter
- pH: 6.5 to 8.5
- sulfate: 250 mg/L
- copper: 1 mg/L
- chloride: 250 mg/L
- fluoride: 2 mg/L
- iron: 0.3 mg/L
- manganese: 0.05 mg/L
- total dissolved solids: 500 mg/l
- zinc: 5 mg/L
- aluminum: 0.05 mg/L
- o-Dichlorobenzene: 0.01 mg/L
- p-Dichlorobenzene: 0.005 mg/L
- ethylbenzene: 0.03 mg/L
- monchlorobenzene: 0.1 mg/L
- pentachlorophenol: 0.03 mg/L
- silver: 0.09 mg/L
- toluene: 0.04 mg/L
- xylene: 0.02 mg/L

See also NATIONAL DRINKING-WATER PRIORITY LIST. Compare NATIONAL SECONDARY DRINKING WATER REGULATIONS.

National Transportation Safety Board   an independent federal agency responsible for the safety of the nation's transportation system. Established in 1975 under authority of the Independent Safety Board Act of 1974, the National Transportation Safety Board (NTSB) investigates accidents, conducts studies and makes safety recommendations to government agencies and the transportation industry. The Safety Board consists of five members that are appointed for five-year terms by the President (with the advice and consent of the Senate). The NTSB investigates all pipeline accidents involving a fatality or substantial damage, all civil airplane crashes and all fatal railroad accidents and those involving significant property damage and selected auto/train crashes. The NTSB also investigates (in cooperation with the Coast Guard) all marine accidents between public and private vessels. The NTSB assures that procedures and equipment for the transportation of hazardous materials is adequate, and that other government agencies handle hazardous materials in accordance with safety guidelines. Address: 490 L'Enfant Plaza, SW, Washington, DC 20594. Phone: 202-382-6600.

Natural Gas   METHANE that is extracted from natural sources. Methane produced from PETROLEUM is also normally called natural gas; About 18.7-trillion cubic feet of natural gas was produced by U.S. wells in 1995. See also DISTILLATION.

Natural Ventilation   see VENTILATION

Negative Ion Generator   a device that drives electrons away from MOLECULES of the common atmospheric gases (see ATMOSPHERE, ION) by passing air through a network of needles or fine wires that are charged with high-voltage electricity. The strong electrical field generated by the charged elements drives an electron from some air molecules, creating a *positive ion*, and the freed electron bonds with another molecule to create a *negative ion*. There are generally only several thousand air ions in a cubic centimeter of air (accompanied by approximately ten million trillion molecules and particles that hold no electrical charge). Water vapor and HYDROGEN and oxygen molecules tend to cluster around negative and positive ions, creating in aggregate what is sometimes called a *small air ion*. Air ions are created naturally by the action of radioactive elements and cosmic radiation, either of which has abundant energy to knock an electron out of its orbit. The shearing energy of water molecules in waterfalls, the pounding of the surf and the abrasive energy of the wind can also create ions.

The exact effect of air ions is a controversial subject. Studies have shown that the concentration of ions in the air can affect microbial growth, and there is some evidence that low levels of air ions, especially negative ions, can increase susceptibility to disease. Air ions can improve air quality by bonding with toxic VOLATILE ORGANIC COMPOUNDS and smaller PARTICULATES. Whether boosting the concentration of negative ions by installing a negative ion generator in a car or a room leads to improved concentration, a sense of well being and improved health, as many adherents claim, is where the controversy begins, but limited evidence does support at least the claims of improved health. See also INDOOR AIR POLLUTION, ELECTROSTATIC PRECIPITATOR.

Nested Well   see MONITORING WELL

Neurotoxin   any substance that damages or destroys nerve tissue. A mild dose of a neurotoxin can cause lethargy, itching and a heavy, dull pain. Numbness, a "pins and needles" sensation in arms or legs, loss of vision, tremors and abnormal muscle jerking are symptoms of more severe poisoning with a neurotoxin. A high dose can lead to permanent brain damage, convulsions, paralysis and death. Because of the general structural similarity of nerve tissue in all life forms, PESTICIDES that kill insects or animals by attacking their nervous systems, including the ORGANOPHOSPHATES, tend to also be neurotoxins in humans. ARSENIC, DIOXIN, LEAD, MERCURY, METHOXYCHLOR, NAPHTHALENE and THALLIUM are well-known neurotoxins.

New Source Performance Standards   emission standards for air and water pollution (see AIR POLLUTION, GROUNDWATER POLLUTION, WATER POLLUTION) from new industrial plants and other STATIONARY SOURCES. New Source Performance Standards (NSPS), which are authorized under the CLEAN WATER ACT, also apply to modifications and new construction at existing plants. The standards require that the BEST AVAILABLE TECHNOLOGY be used to control emissions from such new pollution sources. The act requires that pollution control equipment reduce emissions of pollutants such as SULFUR DIOXIDE, NITROGEN OXIDES and PARTICULATE matter that result from the combustion of fossil fuels to a fixed percentage of what would be emitted with no controls. The remaining criteria pollutants (see CRITERIA POLLUTANT, NATIONAL AMBIENT AIR QUALITY STANDARDS) are governed in terms of the number of pounds

of a pollutant allowed per unit weight of emissions. Plants that emit less than 100 tons of a given contaminant per year are not required to meet the standard for the contaminant. The Clean Air Act specifically states that standards for a given pollutant can be set at zero if it can be shown that complying with such a standard is both technically and financially feasible.

Newsprint   The PAPER on which newspapers are printed. Newspapers constitute about 13% of the volume in a typical LANDFILL according to estimates by the Garbage Project. More than 6.2 billion metric tons of newsprint was produced by U.S. manufacturers in 1991. Newsprint consumes about 13% of world paper production. There has been a surplus of newsprint in recent years because of excess production capacity, but increased newspaper production in eastern Europe, Latin America and Asia has nearly eliminated the surplus. A year's subscription to the *New York Times* typically weighs 520 pounds and takes up 1.5 cubic yards of space. The American Forest And Paper Association estimated that 33.1 million tons of newsprint would be sent to the landfill during 1993, and that 36.7 million tons would be recycled—the first time more newsprint was recovered than dumped. About 15% of the newsprint consumed in New York State was recycled in 1900, but the price soon fell to almost nothing. Rail links to the forests of the Pacific Northwest and innovations in the processing of wood pulp largely eliminated recycled newspapers and rags (see SHODDY) as components of newsprint. Most newsprint recycled in the United States either again becomes newsprint, or is recycled as cereal boxes, gypsum wallboard, cellulose insulation or automobile interiors. Much is also shipped overseas, primarily to Asia. Newsprint and phone books are used to make toilet paper (Forest Green brand) and shred bed, which is a substitute for straw in cattle stalls. In 1988 there were nine newsprint recycling plants in the United States, but by 1994 there were 29.

The California Integrated Waste Management Act requires that at least 25% of the newsprint used by newspaper publishers contains at least 40% post-consumer recycled fibers (see POST-CONSUMER WASTE). The price paid for recycled newsprint in the state quickly went up as a result. Last year, 55,000 tons of recycled newspapers were collected in San Jose alone. Legislation that promoted newsprint recycling that went into effect in New Jersey in 1987 actually had the opposite effect. The newsprint recycling rate climbed from 50% to 62% within a few months, and, as a result, the price paid for recycled newspaper plummeted from $45 per ton to a minus $25 per ton. Many recyclers were therefor forced to pay to have their stock of newsprint hauled to the dump. See also PAPER RECY-CLING, RECYCLING, RECYCLING MARKETS, REFUSE DERIVED FUEL.

Nickel   a silvery-white METAL used in batteries, electroplating, the production of metal alloys including stainless steel, as a catalyst and in the production of coins. More than 36,800 metric tons of nickel were produced by U.S. manufacturers in 1993. More than 60% of the nickel used by U.S. industry is imported. Nickel gets into outdoor air from the combustion of COAL and other FOSSIL FUELS. Nickel gets into surface water from the weathering of rocks and from industrial discharges from electroplating and smelting operations. Nickel and related compounds have high acute TOXICITY to aquatic life (see ACUTE EFFECTS). Nickel's toxicity to aquatic organisms is a function of water hardness—the softer the water, the higher the toxicity. Nickel and its compounds have water solubility ranging from low to high, and are highly persistent in water, with a HALF-LIFE of more than 200 days. Nickel is one of the noble metals.

Nickel DUST and fumes can affect health when inhaled. Skin contact can cause an allergic reaction with itching and redness followed by a rash. An asthma-like lung allergy can result from inhalation. Exposure to airborne nickel can cause soreness or a lesion in the septum, the cartilage that divides the inner nose. Exposure to higher concentrations can cause a cough, shortness of breath and the buildup of fluid in the lungs, which may not occur for up to two days after exposure. Scarring of lung tissues, and disorders of the heart, liver and kidneys can result from a single severe exposure or repeated exposure to lower levels of nickel dust or fumes. Fumes from heated nickel can cause a pneumonia-like illness. Nickel is a probable human CARCINOGEN. It has been shown to cause lung cancer in animals, and there is limited evidence that it causes lung and nasal sinus cancer in humans. Epidemiological surveys have revealed a clear association between nickel refining and increased lung, nasal and throat cancers in humans. Nickel is a possible TERATOGEN.

PUBLIC WATER SYSTEMS are required under the SAFE DRINKING WATER ACT to monitor concentrations of nickel (MAXIMUM CONTAMINANT LEVEL = 0.1 milligrams per liter). The OCCUPATIONAL SAFETY AND HEALTH ADMINISTRATION has established a permissible exposure limit for airborne nickel of 1 milligram per cubic meter, averaged over an eight-hour workshift. It is also cited in AMERICAN CONFERENCE OF GOVERNMENT INDUSTRIAL HYGIENISTS, Department of Transportation NATIONAL INSTITUTE OF OCCUPATIONAL SAFETY AND HEALTH and National Toxicology Program regula-

tions. Nickel dust is a fire and explosion hazard. Nickel is on the Hazardous Substance List and the Special Health Hazard Substance List. Nickel compounds are on the HAZARDOUS AIR POLLUTANTS LIST. It may be necessary to store and dispose of nickel as a HAZARDOUS WASTE, depending on state regulations. Businesses handling significant quantities of nickel or nickel compounds must disclose use and releases of the chemical under the provisions of the EMERGENCY PLANNING AND COMMUNITY RIGHT TO KNOW ACT.

Nicotine   a poisonous alkaloid found in cigarettes and other tobacco products (see ENVIRONMENTAL TOBACCO SMOKE). Pure nicotine is a pale yellow, oily liquid with a slightly fishy odor. It is a chemical relative of caffeine, cocaine, morphine, quinine and strychnine. Scientific evidence indicates that nicotine is the substance to which users of tobacco products become addicted, although the tobacco industry vehemently denies this evidence and contests the addictiveness of nicotine. Jerry Garcia of the Grateful Dead said quitting heroin was easy compared to quitting nicotine. Nicotine is found in lozenges, gums and pills for people who are trying to quit smoking. People who regularly consume alcoholic beverages eliminate nicotine from their body more quickly than do nondrinkers. Although it is little used for this purpose today, nicotine is an effective INSECTICIDE that is powerful by today's standards. Nicotine is sometimes used to tranquilize animals, and is employed in the tanning of leather.

Nicotine can affect health when inhaled or passed through the skin. Tobacco users and those around them—as well as people exposed to nicotine-based insecticides—can receive a high dose of nicotine. People who handle green tobacco leaves can absorb so much nicotine through their skin that they get "green tobacco sickness," characterized by pallor, vomiting and severe cramps. Small doses of nicotine can cause nausea, vomiting, dizziness, headache, sweating, salivation and neurological stimulation. The bodies of regular smokers adapt to the presence of nicotine, however, and such symptoms do not appear. Large doses of nicotine can cause an irregular heartbeat, convulsions and death, although it is difficult for an adult to smoke or ingest enough tobacco for this to occur without first becoming too nauseous to continue. Children, however, occasionally die from nicotine poisoning as the result of eating tobacco products. Nicotine is a known CARCINOGEN for mice, but its ability to cause cancer in humans has not been proven.

The FDA is considering classifying tobacco as a drug because of the addictive nature of nicotine. If tobacco is declared a drug, the FDA could regulate cigarette sales. Skin contact with nicotine in the workplace is regulated by the OCCUPATIONAL SAFETY AND HEALTH ADMINISTRATION. Smoking on domestic airline flights was banned in the United States in 1990. Many states require restaurants to maintain smoking and nonsmoking areas.

Night Soil   human excrement collected to fertilize the soil. See COMPOSTING, FECES, OUTHOUSE.

NIMBY (Not in My Backyard!)   a syndrome associated with the siting of any waste-disposal facility characterized by the reaction, "not in my backyard." The phrase has been repeated with sufficient regularity over the last couple of decades—in relation to the siting of LANDFILLS, GARBAGE INCINERATORS, facilities for the disposal of nuclear or toxic waste (see HAZARDOUS WASTE DISPOSAL, RADIOACTIVE WASTE DISPOSAL), sewage disposal plants and a host of other contemplated projects—that it has earned its own acronym: NIMBY. What has come to be called *nimbyism* carries over into the siting of every conceivable facility right down to disputes about whether a lot should be zoned to allow a gas station in the neighborhood. The ferocity of opposition to such projects tends to increase proportionally as the physical distance to the opposing person's actual backyard (or back door) decreases. While citizens' concerns about the location of such facilities is certainly well founded, given the track record of both industrial polluters and of many of the agencies responsible for enforcing POLLUTION laws, nimbyism has become a serious obstacle to making the public works work. It can, and increasingly does, take years to site a waste disposal facility of any kind, even a "good" one such as a RECYCLING center. Scholarly papers have even been written on the subject.

The prompt disposal of garbage (along with snow removal and a workable sewer system) is one of the primal forces of politics and human nature. Everyone wants the trash to be removed promptly, and they want it to be hauled to a location well out of sight of (and downwind from) their home. Similar expectations surround the flushing of the TOILET. Political mayhem can quickly develop if such basic needs are not promptly met. The bones of the officials who have been dismissed and the mayors who have lost elections over issues such as the quick, efficient disposal of trash, litter the political landscape. So it can be particularly frustrating to an official charged with keeping the stream of waste moving when the best place for the new dump is always "somewhere else."

Vivid examples of the environmental horrors that can result from the dumping of wastes are found in every part of the country. Old municipal dumps that

are now Superfund sites with men in white space suits running around on them, AQUIFERS too tainted to be used or cleaned up, and polluted lakes and streams all have resulted from the virtually unregulated dumping that has occurred in the past. But the pollution threat posed by a modern waste disposal facility—sited and designed with the minimization of emissions in mind—is minimal by comparison. While it is true that such facilities will emit pollutants, this would also be true of potential industrial uses of the same land. The detrimental effects of the increased traffic and resulting increase in danger, noise and pollution along the road to such a facility may be greater than the pollution issuing from the facility itself. But the NIMBY phenomenon shows no sign of abating, and, unpopular though such facilities are, the need for places to dispose of wastes continues to grow. How this conflict will be resolved remains to be seen. See also CITIZENS' CLEARINGHOUSE FOR HAZARDOUS WASTE, NOISE.

Nitrate   SALTS or ESTERS formed by the action of nitric acid (HNO_3) on a METAL or an ALCOHOL. LEACHATE from LANDFILLS contains high levels of nitrates and is a frequent source of contamination of groundwater and surface water. Nitrate is produced in the environment as a BYPRODUCT of the METABOLISM of chemolithotrophic MICROORGANISMS, which oxidize AMMONIA. The conversion of ammonia in nitrate depletes NITROGEN from the soil, reducing its fertility, and the nitrates created can pollute GROUNDWATER (see GROUNDWATER POLLUTION). Nitrates get into surface water from concentrated sources such as sewage plants, industrial effluent and feedlots, and from diffuse sources such as the use of nitrate fertilizers, and from natural sources, in some drainages. Excess nitrates can lead to the EUTROPHICATION of surface water. Nitrates are difficult for conventional SEWAGE TREATMENT plants to process, but can be removed effectively by biological action in artificial marshes. Nitrates get into groundwater from septic tanks and from the application of nitrate fertilizer on overlying land. The most widespread nitrate pollution of aquifers (see WATER POLLUTION) in the United States is found in the Midwest, and is the result of nitrate fertilizer use.

Nitrates and NITRITES are added to cured foods such as smoked meat and fish to preserve color and to protect against *botulism*, which is caused by toxins emitted from a bacterium called *clostridium botulism*. It is generally regarded as the most life-threatening disease associated with food. About 700 tons of nitrites are added to the four million tons of smoked meat and fish produced each year in the United States. Nitrates are converted into nitrites when they enter the stomach. Nitrites and ammonia in the atmosphere are easily converted by natural processes into nitrates. Nitrates are used in fertilizers, explosives and in GLASS making. For Americans living in areas where DRINKING WATER is not polluted with nitrates, 90% of nitrate intake comes from vegetables. Beets, celery, lettuce, parsley, radishes, rhubarb and spinach are among the vegetables with the highest nitrate content.

Nitrates can react with chemicals that may be present in food to form *nitrosamines*, which are suspected human CARCINOGENS. Limited epidemiological evidence has shown that communities with high nitrate levels in drinking water have an increased cancer rate. Nitrosamines have been linked to cancer of the stomach, esophagus and nasal passages, although evidence of their carcinogenity is incomplete. Nitrosamines get into smokers lungs and indoor air from cigarette smoke (see ENVIRONMENTAL TOBACCO SMOKE), and can also find their way into the home in polluted outdoor air. PUBLIC WATER SYSTEMS are required under the SAFE DRINKING WATER ACT to monitor concentrations of nitrates (MAXIMUM CONTAMINANT LEVEL = 10 milligrams per liter) and nitrosamines. The BEST AVAILABLE TECHNOLOGY for the removal of nitrates and nitrites from drinking water is ION exchange and REVERSE OSMOSIS (see WATER TREATMENT).

Nitrite SALTS and ESTERS that contain the nitrite group (NO_{2-}), and that are formed by the reaction of METALS or ALCOHOLS with nitrous acid. Bacterial action on NITRATES also commonly produce nitrites. Nitrites are a health threat because they convert the hemoglobin in blood to methemoglobin, thereby eliminating its ability to carry oxygen. Some nitrites are known or suspected CARCINOGENS. The conversion of nitrates into nitrites in a baby's stomach can result in *methemoglobanemia*, also called the *blue-baby syndrome*. PUBLIC WATER SYSTEMS are required under the SAFE DRINKING WATER ACT to monitor concentrations of nitrite (MAXIMUM CONTAMINANT LEVEL = 1 milligrams per liter).

Nitrogen an inert elemental gas that makes up about 78% of the atmosphere by volume and about 75% by weight. Nitrogen is colorless, odorless and tasteless. It can be condensed into a colorless liquid, and the liquid can, in turn, be compressed into a colorless crystalline solid. Nitrogen serves as a diluent for oxygen in combustion and respiration. Water that is aerated by going over a dam or a waterfall or through a rapids can absorb lots of nitrogen, and nitrogen levels below dam spillways are sometimes high enough to kill fish. NITRATES are a primary source of WATER POLLUTION (see GROUNDWATER POLLUTION). Amines, cyanates, cyanogen, fulminates, nitro compounds, nitric acid,

nitrates, NITRITES, and urea all contain nitrogen. Nitroglycerin is produced from the reaction of nitric acid with ORGANIC COMPOUNDS.

Chemically inert, nitrogen will normally combine with other elements only with high temperature and pressure. Nitrogen can be chemically activated, however, by passing through an electrical discharge at low pressure. For about a minute the resulting "electric nitrogen" is capable of forming numerous compounds by combining with alkali metals to form *azides*; with the vapors of zinc, ARSENIC, MERCURY and CADMIUM to form *nitrides*; and with various HYDROCARBONS to form hydrocyanic acid and CYANIDE. Commercial nitrogen is taken from the atmosphere by drawing air-heated copper or iron. The oxygen is removed in the process, leaving nitrogen and inert gases. Pure nitrogen is obtained by distilling liquefied air. Liquid nitrogen has a lower boiling point than liquid oxygen and can be drawn off first. AMMONIA is made by passing a mixture of atmospheric nitrogen and hydrogen over a metallic catalyst heated to a temperature of 500° to 600°C. Approximately 14 $\frac{1}{2}$ million short tons of nitrogen was produced by U.S. manufacturers in 1993.

The Nitrogen Cycle

The nitrogen molecule (N_2) consists of two nitrogen atoms linked with an extremely strong triple bond, which is very difficult to break. If the bond is broken, however, the individual nitrogen atoms released have the ability to bond with practically any other element. Such nitrogen is called fixed nitrogen. Nitrogen can also be fixed by lightning. Nitrogen-fixing bacteria in the soil convert atmospheric nitrogen into a form such as nitrate that is biologically available to plants. Some nitrogen-fixing bacteria, such as *Azotobacter*, are self-sufficient, while others, such as *Rhyzobium*, are symbiotic, and live in proximity to the nodules on the roots of leguminous plants. Nitrogen in the form of protein and DNA is an essential constituent of living tissue. Once nitrogen has been fixed, it follows a circular pathway known as the nitrogen cycle or the biogeochemical cycle as it passes through an ecosystem. Nitrates and other compounds formed by the free nitrogen atoms are easily taken up by plants, which incorporate the nitrates into their tissues. The animals that eat the plants pass many of the nitrates from the plants as waste, primarily in the form of ammonia and related compounds. As animal wastes and animal bodies decompose—the victims of denitrifying bacteria—elemental nitrogen is released into the atmosphere and nitrates are returned to the soil as ammonia and proteins are broken down. Plant growth is stimulated in the process, an animal eats the plant, and the cycle is complete. See also BIOAVAILABILITY.

Nitrogen Oxides (NO$_x$)   several compounds resulting from different arrangements of nitrogen and oxygen. Nitrogen oxides (NO$_X$) get into indoor and outdoor air primarily as a BYPRODUCT of the COMBUSTION of fuels (see COMBUSTION BYPRODUCTS) such as heating oil, jet fuel, GASOLINE, NATURAL GAS and COAL. Nitrogen oxides are one of the five CRITERIA POLLUTANTS named in the CLEAN AIR ACT. More than half the nitrogen oxides found in the atmosphere come from the combustion of FOSSIL FUELS for the generation of electricity. Most of the rest comes from vehicle emissions. Atmospheric NO$_X$ facilitates the formation of OZONE from airborne VOLATILE ORGANIC COMPOUNDS. Some of the NITROGEN and oxygen molecules in high-temperature exhaust gases combine to form nitrogen oxides. Nitric oxide (NO) and nitric-oxide dimer (N$_2$O$_2$) are formed by the direct union of oxygen and nitrogen at high temperatures or near an electrical discharge. Once formed, the two molecules can undergo a photochemical reaction in which they slowly take on additional electrons to form nitrogen dioxide (NO$_2$) and dinitrogen tetraoxide (N$_2$O$_4$). Nitrogen dioxide, the more common of the two, is the yellowish-brown gas that is largely responsible for the brown tinge associated with most SMOG. The other nitrogen oxides are largely colorless. Nitrogen dioxide is an oxidant (see REACTIVE MATERIALS, OXIDIZING AGENT) that is much more toxic than nitrogen oxide. Nitrogen oxides are readily water soluble, and dissolve in raindrops to form nitrous acid and nitric acid, which contribute to ACID RAIN. They also react with water vapor to form acid PARTICULATES. Nitrous oxide (N$_2$O)—used by dentists to take the pain out of dental work—is called "laughing gas" because of its effects on the user. Nitrogen oxides get into the atmosphere from lightning strikes; the emissions of high-temperature combustion of coal natural gas and PETROLEUM fuels for the production of electricity; and industrial processes. Vehicle emissions (see AUTOMOTIVE AIR POLLUTION, MOBILE SOURCES) are a growing source of NO$_X$ because nitrogen oxides are a byproduct of the catalytic action used to remove CARBON MONOXIDE and HYDROCARBONS from exhaust gases (see CATALYTIC CONVERTER). The average gasoline-fueled automobile emits about 170 pounds of nitrogen oxides in 100,000 miles of travel. In California, 25% of the vehicles sold by 1997 must be low-emission vehicles that emit no more than 0.2 gallons of NO$_X$ per mile traveled.

Nitrogen oxide concentrations tend to peak in urban areas in the morning when automobile exhaust is at its maximum and when sunlight can still penetrate to facilitate its production. NO$_X$ concentrations in urban areas are hundreds of times higher than those found in rural areas. Areas that experience ATMOSPHERIC INVERSION have the highest NO$_X$ levels. The ENVIRONMENTAL PROTECTION AGENCY (EPA) announced plans in April, 1993, to limit NO$_X$ emissions from vehicles other than automobiles. Included were DIESEL-fueled equipment such as tractors, fork lifts and heavy equipment used in road-construction. During 1991, eight northeastern states passed restrictions on nitrogen oxide emissions from oil-, coal- and natural-gas-fired electrical generating stations. The 1990 amendments to the Clean Air Act offer incentives for clean-coal technology that will reduce nitrogen dioxide emissions. A fuel such as natural gas is sometimes added to combustion gases just downstream of the main combustion chamber in a coal fired power plant to reduce NO$_X$ emissions. In May 1992, the SIERRA CLUB sued the EPA for failing to meet deadlines for the establishment of NO$_X$ guidelines imposed under the Clean Air Act.

Nitrogen oxides get into indoor air from gas stoves and ovens, gas and kerosene space heaters, cigarettes (see ENVIRONMENTAL TOBACCO SMOKE), polluted outdoor air (see AIR POLLUTION) and other combustion sources. Nitrogen dioxide, like carbon monoxide, has an affinity for hemoglobin in the blood and it can interfere with the blood's ability to carry oxygen. Nitrogen dioxide can irritate the skin, eyes and mucous membranes. Limited evidence links long-term nitrogen dioxide exposure with chronic respiratory problems and bronchitis. There are no symptoms associated with moderate exposure to nitrogen-dioxide gas, but progressive inflammation can set in five to 72 hours after exposure, which can lead to pulmonary edema and death. NO$_2$ levels in ambient air are not nearly high enough to cause such a reaction, but farm workers sometimes die when exposed to the high concentrations sometimes found in grain silos. Epidemiological studies have concluded that children who live in homes with gas ranges have a higher incidence of reduced lung function when compared to children living in homes with electric ranges. Studies have indicated that exposure to the high concentrations of NO$_X$ found in severe POLLUTION events are at least as damaging as daily exposure to lower levels.

The 1990 amendments to the Clean Air Act call for NO$_X$ emissions to be reduced 2.5 million tons below 1990 levels by the year 2000. The OCCUPATIONAL SAFETY AND HEALTH ADMINISTRATION has established a limit for airborne nitrogen oxides in the workplace of 25 parts per million. The EPA proposed new regulations governing the release of NO$_X$ from power plants in 1993. See also COMBUSTION APPLIANCES, OXIDATION, POLLUTANT STANDARD INDEX.

No Deposit, No Return   bottles intended for a single use followed by disposal or recycling. No-deposit, no-return bottles and other disposable PACKAGING take up a significant portion of LANDFILL space. See CONVE-NIENCE PACKAGING, ONE-WAY PACKAGING, PLASTICS RECYCLING.

Noise unwanted or excessive sound. Much of the noise in a modern city is mechanical, a BYPRODUCT of the industrial revolution. Early factory workers exposed to the shriek of steam engines and the roar of clanging METAL often experienced severe hearing loss. Modern airport workers, loggers, construction workers and other workers exposed to loud noise wear ear protection equipment, but hearing damage from workplace exposure to noise is still common.

An event that makes a sound sends vibrations through the air that are picked up by nerve endings in the inner ear. The vibrations can come from a vibrating object such as a violin string, or from an intermittent event such as the squealing of brakes—a sound caused by the alternating slipping and catching of the brake drum against the brake pads—or the clap of thunder, which is caused when air rushing in from opposite sides of the void created by a lightning strike collides. Sounds vary in pitch and loudness: the more rapid the vibrations, the higher the pitch, and the greater the amount of air moved with each vibration, the louder the sound. Higher pitches must be louder to be heard. Recurring noises such as the clanging of steam pipes in an apartment building are eventually tuned out and are no longer heard by those affected. One of the foremost qualities of wilderness is the lack of noise. Noise is a relative thing. The noise at a football game might be seen as ecstatic, while quiet music and voices can be considered loud in a quiet residential setting, where even relatively mild noises such as the barking of a dog can cause annoyance and loss of sleep.

The *decibel scale* is the most commonly used measure of the loudness of noise. A decibel is a logarithmic ratio between the sound pressure of the source being measured and the sound pressure of a noise at the threshold of audibility. Sound pressure is the slight difference in atmospheric pressure caused by the motion of sound waves. The rustle of leaves registers at 10 on the decibel scale, while a whisper is about 25 decibels (db). Normal conversational levels are about 60 db, a gas lawnmower as heard by the operator emits about 95 decibels, and a jet airplane taking off 200 feet away puts out about 120 db. Sounds start to be physically painful at about 110 decibels; prolonged exposure to more than 85 db can cause ear damage, and 135 db can result in permanent deafness. Perceived loudness doubles each time decibel levels are increased by 10. The physical intensity of sound is measured in watts per square meter, which is a measure of the energy flow associated with the sound wave. A sound with an intensity of 1 milliwatt (one

thousandth of a watt) per square meter corresponds with a decibel rating of 90.

Loud noise can cause damage to the ears that can result in hearing loss or complete deafness. Temporary or permanent deafness can be caused by exposure to a brief but intense noise such as an explosion. Hearing loss, which can follow regular exposure to moderate noise levels, generally occurs at a pitch slightly higher than the one that caused the damage. Damaged auditory nerves, not the physical structure of the ear, are the cause of most hearing loss. The loss of hearing ability is measured in decibels, with a 10 decibel loss meaning that the lowest noise level that can be heard is 10 decibels higher (louder) than before the loss. Factory workers can lose 25 decibels of hearing ability through prolonged exposure to 95 decibels. Steady noise is less damaging than sudden bursts of noise at the same decibel level. Loud noises cause the heart rate and blood pressure to increase, and stimulates the body's production of the hormone ACTH, which is a secretion of the pituitary gland that stimulate the adrenal glands production of steroids that in turn cause elevated blood sugar levels.

Noise pollution is one of the most widespread forms of POLLUTION. Neighborhoods near airports, highways and many industrial facilities are bombarded with constant noise. Noise control measures such as the berming and/or fencing seen along many highways can reduce but not eliminate noise. Thick vegetation is one of the best absorbers of noise. Regulations such as those that govern how commercial jets are allowed to use their engines over a populated area or those that require the use of adequate mufflers on vehicles can also help reduce average noise levels in such settings, but they will remain noisy places. The workplace is often so noisy that workers are threatened with loss of part of their hearing function. The OCCUPATIONAL SAFETY AND HEALTH ADMINISTRATION (OSHA) and the NATIONAL INSTITUTE OF OCCUPATIONAL SAFETY AND HEALTH (NIOSH) are responsible for the regulation of noise in the workplace.

The NOISE CONTROL ACT OF 1972 set goals intended to protect the public from excess noise. The act authorizes the ENVIRONMENTAL PROTECTION AGENCY to set noise limits for commercial products and to work with the Federal Aviation Administration to regulate noise from airplane engines. Most municipalities have public ordinances that limit noise. Outdoor music is frequently regulated in how loud it can be or how late it can be played, and officers ticket hot-rodders because their cars are too loud. Most nations have adopted an average daily exposure limit of 85 decibels. OSHA has established a standard of 90 decibels for eight hours a day, with sudden bursts of noise be limited to 140 db.

The prosecution of violations of workplace noise standards is one of OSHA's most common actions. New regulations that will reduce workplace exposure limits are pending. In 1994, the International Civil Aviation Organization and the U.S. Federal Aviation Administration announced tough new noise regulations for aircraft that weigh more than 34 tons. The regulations, which went into effect in North America in late 1994 and in Europe during 1995, are the first step toward the gradual phasing out of older, noisier aircraft such as the Boeing 707, which is widely employed for hauling freight. Experts say the noisier jetliners will be effectively eliminated by the year 2002.

Noise Control Act of 1972 a U.S. federal law intended to protect the public from excess noise. The ENVIRONMENTAL PROTECTION AGENCY is authorized under the act to set noise limits for commercial products and to work with the federal aviation administration to regulate airplane noise.

Nonattainment Area an area that does not meet the AMBIENT QUALITY STANDARDS for one or more air pollutants, as defined in the CLEAN AIR ACT. Permits are issued for the construction of new stationary sources of AIR POLLUTION or for the significant remodeling of existing sources in a nonattainment area only when it can be shown that pollution from existing sources has reduced sufficiently that overall ambient air quality will not be diminished by the projects. An implementation plan with a timetable for reaching attainment must also be in place for the area in question. New pollution sources must meet the LOWEST ACHIEVABLE EMISSION RATE. CARBON MONOXIDE nonattainment areas must initiate a vehicle emissions inspection program. Failure to meet the federal standards can lead to the withholding of all federal funds for an area.

Noncriteria Pollutant see CRITERIA POLLUTANT

Nondegradable Pollutant a POLLUTANT that is not readily broken down or chemically altered by natural processes. A nondegradable pollutant remains a toxic threat as long as it is present. Compare BIODEGRADABILITY, PERSISTENT COMPOUND.

Nonpoint Pollution see DIFFUSE POLLUTION

Nuclear Reactor a facility in which the heat produced by the controlled FISSION of URANIUM or PLUTONIUM is harnessed to produce heat and decay byproducts. In the United States, 109 nuclear power plants produced approximately 619 billion kilowatt hours of electricity in 1992, a little more than 22% of total U.S. power production. Nuclear reactors produce electricity without the emission of the greenhouse gas, CARBON DIOXIDE (see GREENHOUSE EFFECT), or of other environmental toxins emitted by the COMBUSTION of FOSSIL FUELS, and much less mining is required for the production of fuel. The motive force in a nuclear reactor is provided by a self-perpetuating fission reaction. Part of the heat produced by fission in a nuclear power plant is converted to steam, which spins a turbine that produces electricity.

A typical nuclear reactor consists of a reactor core containing bundles of fuel rods made up of uranium pellets, and control rods, which contain a material such as boron, CADMIUM or COBALT that absorbs radiation. The entire assembly is typically immersed in water, which acts as a moderator and a heat transfer medium. The fuel pellets, which are generally about one-half inch in diameter and one inch high, are enclosed in a hard ZIRCONIUM shell capable of withstanding temperatures up to 1,852°C before melting. Uranium melts at 1,132°C. Control rods are arranged between the fuel rods, and the rate of the nuclear reaction can be regulated by varying how far the control rods are inserted into the core. The reaction can be stopped—or *scrammed*—by lowering the control rods all the way down. The moderator controls the speed of the free neutrons, thereby controlling the rate at which the nuclear reaction occurs. Graphite rather than water is used as a moderator in some of the older commercial reactors and in nuclear weapons production facilities (see HANFORD NUCLEAR RESERVATION, SAVANNAH RIVER). The CHERNOBYL disaster ended the commercial use of graphite cores in most countries. Lowering the control rods in a reactor where graphite is used as a moderator does not entirely stop the nuclear chain reaction.

Types of Reactors
A *light-water reactor* uses ordinary, "light" water as a moderator and coolant, while a *heavy-water reactor* uses HEAVY WATER, which has molecules built around the hydrogen isotope, DEUTERIUM. In a *boiling-water reactor*, water is heated to about 550°F in the core while under about 1,000 pounds per square inch (psi) of pressure. The water is converted to steam, which drives the turbines, and is then recondensed and circulated back to the core for reheating. In a *pressurized-water reactor*, which were initially developed for use in nuclear submarines, super-heated water in the primary reactor cooling system is prevented from boiling because it is under high pressure. Water temperatures of 580°F and a pressure of about 2,000 psi are typical. Heat from the high-temperature, high-pressure primary cooling system flashes water in a secondary cooling system to

steam, which in turn drives a power turbine. One advantage of such a system is that radioactivity is confined primarily to the reactor core and the primary cooling system, making maintenance easier and making the containment of radiation in the power-production area unnecessary. About 60% of U.S. commercial nuclear installations use pressurized-water reactors (see THREE MILE ISLAND). In a BREEDER REACTOR, some of the neutrons liberated by the chain reaction are used to convert nonfissionable U^{238} into fissile U^{235}, or plutonium, thereby creating more fuel than is consumed. A *fast reactor* dispenses with the moderator. An *advanced gas reactor* uses a pressurized inert gas such as argon as a coolant and a propellant for turbines. A *graphite-core reactor* uses graphite rather than water as a moderator (see RBMK-1000 REACTOR).

Nuclear Safety

In contrast to the explosive runaway chain reaction that occurs when a critical mass of uranium^{235} or plutonium^{239} is brought together in a bomb, the fission inside a nuclear reactor is controlled. The chances of a catastrophic loss-of-coolant accident and resulting explosions and core damage are very remote, but the consequences of such an event—as demonstrated in the core meltdown at Chernobyl—are so severe that every precaution must be taken to prevent them. Accidents occur even at responsibly managed, modern nuclear plants, but nuclear installations are designed, operated and regulated with accident-prevention and the minimization of impacts the foremost consideration. A typical modern reactor has redundant safety measures, and there are one or more backups for essential pumps, valves and controls. An emergency core cooling system (a tank that often holds more than one million gallons of water that can be dumped into the reactor core on a moments notice) and diesel generators to provide backup power for pumps and controls in the event of a power outage (usually with one or more backups, especially in cold climates) are among the measures taken to minimize the risk of a nuclear accident. A containment structure—designed to withstand the heat and pressure of a runaway nuclear reaction, and to survive an accident such as the crash of jumbo jet directly into the building—surrounds the reactor core. The air pressure inside the containment structure is kept lower than that outdoors to assure that no radioactive gases will escape the containment structure into the atmosphere. In terms of public safety, the safe storage and disposal of the fission products created in the nuclear reaction is actually a greater concern than the remote chance of a nuclear catastrophe (see HIGH-LEVEL WASTE, LOW-LEVEL WASTE, REACTOR WASTE).

Cooling

Because of the intense heat that is continually generated by nuclear fission, and because the fission process does not stop immediately when the control rods are lowered into the reactor core—it is essential that the circulation of coolant through the reactor core not be interrupted. For this reason, most reactor designs feature redundant components throughout the cooling system and a large reservoir of coolant that can be used to flood the core in the event of a loss-of-coolant accident. The most notable features of nuclear installations are the huge cooling towers that are used to recondense the steam used to drive power turbines. Compare FUSION. See also OAKRIDGE NATIONAL LABORATORIES, PRICE ANDERSON ACT, ROCKY FLATS.

Nuclear Regulatory Commission an independent federal agency that licenses and regulates civilian users of nuclear energy. Established under the Energy Reorganization Act of 1974, the Nuclear Regulatory Commission took over the regulatory responsibilities of the ATOMIC ENERGY COMMISSION. An NRC license must be obtained before a private company can build or operate a NUCLEAR REACTOR or can possess, process or dispose of nuclear materials (see PLUTONIUM, URANIUM, RADIOACTIVE WASTE). Measures for the security of nuclear facilities and nuclear fuel and the proper storage and handling of nuclear waste are specified by the NRC in a facility's operating permit. Once a permit has been obtained, the NRC periodically inspects the plant for compliance and assesses fines and/or specifies corrective measures for facilities found to be in violation. The agency also investigates incidents and accidents involving civilian nuclear facilities, and works with the states in the regulation of nuclear fuels and wastes from commercial nuclear facilities.

Principal divisions within the NRC are the Office of Nuclear Reactor Regulation, the Office of Nuclear Material Safety and Safeguards and the Office of Nuclear Regulatory Research. The Advisory Committee on Nuclear Waste and the Advisory Committee on Reactor Safeguards make recommendations concerning the use and handling of nuclear fuels to the commissioners. The NRC maintains five regional offices, which are responsible for monitoring and working with licensees in their region. Agency documents are available at the NRC Public Document Room at 2120 L Street, NW, Washington, DC 20555. The commission also maintains approximately 90 public document collections around the country, typically located at a public library near a nuclear installation regulated by the agency. Information on the availability of specific documents at local public document centers can be obtained at 800-638-8081.

Address: U.S. Nuclear Regulatory Commission, Washington, DC 20555. Phone: 301-951-0550. See also RADIOACTIVE WASTE DISPOSAL, THREE MILE ISLAND.

Nuclear Waste   see RADIOACTIVE WASTE

Nuclide   each of the different forms in which an ELEMENT can occur. The characteristics of an element are determined by the number of protons in the nucleus of its atom. Additional neutrons can be added to the nucleus, but the atom will exhibit the characteristics of the same element as long as the number of protons remains constant. For example, hydrogen-1 is a nuclide of hydrogen called *protium* that consists of a single proton with an electron orbiting it. Hydrogen-2 is a hydrogen nuclide called DEUTERIUM, which has a nucleus containing one proton and one neutron with one electron orbiting the nucleus. Hydrogen-3, the radioisotope TRITIUM (see ISOTOPE, RADIOACTIVE MATERIAL), is a hydrogen nuclide that occurs naturally only rarely, but that is a decay product commonly associated with nuclear FISSION. Tritium consists of one proton, two neutrons and one electron. See also RADIONUCLIDE.

Nuisance   in law, the use of a property in a manner that substantially interferes with the use of neighboring properties. A *public nuisance* causes damage to a large number of people or to the public in general. A *private nuisance* damages one person or a small group of people significantly more than others. Individuals commonly sue to stop or to be compensated for a private nuisance, but generally only a government body can file suits claiming a public nuisance. Suits involving AIR POLLUTION, NOISE pollution and the accumulation of trash or properties deemed to be an eyesore are common. Injunctions against the continuation of the act in question, and cash awards, can be awarded to the winner of such a suit. In order to have grounds for a nuisance suit, the initiators must generally establish that their use of the property predated the existence of the nuisance. To be successful, a nuisance suit must establish that the damage being caused to neighboring property exceeds the cost to the owner of removing the nuisance. In the case of a public nuisance, establishing standing (the right to sue) can be difficult.

Nuisance Particles   DUST that has little adverse effects on the lungs, and that does not produce a disease or toxic effect when exposure is kept under reasonable control. The inhalation of sufficient nuisance dusts can cause lung irritation, so the term "inert dust," which is sometimes used to describe nuisance particles is not literally correct. Even when excessive nuisance particles are inhaled, the architecture of the air spaces in the lungs remains the same, no significant scar tissue is formed and what tissue damage there is is repairable. FIBROGENIC DUSTS, in contrast, leave scar tissue when inhaled in excessive amounts. High levels of nuisance dust in the workplace can reduce visibility, cause irritation to the eyes, ears and nasal passages and damage the skin and mucous membrane.

Nutrient   an ELEMENT or COMPOUND that builds living tissue and fuels METABOLISM. Elements and simple inorganic compounds are nutrients in plants. They are converted to proteins and carbohydrates by photosynthesis and other processes. Animals rely primarily on the proteins and carbohydrates built by plants as nutrients, usually using a digestive process that breaks them down into amino acids and sugars, which are in turn reassembled into the proteins and carbohydrates necessary for the animal's metabolism. NITROGEN (see AMMONIA, NITRATE) and PHOSPHOROUS discharged into streams by EFFLUENT outfalls and RUNOFF stimulate aquatic life, and are one of the factors that can lead to EUTROPHICATION. The concentration of nutrients in water and in SEDIMENTS both influence the rate of biological productivity of a body of water. DETERGENTS based on trisodium phosphate produce phosphorous, and bans on phosphate-based detergents have been instituted in many areas with lakes and streams suffering from eutrophication. Nutrients in the form of anhydrous ammonia and phosphoric acid are sometimes added to industrial liquid waste that contains little nitrogen and phosphorous to support microbial action in wastewater treatment plants.

O

Oakridge National Laboratories a federal facility where uranium parts for nuclear weapons are produced. The Oakridge National Laboratories were built on the federal 37,000 acre Oakridge Reservation in Tennessee's Cumberland Mountains in 1943 as part of the war effort to build the first atomic bomb. On-site disposal of MERCURY between 1950 and 1966 resulted in the release of an estimated 2.4 million pounds of the metal into the surrounding environment, polluting GROUNDWATER and nearby streams. Surface and groundwater in and around the site are also contaminated with a variety of radioactive materials. Martin Marietta now manages Oakridge for the government.

The Department of Energy announced in 1983 that as much as 200,000 pounds of mercury, which had been used to process lithium at the plant, was dumped in unlined pits and otherwise mishandled between 1955 and 1963, and that the East Fork of Poplar Creek, which runs through the town of Oakridge, was polluted with mercury as a result. Researchers found that creek-bed SEDIMENTS were loaded with mercury, and that some of the sediments had been used as construction fill. The lawn of the Oakridge Civic Center had to be dug up after it was found to be contaminated. By 1989, $838 million had been spent and the seepage of radioactive and chemical wastes into groundwater and nearby streams had been reduced, but were still not stopped. A toxic waste incinerator, a water-purification plant and five installations for treating liquid chemical waste were built as part of the cleanup effort that it was estimated in 1989 would take five years and cost $19.5 billion. A $66 million effort to cover a 92-acre chemical waste dump with a cap composed of a plastic liner and clay was initiated in 1990. Waste from several unlined lagoons was mixed with concrete and stored in 60,000 82-gallon barrels, which the DOE is trying to get permission from the ENVIRONMENTAL PROTECTION AGENCY to dispose of in a nearby low-level radioactive waste site that is under construction nearby.

Cesium and strontium, which are byproducts of PLUTONIUM production, were dumped into seven unlined pits on the site between 1945 and 1963. Water in a stream that runs near the pits was found to be radioactive. The pits still emit considerable radiation even though they are covered with a thick layer of asphalt.

Occupational Safety and Health Administration The federal agency—part of the U.S. Department of Labor—that is responsible for health and safety in the workplace. Substances that are hazardous because of their TOXICITY, their reactivity (see REACTIVE MATERIALS), their flammability or their explosivity (see EXPLOSIVE MATERIAL) are regulated under the terms of the Occupational Safety and Health Act. The federal Coal Mine Health and Safety Act of 1969 also puts authority for mine health and safety under federal management, although much of this authority is usually passed on to state governments. OSHA runs the Toxicology Information Response Center in Oak Ridge, Tennessee. OSHA regulations apply to any workplace with one or more employees who are not family members of the business's owner. The NATIONAL INSTITUTE FOR OCCUPATIONAL SAFETY AND HEALTH (NIOSH) conducts research and testing to identify workplace hazards and develop workplace safety standards.

Occupational Safety and Health Review Commission an independent, quasi-judicial agency established to oversee the operation of the OCCUPATIONAL SAFETY AND HEALTH ADMINISTRATION. The commission considers appeals of OSHA decisions relating to worker health and safety. Address: 1120 Twentieth Street, NW, Washington, DC 20036. Phone: 202-606-5100.

Ocean Disposal the intentional disposal of wastes such as SEWAGE, GARBAGE and RADIONUCLIDES in the ocean. For more than 50 years, New York City and other municipalities dumped a total of more than 8 million tons of sewage sludge per year into the ocean at a site 106 miles off the New Jersey coast until the ENVIRONMENTAL PROTECTION AGENCY (EPA) forced them to switch to land disposal by 1992. New York City, which alone was annually dumping more than 4.5 million tons of sludge at the site, must come up with a permanent site for disposal of its sludge by the end of 1998, according to the terms of its agreement with the EPA at an estimated cost of $700 million. Congress originally outlawed the ocean disposal of sewage SLUDGE in the starting in 1981, but the city sued the EPA that year, claiming that its sludge was not that degrading to the ocean environment. A federal court ruled in New York's favor, and the city, Westchester and Nassau counties in New York state and six New Jersey municipalities continued to dump sludge at sea under the court order until the ocean dumping ban went into effect in 1992. They complied with the EPA's order in 1987 that they abandon a site 12 miles of Sandy Hook, New Jersey, because of its proximity to the coastline and because it had turned the relatively shallow dump site into a virtual dead zone. Congress passed the Ocean Dumping Act, which banned the ocean disposal of sludge after the end of 1989 except for cities that had entered into an agreement with the EPA to end the practice. New York City is drying its sewage sludge and paying independent contractors to dispose of it as an interim solution.

Boston and Los Angeles are the only remaining major cities that dump raw sewage into the ocean, and Boston is in the process of building a new sewage treatment plant that will end the practice. The pollution of Boston Harbor was a point of contention in the 1988 presidential campaign when George Bush pretty successfully laid the blame for the harbor's condition on Michael Dukakis' tenure as governor of Massachusetts, although the charge was far from accurate. Boston is now under court order to end the emission of EFFLUENT into the harbor. A $6-billion sewage treatment plant—that will be the second largest in the nation when it is completed—will treat wastewater, and the treated water will pass through a tunnel that is 24 feet in diameter and that will run for nine miles through solid bedrock 400 feet under the surface of Boston Harbor to carry effluent from the new treatment plant out to sea.

RADIOACTIVE WASTE was hauled out to sea and dumped by the United States and other nations in the early days of the nuclear era. A study by the EPA disclosed that most of the barrels holding the waste had imploded or had been crushed, releasing radioactive materials into the sea water and ocean SEDIMENTS. A study by the International Atomic Energy Agency of one dumping site in the Atlantic found that radioactive decay products were, in fact, spreading through the marine ecosystem in the area. Fear that the RADIONUCLIDES released could work their way up the food chain and contaminate commercial seafood led to the London Dumping Convention of 1972, which prohibits the disposal of high-level nuclear waste in the ocean. Parties to the convention voted in 1990 for a nonbinding resolution to stop the scuttling of nuclear submarines at sea. After protests from Japan and the United States, Russia agreed in October 1993 to stop dumping liquid radioactive waste into the Sea of Japan, but warned that it would resume the practice unless richer countries help it process the waste for underground burial. The Russians were dumping more than 20,000 cubic meters of LIQUID WASTE produced by Russia's nuclear subs each year into the sea in violation of a 1983 international moratorium on such dumping.

Offal   a material that is left as waste or that is a byproduct of manufacturing or preparation or a dead animal or its parts. Offal is a much-used term that can refer to:

- the stalks and particles left over from the processing of a tobacco leaf;
- the least valuable portions of a hide;
- the material that is removed from a grain when it is milled (which is used as stock feed);
- the parts of an animal such as the brains, heart and liver that are removed during butchering;
- fish that are too small to be marketed.

See also GARBAGE, REFUSE, RUBBISH, TRASH.

Ogallala Aquifer   an AQUIFER that underlies a large part of the United States High Plains, including all of Nebraska and parts of Colorado, Kansas, South Dakota, Oklahoma, New Mexico and Texas. The aquifer is in a region that generally receives less than 20 inches of water per year, and much of it is overlaid with an impenetrable layer of caliche that prevents recharge. As a result, the aquifer contains primarily three-million-year-old "fossil water" that will not be recharged after it has been used. The aquifer has been tapped for IRRIGATION WATER and DRINKING WATER across much of the region, and it is quickly becoming depleted as a result. It is estimated that 21 million acre feet of water are withdrawn from the Ogallala Aquifer annually while the recharge rate is only about 100,000 acre feet per year, resulting in an annual drop in the WATER TABLE of 10 feet. The situation is especially critical in the Texas High Plains, where a billion dollar agricultural industry has been built around water drawn from the aquifer in an area without other sources of water to turn to when it is used up. About a quarter of the U.S. cotton and grain sorghum and 15% of its cattle feed are raised in the area. The falling water table has already led to cutbacks and elimination of irrigation on a substantial portion of the area's farm acreage. See also GROUNDWATER.

Oil Shale shale that contains rich organic plant remains, which occur primarily as *kerogen*, an oily solid that is a precursor to PETROLEUM. Kerogen can be converted into crude petroleum for refining purposes. Rocks in some of the most kerogen-rich shale deposit will burn when lit, but most deposit are considerably less rich. There are an estimated 1.43 trillion barrels of oil locked in oil shale, of which about 80% are considered rich enough and close enough to the surface to be economically developed. The refining process for kerogen is much more complex than that used with liquid petroleum. Since crude oil has remained relatively cheap and readily available, none of the large tracts of oil shale land leased by oil companies in western states have been developed. Protests of the environmental effects of strip mining vast tracts for oil have slowed lease development in some areas. In a typical oil-shale operation, approximately one ton of rock must be processed to yield the equivalent of 25 gallons of petroleum. Since the processed rock takes up considerably more room than the original, disposing of it is a problem of major proportions.

Oils substances that are soluble in ether, CHLOROFORM or other nonpolar SOLVENTS (see POLAR LIQUID), but that are not soluble in WATER. Oils are liquids or soft, malleable solids with a greasy surface (see GREASE). There are three basic types of oils: *mineral oils*—natural oils derived from PETROLEUM that are used as engine lubricants (see ENGINE OIL); *fixed oils*—nonmineral oils that do not volatilize easily (see VOLATILITY); and *essential oils*—nonmineral oils, such as turpentine, that evaporate easily. Liquid oils spread in a thin sheet on top of the water when spilled. An *oil film* can be as thin as the diameter of one molecule (about 2.5 microns or 0.0001 inch). An oil layer thicker than 2.5 microns is called an *oil slick*. Oil is one of the most visible kinds of WATER POLLUTION because it does not mix with water and floats to the surface. Oil spills can devastate water quality and wildlife and lead to long-term pollution of SEDIMENTS (see OCEAN DUMPING). Heavy petroleum oils tend to saturate soil without much further movement when spilled, and removing them can be difficult. A technique called *land farming* enhances microbial action to neutralize contaminants. The soil is excavated and placed on a liner to prevent the spread of contamination. Fertilizer

and carbon are added, and the soil is aerated (see AERATION) by tilling. Oils and greases of animal origin are generally biodegradable and can normally be processed at a sewage treatment plant, but petroleum-based oils generally cannot be removed from the wastestream through biological processing. Since it is difficult to distinguish between the two types of oils and greases, however, both are generally removed.

On-Site Disposal the disposal of wastes on the site at which they are generated. Incinerators such as those used by medical facilities (see MEDICAL WASTE INCINERATOR) and many industrial installations are an example of on-site disposal.

One-Way Packaging PACKAGING that is designed to be used once and discarded. PAPER and PLASTICS are used in the production of most one-way packaging. Some plastics and much of the paper and CARDBOARD used in disposable packaging can be, and is sometimes designed to be, recycled (see MCTOXICS CAMPAIGN, PAPER RECYCLING, PLASTICS RECYCLING). Composite packaging that contains a combination of paper and plastic or a combination of plastics is virtually impossible to recycle. See also CONVENIENCE PACKAGING.

Opacity the transparency or translucency of an object. Air or water's opacity is its transparence to visible light. Airborne matter (see PARTICULATES) reflects light and obscures the view while pure air is almost totally transparent. The Ringlemann number is a gauge of the opacity of water. See AIR POLLUTION, GROUNDWATER POLLUTION, VISUAL POLLUTION, VISIBILITY, WATER POLLUTION.

Open Burning burning in the open ATMOSPHERE. Open burning includes backyard grass and brush burning, trash-burning drums, slash burning, field burning, PAPER and WOOD combustion at an OPEN DUMP, forest fires and brush fires. Access to oxygen in a pile of burning brush or logging slash is uneven, materials are often damp and combustion temperatures are low—conditions that lead to incomplete combustion and the atmospheric emission of toxic gases such as CARBON MONOXIDE and PARTICULATES such as BENZO-A-PYRENE (see AIR POLLUTION, COMBUSTION, COMBUSTION BYPRODUCTS, SMOKE). The smelting of copper ore was once accomplished by open burning in the early days of the mining town of Butte, Montana. Huge heaps consisting of layers of ore sandwiched between layers of wood and COAL, were built up and then lit to smolder for weeks. The practice was discontinued in favor of a smelter with a smokestack after severe health problems resulted among the people in town. Most towns once used open burning at the dump to reduce the volume of waste, and until the 1970s the practice was still common. Fatal or destructive fires such as the one at the KENILWORTH DUMP were a common occurrence, and the practice is now illegal.

Open Dump   garbage dumps in which refuse is not regularly covered with soil, as is the case in a modern sanitary LANDFILL. Open dumps, which were common in the United States until the passage of the RESOURCE CONSERVATION AND RECOVERY ACT in 1976, usually practiced OPEN BURNING, and were a perennial source of smoke and odors. Such dumps had no liner or other provisions to deal with the production of leachate, and therefore polluted GROUNDWATER and surface water.

Organic Compound   a COMPOUND that contains carbon. Organic compounds were believed until about 1830 to always be a product of living tissue, hence the name. Organic chemicals are further defined as having a carbon-carbon bond, a carbon-hydrogen bond, or both—a definition that excludes simple carbon compounds such as CARBON MONOXIDE and CARBON DIOXIDE. Approximately 10 million organic compounds are now known—about 50 times the number of all other known compounds combined—and about 500,000 new manmade organic chemicals are created each year. See also AROMATIC HYDROCARBON, ALIPHATIC HYDROCARBON, HYDROCARBON, VOLATILE ORGANIC COMPOUND.

Organic Material   material such as dead vegetable matter, animal waste and PAPER that is derived from living things. Many municipalities have set up large-scale composting programs or have encouraged residents to do there own backyard composting to reduce the volume of leaves and grass clippings hauled to landfills. Grass clippings can contain residues of lawn-care chemicals.

Organochloride   an ORGANIC COMPOUND that contains CHLORINE. Organochlorides include some very dangerous chemicals such as DDT VINYL CHLORIDE, and PCBs (see POLYCHLORINATED BIPHENYLS). Most of the organochloride PESTICIDES have either been banned or are tightly regulated because of their TOXICITY and persistence in the environment.

Organophosphate   an ORGANIC COMPOUND than contains at least one PHOSPHATE group. Malathion and PARATHION are organophosphate INSECTICIDES.

Outfall   the point where LIQUID WASTES are dumped into a body of surface water (called the *receiving waters*) from a sewage plant or industrial installation.

Outgassing   the passage of a gas into the ATMOSPHERE as the result of the curing and aging of building products and furnishings that contain synthetic resins. The outgassing of FORMALDEHYDE from glued-wood products bound together with urea-formaldehyde resins is one of the best known examples of the phenomenon. Formaldehyde from the outgassing of glued building materials is a form of GASEOUS WASTE, most of which is released within five years of the original manufacturing process.

Outhouse   a small structure built over a hole in the ground used as a TOILET in homes where there is no indoor plumbing. High-tech outhouses that break down waste through aerobic and anaerobic decomposition are used in rural China and other areas. An outhouse constructed over porous soils can cause GROUNDWATER POLLUTION. See also FECES.

Oxidation   the loss of electrons by an element during the course of any chemical reaction. Oxidation was formerly defined as any chemical reaction involving oxygen, hence the name. The oxidation of a compound usually involves the release of energy. The oxidation of carbon compounds during respiration releases energy and creates many of the compounds found in living tissue. Powerful oxidants such as OZONE are dangerous air or water POLLUTANTS (see AIR POLLUTION, GROUNDWATER POLLUTION, WATER POLLUTION) because they can damage living tissue and combine with other pollutants to form, new, more toxic compounds. The oxidizing abilities of CHLORINE are the basis of its use both as a pesticide and disinfectant and as an industrial process chemical. Oxidation is used to neutralize waste in HAZARDOUS WASTE TREATMENT and SEWAGE TREATMENT.

Oxidation Pond   see POLISHING POND

Oxide   a binary COMPOUND with oxygen as one of its two constituent elements. Almost all elements can form oxides, the exceptions being the *noble gases*: helium, argon, RADON, neon and krypton. Oxides of metals are usually BASES, while oxides of nonmetals are generally ACIDS. Commonly occurring oxides include WATER (H_2O), CARBON DIOXIDE (CO_2) and IRON rust (Fe_2O_3). See also CORROSION, CORROSIVE MATERIAL, OXIDIZING AGENT.

Oxidizing Agent   an ELEMENT or COMPOUND that receives electrons during a chemical reaction, thereby causing the OXIDATION of the element or compound from which the electrons are removed. To be a good oxidizing agent, a compound must react with a wide variety of materials. A material's ability as an oxidizing agent, or *oxidant*, is dependent on its *electronegati-*

vity—its ability to attract electrons. See also HAZ-ARDOUS WASTE, REACTIVE MATERIALS.

Oxygenated Fuels   FUELS to which ALCOHOL (see ETHANOL, METHANOL) or METHYL-TERTIARYBUTYLETHER (MTBE) have been added to increase its oxygen content and therefore its combustion efficiency; also called *oxyfuels*. The oxygenated fuels program was originally developed to reduce the emissions of CARBON MONOXIDE. Communities that exceed federal guidelines for carbon monoxide and OZONE are required under the CLEAN AIR ACT to use oxygenated fuels. The Reformulated Gasoline Program that gets underway in 1995 will go a step further than oxyfuels by changing the formulation of GASOLINE to control contaminants such as sulfur, nitrogen dioxide, particulate matter and low-level ozone.

Ozone   an oxygen molecule that contains three—rather than the usual two—atoms per MOLECULE. Ozone (O_3) is a pale-blue gas with a penetrating odor, and is formed when VOLATILE ORGANIC COMPOUNDS and NITROGEN OXIDES react with other atmospheric gases and sunlight. Ozone is used in the purification of air and water and in the bleaching of some foods. An increasing number of public swimming pools are drastically reducing the amount of CHLORINE used to disinfect water, and are relying instead on ozone to handle the bulk of the microbial contaminants found in pool water. The result is water without the eye-stinging ability of chlorine.

Ozone is one of the five CRITERIA POLLUTANTS named in the CLEAN AIR ACT. Ozone, also called *trivalent oxygen*, is also produced when a spark passes through oxygen. Its odor can sometimes be detected near electrical machinery, after a lightning strike or near electrical transmission lines. Liquid ozone is deep blue and highly magnetic. When found in the stratosphere (see STRATOSPHERIC OZONE), ozone is beneficial because it absorbs some of the sun's ultraviolet radiation. But ozone is a dangerous POLLUTANT when found at high concentrations in the lower atmosphere because it is a powerful oxidant (see OXIDATION, OXIDIZING AGENT) capable of irritating respiratory systems and altering cellular metabolism. Concentrations as small as a few parts per million can cause damage to living tissue. Lower atmospheric concentrations, however, may play a role in maintaining the freshness of fresh air.

All that is required to support the photochemical reaction that can create ozone out of the chemical soup of the *ozone precursors*, volatile organic compounds and nitrogen oxides that exist in most urban areas, is hot, dry weather with lots of sunshine. Ozone concentrations will be highest when there is little air movement to disperse the pollution (see AIR POLLUTION, ATMOSPHERIC INVERSION). INTERNAL COMBUSTION ENGINES and the fuels used to run them are the primary source of ozone-creating chemicals, and the gasoline-fueled automobile, by itself, contributes more than half the burden in many areas. While the concentration of other PRIORITY POLLUTANTS declined during the 1980s, ozone has been harder to control. The ENVIRONMENTAL PROTECTION AGENCY announced in 1992 that 98 U.S. cities did not meet federal standards for ozone. Only in Los Angeles, where tremendous progress has been made in reducing ozone levels, is ozone pollution in the extreme category, with nearby San Diego County rated as severe. Six of the seven cities with the highest ozone readings in the country are located in California.

Residents of areas that exceed federal limits for airborne ozone are advised during pollution season:

- to avoid extended engine warm-up and idling;
- to avoid using mowers or other gasoline-powered equipment;
- to consider using alternative fuels, such as methanol or propane;
- to avoid using charcoal lighter fluid if barbecuing (gas grilling is okay);
- to use low vapor pressure fuels if possible, to avoid overfilling their tank, to prevent spillage and resulting EVAPORATION and to be certain the gas cap is tightly sealed when finished.

Ozone can affect health when inhaled. It is a severe irritant to the respiratory tract, and rapid, labored breathing and congestion are symptoms of exposure. Symptoms are displayed by some individuals at concentrations only slightly higher than federal ozone standards. Laboratory studies using human volunteers have demonstrated that intermittent exercise in air with ozone levels typical of urban air reduces lung function in adults, making it impossible to inhale deeply. Although some lung capacity returns if ozone exposure ceases, permanent scarring of the lungs can prevent complete recovery. An increased rate of asthma attacks and of infections of the upper respiratory system (due to damage to the lining of the airways) also result from prolonged exposure to atmospheric ozone. Ozone and other oxidants can also cause severe damage to crops and forests. Officials in Sequoia and Yosemite National Parks announced in 1989 that airborne ozone had damaged from 80% to 90% of the pines in the two parks. Ozone pollution has caused the loss of more than 20% of tomato and beancrops in some instances.

Many believe the current federal limit for ozone of 0.12 parts per million (ppm) is inadequate. The American Academy of Pediatrics issued a statement in

June, 1993, stating that the standards for ozone pollution must be tightened if it is to protect children from the effects of ozone. In June 1994 the American Lung Association called the existing U.S. standard for ozone inadequate, saying that nearly 15 million American adults and children with respiratory problems are at increased risk because of ozone. Studies in Canada suggest that even their 0.08 ppm limit may be high enough to protect human health. But the emphasis in the CLEAN AIR ACT is to attain compliance with the old standard before adopting a new, more-stringent limit.

The Clean Air Act requires cities that exceed federal limits for ozone to use special fuels during the pollution season. The act also calls for reducing volatile organic compounds (VOCs) passed into the atmosphere by switching to PAINTS with fewer SOLVENTS. Industries that apply paints and other coatings must either switch to low-solvent products or collect VOC-laden fumes and incinerate them. See also AUTOMOTIVE AIR POLLUTION, POLLUTANT STANDARD INDEX.

Ozone Layer   see STRATOSPHERIC OZONE

P

Packaging materials used to enclose, contain and/or protect a product. From the ubiquitous "tin" can (see STEEL) to the infamous McDonald's foam clamshell (see MCTOXICS CAMPAIGN), packaging is an inescapable part of everyday life. Commonly encountered forms of packaging include CARDBOARD boxes; ALUMINUM and steel cans and foil; PAPER; GLASS bottles and jars; PLASTIC wrap, rigid plastic foam packing extrusions and plastic foam peanuts (see EXPANDED FOAM); and infinite combinations of the above materials. Containers made of plastic, cardboard, aluminum, steel and glass are often recycled (see RECYCLING), but combinations of different materials (see COMPOSITE PACKAGING) usually go to the LANDFILL or the incinerator. Plastics used in packaging other than beverage containers and paper used in packaging are seldom recycled.

Almost 60% of the packaging used in the United States contains food and beverages. FAST-FOOD PACKAGING often gets tossed out the car window to become LITTER. Because it is highly visible, this easily identifiable form of TRASH, undoubtedly has much to do with the American public's disdain for such GARBAGE. Packaging protects food from contamination, excessive moisture or dryness and postpones decay. The United Nations estimates that improved packaging of food in Third World countries could reduce crop losses by 5%. In industrialized nations, it is the age of fast food, however, and even foods that are cooked at home are marketed as much for their convenience (see CONVENIENCE PACKAGING) as for taste and nutrition: "Just pop it in the microwave." More than 40% of food sold at grocery stores can be prepared in less than five minutes.

The average respondent to a Garbage Project survey conducted at a 1989 meeting of the National Audobon Society said that expanded POLYSTYRENE of the type used in cups and sandwich packaging made up 20% to 30% of total volume of a typical landfill, and that fast-food wrappers constitute an additional 25% to 40%. Garbage Project statistics based on the examination of 14 tons of garbage from nine municipal landfills, however, found that only about *one-third of one percent* of average landfill volume is devoted to fast-food packaging, with the total for *all* expanded foam products, of which the expanded polystyrene used in fast-food packaging makes up only a small part, being about 1%. So discarded packaging is not nearly as great a concern as the public often perceives. The production of the metals, plastics and paper used in packaging are a major source of air and water pollution, and HEAVY METALS are found in some inks and adhesives used in packaging.

The packaging industry grosses $75 to $80 billion a year, and employs two million people in the United States. Primary consumer packaging—the cans, bottles, cardboard cartons and plastic wrappings in which products are found in stores—actually comprises only about half the packaging associated with a typical product. The other half is distribution packaging, which is used to ship products from manufacturers to distributors, stores and consumers. Distribution packaging makes goods easier to handle, and protects them from moisture and physical abrasion. Items that will be on display in stores are often encased in rigid plastic or plastic shrink-wrap and mounted on bright cardboard to prevent shoplifting and promote sales with a bright, attractive package covered with advertising. Packaging for over-the-counter drugs and some personal care products now come in "tamper-evident" packages, to prevent people with nothing better to do from putting poisons inside. Child-proof packaging that (in theory at least) makes it more difficult for the kids to eat the birth-control pills is also common. People with disabilities such as impaired eyesight or arthritis, who may have a difficult time opening even the most straightforward packaging, can be completely stumped by tamper-proof or child-proof packaging.

Medical equipment often comes sealed in a sterile plastic package that is discarded when opened. Most medical professionals say the convenience and guaranteed sterility of such products is worth the investment in throwaway packaging. Packaging is an especially useful and reliable DEMOGRAPHIC MARKER that is used in the science of GARBOLOGY. The labels on food packaging reveal exactly what and how much food people buy, and the labels on other products can similarly be used to trace consumption patterns. The amount of discarded toy packaging in a household garbage can each week can be used to estimate the number of children in a community, and packaging from feminine hygiene products can give a good approximation of the number of women in a household. See also CONSUMERISM, CONSTRUCTION WASTES, COUNCIL ON PACKAGING IN THE ENVIRONMENT, NO DEPOSIT NO RETURN, ONE-WAY PACKAGING, PLASTIC RECYCLING, PLASTIC WASTE, REUSABLE PACKAGING.

Packaging Research Foundation   an independent, nonprofit organization formed in 1985 as the Plastics Recycling Foundation (PRF) by members of the plastics packaging industry interested in increasing the post-consumer recycling of plastics. The foundation is the largest U.S. sponsor of university research related to PLASTICS RECYCLING and PACKAGING, and publishes technical papers relating to the subjects. Address: 1275 K Street, NW, Washington, DC 20005. Phone: 202-371-5200. The Center for Plastics Recycling Research was established by the PRF—Address: Rutgers, The State University of New Jersey, Busch Campus Bldg. 3529, Piscataway, NJ 08855. Phone: 201-932-3683.

Paint   color or pigment, either dry or mixed with a viscous fluid that is used in decorative and protective coatings for WOOD, METAL and other materials. Pigment is mixed with a *vehicle* that ultimately forms the adherent, flexible skin-like coating associated with paint. The vehicle contains SOLVENTS or thinners that allow the paint to flow, then evaporate to allow the paint film to harden. It is the pigment that gives the hardened paint its color and opaqueness. The vehicle may be composed of POLYMERS, or may consist of a drying oil (usually an ESTER) mixed with a viscous ALCOHOL such as glycerin. The disposal of used paint at municipal landfills can cause toxic solvents to get into LEACHATE. Used paint should be treated as HAZARDOUS WASTE and disposed of accordingly (see HAZARDOUS MATERIAL, HOUSEHOLD HAZARDOUS WASTE). Many of the numerous VOLATILE ORGANIC COMPOUNDS (VOCs) that can be released as paints dry are precursors of OZONE, and others are toxic in their own right. The VOCs that can escape from paint for years after it is applied are a source of INDOOR AIR POLLUTION. Latex paints are based on a fluid binder that resembles the

white latex sap found in rubber trees, which is composed of gum resins and fats. In paint, the latex is an emulsion in water of synthetic rubbers or PLASTICS obtained by polymerization. While latex paints are based on water, oil-base points rely on a mix of oils such as linseed oil and volatile solvents to act as a transport mechanism for the pigments being applied. The proportion of latex paints on the market has increased dramatically as a result of rules governing the release of VOCs from paint. The painting process is hazardous from the standpoint of indoor air quality. Scraping and sanding can release toxic substances such as LEAD and ASBESTOS into the air, and paint strippers can also emit a variety of toxic gases and PARTICULATES into indoor air. Before 1940, lead carbonate or white lead was the primary white pigment used in paints, and more than half the volume of many products in use at the time was lead. Outdoor paints based on red lead were also universally used as a metal primer at the time. After decades of growing concern about the health effects of lead-based paint, the Lead Based Paint Poisoning Prevention Act was passed in 1971. The law called for the use of lead in paints to be ended by 1978. The ENVIRONMENTAL PROTECTION AGENCY (EPA) estimates that paint in more than half of U.S. homes contains lead. The principle white pigments found in modern paints are zinc oxide, lithopone and titanium oxide. A variety of other toxic ingredients are added to paint including the solvents BENZENE and TRICHLOROETHYLENE; COBALT, which is used as a drier in paints and varnishes; and PENTACHLOROPHENOL (PCP or penta), which is used as a FUNGICIDE and BACTERICIDE in paints and stains designed for outdoor use. The EPA ruled that MERCURY, because of its known TOXICITY, could not be added to paints after the end of August, 1990. Before the ban, about one-third of interior house paints used mercury to control bacteria that can thin paint and turn it rancid. Mercury is still allowed in exterior paints.

VOCs

The principal reason for the changes in paint formulations is the fact that VOCs are a principal precursor of ozone. While some progress has been made in controlling emissions of toxins such as particulate and CARBON MONOXIDE over the last decade, average ozone levels have continued to climb along with the number of urban areas in violation of federal ozone standards. Oil-base paints are impacted by 1990 amendments to the CLEAN AIR ACT (CAA) because they contain HYDROCARBON solvents such as TOLUENE, XYLENE and naptha that emit VOCs as the paint dries. The National Paint and Coatings Association favors national adoption of California standards that limit the VOCs emissions from oil-based paint sold in the state to 250 grams per liter (as opposed to the 400 grams per liter emitted by a traditional oil-based paint). The association sees national regulation as a way of ending the hodgepodge of regulations that now govern the production of paint.

Paint Formulations

Traditional oil-based paint is about 60% liquid and 40% solid. The liquid in a typical alkyd paint is a mixture of linseed oil and volatile solvents that helps the paint spread and dry, and solids including pigments and the alkyd resins that act as binders. The alkyd resins in the paint polymerize to form the paint film. Traditional alkyds are long-chain polymer MOLECULES that readily form chemical bonds with other long-chain molecules. The new clean-air alkyds have a shorter chemical chain that makes them easier to apply in a paint containing fewer solvents, but the resulting chemical bond in the paint film is not as strong and takes longer to develop. Dry painted surfaces are not as strong as a result. Paint manufacturers have, in many cases, found it easier to discontinue traditional oil base paints rather than to reformulate them. Oil-based gloss enamel has been especially hard to replace because of its smoothness and high initial gloss, although some acrylic enamels come close and actually retain their glossy surface longer. The solid-content of clear coatings, which have contained 38% to 45% solids in the past, increases to 63% to 65% solids in the new paints. Stains, which had been about 85% liquid, contain 65% to 70% solids in clean-air versions. The new stains are thick enough that they are best wiped on with a rag rather than being applied with a brush. Paint manufacturers added more solids to oil-base clear finishes rather than switch to a latex paint because sunlight passes through water-base sealers into the wood surface below where ultraviolet light can degrade the wood's surface. The resins in traditional clear finishes absorb ultraviolet radiation, and therefore the sealer, not the underlying wood finish, eventually shows damage from exposure to sunlight. Clean-air paints go on in thicker coats because they contain fewer solvents, and can take about twice as long to dry.

While many of the latex replacements for traditional oil-base favorites have been successful, latex semi-gloss, gloss and eggshell enamels are not as effective as their predecessors, and tend to have an uneven, ropy surface when dry. Among the qualities lacking in the new paints are *brush out*, the ability to smooth out paintbrush or roller marks, because the paints do not flow and level out as well; and *holdout*, the ability to resist penetration when applied over a porous surface, which could be poorer because the paint does not dry as quickly and has more of an opportunity to penetrate into the underlying surface. The new paints have less mildew resistance because there is more organic material in the paint for the mildew to grow on. The advantages of the new paints include the fact that they are fast-drying and have a tendency to retain their color. They also emit fewer odors and toxic vapors and represent a reduced fire hazard. Because they dry more slowly, the new paints have a better *wet edge retention*, meaning that a painter can more easily go over the edge of an area that has already been painted without leaving lap marks. Because the new paints have a more flexible paint film, their resistance to cracking and peeling is greater.

Paper a thin sheet usually composed primarily of cellulose or other vegetable fibers. The ancient Egyptians used the pith of papyrus (the source of the word "paper"), a reed-like plant that grows in the Nile Valley, to produce paper. The first known paper was produced in China with fiber from rags, tree bark and hemp, bamboo, jute, and straw. In addition to paper's original use as a medium on which to write or draw, paper and paper products are used in books, magazines, catalogs; in a variety of tissues for cleaning, wiping and nose-blowing; in PACKAGING; in the production of BUILDING MATERIALS and in air and water filters. An increase in world literacy rates—from 56% in 1950 to 74% in 1992—and increased use of paper for packaging have contributed to a worldwide increase in the use of paper products. Futurists predicted the demise of the printed word with the advent of the electronic information age in the 1970s, but as William Rathje and Cullen Murphy say in their book *Rubbish*, "with respect to paper, advancing technology is not a contraceptive but a fertility drug." The estimated 19 million photocopy machines now in use in the United States alone, and countless computers in homes and offices, have dramatically increased the use of paper. Total world production of paper products in 1993 was about 247 million tons, a six-fold expansion over total production in 1950. More than 64 million tons of paper and paper-board are produced annually at U.S. paper mills, and American consumers use nearly 71 million tons of paper and CARDBOARD a year. The United States, Canada, Japan and China are the world's four leading producers of paper, accounting for more than half of total production. The Garbage Project estimates that more than 40% of the volume of a typical LANDFILL is consumed by paper products, with newspapers alone constituting about 13% and magazines about 1.2%. The Forest Service estimates that more than one-fourth of the trees logged in the U.S. are used to make paper or paperboard. An estimated 52 billion pieces of advertising printed on paper—often referred to as

JUNK MAIL, including 14 billion mail-order catalogs—are handled by the U.S. Postal Service annually, according to estimates by the WORLDWATCH INSTITUTE, and much of it is thrown away without being opened.

Paper mills are major polluters of air and water. The Worldwatch Institute estimates that 950,000 tons of EFFLUENT, which includes numerous toxic organochlorine substances (see ORGANOCHLORIDE), are dumped into the world's rivers each year from paper mills, and that 100,000 tons of SULFUR DIOXIDE is emitted in the paper-production process. WATER POLLUTION from pulp mills has given streams the appearance of a frothy sewer with an odor like rotten eggs, often killing most of the aquatic life for miles downstream, but emissions in the United States have been greatly reduced.

Most paper in industrialized nations is made from wood pulp, but other sources are used to produce about 10% the paper made worldwide. Plant fibers from agricultural wastes such as sugarcane stalks (called *bagasse*) and straw are widely used to produce paper in developing countries. Fiber crops such as hemp and kenaf that are seeing increased use in paper production can produce pulp more quickly and require less space to grow than wood. It is estimated that kenaf produces four times as much fiber per acre as southern pine. About 60% of Chinese paper is produced from fibers other than wood. More energy is required to process wood pulp in a paper mill than that required for such alternatives.

Making Paper

The basic paper-making process involves suspending a fibrous material in water, spreading the suspension over a porous surface through which water can drain, pressing most of the rest of the water out of the fibers and drying the resulting piece of paper. Lower grades of paper are made through a process called *mechanical pulping*, also known as *groundwood pulping*, in which wood chips are converted to pulp in an attrition mill, known as a disk refiner. Groundwood pulp contains *lignin*, which is what gives trees their structure and about half their weight. Groundwood pulping converts a higher percentage of the pulp found in a tree into paper than chemical processes. *Chemical pulping* increases the strength of wood fibers and produces a pulp capable of making a finer grade of paper. To produce the chemical pulp, wood chips are dumped into a digester and cooked with steam in a chemical SOLUTION called a cooking liquor. The *kraft process*, which yields the strongest fiber and is how most chemical pulp is treated, breaks down the resins, waxes, and fats found in some pines that are hard to pulp with other methods. Paper mills that process virgin fibers (see VIRGIN MATERIAL) to produce kraft paper emit sulfur compounds, acetone, METHANOL and chlorine compounds into the atmosphere.

Paper is made in *printing grades* and *industrial grades*. Printing-grade paper includes NEWSPRINT and catalog, *rotogravure*, and magazine papers, which are made from mechanical and thermomechanical pulps. Industrial grades are used in paper bags, liner board, corrugated cardboard, tissues, building materials, food packaging and DISPOSABLE DIAPERS. About three-fifths of the weight of a typical disposable diaper is cellulose fiber. Newsprint, phone books and the paper in a paperback book contain mechanical pulp, which deteriorates rapidly when exposed to light and heat because the lignin it contain turns yellow and brittle as it ages. Hardcover books are made from fully bleached pulps that are less acidic and much more durable. The highest quality paper used in books and for writing and printing papers contains fibers that are derived from processed cotton and linen rags. Brown kraft paper is used to make paper bags and cardboard boxes, and bleached kraft paper is used in books, magazines, and writing and wrapping paper. Most writing and magazine papers contain up to 30% mineral filler (usually calcium carbonate or clay), which makes the paper smoother, less absorbent and more opaque, thereby improving its optical and printing qualities.

Chlorine dioxide is used to bleach paper, but the discovery during the 1980s that WASTEWATER from paper mills that use chlorine as a bleaching agent contains FURANS, DIOXINS and CHLOROFORM has led some mills to switch to less-polluting bleaching agents including OZONE, oxygen and hydrogen peroxide for the majority of the bleaching process. A chlorine compound is used only for a finish bleaching. Bleaching is required to remove lignin from pulp that will be used to produce higher grades of paper, and it is primarily the reaction of chlorine with lignin that creates toxins. Hydrogen peroxide is used as a bleach in most paper mills in Germany, Austria, Switzerland and the Nordic countries. Bleaching removes the residual lignin and brightens the pulp.

The cellulose fibers in wood are bound together by a chemical called lignin. Paper makers have in the past burned millions of tons of lignin every year because there was no practical use for it. Lenox Polymers Ltd. in Port Huron, Michigan, has developed a new lignin-based resin that can be substituted for phenolic resins in the production of sand molds for metal castings. Phenolic resins contain FORMALDEHYDE.

Paper or Plastic? a choice now being offered to grocery shoppers at many supermarkets every time they go through the checkout stand. The answer—in terms of which is more environmentally friendly—is definitely PLASTIC. A big PAPER grocery bag weighs almost 65 grams, while the biggest plastic bag weighs a little less than 7 grams. And the POLLUTION generated per gram of bag is probably actually more for the paper. But even assuming that they are equal, the plastic bag wins because it weighs only a tenth as much. Both the paper and the plastic bags are easy to recycle (see PAPER RECYCLING, PLASTIC RECYCLING, RECYCLING), and both are pretty likely to have a fairly high percentage of post-consumer recycled material (see POST-CON-SUMER WASTE) in them. It is difficult to compare the resource depletion cost of the PETROLEUM pumped out of the ground to make the plastic bag to the trees cut to make the paper (something on the order of 10 to 31 trees per ton), but the effects of cutting the trees *is* more obvious. The plastic bag has about the same cross-section (in fact, some stores stuff their paper bags inside a plastic bag to make them more convenient to carry), but more groceries can probably be safely carried in one paper bag than in one plastic bag. The plastic bags look flimsy, but, if anything, the paper bag is more likely to tear, which is why the bag-person uses double paper bags for the heavier items. Even then, it just *seems* like the paper bag should be better. But grocery consumers concerned about the environmental correctness of their bag choice are probably going to bring their own reusable grocery bags to the store anyway. All of the above demonstrate the complexities involved in making such comparisons.

A study that compared the energy consumption and production of pollution of paper and EXPANDED POLY-STYRENE hot drink cups by a chemist at the University of Victoria, British Columbia reached the following conclusions:

- An equivalent amount of HYDROCARBONS are consumed in the production of the two types of cup. The hydrocarbons are burned in the paper-production process, while they become part of the polystyrene cup.
- Making the paper cup consumes between 160 and 200 kilograms of PROCESS CHEMICALS per metric ton of wood pulp processed, while manufacturing the foam cup requires only 33 kg per metric ton of polystyrene.
- Because it is a much denser material, there is six times as much wood pulp in a paper cup as would be found in a comparable foam cup.
- The production of a paper cup consumes 12 times as much steam, 36 times as much electricity and twice as much cooling water as is required for a foam cup.
- The wastewater produced in the paper-cup production process has 10 to 100 times the contaminants found in the wastewater stream associated with the production of the Styrofoam cup.

- The emission of air pollution from production of one polystyrene cup is 60% less than that for the paper cup.
- Paper cups are not recyclable because they are held together with a hot-melt glue or a SOLVENT-based adhesive, and because they are coated with wax or a plastic film.
- The foam cups cost less.

And one additional note not in the study: Foam cups are easier to hold when filled with a hot beverage (and keep the beverage hotter for longer), and a beverage consumer is unlikely to use two cups to avoid burning his or her fingers, while double-cupping is fairly common with paper cups. See also MCTOXICS CAMPAIGN.

Paper Recycling the incorporation of wastepaper into new PAPER products. Paper recycling reduces the volume of GARBAGE that is dumped in LANDFILLS, which is important since an estimated 40% of the volume of TRASH in a typical landfill is taken up by paper products.

Toronto's curbside recycling program cut the volume of newspaper arriving at the city dump from 12.9% to 6.7%. Reprocessing recycled paper also significantly reduces the AIR POLLUTION and WATER POL-LUTION associated with the treatment of virgin wood pulp. The only atmospheric emissions associated with the processing of recycled paper are those associated with the COMBUSTION of fuel necessary to run the paper mill. It has been estimated that the fibers that make up paper can be reused up to a dozen times without experiencing significant deterioration. Recycled paper from the United States has traditionally been used to produce many of the paper products manufactured in Asian countries, but more is being used domestically as more paper mills purchase the necessary processing equipment. The recovery rate for paper products was 22% in 1970, but it had increased to 40% by 1994. The British recycle about 35% of their paper, while Germany recycles about 50%. The composite paper/plastic products often used in PACKAG-ING are nearly impossible to recycle (see COMPOSITE PACKAGING).

President Clinton signed an executive order in October 1993, requiring federal government agencies to buy paper containing a minimum of 20% post-consumer fiber (see POST-CONSUMER WASTE) starting in 1995, and to switch to paper with a 30% post-consumer content in 1998. Although the federal government uses only 2% of the 22 million tons of printing and writing paper sold annually in the United States, the order is expected to set a precedent and to boost the production of recycled office paper. The government buys about $150 million worth of paper annually, primarily computer and copier paper. The recycled paper is expected to cost 3% to 4% more than virgin paper initially, and federal agencies will reportedly be told to absorb the increased cost by decreasing their overall use and waste of paper. Less than 10% of the printing and writing paper purchased by the government before the order went into effect contained recycled material. The paper industry had argued that it would be more effective to start with a 10% post-consumer fiber requirement than a higher one because the larger paper mills would be able immediately to start production of recycled paper with that content. Calling for a higher percentage would mean that only smaller mills would gear up to produce the paper, according to industry officials.

The paper industry has been wary of making a large investment in de-inking equipment given uncertainty about both the market for recycled office paper and the supply of high-quality office paper from recyclers. Industry officials say that even though the demand for reprocessed office paper is increasing, they are still confronted with an inadequate supply of high-quality recycled paper. Paper used for printing and writing, which comprises 28% by weight of all paper and paperboard manufactured in the United States, contained an average of only 9% wastepaper in 1992. Paper tissues, by comparison, had a 60% recycled paper content and NEWSPRINT 36%. If wastepaper is not well sorted, it can easily foul the expensive de-inking equipment in which it is processed. Recyclers say that even paper from large offices is generally poorly sorted. Separating different types of office paper out of a mixed batch is "like trying to unscramble an egg," according to one recycling industry executive. Industry sources say that it is almost impossible to get the well-sorted supply of office paper needed for recycling from municipal waste collection systems. Poorly sorted office paper that is recycled will probably end up being used for lower grade products such as fiberboard and CARDBOARD, if it is used at all.

If high-quality paper is to be produced from recycled paper that has printing or photographs on it, the ink, fillers and coatings that it contains must be removed. This is accomplished by first converting the used paper into a pulp in a vat where mechanical, thermal and chemical processes are used to swell cellulose fibers and disperse ink. The pulp is then routed through a screen that intercepts foreign material such as staples or paper clips that escaped the initial sorting, and then through a (nonchlorine) bleaching process that removes part of the remaining inks and dyes. Detergents and water are added to the pulp stock and it is then agitated in a machine resembling a giant washing machine. Most of the remaining ink particles become *hydrophilic* in the process, and become chemically bonded to water molecules, which are then drained out of the pulp. Foaming agents and other chemicals that make the remaining ink particles *hydrophobic* are then added to the stock, and air is bubbled through the mixture. The remaining ink particles lock on to the air bubbles and are carried to the surface. The pulp then is spun at high speed in a centrifugal cleaner that removes any synthetic compounds and hot-melt glue that remain in the stock. The processed pulp is then ready to be made into paper using the same techniques used for virgin pulp. More than 20 pounds of SLUDGE is typically produced for every 100 pounds of wastepaper run through the de-inking process. The sludge is generally cleaner than that created by most secondary SEWAGE TREATMENT plants (see SECONDARY TREATMENT), and most of it is disposed of in private landfills, although some paper companies have received permission to dispose of it in municipal landfills. In some cases the sludge contains toxic materials, and must be disposed of as a HAZARDOUS WASTE.

Parathion an organophosphate insecticide (see ORGANOPHOSPHATE, INSECTICIDE) widely used on commercial fruit, nut, vegetable and field crops. Parathion gets into air and water primarily from treated agricultural land. It is moderately soluble and moderately persistent in water, with a HALF-LIFE of 20 to 200 days. Parathion has high chronic TOXICITY (see CHRONIC EFFECTS) to aquatic life. Parathion can affect health when inhaled or passed through the skin. Skin contact can cause organophosphate poisoning, with headaches, sweating, nausea, vomiting, diarrhea, loss of coordination and, in extreme cases, death. Repeated exposure may cause personality changes including depression, anxiety or irritability. Parathion is a possible TERATOGEN.

The OCCUPATIONAL SAFETY AND HEALTH ADMINISTRATION has established a permissible exposure limit for airborne parathion of 0.1 milligrams per cubic meter, averaged over an eight-hour workshift. It is also cited in AMERICAN CONFERENCE OF GOVERNMENT INDUSTRIAL HYGIENISTS, Department of Transportation NATIONAL INSTITUTE OF OCCUPATIONAL SAFETY AND HEALTH and ENVIRONMENTAL PROTECTION AGENCY regulations. Parathion is on the Hazardous Substance List and the HAZARDOUS AIR POLLUTANTS LIST. It is a reactive chemical (see REACTIVE MATERIAL) and is an explosion hazard. Businesses handling significant quantities of parathion must disclose use and releases of the chemical under the provisions of the EMERGENCY PLANNING AND COMMUNITY RIGHT TO KNOW ACT.

Particulate a particle of solid matter or droplet of liquid small enough to remain suspended in the air, at least temporarily. Airborne particles with a diameter larger than 10 microns are subject to the pull of gravity, and—in the absence of sufficient turbulence to keep them suspended—quickly settle out of the atmosphere. Particles with a diameter of less than one micron, which comprise 99% of the total particle count in a typical air sample, can remain aloft indefinitely. Particulates can be as large as 1,000 microns. Even in clean rural air there are typically more than one million particulates suspended in each cubic foot of air. Such particles are kept aloft by *Brownian Motion*—the irregular movement of particles suspended in air as the result of collisions between atoms, molecules and small particles. Although defined as liquid or solid particles, the word *particulates* is more often used to denote solids, while AEROSOLS, which are also technically liquids or solids, refers to liquid particles.

Airborne particles and gases that act together as contaminants are called *dispersoids*. The reaction of SULFUR DIOXIDE with oxidants (see OXIDATION, OXIDIZING AGENT) and particulates to form sulfates and sulfuric acid is one such process (see PHOTOCHEMICAL SMOG). DUST, ASH, SMOKE, soot, metals and liquid droplets released by burning fuel are all particulates (see COMBUSTION BYPRODUCTS). Wood stoves, vehicle exhaust, dirt and gravel roads, farms, forests and industrial emissions are among the countless sources of particulate POLLUTION. AIR FILTERS, SCRUBBERS and ELECTROSTATIC PRECIPITATORS are used to take samples of particulates for laboratory analysis and to remove particulates from air (see PARTICULATE MEASUREMENT, PARTICULATE REMOVAL). Airborne POLLEN and MICROORGANISMS come in every imaginable shape and size ranging from about 1 to 100 microns. Particulates are a major source of water pollution (see ACID RAIN, AERIAL DEPOSITION). Many of the toxins emitted by tobacco products are particulates (see ENVIRONMENTAL TOBACCO SMOKE).

The upper respiratory system filters out particles larger than approximately 10 microns in diameter. Particles smaller than 6 microns are called *respirable suspended particulates* because they are small enough to be inhaled deep into the lungs. Suspended particulates are what makes a beam of sunlight visible (by reflecting light). Only about 1% of airborne particulates—those larger than 10 microns—can be detected by the human eye. A report released in 1994 by the American Lung Association estimated that nearly 23 million Americans live in areas where suspended particulates exceed federal standards. The report, which was published by the federal Centers for Disease Control and Prevention, estimated that an additional 92 million people live in areas that do not meet the stricter California particulate standards. Seventy-one areas exceeded federal particulate limits in 1991. Rough estimates by the ENVIRONMENTAL PROTECTION AGENCY (EPA) suggest that between 50,000 and 60,000 deaths each year are caused by particulates, which would make particulates the most deadly form of air pollution if correct. There has been relatively little research into the TOXICITY of particulates. Even nontoxic particulates can pose a health hazard, as they can become the vehicle by which a toxic hitchhiker gets into the lungs. Even if the lungs expel the particle, the toxin is often left behind. Particulates get into indoor air from combustion appliances and microbial growths. BENZO-A-PYRENE, the result of incomplete combustion, is one of the most dangerous particulates found in indoor air.

DIESEL engines are one of the most obvious sources of particulate pollution, and fumes from diesel-powered trucks, buses and cars are especially dangerous because they are emitted in urban areas and in heavy traffic where many people are exposed to them. The particulates in diesel exhaust absorbs sulfuric acids and VOLATILE ORGANIC COMPOUNDS (VOCs) once emitted, and can carry them into the lungs. Epidemiological studies of truck drivers, railroad workers and heavy-equipment operators, all of whom are exposed to particulates from diesel engines regularly, have shown increased incidence of liver, lung and bladder cancer. The EPA classified particulate matter as a CARCINOGEN in 1983. Congress ordered the EPA in 1977 to have stricter particulate standards in effect for buses by 1981. In 1980, the agency proposed new standards to be implemented in 1983, having missed the goal of imposing the tougher standards on 1981 buses, but the Reagan Administration's new EPA administrator shelved the rules, citing the hardship they would impose on the ailing auto industry. The 1990 amendments to the CLEAN AIR ACT called for new standards that required a 95% reduction in particulate emissions from trucks and buses produced from 1994 on, and required the EPA to institute new particulate standards by 1995. See also AEROSOL, AIR POLLUTION, ASBESTOS, AUTOMOTIVE AIR POLLUTION, BIOLOGICAL AEROSOL, CALIFORNIA CLEAN AIR ACT, DANDER, FIBROGENIC DUST, FLY ASH, MINERAL FIBERS, MOBILE SOURCES, NITROGEN OXIDES, SUSPENDED SOLIDS, TOTAL SUSPENDED PARTICULATES.

Particulate Measurement the sampling and measurement of airborne particles. Several principal techniques are used:

- *Direct observation* of the opacity of smokestack emissions is a quick way of estimating the particulate content of exhaust gases.

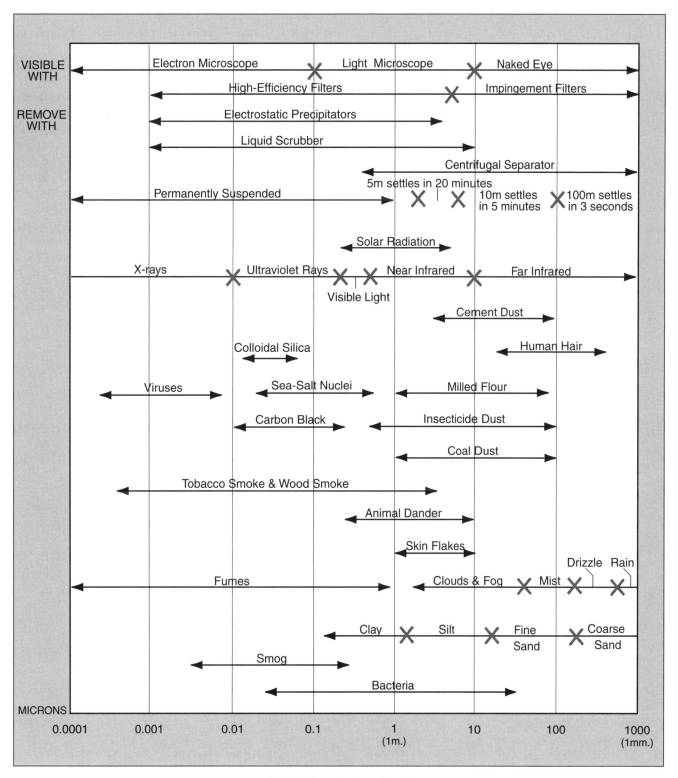

Particle diameter/wavelength

- DUSTFALL—the settling of particulates out of the atmosphere due to gravity—can be measured with a *dustfall jar* or an *adhesive sheet*.
- A *molecular correlation spectrometer* is an instrument that can be used from a distance to measure the concentration of gases, especially nitrogen dioxide and SULFUR DIOXIDE, in a pollution PLUME. The instrument reads the spectrum of incoming solar radiation that has passed through the pollution plume being measured. The spectrum of the pollutant being measured is stored electronically inside the instrument, and the percentage of radiation coming from the direction in which it is pointed that matches the pattern is revealed.
- *Photometric analysis* is used primarily to determine the mass or concentration of particles. When a beam of light hits a particle, it is scattered, and by analyzing the resulting pattern of reflected light, it is possible to come up with a good estimation of a particle's size. A *single-particle photometer* is used to determine the distribution of particle sizes in an air sample. An *aerosol photometer* collects the light scattered by numerous particles in an airstream, and analyzes it to produce an estimate of the total cross section of particles in the sample. A beam of light from a pulse laser can be reflected off a smoke plume and the reflected light can be analyzed to get an estimate of the particle concentration.
- *Inertial particle collectors* use the inertia of particles moving in an airstream to capture them by putting an obstruction with a sticky surface in the airstream. The air moves around the obstruction, but particles that are large enough hit the adhesive surface and stick. A *single-stage impactor* forces the air to be sampled through an orifice with a sticky substrate directly in front of it. The airstream is accelerated through the orifice, giving suspended particles larger than about 0.5 microns sufficient inertia to hit the collection strip and stick. *Multi-stage impactors* or *cascade impactors* consist of two or more single-stage impactors operating in series. The first impactor has the lowest velocity, and collects the largest particulates, with succeeding stages having successively higher velocities and capturing smaller particles. A *cyclone separator* removes particulates by forcing the airstream to move in a tight vortex, which causes particulates to be flung to the outside of the chamber onto the collector walls.
- *High-volume samplers* measure airborne particulates. A suction pump pulls the air to be sampled through a filter of known weight, which typically filters 70,000 cubic feet of air during a period of 24 hours. After the test run, the filter is removed and weighed again to determine the total weight of the particles collected. The concentration of airborne particulates can then be computed since the capacity of the pump and the amount of time it was run are known. Electronic high-volume sampling systems are now available for computerized monitoring. Particulate standards in the CLEAN AIR ACT and most other air quality regulations are based on air sampling with high-volume samplers.
- TAPE SAMPLER—a tube through which air to be tested is drawn by a vacuum pump and contaminants are intercepted on a filter tape mounted at one end.
- An ELECTROSTATIC PRECIPITATOR can be used to collect relatively large samples of particulates with high efficiency for all particle sizes, or to deposit small samples of very small particulates directly onto the grid of an electron microscope.
- A *thermal precipitator* removes a relatively small quantity of precipitates from the airstream through the *thermoforesis force* experienced by a particle in a temperature gradient, which drives the particle towards a cool surface. Particle motion is caused by the difference in molecular bombardment on different sides of the particle. A thermal precipitator collects particulates from microscopic size up to 3 to 5 microns in diameter.

The analysis of particulate samples can be very difficult because of the number and variety of particles that can be involved. The concentration and mass and particles in a given air sample can be computed relatively easily by standard analysis techniques (see AIR POLLUTION MEASUREMENT), and the composition of individual particles can similarly be analyzed. But producing an exact breakdown of the millions of particles that can be suspended in even a closet-sized space is next to impossible. Microscopic analysis of representative samples can give a researcher an estimate of the makeup of the particles, and computer analysis can give a far clearer picture of particulate mix than has been available in the past.

Particulate Removal   the removal of particulates from air. AIR FILTERS, centrifugal collectors, (see CYCLONE), SCRUBBERS and ELECTROSTATIC PRECIPITATORS are some of the principal methods of particle removal:

- *Wet collectors*, called scrubbers, treat polluted air by spraying it with a liquid, usually water.
- A *gravity collector* is a settling chamber where large particles (generally larger than 50 microns in diameter) are pulled out of an airstream by gravity.
- A *momentum collector* forces the airstream to be cleaned through sudden changes in direction. Particles larger than about 20 microns in diameter are unable (because of their inertia) to make the turn

and are impacted on baffles or louvers. Gravity and momentum collectors often provide pretreatment of an airstream destined for treatment in a higher efficiency filter.

- A *cyclone* or *dry centrifugal collector* whirls contaminated air in a vortex inside a tapered, cylindrical chamber, and centrifugal force throws particles to the outside of the collector, where they settle to the bottom for removal. A centrifugal collector can remove up to 70% of the particles five microns and larger in diameter from an airstream, and are used as primary collectors for FLY ASH, WOOD, DUST and PLASTICS dust.
- A BAGHOUSE is a network of tubes or bags that filter contaminants out of an airstream.
- In an *air classifier*, air containing particulates is injected into a forced stream where particles are separated according to the size, density and aerodynamics.

See also AIR POLLUTION, CLEAN AIR ACT, INDUSTRIAL AIR POLLUTION TREATMENT.

Pathogen an organism that causes disease in other living things. Most pathogens are VIRUSES or BACTERIA. To infect another organism, a pathogen must be capable of invading the host organism, disabling or combating the host's immune system, reproducing and spreading, and causing symptoms of infection in the host. Pathogens cause disease by producing toxins and ALLERGENS, disrupting cell metabolism and other means. Airborne pathogens that are usually present in either outdoor or indoor air include bacteria, viruses, spores, molds, mildews and POLLENS. A *virulent pathogen* causes relatively major damage to the host with relatively few pathogens present. See also AEROALLERGENS, PATHOGENIC WASTE.

Pathogenic Waste liquid, solid or gaseous waste that contains BACTERIA that can be harmful to health. Principal sources of pathologic waste are animal waste from livestock production, laboratory animals, slaughterhouses and tanneries; pharmaceuticals manufacturing waste; and food-processing waste. The concentration of FECAL COLIFORM BACTERIA is monitored in drinking water and wastewater treatment systems to estimate the concentration of more pathogenic bacteria such as SALMONELLA. Chlorination is the most common way of killing pathogenic bacteria found in wastewater (see CHLORINE). See also MICROORGANISMS.

Pentachlorophenol a chlorinated AROMATIC HYDROCARBON biocide widely used in paints and stains as a fungicide and bactericide; also called *PCP* or *penta*.

Pentachlorophenol is a light brown solid that is normally used in SOLUTION. Installations that treat posts, poles and railroad ties with penta, often in solution with DIESEL oil, have been a common source of WATER POLLUTION in the past (see GROUNDWATER POLLUTIONS), and many have become Superfund cleanup sites. Penta also gets into air and water from already treated wood, industrial discharges, municipal waste treatment plants, spills and agricultural RUNOFF. It is slightly soluble and moderately persistent in water, with a HALF-LIFE of 20 to 200 days. Pentachlorophenol has high acute and chronic TOXICITY to aquatic life (see ACUTE EFFECTS, CHRONIC EFFECTS). The toxicity of pentachlorophenol increases as the water's acidity increases. DIOXIN is often present as an impurity in penta.

PCP can affect health when inhaled or passed through the skin. It can cause severe irritation of the skin, eyes, nose and throat. PCP poisoning, characterized by sweating, trouble in breathing, high fever, rapid pulse and pain in the chest or abdomen, can quickly cause death if not promptly treated. Repeated contact can produce an acne-like skin rash, and can damage the liver and kidneys. Penta can cause bronchitis, with cough and phlegm. Penta may cause genetic changes in cells. Limited evidence suggests that it may be a TERATOGEN in animals, and that it is a possible teratogen in humans. Penta is still used as a wood preservative, but all other uses were banned in 1987.

PUBLIC WATER SYSTEMS are required under the SAFE DRINKING WATER ACT to monitor concentrations of PCP (MAXIMUM CONTAMINANT LEVEL = 0.001 milligrams per liter). The BEST AVAILABLE TECHNOLOGY for the removal of penta from DRINKING WATER is granular ACTIVATED CARBON. The OCCUPATIONAL SAFETY AND HEALTH ADMINISTRATION has established a permissible exposure limit for airborne PCP of 0.5 milligrams per cubic meter, averaged over an eight-hour workshift. It is also regulated by AMERICAN CONFERENCE OF GOVERNMENT INDUSTRIAL HYGIENISTS. It is on the Hazardous Substance List and the HAZARDOUS AIR POLLUTANTS LIST. Businesses handling significant quantities of PCP must disclose use and releases of the chemical under the provisions of the EMERGENCY PLANNING AND COMMUNITY RIGHT TO KNOW ACT.

Perc Test a test used to determine the approximate PERCOLATION rate of a soil. A perc test is required in many areas before a permit to install a septic tank (see SEPTIC SYSTEM) can be approved. To perform a perc test, a hole about three feet across is dug in an area with soils typical of the area to about the same depth that drain fields will be run. The bottom of the pit is filled with gravel, and the pit is then filled with water, which is allowed to settle overnight. The next day, water is poured into the pit to a level about a foot above the top of the gravel bed. The amount of water in the pit and the amount of time it takes to drain a given distance is then computed. See also GROUNDWATER, PERMEABILITY.

Perched Aquifer   a zone saturated with GROUNDWATER that sits on top of an impermeable material such as rock or clay and that has a surface that is generally higher than the region's water table. Perched aquifers are separate from other AQUIFERS and often have a different water source than other bodies of groundwater in the area. A perched aquifer is generally small, and is subject to influence by seasonal fluctuations of rainfall and is vulnerable to infiltration by surface contaminants. Perched aquifers are generally not a reliable source of DRINKING WATER. See also PERCOLATION.

Perchloroethylene (PCE)   a clear liquid SOLVENT with a sweet CHLOROFORM-like odor that is used primarily for dry cleaning and metal-degreasing. Perchloroethylene—also called *tetrachloroethylene* and *ethylene tetrachloride*—is also used as a heat-transfer fluid and in the production of fluorocarbons. It gets into air and water from discharges from dry cleaners and metal processors, from municipal waste-treatment plants and from spills. A New York State study concluded that fumes from dry cleaners, especially PCE, may be affecting as many as 160,000 people in the state, including 100,000 who work or live near dry cleaners in New York City. Perchloroethylene is moderately soluble and nonpersistent in water, with a HALF-LIFE of less than two days. Concentrations may build to dangerous levels in GROUNDWATER.

PCE can affect health when its fumes are inhaled or passed through the skin. Inhaling the vapor can irritate the lungs, causing coughing or shortness of breath. PCE fumes can also irritate eyes, nose, mouth and throat. Exposure to higher concentrations can cause dizziness, lightheadedness and unconsciousness, and can cause fluid to build up in the lungs. Contact with liquid PCE can cause severe skin and eye burns. Long-term exposure can cause drying and cracking of the skin. Severe exposure can cause the heart to beat irregularly or to stop, and can damage the liver and kidneys severely enough to cause death. Perchloroethylene is a possible human CARCINOGEN that has been shown to cause liver cancer in animals in tests. It is also a possible human TERATOGEN.

PCE is one of the contaminants for which regulation is required under the 1986 amendments to the SAFE DRINKING WATER ACT. The BEST AVAILABLE TECHNOLOGY for the removal of PCE from drinking water is granular ACTIVATED CARBON or stripped-tower AERATION.

PUBLIC WATER SYSTEMS are required under the Safe Drinking Water Act to monitor for PCE (MAXIMUM CONTAMINANT LEVEL = 0.005 milligrams per liter). The OCCUPATIONAL SAFETY AND HEALTH ADMINISTRATION has established a permissible exposure limit for airborne PCE of 25 parts per million, averaged over an eight-hour workshift. It is also cited in NATIONAL INSTITUTE OF OCCUPATIONAL SAFETY AND HEALTH, AMERICAN CONFERENCE OF GOVERNMENT INDUSTRIAL HYGIENISTS and Department of Transportation regulations. It is on the Hazardous Substance List and the Special Health Hazard Substance List. Businesses handling significant quantities of perchloroethylene must disclose use and releases of the chemical under the provisions of the EMERGENCY PLANNING AND COMMUNITY RIGHT TO KNOW ACT.

Percolation   the passage of a liquid through a permeable solid (see PERMEABILITY) or through a mass of unconsolidated material such as gravel, soil or sand. Percolation occurs very slowly, with the force of gravity and of capillary action both playing a role in the movement of the liquid from pore to pore through the medium. The percolation rate of an AQUIFER is a measure of the ease and speed with which GROUNDWATER can move through it. Water will flow readily through coarse alluvial and glacial soils composed of rocks and coarse sand, but can only seep slowly through finer materials. The percolation rate of the materials above the WATER TABLE determine how easily surface contaminants can make their way into the aquifer (see LEAKING UNDERGROUND STORAGE TANKS, GROUNDWATER POLLUTION, WATER POLLUTION). The shape of the CONE OF DEPRESSION that forms on the surface of a water table when a pump sucks water out of an aquifer and the pollution PLUME that forms in an aquifer downstream from a pollution source are a function of percolation. Soils must allow adequate percolation for a septic tank drainfield to function properly (see PERC TEST, SEPTIC SYSTEM). See also INFILTRATION CAPACITY, POROSITY.

Permeability   the ability of a material to allow the passage of a liquid, usually water, from one part to another. A material through which liquids flow very easily is said to be highly permeable, while a material through which no liquid will pass is said to be impermeable. Permeability is determined more by the nature of the connection between openings within the material than by the size or shape of the openings themselves. Compare POROSITY. See also INFILTRATION CAPACITY, PERCOLATION, PERMEABLE LAYER.

Permeable Layer   a layer of rock, gravel, sand or a mixture of materials through which water can flow.

Solid rocks such as sandstone can be permeable, and are called a *consolidated permeable layer* when this occurs. Fractured rocks, sand, soil, gravel and the like are called an *unconsolidated permeable layer*. See also GROUNDWATER POLLUTION, PERMEABILITY, POROSITY,WATER POLLUTION.

Permeable Pipe a pipe constructed of a material through which water can easily percolate (see PERCOLATION). METAL or PLASTIC pipe that has been perforated, and unfired clay and composite materials that allow the passage of water, are used to make permeable pipe, which is used primarily to drain water from saturated soil. The footing drains that remove groundwater from a building's foundation and the LEACHATE drains that collect GROUNDWATER around a landfill or a hazardous waste dump are common uses for permeable pipe.

Persistent Compound a substance that will not break down chemically for many years when exposed to the elements. Because they do not break down easily, persistent compounds can accumulate to toxic levels in living tissue over a period of time. The process, called BIOACCUMULATION, means that even nontoxic concentrations of a contaminant can lead to poisoning. Life-forms at the bottom of the food chain may have drastically lower levels of accumulated toxins than do beings at the top of the food chain, which may have concentrations 100,000 times as high. For this reason, the environmental accumulation of a toxin is frequently discovered in raptors and other carnivores. CHLORINATED HYDROCARBONS such as DDT, POLYCHLORINATED BIPHENYL, CHLORDANE, PENTACHLOROPHENOL and HEPTACHLOR and metals such as MERCURY and LEAD are especially persistent toxins. Compare BIODEGRADABILITY.

Pesticide any substance that is used to control pests. INSECTICIDES, FUNGICIDES, RODENTICIDES and HERBICIDES are commonly used pesticides. Approximately 2.5 billion pounds of pesticides are applied each year in the United States, and many developing nations use much more per capita. More than 35,000 pesticide formulations with more than 1,400 active ingredients are available in the United States.

The widespread use of pesticides has lead to extensive air and water pollution (see AIR POLLUTION, GROUNDWATER POLLUTION, WATER POLLUTION) and associated disease and ecological damage. Pesticides applied by *broadcast spraying*—spraying plants with a fine mist, usually from an airplane or farm vehicle—can easily blow over onto neighboring property or into nearby watercourses, and for this reason the conditions under which broadcast spraying is allowed are tightly regulated. With *topical application*, pesticides are sprayed directly onto the plant to be treated, usually from a hand-held sprayer with a backpack tank or a small spray system mounted on a tractor. There is less overspray with topical application, but non-targeted life forms can still be affected. Animals have a tendency to eat vegetation that has been sprayed selectively, and can get a dangerously high dose of herbicide if allowed in a pasture too soon after spraying. Pesticides get into surface water and GROUNDWATER primarily from non-point pollution and from RUNOFF from farms, forests, gardens and lawns in which pesticides are used. Spills, industrial effluent and HAZARDOUS WASTE DISPOSAL are also significant sources of pesticide pollution in water.

Pesticides get into outdoor air from industrial emissions, spills and from the application of pest-control products. Rainwater often contains traces of pesticides picked up from contaminated air, often from halfway around the world (see GLOBAL POLLUTION). As much as 80% of Americans' exposure to pesticides occurs indoors, according to ENVIRONMENTAL PROTECTION AGENCY (EPA) estimates. Cleansers, plant food, pet collars, insecticides, pesticides tracked or blown in from outside and those picked up on clothes and shoes at work can all contaminate indoor air (see INDOOR AIR POLLUTION).

Some of the principal types of pesticides are:

- *arsenical compounds* that contain ARSENIC or an arsenic compound;
- ORGANOPHOSPHATES—ORGANIC COMPOUNDS that contain phosphorous;
- *biological pesticides*, which contain pathogens of the targeted pests;
- ORGANOCHLORIDES—organic compounds that contain chlorine.

Organochloride pesticides including the CHLORINATED HYDROCARBONS DDT, dieldrin, endrin, CHLORDANE, ALDRIN and kepone tend to be very persistent in the environment, and most uses have been either banned or sharply curtailed in the United States. *Hard pesticides* are PERSISTENT COMPOUNDS, which may exist in their original form for years or decades once they have been introduced into the environment. *Soft pesticides* such as the organophosphates malathion and PARATHION degrade more quickly once released into the environment, although their acute TOXICITY may actually be higher (see ACUTE EFFECTS).

Continued reliance on pesticides to control the population of undesirable species can initiate a deadly circle of cause and effect commonly called the *pesticide treadmill*, in which pests develop immunity to pesticides through natural selection—which in turn leads to heavier applications of pesticides that can destroy not

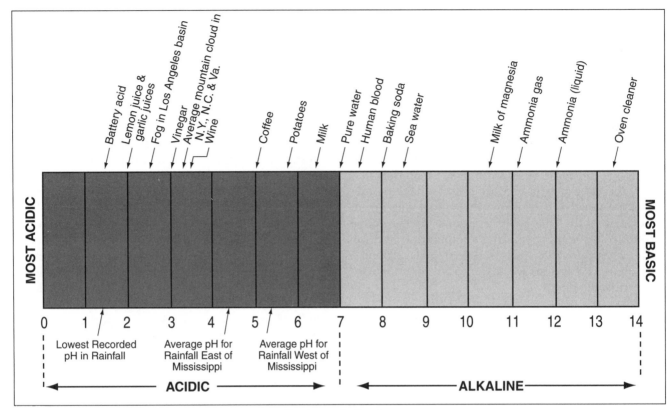

Spectrum of pH scale

only the targeted species but may also eliminate the pest's natural enemies. Decline of the species that would ordinarily control the pest's population means that even higher levels of the pesticide must be applied to keep down the pests. An increasing number of farmers are resorting to INTEGRATED PEST MANAGEMENT techniques in an attempt to reduce pesticide use, restore damaged land and save money.

Pesticides are regulated under the FEDERAL INSECTICIDE, FUNGICIDE AND RODENTICIDE ACT (FIFRA), which originally became law in 1947. The majority of the pesticides now in use were registered before 1972, and have not been subjected to the rigorous health tests required for newer formulations. A 1988 law required pesticide manufacturers to submit health data on the 620 active ingredients in older pesticides. The EPA issued rules during 1994 to protect workers on farms, forest lands and in nurseries and greenhouses from exposure to pesticides. The EPA estimates that one or more pesticide is found in the water of 10% of U.S. community drinking-water systems (see PUBLIC WATER SYSTEM) and 4% of rural domestic wells at levels higher than federal guidelines. The National Coalition Against the Misuse of Pesticides and the Northwest Coalition for Alternatives to Pesticides sued the EPA in May 1994, claiming that the agency was not properly

reviewing manufacturer's claims about the toxicity of the more than 2,300 inert ingredients that are commonly part of pesticide formulations.

Pesticide Manufacturing Wastes   BYPRODUCTS of the chemical processes used to produce pesticides. Some of the most polluted U.S. Superfund toxic waste sites, such as LOVE CANAL, were dumping sites for PESTICIDE manufacturing wastes. Toxic materials found in the waste products created in pesticide production include traces of the product being produced, and byproducts of the chemical processes being used. There was little control of on-site dumping of pesticide-production residues as recently as the 1970s, and soil and GROUND-WATER around pesticide factories were frequently seriously contaminated as a result. Workers in many plants were routinely exposed to pesticides and their byproducts, with many contracting cancer and other disorders as a result. See also HAZARDOUS WASTE, HAZARDOUS WASTE DISPOSAL.

PET Plastics   see POLYETHYLENE TEREPHTHALATE PLASTICS

Petrochemical   any of the HYDROCARBONS produced from PETROLEUM. Petrochemicals are also called petroleum distillates (see DISTILLATION). The lighter petro-chemicals are VOLATILE ORGANIC COMPOUNDS.

Petroleum a complex mixture of HYDROCARBONS containing varying amounts of compounds of oxygen, SULFUR and NITROGEN. PLASTICS, pharmaceuticals, FER-TILIZERS, PESTICIDES, GASOLINE, DIESEL, MOTOR OIL, industrial PROCESS CHEMICALS, BUILDING MATERIALS, FUELS and even some foods are all petroleum-based products. Buckminster FULLER compared the world's supply of petroleum to the nutrient in the white of an egg—a source of sustenance for developing life—and likened the necessity of eventually ending human dependence on FOSSIL FUELS and converting instead to the use of RENEWABLE RESOURCES to the instinctual compulsion of the chick to break out of the egg and fend for itself in the world outside. Petroleum is a liquid that is found primarily in deep sandstone and limestone formations, often mixed with brine and salts. It is apparently formed when the fossilized remains of MICROORGANISMS deposited on the beds of ancient seas are buried by tectonic activity. These petroleum-bearing formations tend to move toward the surface until trapped by an impermeable layer of rock. Petroleum is tapped for human use by drilling through the layer under which it is trapped. The availability of petroleum has molded the manmade environment by making the INTERNAL-COMBUSTION ENGINE and related developments possible. Petroleum products play a role in nearly every human endeavor. The use and control of petroleum reserves plays a dominant role in world political and military activity. More than 2.5 billion barrels of crude petroleum were produced by U.S. wells in 1993, which was a little less than half of the nation's total consumption of crude oil (a barrel holds 42 gallons—see MEASUREMENTS). A record high 52% of the crude used in the United States in 1993 was imported.

The mix of hydrocarbons present in petroleum varies widely in oils from different localities, or even in petroleum from different wells that are in the same region. A variety of air and water POLLUTANTS (see AIR POLLUTION, GROUNDWATER POLLUTION, WATER POLLUTION) are generated by the extraction and refinement of petroleum. Most crude oils contain a variety of AROMATIC HYDROCARBONS such as BENZENE and TOLUENE. The petrochemicals present in a given crude oil have a wide range of boiling points, so various hydrocarbon groups can be removed by gradual heating process called *fractional distillation* (see DISTILLATION). As petroleum is heated, the gaseous hydrocarbons METHANE and BUTANE are the first to be released. Petroleum ethers, *pentanes* and *hexanes*, evaporate next, at temperatures from 20° to 60°C. Gasoline—a mixture of hydrocarbons that evaporate between 38° and 205°C—consists of mixed pentanes and *nonanes*. As the temperature is increased, kerosene, then fuel oil and lubricating oil are evaporated from the crude oil. Jet fuel, a derivative of kerosene and fuel oil, makes up about half of every barrel of crude oil, depending on the crude being refined and the refinement process used. The residue of the distillation process is either asphalt or coke. Petroleum distillates can be further refined through a variety of processes. *Thermal cracking* is used to split some of the hydrogen molecules in heavier petroleum distillates by subjecting them to very high temperature and pressure. *Catalytic cracking* employs a catalyst in conjunction with heat and pressure to split the hydrogen molecules. Processes such as hydrogenation, alkylation, hydro-forming, polymerization (see POLYMER) completely modify the chemical structure of petrochemical molecules. It is expensive to change oil refineries to produce new, cleaner fuel products. In 1993 the federal government estimated that by the year 2000 the petroleum refining industry would spend about $37 billion retooling to meet new environmental laws, and that by the end of the decade consumers would pay about $18 billion a year more for petroleum products such as gasoline and heating oil because of the cost of the new laws.

The recovery, transportation and refinement of petroleum, which is commonly referred to as *raw crude oil*, is a major source of air and water pollution, as is the processing of petrochemicals. The rupture of a major oil pipeline—such as occurred in Houston during flooding in November 1994—not only pollutes the environment and threatens human life, but also threatens the economic well-being of the country. The BRINE that is removed along with the petroleum is disposed of in shallow ponds designed to allow the liquid to seep into the ground, or is pumped into the ground in an INJECTION WELL. Both methods of brine disposal often pollute groundwater.

pH the balance of ACIDITY and basicity (see ALKALINITY) of a liquid. The pH of a liquid is a function of the number of HYDROGEN IONS present. Pure water has a pH of 7. A liquid with a higher concentration of hydrogen ions than pure water is termed an ACID, and will have a pH of less than 7. A liquid with fewer hydrogen ions than pure water is termed a BASE, and will have a pH of more than 7. Since pH is a logarithmic scale, each full step on the scale represents an increase or decrease by a factor of 10. Thus an acidic SOLUTION with a pH of 6 will contain 10 times as many hydrogen ions as pure water with a pH of 7. Similarly, a basic solution with a pH of 8 will contain one-tenth as many hydrogen ions as pure water. A pH meter measures hydrogen ion concentration by gauging the difference in electrical potential between two electrodes placed in a liquid. An approximate measurement of a liquid's pH can be obtained with litmus paper, which employs chemicals that change color depending on a solution's ACIDITY. The acidity of a body of water or of LIQUID WASTE that is dumped into water is important because many biological functions of aquatic organisms are affected by acidity. The TOXICITY of many pollutants varies according to water's pH. A sudden change in pH can destroy the MICROORGANISMS that process waste in a SEWAGE TREATMENT plant.

Phenol an aromatic compound that consists of a BEN-ZENE RING with an attached hydroxl group (OH–). Phenol is a colorless to pink solid or thick liquid with a sweet, tar-like odor. It is used primarily in the production of other chemicals, and also in the manufacture of plywood, pharmaceuticals (including aspirin), PLAS-TICS and rubber. Oil refineries and operations that convert COAL to gaseous or liquid fuels or metallurgical COKE produce phenol as a BYPRODUCT. Discharges from these industries, municipal landfills and SEWAGE TREATMENT plants and spills are the primary ways phenol gets into air and water. Phenol is highly soluble and slightly persistent in water, with a HALF-LIFE of two to 20 days. Phenols are also a class of compounds formed by adding one or more functional groups to the phenol molecule. Biphenyl, phenolics, trichlorophenol and PENTACHLOROPHENOL are examples. Phenols get into indoor air from household disinfectants, antiseptics, perfumes, mouthwashes, polishes, waxes, glues and air fresheners.

Phenol can affect health when inhaled or passed through the skin. Inhaling phenol vapor can irritate the mouth, nose and throat. Phenol can severely burn the skin and cause permanent damage to the eyes. Long-term or high-level exposure can damage the liver, kidneys and heart. Exposure to high concentrations can cause poisoning with vomiting, difficulty swallowing, diarrhea, lack of appetite, headache, dizziness and fainting. Significant skin contact or inhalation can lead to death within minutes. Skin contact is not immediately painful, but deep damage to skin and even local gangrene can result. Phenol is a MUTAGEN.

The OCCUPATIONAL SAFETY AND HEALTH ADMINIS-TRATION has established a permissible exposure limit for airborne phenol of 19 milligrams per cubic meter averaged over an eight-hour workshift. It is also cited in AMERICAN CONFERENCE OF GOVERNMENT INDUS-TRIAL HYGIENISTS, NATIONAL INSTITUTE FOR OCCUPA-TIONAL SAFETY AND HEALTH and ENVIRONMENTAL PROTECTION AGENCY regulations. Phenol is on the Hazardous Substance List, the Special Health Hazard Substance List and the HAZARDOUS AIR POLLUTANTS LIST. Its odor threshold is 0.15 mg/m³. Businesses han-dling significant quantities of phenol must disclose use and releases of the chemical under the provisions of the EMERGENCY PLANNING AND COMMUNITY RIGHT TO KNOW ACT.

Phosgene a colorless, highly toxic gas at room temperature, or a clear to yellow volatile liquid when below the freezing point of water. Phosgene is used primarily as an intermediate to produce other chemicals. It is used in the production of polyurethane, isocyanates, polycarbonates, pharmaceuticals, dyes, PESTICIDES and perfume. It gets into outdoor air primarily from industrial discharges and spills. Carbon monoxide can react with CHLORINE under the influence of heat or light to form phosgene, which often is emitted when PLASTICS and other synthetic materials are burned in a confined, unventilated space. Phosgene was used as a poisonous gas during World War I. Phosgene affects health when inhaled. Exposure may initially cause only minor irritation of the eyes, nose and throat, but severe lung burns and a fatal build up of fluid in the lungs can occur within hours. Repeated exposure to even very low levels can result in emphysema and bronchitis.

The OCCUPATIONAL SAFETY AND HEALTH ADMINISTRA-TION has established a permissible exposure limit for airborne phosgene of 0.1 parts per million, averaged over an eight-hour workshift. It is also cited in AMERICAN CONFERENCE OF GOVERNMENT INDUSTRIAL HYGIENISTS, Department of Transportation and NATIONAL INSTITUTE OF OCCUPATIONAL SAFETY AND HEALTH regulations. Phosgene is on the Hazardous Substance List and the HAZARDOUS AIR POLLUTANTS LIST. Its odor threshold is 0.9 parts per million. Businesses handling significant quantities of phosgene must disclose use and releases of the chemical under the provisions of the EMERGENCY PLANNING AND COMMUNITY RIGHT TO KNOW ACT.

Phosphate a COMPOUND that contains the phosphate group (PO₄). Phosphates are primarily salts of phosphoric acid. Phosphates are essential to METABO-LISM in all levels of life from VIRUS and BACTERIA through animals and human beings. Because of the high level of biological activity caused by phosphates, they can also act as POLLUTANTS. Excess phosphates and nitrates in a stream can lead to EUTROPHICATION. For this reason, bans on the use of phosphate DETERGENTS have been instituted in many parts of the nation as a way of reducing the emission of phosphate from sewage treatment plants into lakes and streams.

Phosphates are used in fertilizer, most commonly in the form of *superphosphate* (monocalcium phosphate). Almost 9 million short tons of phosphate were produced in 1991. GYPSUM is a BYPRODUCT of the processing of superphosphate. *Phosphate rock* is a sedimentary mineral composed primarily of *fluorapatite* mixed with sand and clay. It is slightly radioactive, and imparts some of its RADIOACTIVITY to GROUNDWATER. Some phosphate rock can be used directly as a FERTILIZER, but most is first chemically treated with a sulfuric-acid solution. About 47 million metric tons of phosphate was mined by U.S. companies in 1992. A byproduct of the processing of phosphate rock is *phosphate slag*, which is a low-level RADIOACTIVE WASTE. The use of phosphate slag as a fill material for buildings and as an aggregate for concrete has caused elevated radiation levels in indoor air. See also NUTRIENT, PHOSPHOROUS.

Phosphorous an ELEMENT found in igneous and sedimentary rocks. Commercial-grade phosphorous, a white to yellow, crystalline solid with a waxy appearance, is processed from phosphate rock. Phosphorous is used to make FERTILIZER (see PHOSPHATE), explosives, detergents, food and beverages, roach and rodent poisons, match heads and smoke bombs. Phosphorous is an essential constituent of protoplasm, nerve tissue and BONES, and as such is an essential NUTRIENT. Phosphorous helps remove lead from the bloodstream. It gets into water from industrial discharges, agricultural RUNOFF, municipal waste treatment plants and spills. Pollution of water (see GROUNDWATER POLLUTION, WATER POLLUTION) by phosphorous compounds is a primary cause of EUTROPHICATION. Commercial phosphorous, known as yellow or white phosphorus, ignites in the presence of air and moisture.

Phosphorous can be inhaled or passed through the skin. Exposure to fumes is extremely irritating to eyes, nose, throat and lungs. Repeated exposure to low concentrations of phosphorous can destroy bone, especially the jaw bone and the eye sockets, and can cause anemia, weight loss and bronchitis. It may ignite on contact with eyes or skin, causing severe burns. Sudden death may occur after relatively minor phosphorous burns covering 10% to 15% of the body. Phosphorous exposure can damage the liver, kidneys and nervous system.

The OCCUPATIONAL SAFETY AND HEALTH ADMINISTRATION has established a permissible exposure limit for airborne phosphorous of 0.1 milligrams per cubic meter averaged over an eight-hour workshift. It is also cited in AMERICAN CONFERENCE OF GOVERNMENT INDUSTRIAL HYGIENISTS, and Department of Transportation and ENVIRONMENTAL PROTECTION AGENCY regulations. Phosphorous is on the Hazardous Substance List, the Special Health Hazard Substance List and the HAZARDOUS AIR POLLUTANTS LIST. Businesses handling significant quantities of yellow or white phosphorous must disclose use and releases of the chemical under the provisions of the EMERGENCY PLANNING AND COMMUNITY RIGHT TO KNOW ACT.

Photochemical Smog a phenomenon in which atmospheric NITROGEN OXIDES and VOLATILE ORGANIC COMPOUNDS combine in the presence of sunlight to produce OZONE. The principal source of both HYDROCARBONS and nitrogen oxides is emissions associated with the automobile (see AIR POLLUTION, AUTOMOTIVE AIR POLLUTION, CLEAN AIR ACT, SMOG).

Phytotoxicity the ability of a compound to kill or seriously injure plant life. See also HERBICIDE, LD$_{50}$, PESTICIDE, TOXICITY.

Picocurie a measure of RADIOACTIVITY which represents one-trillionth of a curie. A substance that is emitting one picocurie is losing about one atom every 27 seconds to radioactive decay.

Placer Mine a mine in a placer deposit (a deposit of SEDIMENTS that have eroded from a mineral lode). GOLD is the valuable mineral most often found in placer deposits, but platinum, tin and gemstones are also sometimes extracted. Placers are usually mined by removing the sands and gravels that contain the minerals of values, washing away the lighter materials with water, and catching heavier minerals such as gold on the ridged bottom of a sluice box.

The simplest form of placer mining—one still used by prospectors—involves the *gold pan*. Sands and small gravels from a placer deposit are placed in the flat pan, water is added and the mixture is swished around inside the pan. Lighter minerals are carried up the gently sloping sides of the pan by the gentle circular motion, and they spill over the side of the pan along with some of the water, leaving heavier sands and gravels (and hopefully a gold nugget!) behind.

The banks and beds of many western streams that drain gold country contain mineral-bearing placer deposits, and the beds of thousands of miles of western creeks and rivers were removed with dredges and the banks were blasted away with high-pressure hoses to get at the gold and silver they contained. Massive amounts of sediments were washed downstream by the placer operations, leading to SEDIMENTATION in gravels that are essential to fish and other aquatic organisms. When the mining operation moved on, watercourses were lined with sterile spoils piles composed mostly of sand and gravel. Rarely was any attempt made at stream restoration, and the bare piles of rock that still line many such streams look like they might have been deposited less than a decade ago, rather than more than a century. Most drainages have yet to fully recover from the resulting disruption of the aquatic ecosystem. See also HARD ROCK MINE, STRIP MINE, TAILINGS.

Planned Obsolescence   the practice of periodically changing the design of a product, often in trivial or cosmetic ways, with the intention of stimulating the sales of replacement items, which are advertised as "new," and "improved." Perhaps nowhere is the concept of planned obsolescence as obvious as with children's toys. Kids are bombarded with television ads telling them that they simply must have the latest plastic replica of the latest movie hero or video game, with a new "must have" appearing every year. The patterns of consumption established in childhood—always "needing" the latest toy or game, being unsatisfied with last year's purchases—are thought by many to lead to similar patterns of CONSUMERISM when the children grow up. Automobiles are another obvious example of planned obsolescence. When Henry Ford introduced the Model T, his business plan was to provide a durable machine that was easy to fix and to reduce its cost through mass production. The Model T was produced with only minor changes for 15 years, and the price declined from $780 to $290 during that period. Ford's competitors found it increasingly hard to compete, and General Motors came up with the idea of using changes in style to sell new cars, an approach that has dominated automobile sales since that time. While real advances are constantly occurring, especially in a volatile field such as electronics, replacing a serviceable item just to get something new is a needless use of resources (see RESOURCE DEPLETION COST) and energy.

The sale of clothes, where technical advances and improved quality are relatively minor selling points, relies even more heavily on changes in style to bring consumers into the stores. The fashionable dresser wouldn't be caught dead in last year's colors or tailoring cut. Similar sales strategies are employed with breakfast cereals, household cleansers, furnishings and a host of other items. The consumer is constantly being encouraged to buy the "new, improved" version of just about everything. One solution to the garbage crisis is a return to the values of Henry Ford—selling products on the basis of how durable they are and how easily they can be repaired, if needed.

Compare SUSTAINABLE AGRICULTURE, SUSTAINABLE SOCIETY. See also ENVIRONMENTALISM, GARBOLOGY, MATERIALISM.

Plastics   a broad definition includes any material that can be molded, cast or spun into the desired shape when fluid, and that will maintain its shape under normal temperature and pressure. The term "plastic"

more specifically refers to any of a large group of primarily synthetic compounds composed of organic molecules arranged in certain complex patterns. Plastics are used in BUILDING MATERIALS, PAINTS and coatings, adhesives, floor coverings and resin-bonded WOOD products. They are produced by chemically modifying a MONOMER by forcing it to combine at each of its ends with other monomers to form a long chain, or POLYMER. Plastics and plastic resins are made from a variety of reactive molecules (see REACTIVE MATERIAL) including acrylonitrile, ethylene, ethylene oxide, propylene oxide, STYRENE, vinyl acetate, VINYL CHLORIDE, FORMALDEHYDE, methyl methacrylate and methyl dichlorosilane. The plastics manufacturing industry is a major source of air pollution and water pollution (see AIR POLLUTION, GROUNDWATER POLLUTION, WATER POLLUTION). CARBON MONOXIDE can react with CHLORINE under the influence of heat or light to form PHOSGENE, an extremely poisonous gas that often forms when plastics and other synthetic materials burn in confined, unventilated spaces. Volatile chemicals that evaporate from plastics and plastic resins are a source of INDOOR AIR POLLUTION. Some individuals with multiple chemical sensitivities are allergic to many of the fumes emitted by plastics (see OUTGASSING). DRINKING WATER that stands in plastic pipe for several hours or more can become contaminated with METHYL ETHYL KETONE and other toxins. See BIODEGRADABLE PLASTICS, POLYETHYLENE TEREPHTHALATE (PET).

History

Development of plastics was accelerated in the 1860s when a company that manufactured billiard balls offered a prize of $10,000 to anyone who could provide a suitable substitute for ivory, which was becoming scarce and expensive as a result of the decimation of wild elephant herds. In an effort to win the prize, John Wesley Hyatt developed a material called *celluloid*, a product of the action of camphor on the cellulose nitrate *pyroxylin*. Although Hyatt didn't win the contest, and in spite of the fact that celluloid was flammable and tended to break down when exposed to light, the new plastic material was a commercial success. Cellulose acetate has since been used to produce a plastic material that is less flammable and more resistant to weathering and the effects of mechanical impact. In 1895, the German-American inventor Emile Berliner developed a plastic material bound together with shellac that was used to mass-produce phonograph records. *Cold-molded plastics*, which consisted of an asbestos filler with asphalt, pitch or coal tar, used as a binder, were introduced in the United States in 1909. The resulting plastics, which were used in electrical insulators and heat-resistant handles, were molded into the desired shape and then heated in an oven to drive off solvents and dry oils or to set the resins in the binder. In 1909, the Belgian-American inventor Leo Hendrick Baekeland introduced the first thermosetting synthetic resin, which was based on a reaction between PHENOL and FORMALDEHYDE. The resulting material, which was marketed under the tradename "Bakelite," was strong and a good electrical insulator. Urea-formaldehyde plastics were introduced in 1929. The product's light weight, high resistance to shock and chemical inertness made it popular for use in plastic dishes. Resins produced from VINYL CHLORIDE and vinyl acetate were introduced at about the same time, and were soon adapted for use in laminated safety GLASS, phonograph records and in clothing such as belts, suspenders and shoes. POLYSTYRENE resins were first produced in 1937, and gained quick popularity because of their resistance to chemical or mechanical alteration at low temperatures and the fact that they are almost completely waterproof. Large-scale production of a plastic called *nylon*, which is based on polyamide resins, started in the 1940s. The resins yield FIBERS with exceptional strength and elasticity that are ideal for use in fabrics. A shortage of silk created by the cessation of exports from Japan during World War II stimulated the production of nylon fabrics.

Vinyl copolymers and the related *acrylic resins* are used today to produce yarns, fabrics, fishnets and tents. The acrylic resin, *methyl methacrylate*, was first produced for use as a binder in laminated safety glass, and has been marketed under the trade names "Lucite" and "Plexiglass" since 1937. Plastics based on casein from milk and on soybeans, including the acrylic fiber *Orlon* and the polyester fiber *Dacron*, are widely used in fabrics because of their warmth, lightness and resistance to wrinkling. *Rayon*, which is based on cellulose fiber, is also widely used in textiles. *Phenolic plastics* are used as electrical insulators, in industrial adhesives, as the bonding agents in laminated PAPER products and marine plywood and in fabrics. *Alkyd resins* are widely used as a base for paints. *Polyethylene* is a flexible plastic that is used in containers for liquids (see HIGH-DENSITY POLYETHYLENE). Tetrafluoroethylene resins, sold under the tradename "*Teflon*," are used in electrical insulation, gaskets, as a nonstick lining for cookware and in other applications where resistance to heat and CORROSION are critical. *Silicone plastics* are highly resistant to high temperatures and penetration by water, and are used in water repellents and as electrical insulators. Replacements for internal body parts such as blood vessels, heart valves, tendons and bones are commonly made of silicone-base plastics. *Polysulfones* are notable for their resistance to the effects of extreme heat and cold, and can withstand temperatures from –150°F to +300°F. They are used to make parts for computers, automobiles, electrical circuit breakers and wire coatings. *Polyimides* can resist temperatures up to 750°F, and are used to make parts for airplanes and space craft and bearings. *Melamine* (polytriazine) is formed into laminated sheets and sold under the tradename "Formica." Other modern plastic products include polychlorobutadiene (neoprene); polyisobutylene (butyl rubber); polyphenylisocyanate (urethanes); and polyester (alkyds). Filters and conditioners for automobile radiators and other cooling systems rely on plastic ion-exchange resins to clean water. Such resins are also used: in antacids and intestinal absorbents; in the treatment of peptic ulcers, migraine headache, heart disease and edema; in the preparation of blood plasma; and for diarrhea control. Plastic products have replaced many items once made of wood, metal or paper over the last few decades. In 1960, the average automobile contained less than 1% plastic by weight, but by 1984, the average percentage had increased to 6.4%, or about 200 pounds per car (see JUNKYARD).

The majority of modern plastic products are made by introducing liquid plastic into a mold. Blow molding is used to make plastic bottles. Compression molding, injection molding, transfer molding, extrusion molding and high pressure laminating are also used to produce plastic products. Calendering and sheet forming is used to produce plastic films. See also ACETALDEHYDE, AMMONIA, BENZENE, CADMIUM, EPICHLOROHYDRIN, NAPTHALENE, PETROLEUM, POLYVINYL CHLORIDE, VINYL.

Plastics Recycling   the use of resins derived from recycled plastic materials to produce new products. The RECYCLING of PLASTICS—with the exception of PET Plastics (see POLYETHYLENE TEREPHTHALATE PLASTICS) and HIGH-DENSITY POLYETHYLENE (HDPE)—has so far met with limited success. The different, non-mixable formulations used to produce different kinds of plastic resins, the relatively low price paid for most virgin resins (see VIRGIN MATERIAL) and the high degree of purity that is generally required before a batch of plastics can be recycled have worked against the creation of a vigorous plastics recycling industry. Many kinds of resins are generally not recycled at all, and composite products such as those containing more than one kind of plastic or CARDBOARD or PAPER bonded to plastic (see COMPOSITE PACKAGING) are seldom recyclable. Many of the plastics recycled in the United States are shipped to Asia for reprocessing and remanufacture or, too frequently, are hauled to the dump because they cannot be sold.

Opinion surveys conducted by Procter and Gamble and others reveal that the public expects their plastic

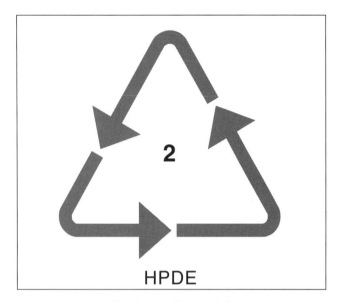

Plastics recycling symbol

GARBAGE to be recycled, but that people usually are not willing to pay more for products or containers that contain recycled plastic. Insiders agree that consumer education is essential if the public is to embrace plastics recycling in fact as they do in theory. In 1993, the recycling centers in nearly 7,000 communities accepted some plastics—usually PET plastics and HDPE plastics. The estimated one billion pounds of plastics recycled during 1993 included 450 million pounds of HDPE, 448 million pounds of PET, 88.4 million pounds of LDPE (low-density polyethylene) and 5.5 million pounds of PVC (POLYVINYL CHLORIDE).

Plastic manufacturers generally prefer to use virgin resins because of their chemical integrity and lack of foreign materials. In addition, oil- and natural-gas depletion allowances permit U.S. plastics manufacturers to write off the natural resources they extract and use. Representatives of small recycling businesses say that large industrial producers of plastic resins such as Dow Chemical, DuPont and Union Carbide—even though they have invested in plastic recycling programs of their own (usually amid considerable fanfare from their respective public relations departments)—actually see a viable plastics recycling industry simply as competition for the virgin plastic resins that are their bread and butter. Detractors say that the large chemical companies recognized the popularity of recycling plastics, and felt it was essential to promote plastics as recyclable to combat the anti-plastics ordinances that emerged during the mid-1980s. So while the giants of the plastic industry have publicly supported recycling, they have at least as often worked behind the scenes to ensure its failure. The American Plastics Council spends an estimated $3 to $8 million dollars lobbying against recycling legislation, using the "free market" argument that holds that the government should not be involved in the regulation of PACKAGING or the subsidization of recycling. The makers of plastic resins and plastic products have lobbied against laws mandating the use of recycled resins in preference to virgin resins, set tough standards governing the purity and cleanliness of recycled plastics and occasionally dropped the price paid for plastic resins in an apparent attempt to force smaller companies out of the business. Because of such industry tactics and the relatively weak market for recycled plastic, many plastic recycling businesses and programs have gone broke, sometimes after losing millions of dollars. The "green-dot" plastics recycling program in Germany was forced to convert $533 million of its debts into loans during 1993 as the result of unanticipated processing costs, difficulty selling the recycled materials and problems with collecting the recycling tax imposed on all German companies.

Different kinds of plastics are now generally hand-sorted, but that job grows more difficult as the variety and volume of recycled plastics grow. Technology to improve the speed and efficiency of hand sorting or to replace it with automated techniques is available or under development, but generally remains too expensive for use at most recycling facilities. Ultraviolet radiation (see ELECTROMAGNETIC SPECTRUM) causes certain plastic resins to emit visible light of different colors, a fact that can be used to increase the speed and efficiency of manual sorting. Radiation in the near infrared and x-rays can similarly be employed either to assist manual sorting or as the eyes of an automatic sorting system. *Molecular marking*, which is being developed by Eastman Kodak, holds perhaps the most promise for the efficient and automated sorting of plastics. A small amount of a material with a molecular structure that is easily identified with inexpensive electronic sensors is added to a batch of plastic when it is prepared, with each variety of plastic having its own distinct marker molecule. All the manufacturers of plastic resins would have to add the appropriate molecular marker in order for the system to work. A variety of other sorting methods that utilize differences in density, the ability to hold an electrical charge and other chemical and physical properties of plastics are also under development.

The Plastic Bag Association estimates that 17,000 of the 31,000 grocery stores in the United States have recycling bins for collecting plastic grocery bags, which are used to make plastic lumber, irrigation pipe and containers. Sonoco Products Co., which produces about 30% of U.S. plastic grocery bags, collects approximately 300 tons of bags per month from the recycling bins it maintains in about 8,000 stores, and uses the recycled bags to produce new plastic bags with about

10% post-consumer content. If a batch of bags (or other recycled plastics) contains debris such as receipts, coupons or bits of food, or if it contains other types of plastics, it is prohibitively expensive to sort the unwanted material out, and the entire batch will likely be rejected and end up in the dump.

A Pontiac, Michigan company, Waltom Services, had developed a product called Dri-Strip, tiny grains of plastic that are used in high-pressure blasting equipment. The company grinds up recycled plastic for use in the process, which is an alternative to sandblasting and chemical stripping. After Dri-Strip is used, most of the plastic is separated from the material it has removed in an air cyclone, so that it can be reused.

Wisconsin and a few other states have considered legislation that would make the dumping of recyclable plastics illegal. In Oregon, "rates and dates" legislation was enacted in 1991 that makes plastics manufacturers responsible for including a certain percentage (the rate) of recycled plastics in products sold in the state by a certain date. A proposed law in New York State would require manufacturers to attain a "threshold level" of *reusability* (returning a container for refilling), *recyclability*, SOURCE REDUCTION (products would contain less plastics than previous versions or would meet a product-to-package efficiency ratio), or *efficient packaging* (packages could use no more than a specified amount of materials).

The American Plastics Council distributes a booklet, a Recycled Plastic Products Source Book, meant to assist local governments and the recycling industry to find markets for recycled plastic materials.

See also MCTOXICS CAMPAIGN, NATIONAL POLYSTYRENE RECYCLING COUNCIL.

Plastics Recycling Symbol   a coding system for some of the most commonly used plastic resins introduced by the Society of the Plastics Industry in 1988. The plastics recycling symbol—a number from 1 to 7 surrounded by a triangle of chasing arrows—is cut into the dies that produce bottles and other plastic containers. The seven codes are: 1 = PET plastics (see POLYETHYLENE TEREPHTHALATE PLASTICS); 2 = HPDE (see HIGH DENSITY POLYETHYLENE); 3 = POLYVINYL CHLORIDE; 4 = low-density polyethylene; 5 = polypropylene; 6 = POLYSTYRENE; 7 = other (including multi-layer plastics). Representatives of RECYCLING businesses have complained that although the numbering system does make it easier for consumers to sort plastics for recycling, the use of the symbol leads people to believe that any plastic product that bears the symbol can be recycled, when, in fact, only PET and HPDE plastics are widely accepted at recycling centers. The mixing of even a small amount of a different plastic resin into a

batch of plastics bound for recycling can ruin the entire batch of plastic produced, so even a few consumers who throw every piece of plastic with a recycling symbol on it into the HPDE plastics bin can cause considerably havoc further down the plastics-recycling stream.

The NATIONAL RECYCLING COALITION (NRC) voted in 1994 to replace the recycling symbol with a square in which a code consisting of numbers and letters would more accurately identify recyclable products, making sorting easier. The plastics industry has objected to changing the markings, claiming that it would cost about $80 million to modify molds to completely change the symbol or $20 million to modify it. Under the NRC recommendations, the industry would have four years after laws are amended in the 39 states that require the recycling symbol on plastic products to change the molds.

Plastics Waste   waste from the production, manufacturing and use of plastic products. Plastic makes up about 16% of the contents of a typical LANDFILL, according to estimates by the Garbage Project. While the number of plastic items hauled to landfills in recent years has increased, the overall volume and weight has stayed about the same because plastic packaging and containers are constantly getting lighter (see LIGHTWEIGHTING). The chemical plants that produce plastic resins can emit METHYLENE CHLORIDE and a variety of toxic VOLATILE ORGANIC COMPOUNDS into outdoor air. Any of the numerous toxic compounds used in plastics production, including BENZENE, CADMIUM, CHLORINE, NAPTHALENE, VINYL CHLORIDE, STYRENE, and VINYL, can pollute local water sources.

The manufacture and use of plastic products has received so much negative publicity, and plastic debris are such a long-lasting and obvious component of LITTER that the public has formed a distorted idea of quantity and toxicity of plastic waste. The plastic six-pack holders that can snare and kill wildlife and the white plastic foam that commonly line the shores of oceans and lakes do little to allay public suspicion of plastic waste. William Rathje, director of the Garbage Project said in his book, *Rubbish*, that "plastic is surrounded by a maelstrom of mythology; into the very word Americans seem to have distilled all of their guilt over the environmental degradation they have wrought and the culture of consumption they invented and inhabit. Plastic has become an object of scorn. . . . Plastic is the Great Satan of garbage." Representatives of the Garbage Project conducted a survey at a 1989 meeting of the National Audobon Society in which attendees were asked what percentage of the total volume of a typical landfill was taken up by the expanded POLYSTYRENE foam (see EXPANDED

FOAM) that is used to make disposable coffee cups and fast-food PACKAGING. Respondents estimated that expanded polystyrene would typically comprise 20% to 30% of total volume and that fast-food wrappers (see MCTOXICS CAMPAIGN) would make up another 25% to 40%. But according to Garbage Project statistics compiled through the careful examination and sorting of more than 14 tons of garbage from nine municipal landfills, only about one-third of one percent of average landfill volume is devoted to fast-food packaging, with the total for *all* expanded foam products, of which the expanded polystyrene used in fast-food packaging makes up only a small part, being about 1%. The Garbage Project dump digs found the total volume of plastics found in landfills to be less than 16%. Further, the majority of the plastic products found in landfills are nearly perfectly preserved, and pose no threat of breaking down and releasing pollutants. A process called lightweighting—using the minimum amount of plastic resins necessary for the item in question to maintain its necessary functional characteristics—has kept the volume of landfilled plastics constant or slightly smaller, even though the number of plastic items in garbage has increased constantly in recent decades.

Plastics' high combustion temperature makes them welcome at trash-to-energy incinerators, although DIOXIN, vinyl chlorides, FORMALDEHYDE and a variety of other toxins can be emitted when plastic products are burned. The use of plastics in the medical field has expanded exponentially in the last few decades. Medical equipment, which is often made of plastic or a plastic composite, comes sealed in a sterile plastic package, with the whole works designed to be discarded after a single use. While such practices may seem wasteful, most medical professionals agree that the convenience and guaranteed sterility such products provide is worth the investment in throwaway plastic materials. At any rate, plastics have come to compose a significant portion of medical waste, most of which is treated as infectious waste that is usually burned in an incinerator. See also PLASTICS RECYCLING.

Plume the shape taken by a body of waste emissions into the water or air. The characteristics of the POLLUTANT and of its source, and the properties of the medium into which the pollutant is being dumped, determines the shape and size of a POLLUTION plume. Only part of a plume is normally visible, and a plume may be completely invisible, as is often the case with colorless contaminants dumped into surface water or GROUNDWATER. Much of the visible component of atmospheric emissions is often water vapor, with the more toxic gases and fine PARTICULATES being invisi-

ble. The shape of a plume created by the emission of contaminants into the atmosphere depends first on weather conditions. Wind speed and direction, temperature, humidity, barometric pressure, precipitation and general atmospheric conditions all play a role in determining the size, shape and location of a plume. An ATMOSPHERIC INVERSION can lead to dangerous concentrations of airborne toxins at ground level. The size and height of the smokestack or vent from which emissions are issuing and the geography of the area around the pollution source also play a major role in dictating plume formation. The buoyancy of the material being emitted, which is dependent on the density and temperature of gases and the size of particulates, also influences the nature of the plume.

Atmospheric plumes are classified according to their shape. A *coning plume* spreads downwind in a cone shape with its apex at the emission source. A *looping plume* first dips toward the ground, and then climbs rapidly upward. Severe pollution incidents can occur when a looping plume touched the ground. A *fanning plume* is a coning plume that has been flattened out into a relatively narrow band by an inversion or other atmospheric conditions that cause distinct air layers to develop. Fumigation occurs when a fanning plume is pushed down to ground level in cold, still weather, and is one of the most hazardous conditions associated with air pollution.

The characteristics of a plume of pollution issued into a lake or stream are also determined in part by the heat and density of the material being emitted and also by the diffusion rate of any water-soluble pollutants that are emitted, but currents and water temperature are the principal determinants of the shape and size of the pollution plume. A plume of pollution in surface water is generally less visible than an atmospheric plume. A groundwater pollution plume forms according to the density, water solubility and volume of contaminants, and on the rate and flow pattern of water within the receiving AQUIFER (see GROUNDWATER POLLUTION, GROUNDWATER TREATMENT). The degree of capillary action around the fringes of an aquifer can also influence plume formation. Determining the shape of an underground pollution plume can be an extremely expensive and time consuming operation that can cost millions of dollars. Numerous test wells must normally be drilled, and water samples must be taken at different depths over a period of months or even years before a groundwater pollution plume can be accurately mapped. See also COLATION.

Plutonium a RADIONUCLIDE created through the fissioning of URANIUM. Plutonium undergoes FISSION when exposed to free neutrons (see RADIOACTIVITY). It

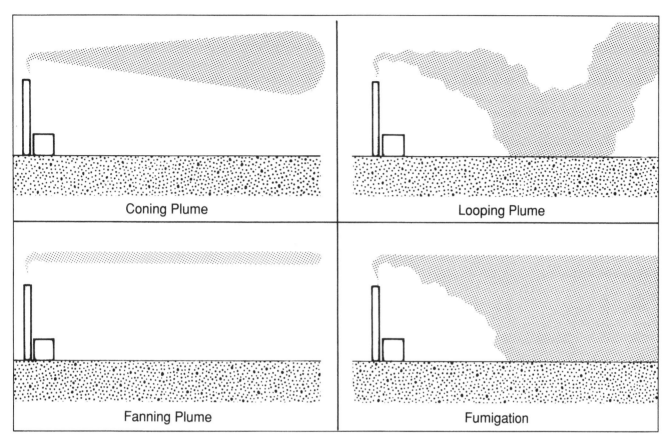

Plumes—coning, looping and fanning (*Encyclopedia of Environmental Studies*)

is used primarily in the production of nuclear weapons, although small quantities are used as fuel in nuclear reactors. When uranium-238 (U_{238}) is exposed to ionizing radiation, it can pick up an extra neutron to become U_{239}. Rapid decay and the emission of a BETA PARTICLE can change U_{239} into neptunium-239. Emission of an electron transforms the neptunium into fissile plutonium-239. Plutonium-239 has a HALF-LIFE of 24,000 years.

The chemical and nuclear processes for extracting plutonium from fission byproducts has been perfected through the production of nuclear weapons for the Cold War, and disposal of the world arsenal of plutonium bombs and the production of plutonium as a byproduct of the operation of nuclear power plants has led to a growing world stockpile of plutonium and associated security problems. Destroying the plutonium by using it as a fuel in nuclear power plants is one disposal option that has been discussed. The production of plutonium for nuclear weapons during the 1940s through the 1970s caused major environmental contamination in the United States (see HANFORD NUCLEAR RESERVATION, OAKRIDGE NATIONAL LABORATORIES, ROCKY FLATS, URAVAN URANIUM MILL).

Point Source a source of POLLUTION that enters the environment from a single, stationary location such as an outfall from a sewage plant or industrial installation, or from a factory smokestack. Pollution from a point source is easier to control than DIFFUSE POLLUTION because the origin of the pollution PLUME emitted from the source can be easily traced, and because contaminants are already concentrated in a smokestack or effluent pipe making remediation easier. Pollution from a diffuse source, by contrast, can be extremely difficult to trace to its source, and it can be difficult and expensive (or impossible, in many cases) to control the source of emissions. While point sources are governed by numerous state and federal regulations including the CLEAN AIR ACT and the CLEAN WATER ACT, diffuse sources are, in many cases, not covered by pollution law. Although tailpipe emission from automobiles have many of the characteristics of a diffuse source of pollution, they are often treated as a mobile point source because pollutants are entering the air from a single point that happens to be mobile. All the vehicles in a given AIRSHED are often treated as a single point source for the purposes of air-pollution monitoring and regulation. See also BUBBLE, STATIONARY SOURCE.

Poison a substance that has the ability to damage living tissue or to cause disease or death to a living organism. Poisons, which are also called *toxins*, may be solids, liquids or gases, and they may have their origins as animal, vegetable or mineral matter. Any substance that interferes with any kind of complex chemical reaction is sometimes referred to as a poison. Poisons can enter the body by being inhaled, ingested or passed through the skin. The effect of poisons that get inside the body to impact internal organs is generally more serious than those that affect only the skin and external organs. The TOXICITY of a poison is a measure of its ability to kill (see LD_{50}) or injure an organism. About half of human poisoning occurs as the result of ingesting common drugs such as barbiturates or household products such as insecticides or cosmetics. A poison may interfere with METABOLISM and damage living tissue in a variety of ways: by preventing the adsorption of food in the digestive tract; by interfering with the production of essential enzymes and hormones; or by damaging the nervous system or the immune system. An *asphyxiant* such as CARBON MONOXIDE restricts the flow of oxygen to the cells by taking its place in red blood cells. Many poisons are designed to kill or control a specific group of organisms such as insects (INSECTICIDE), rodents (RODENTICIDE), plants (HERBICIDE), soil organisms called nematodes or fungal growths (FUNGICIDE). A poison such as a CARCINOGEN, a TERATOGEN or a MUTAGEN may cause a serious chronic disorder (see CHRONIC EFFECTS) without having noticeable ACUTE EFFECTS.

The basic types of poisons are corrosives (see CORROSION, CORROSIVE MATERIAL), IRRITANTS and narcotics (see NEUROTOXINS). A corrosive is a material that can chemically or electrochemically alter other substances. Most corrosives contain strong ACIDS or alkalis that are capable of causing chemical burns on external or internal tissue. An irritant is a toxic material that damages the mucous membrane causing gastrointestinal irritation or inflammation often accompanied by stomach pain and vomiting. ARSENIC, MERCURY, iodine and laxatives are commonly occurring irritants. Irritants include *cumulative poisons* that can be taken up by the body gradually with no apparent ill effects, and then can cause systemic damage when a threshold level is reached. A narcotic acts on the central nervous system or on major organs such as the heart, liver, lungs or kidneys until they affect the respiratory or circulatory systems. Coma, convulsions and delirium are symptoms of narcotic poisoning. Ethyl alcohol (see ETHANOL), opium and opium derivatives such as heroin, CYANIDE, CHLOROFORM and strychnine are narcotics. One of the most dangerous poisons known, the botulin toxin that causes the acute form of food poisoning known as botulism, is

a narcotic. Gaseous poisons may be corrosives, irritants or asphyxiants. *Blood poisoning* occurs when virulent MICROORGANISMS invade the bloodstream through a wound or infection. See also PESTICIDE, TOXIC WASTE.

Polar Liquid a liquid in which individual MOLECULES are electrically polarized with one end of each molecule having a positive charge and the other end a negative charge. A molecule with covalent bonds exhibits polarity when one of its atoms attracts electrons more strongly than the other. The electrons spend more time orbiting near the attracting atom, causing that side of the molecule to have a negative charge and the other side to have a positive charge. A polar liquid has the ability to chemically pull apart a polarized molecule that is immersed in it, turning the molecules into IONS, a process called *dissociation*. WATER (H_2O) is a strongly polar liquid, with the oxygen in each molecule being the negative pole and the two hydrogen atoms the positive pole. Water's ability to dissociate a wide range of compounds is the reason it is called the universal SOLVENT.

Pollen the fine DUST emitted by seed-bearing plants as a means of carrying the male gametophyte, or sperm, to the female gametophyte, or egg. Each pollen grain consists of a tough outer shell called an *exine* that surrounds an inner coating called the *intine*, which in turn surrounds the sperm. There are usually two sperm in each pollen grain. Airborne pollens cause allergic reactions in many people. See also ALLERGEN.

Pollutant a contaminant that is harmful to the environment. Some pollutants are simply resources that are out of place. For instance, a PESTICIDE is a resource when it does its job of controlling an undesirable life form, but can become a pollutant if it later gets into air or water. Similarly, an element such as CHROMIUM is a resource when it is used in the metals or photography industries (and could be used again as a resource if recovered from the waste stream), but is a pollutant when it enters the environment. The transformation of a resource into a pollutant usually often involves a degree of economic expediency. It might be considered economic, for instance, to flush an industrial PROCESS CHEMICAL down the drain and replace it with a new supply rather than to reclaim the chemical from EFFLUENT, especially if the environmental costs of POLLUTION are not considered. Other pollutants, such as DIOXIN and FLY ASH, are simply waste products that are released into the environment.

Pollutant Standard Index an air-quality measurement method required by the CLEAN AIR ACT that expresses the average concentration of the five CRITE-

RIA POLLUTANTS measured in AMBIENT AIR. To compute the PSI, each of the criteria pollutants—nitrogen dioxide (see NITROGEN OXIDES); SULFUR DIOXIDE; CARBON MONOXIDE; OZONE; and TOTAL SUSPENDED PARTICULATES—is assigned a number based on what percentage of the concentration considered under federal law to be harmful to human health is present in ambient air. For example, if carbon monoxide (CO) is found to average 7.2 parts per million in ambient air samples, or 80% of the nine parts per million federal CO limit, it would be given the number 80. This would be added to the average numbers for the other five criteria pollutants to arrive at the PSI number for the AIRSHED in question. A PSI of less than 100 means ambient air quality is healthful, a reading of 100 to 199 is classed unhealthful, a total of 200 to 299 is termed very unhealthful, and totals over 300 are considered hazardous. The highest possible PSI is 500 (meaning that the concentration of all five pollutants are at or above federal limits).

Pollution the tainting of the environment with one or more CONTAMINANTS. Pollution is usually toxic (see TOXICITY) or otherwise harmful to living things. The absorptive abilities of water and air plays a central role both in their pollution and cleanup (see ABSORPTION, AIR POLLUTION, GROUNDWATER POLLUTION, WATER POLLUTION). Pollution comes from virtually every sector of society. Transportation, agriculture and industry all contribute at least some degree of environmental contamination. Centralized sources of pollution such as smokestacks, effluent discharges, underground storage tanks, garbage dumps and toxic waste dumps are called POINT SOURCES of pollution. The RUNOFF from farms, fields and city streets are diffuse sources of pollution (see DIFFUSE POLLUTION). Pollution issuing from vehicle exhaust pipes, although coming from many, diffuse sources, is treated as a *mobile point source* for purposes of regulation, and the cumulative emissions in a given airshed may be treated as a single point source. For details on pollution sources, see AUTOMOTIVE AIR POLLUTION, CROSS-MEDIA POLLUTION, MOBILE SOURCE, NOISE, THERMAL POLLUTION, VISUAL POLLUTION. See also POLLUTANT.

Pollution Measurement the sampling and analysis of CONTAMINANTS in air, water and soil. The measurement of contaminants in lakes, streams, AQUIFERS, DRINKING WATER, PROCESS WATER, SEWAGE, AMBIENT AIR and pollution PLUMES in air and water is often mandated under federal laws such as the CLEAN AIR ACT and the CLEAN WATER ACT (see AIR POLLUTION MEASUREMENT, INDUSTRIAL AIR POLLUTION MEASUREMENT, WATER POLLUTION MEASUREMENT). Collecting a sample that is representative of the air, water or soil of which it is a part can be difficult (see INDOOR AIR POLLUTION). The location(s) in which samples are taken, the time of day and season, air and water currents, the characteristics of the pollutants and numerous other factors must be taken into account if the monitoring of a plume of pollution, or even of a large building, is to be successful. Contaminants are unique in the way in which they respond to heat, light, temperature or chemicals, and can generally be identified by these characteristic responses. Laboratory tests, instruments and monitoring equipment rely on physical, chemical and electrical characteristics that are unique to the atom or molecule of interest, to identify and quantify the substance.

Analysis Techniques

Samples may be taken manually for later analysis in a lab, or readings may be taken directly with a handheld instrument. Passive monitors such as those used for RADON and PARTICULATE monitoring are exposed to air for a specified period, then analyzed in a lab. DETECTOR TUBES contain chemicals that change color or extend a stain in proportion to the amount of a specific airborne contaminant present in air that is drawn through the tube. Badges that change color in response to radiation or the presence of toxic substances must be worn in some industrial and military installations. Visual observations of factors such as the OPACITY of an air pollution plume, the extent that haze is limiting visibility (see VISUAL POLLUTION) or of the presence of COLOR in WASTEWATER (see WATER POLLUTION) are usually part of a pollution monitoring effort. Automated monitoring systems capable of transmitting data to a computer for storage and analysis are seeing increased use in industrial installations and in the monitoring of urban SMOG. *Spectral analysis*—the investigation of the unique spectrum generated when a substance's ability to transmit or absorb radiation is plotted as a function of its wavelength—is used to identify and analyze many contaminants.

In *mass spectrometry*, molecules are bombarded with electrons, forming ions, which in turn may be converted into *daughter ions*. The ions are accelerated, separated and focused on an ION detector through the use of magnetic fields or a *quadrupole mass analyzer*. Mass spectrometry is used to measure the content of NITROGEN, oxygen and other elements in MOLECULES, and to monitor VOLATILE ORGANIC COMPOUNDS (VOCs), CARBON MONOXIDE, SULFUR DIOXIDE, NITROGEN OXIDES, OZONE, CHLORINE and chlorides, CYANIDES and hydrogen fluoride and other FLUORIDES, and organic and inorganic vapors. The *fragmentation pattern* resulting from electron bombardment produces a unique molecular fingerprint for many compounds, which can be read through *infrared* and *nuclear magnetic resonance spectra analyzers*. However, if a mixture is introduced into a mass spectrometer, the fragmentation patterns produced by electron bombardment overlay one another, making analysis difficult or impossible.

Gas chromatography is a technique for separating complex mixtures of gaseous VOCs and inorganic compounds with boiling points less than 480°F (250°C) into their component elements. It is often used to separate gases for analysis by mass spectrometry. *Gas-liquid chromatography* employs a film composed of a nonvolatile liquid SORBENT that adheres to the surface of a granular solid such as inert diatomaceous earth. Liquid chromatography occurs in a *packed column*, a glass or metal tube 0.125 or 0.25 inches (6 or 13 mm) in diameter that is filled with inert diatomaceous earth. A nonvolatile liquid sorbent (see SOLUTION) coats the inert medium. In *gas-solid chromatography*, a granular solid with a large surface area is used as a sorbent, and the *carrier gas* is an inert gas such as nitrogen or helium. The speed at which a compound moves through the column is a function of its solubility in a gas-liquid system, and of its rate of adsorption in a gas-solid system. Since it takes different amounts of times for the gases in a mixture introduced into a chromatographic column to reach the other end, it is possible to separate them for further analysis (a process called *elutriation*—the removal of an absorbed material from a solvent). Liquid samples, which generally range in size from 0.5 to 5 microliters (1 microliter is about $1/50$th of a normal sized drop), are injected into a chromatograph with a calibrated hypodermic syringe through a silicone rubber septum. The liquid is vaporized in the injection port, which is kept at a temperature greater than the liquid's boiling point. Gaseous samples are admitted through a valve. Thermal conductivity and flame ionization are the most commonly employed methods of identifying separated gases in a gas chromatography.

Thermal conductivity measurement can be used to measure the concentration of many gases. A metal-wire filament or thermistor positioned in the gas stream at one end of the column is heated by an electrical element. Organic compounds have a conductivity that is lower than that of the helium carrier gas, so the rate at which energy moves away from the heated filament will be greater when they are present. Just how much warmer depends on the makeup of the organic gases in the chamber. The electrical resistance of a metal filament increases as it gets warmer, and the resistance of a thermistor decreases with increased temperature. The readings obtained for the sample can be compared to those of a known reference gas to interpret the results of the test.

A *flame ionization detector* is used to detect HYDROCARBONS and other organic vapors by measuring the number of carbon atoms present. The sample is burned in a hydrogen flame as it is eluted from the column of a gas chromatograph, creating charged carbon particles that are collected by a polarized collector electrode. Flame ionization is used to detect ORGANIC COMPOUNDS, nitrogen and sulfur oxides, but can not be used for carbon monoxide or CARBON DIOXIDE. *Flame-photometric detectors* are used to measure compounds of SULFUR and PHOSPHOROUS and volatile metallic compounds.

In *spectography*, the chemical composition of a compound is analyzed by using the electromagnetic spectrum displayed on a spectroscope to produce a photograph. Spectroscopy is used to identify and analyze materials on the basis of their interaction with different parts of the ELECTROMAGNETIC SPECTRUM. In *spectrophotometric analysis*, the shape and intensity of a compound's characteristic spectra are analyzed to identify the compound and its characteristics. *Ultraviolet spectrometry* is used to measure ozone. Carbon monoxide is measured via *infrared spectrometry*.

Other analysis methods include:

- *Coulson Conductometric Detectors*—used to measure AMMONIA and organic nitrogen compounds;
- *electron-capture detectors*—for halogenated compounds, POLYAROMATIC HYDROCARBONS and PESTICIDES;
- *x-ray fluorescence spectroscopy* and *neutron activation analysis*—for particulates;
- *chemiluminescence*—used to measure ozone and nitrogen oxides;
- *pulsed fluorescence*—used to measure sulfur dioxides.

Polyaromatic Hydrocarbons an AROMATIC HYDROCARBON that contains two or more BENZENE RINGS or other aromatic ring structure—also called *polynuclear aromatic hydrocarbons*. Polyaromatic hydrocarbons (PAHs) are persistent once they have entered the environment, and many are CARCINOGENS. NAPTHALENE and BENZO-A-PYRENE are both PAHs. The COMBUSTION of FOSSIL FUELS, solid-waste incinerators and discharges of oil and grease from industrial sources often contain PAHs.

PUBLIC WATER SYSTEMS are required under the SAFE DRINKING WATER ACT to monitor concentrations of benzo-a-pyrene (MAXIMUM CONTAMINANT LEVEL = 0.0002 milligrams per liter), and the ENVIRONMENTAL PROTECTION AGENCY is considering instituting MCLS for six other PAHs that are classified as probable human carcinogens: benzo(b)fluoranthene, benzo(k)-fluoranthene, chrysene, dibenz(a,h)anthracene and indenopyrene.

Polychlorinated Biphenyls a family of more than 200 synthetic ORGANIC COMPOUNDS that contain two or more CHLORINE atoms attached to a bonded pair of BENZENE RINGS. Polychlorinated biphenyls (PCBs) are clear to yellow oily, odorless liquids and solids that vary according to the percentage of chlorine in the mixture. Those with less chlorine are colorless mobile liquids. Increased chlorination produces thicker liquids, and yet further chlorination produces sticky resins and white powders. PCBs get into air and water from industrial and electrical equipment, leaching from municipal landfills, and from previously contaminated SEDIMENTS (see INTERNAL LOADING). Polychlorinated biphenyls are slightly soluble and highly persistent in water, with a HALF-LIFE of more than 200 days. PCBs are almost chemically inert, are stable when heated and are good electrical insulators, all of which led to their use as heat-transfer fluids in electrical transformers and capacitors and in industrial applications. PCBs were also used in the carbons in copy PAPER and in NEWSPRINT ink. The DIOXIN that is frequently found as an impurity in PCBs, and the fact that PCBs are only slightly biodegradable (see BIODEGRADABILITY, PERSISTENT COMPOUND) mean that they can accumulate to dangerous levels in sediments and living tissues after entering the environment. PCBs have high chronic and acute TOXICITY to aquatic life (see ACUTE EFFECTS, CHRONIC EFFECTS). They were banned in the United States in 1979 under the terms of the TOXIC SUBSTANCES CONTROL ACT of 1976.

Polychlorinated biphenyls can affect health when inhaled or by passing through the skin. They can be passed to a child through mother's milk. Most PCBs have a relatively low acute toxicity. It is their virtual indestructibility, and the fact that they tend to accumulate in living tissue that makes them dangerous. Exposure can cause an acne-like skin rash called chloracne. Exposure to higher concentrations can damage the nervous system, causing numbness, weakness and tingling in the arms and legs, and can damage the liver. PCB vapor can irritate the eyes, nose and throat. PCBs are probable human CARCINOGENS. There is limited evidence that they cause skin cancer in humans, and PCBs have been shown to cause liver cancer in animals. PCBs are possible human TERATOGENS, and have been shown to be teratogens in animals. Repeated exposure to PCBs can cause liver damage.

PUBLIC WATER SYSTEMS are required under the SAFE DRINKING WATER ACT to monitor concentrations of PCBs (MAXIMUM CONTAMINANT LEVEL = 0.0005 MILLIGRAMS PER LITER). THE occupational safety and health administration has established a permissible exposure limit for airborne PCBs of 1 milligrams per cubic meter for 42% chlorine PCBs and 0.5 mg/m^3 for 54% chlorine

averaged over an eight-hour workshift. They are on the Hazardous Substance List and the Special Health Hazard Substance List. Businesses handling significant quantities of PCBs must disclose use and releases of the chemical under the provisions of the EMERGENCY PLANNING AND COMMUNITY RIGHT TO KNOW ACT. The BEST AVAILABLE TECHNOLOGY for the removal of PCBs from drinking water is granular ACTIVATED CARBON.

Polychlorinated Dibenzofurans see FURANS

Polycyclic Organic Matter (POM) organic compounds that contain two or more ring structures (see AROMATIC HYDROCARBON, BENZENE RING) that are found in animal or plant wastes or in the decomposing remains of plants or animals. POM from decaying leaves, flower petals, twigs, sawmill wastes, animal urine and feces is often found in water (see GROUND-WATER POLLUTION, WATER POLLUTION). When DRINKING WATER that contains POM is treated with CHLORINE (see WATER TREATMENT), TRIHALOMETHANES (THMs) are formed. Although POM is not especially toxic, THMs such as the probable CARCINOGEN and suspected TERATOGEN, CHLOROFORM, are dangerous toxins, so the concentration of POMs in drinking water that is to be chlorinated must be carefully controlled.

Polyethylene Terephthalate Plastics the stiff, flexible clear plastic used in soda-pop bottles and containers for salad dressing and household cleansers. PLASTICS made from the polyethylene terephthalate POLYMER—commonly called *PET plastics*—are recycled more frequently than other plastics because the recycled resin is less expensive than virgin resin (see PLASTICS RECYCLING, VIRGIN MATERIAL). More than 30% of the PET produced in the United States during 1993 was recycled, the highest recycling rate for any plastic. The presence of even traces of POLYVINYL CHLORIDE (PVC) can ruin a batch of recycled PET plastics by forming acids that alter the PET's chemical and physical, producing a brittle plastic with a yellow tint. Breakdown can result from concentrations of PVC as small as 50 parts per million (less than one PVC container in an 800 pound container of recycled PET flake). Efficient sorting of recycled PET products is, therefore, a necessity.

Old soft-drink bottles made out of PET plastic can be pulverized, treated with chemicals and reconstituted into synthetic fibers that look and feel like cotton. The pile jackets, sleeping-bag insulation and other clothing made from recycled PET plastics have become quite popular—at least in part because they are an end product of the recycling of household throwaways—and the market for recycled PET resins has been very strong as a result. It takes the plastic in 5 two-liter PET soda bottles to produce the fiber required to make one T-shirt or a ski jacket, and it takes 25 bottles to make a sweater. Recycled PET plastics are also used to make new containers; carpet; automotive parts such as headliners, luggage racks, fuse boxes, bumpers, grills and door panels; and food-service trays, egg cartons, containers for baked goods and industrial strapping. See also PLASTICS RECYCLING SYMBOL.

Polymer   a large MOLECULE that is made up of small, repeated units called MONOMERS. A *polymer chain*—which may contain thousands of monomers—consists of a so-called *backbone* made up of groupings of HYDROCARBONS or similar materials to which a variety of monomer *side groups* are attached. The nature and arrangement of the monomers in the polymer's molecule determine its characteristics. The exact number of monomers in molecules of the same polymer may vary without changing its chemical characteristics. The rigidity and strength of a polymer can be increased by using reactive side groups that can undergo chemical reaction with the backbones of adjacent polymer chains, thereby cross-linking the polymers into a three-dimensional network.

Most chemicals with the prefix "poly" are polymers. A *copolymer* results when a mixture of two different monomers is polymerized with the two materials alternating in the polymer's backbone and its side groups. PLASTICS and synthetic rubbers are polymers, as are feathers, hair and woody fiber. An *addition polymer* is formed by linking individual monomers together. A *condensation polymer* is formed by linking monomers in such a way that each breaks off at the point of linkage creating a polymer and a BYPRODUCT, which is usually WATER. A copolymer contains more than one type of monomer, and may be formed as either addition or condensation polymers. Polymerization is used to produce all plastics and plastic resins, and is used in the production of GASOLINE, adhesives and PAINTS.

Polystyrene   a POLYMER created when the STYRENE MONOMER is heated. Polystyrene is a clear plastic with excellent insulating properties, high resistance to chemical and mechanical alteration at very low temperatures and water ABSORPTION of practically zero. It is used in the production of parts for electronics equipment and equipment such as refrigerators and airplanes that must operate in very low temperatures. It was first produced commercially during the latter 1930s. Expanded polystyrene foam is used in "styrofoam" building insulation, in disposable cups for hot beverages, plates, egg cartons, supermarket meat trays and in foam packing peanuts. McDonald's Corporation used expanded poly-

styrene "clamshell" packages to wrap hamburgers until forced by public complaints to switch to PAPER/PLASTIC COMPOSITE PACKAGING in 1990 (see MCTOXICS CAMPAIGN, PLASTIC WASTE). Polystyrene foam was produced through a process that emitted CHLOROFLUOROCARBONS (CFCs) until the early 1990s, at which time the CFCs were largely replaced with hydrofluorocarbons. More than a dozen cities and counties banned the use of expanded polystyrene in fast-food coffee cups and hamburger boxes in the late 1990s. Governor Deukmejian of California vetoed legislation that would have outlawed the use of expanded polystyrene made with CFCs two years in a row (in 1989 and 1990), saying that the industry was correcting the problem without the need of a ban. Expanded polystyrene foam is valued at waste-to-energy incinerators (see WASTE-TO-ENERGY PLANT) because of its high heat content and because the polystyrene helps to break down complex molecules found in other GARBAGE. Researchers at the Garbage Project estimate that polystyrene foam made up about 1% of total LANDFILL volume between 1980 and 1989. See also EXPANDED FOAM, PLASTICS RECYCLING.

Polyvinyl Chloride   a widely used PLASTIC that is a POLYMER of VINYL CHLORIDE. Polyvinyl chloride resin, the raw ingredient of polyvinyl chloride (PVC) plastic, is produced from vinyl chloride. PVC is used in the production of plastic pipe, film, beverage containers and numerous other plastic products. Containers made of PVC are marked with the triangular PLASTICS RECYCLING SYMBOL with a "3" in the middle. When burned, PVC can emit DIOXIN, hydrochloric acid and HEAVY METALS, and tends to corrode inside of incinerators and SCRUBBERS. *Meat wrapper's asthma*, which is associated with the PVC film used to wrap meat, is also suffered by workers in PVC production. An estimated 5.5 million pounds of PVC were recycled in 1993. Even concentrations of PVC as small as 50 parts per million can ruin a batch of recycled PET plastics by forming acids that alter the PET's chemical and physical structure, so thorough separation of the two plastics is essential. CADMIUM is used as a stabilizer in PVC pipe.

Ponding   the treatment of biological wastes by BACTERIA in a SEWAGE lagoon or the accumulation of water on the surface of a TRICKLING FILTER resulting when the filter becomes clogged with bacteria or other insoluble materials.

Population Growth   an increase in human population. Population growth is linked to worldwide problems such as the depletion of natural resources, the pollution of air and water and the growing difficulty of disposing of wastes. World population reached 1 billion in the early 1800s and 2 billion in 1930, and at the current rate of growth it increases by 1 billion every 12 years. The United Nations predicted in 1982 that world population would stabilize in the year 2100 at 10.2 billion, but the agency now predicts that human numbers will not stabilize until sometime after the year 2200, at 11.6 billion. World population increased to more than 5.5 billion in 1993—87 million more people than in 1992—with 94% of the growth occurring in the Third World.

The 89-million-person increase in population in 1989 was a record. The annual rate of population growth has slowed from 1.75% in 1986 to 1.56% in 1993. The rate declined in 1993 due primarily to a lower-than-expected reproduction rate in China, where one-fifth of the world's population lives. Chinese officials attributed the decrease from an average of 2.3 children per woman in 1991 to 1.9 in 1993 to the nation's strictly enforced one-couple, one-child policy, to improved birth-control methods and to a rising standard of living. Critics say China's decrease in growth rates involves coercion and involuntary sterilization. The fertility rate is just under four in developing countries and averages two in industrialized countries, with a few western European countries having attained zero or negative population growth (see ZERO POPULATION GROWTH). Residents of developing nations are more likely to want a large family because many children die from war, famine and disease. The rate of U.S. population increase is now 1.04 and Canada's is 1.5.

The science of demography, the study of human population, defines fertility as the average number of births per woman, and *replacement-level fertility* refers to the number of children average parents must have to replace themselves. The number is a little more than two because some children will die before they bear children. *Migration*, which occurs primarily within a country's boundaries, is defined as a movement of primary residence. *Emigration* is the movement of people out of a country, while *immigration* is the movement of people into a country. Population growth is determined by subtracting the number of people that die in or emigrate from a country from the number of newborn and immigrants, and then comparing the resulting number with the previous population size. Even if average fertility rates immediately dropped to two children per woman, population would continue to grow by another 3 billion before stabilizing because the average human age is now so young that an unprecedented number of young women are approaching child-bearing age (a phenomenon called *population momentum*). The World Bank estimates that if replacement-level fertility were to be attained in India by 2015, population momentum would still cause the country's population

to grow from an estimated 1.2 billion that year to 1.9 billion before it would level off. Demographers define a *stable population* as one in which the ratio of individuals of different ages and sexes remains constant, even if the total population is increasing. *Stabilized population* is a popular term meaning that the population of a given country is no longer growing. *Carrying capacity* is a term used by biologists to denote the number of animals that can be supported by a given ecosystem without depleting available sources of food and suffering population collapse. Carrying capacity does not necessarily apply to humans, however, because they are more adaptable and can come up with innovations to increase apparent carrying capacity. The mass famines predicted by some to occur during the 1970s and 1980s did not occur for this reason.

The world's big cities are growing by 1 million people a week, and will hold more than half the Earth's population within a decade, according to estimates in a World Bank study. The study predicted that by 2020, 3.6 billion people will inhabit urban areas while about 3 billion will remain in rural areas. In 1990, there were 1.4 billion people living in the world's urban areas, compared with about 2.7 billion in rural areas. The study projected that by 2000, there will be 391 cities with more than 1 million residents, up from 288 in 1990, and that 26 cities will be "megacities" with more than 10 million people. The growth of urban populations is linked to increases in air and water pollution (see AIR POLLUTION, GROUNDWATER POLLUTION, WATER POLLUTION) and corresponding decreases in sanitation.

Stanford University population biologist Paul Ehrlich stated his opinion on the impacts of population growth in the formula: I = PAT, where "I" is environmental impact, "P" is human population, "A" is affluence or per capita consumption and "T" is the output of technology used. Applied to automotive emissions in an urban area (see AUTOMOTIVE AIR POLLUTION), the formula demonstrates that even if per-capita consumption (A) reduces because people drive their cars less, and improved emissions control technology (T) reduces the number of POLLUTANTS produced per mile traveled, that a sufficient increase in population (P) and a corresponding increase in the total number of cars will mean that the environmental impact (I) of total automotive emissions produced will increase. The validity of Ehrlich's theory is demonstrated in California, where the population tripled between 1950 and 1993 to a current total of 32 million. The number of automobiles in California swelled from 12 million in 1971 to an estimated 22 million in 1993. Despite having the strictest automotive emissions standards in the United States, the state is struggling to improve air quality to an acceptable level.

The United Nations' International Conference on Population and Development was held in September 1994 in Cairo, Egypt. At the meeting, delegates endorsed a Population Action Plan that must go to the U.N. General Assembly for approval. The plan calls for spending $17 billion annually by the year 2000 to control world population growth, with spending to increase to more than $21.5 billion by the year 2015. Western nations are expected to pay one-third of the cost, and developing nations, two-thirds. Current spending on world population control is less than $6 billion a year. The action plan would control population primarily through the empowerment of women through education, better health care and political equality. The rationale is that educated and secure women will produce fewer children and contribute to economic development. Even the Vatican, which generally opposes any measure aimed at controlling world population through contraception, gave a limited endorsement to the plan. The Clinton administration restored funds for the U.N. Population Fund and the International Planned Parenthood Federation in 1993.

The exponential growth in world population can be seen as a human triumph over traditional sources of population control such as disease and child mortality. But the concept of sustainable development, which holds that we must consider the well-being of future generations, becomes more difficult in the face of growing population. See also DESERTIFICATION.

Porosity   the total amount of water that an AQUIFER is capable of holding. Porosity is dependent on the size and number of spaces, or pores, that exist between pieces of impermeable material in the aquifer. PERMEABILITY, the ability of water to move through earth or other materials, is not considered when measuring an aquifer's porosity. See also INFILTRATION CAPACITY, GROUNDWATER, PERCOLATION.

Post-Consumer Waste   materials that have been manufactured, sold to the public, used and discarded. The post-consumer waste content of new products that contain recycled materials is sometimes listed. A post-consumer waste content of more than 10% is significant. The post-consumer waste content of materials such as newsprint is dictated by law in many states. See also RECYCLING. Compare PRE-CONSUMER WASTE, VIRGIN MATERIAL.

Pre-Consumer Waste   material that is discarded during the manufacture of a product. The RECYCLING of pre-consumer waste is generally much easier than recycling materials that have been used by consumers, because the material is more pure and because a large

volume is present at one location. Such recycling of manufacturing wastes has been widespread since long before recycling of used waste products became common simply because of the economic benefit of reusing such materials rather than discarding them. Many products advertised as "recycled" contain only pre-consumer waste. Compare POST-CONSUMER WASTE, VIRGIN MATERIAL.

Precipitation   liquid or solid WATER that has become too heavy to remain suspended in the ATMOSPHERE and falls to earth. Rain, sleet, snow and hail are some of the forms taken by precipitation, which is the source of all fresh water and surface RUNOFF. Condensation causes tiny droplets of water to form (or solid ice crystals, if temperatures are below freezing) in air that is saturated with water vapor (at 100% relative humidity). The droplets or ice crystals gather to form fog or clouds, which are carried by the atmosphere until the growing particles of liquid or solid water get too heavy to be supported any longer. The liquids or solids that fall from clouds are called precipitation because the water is a PRECIPITATE, and precipitation also refers to the process of precipitate formation. See also ACID RAIN, AERIAL DEPOSITION, TOXIC PRECIPITATION.

Precipitate   a solid, liquid or gas that is dispersed throughout a material of a different, usually higher, phase as the result of a chemical or physical reaction. The formation of PRECIPITATION from atmospheric water vapor is the best-known form of precipitation. When a SOLUTION is supersaturated, part of the solute is said to precipitate—it becomes a solid suspended within the liquid solution, and may settle to the bottom of the container holding the liquid. Many precipitates that form as the result of chemical reactions are solids that form within a liquid. Two dissolved substances can react chemically to produce a third, insoluble, precipitate, which can, in turn be removed from the liquid with a WATER FILTER, a centrifuge, or by allowing it to settle out in a SEDIMENTATION basin. It is possible for a precipitate and the substance it is derived from to be in the same phase. Certain metal alloys, for instance, are formed by the precipitation of solids within other solids. Precipitation is widely used in the purification of DRINKING WATER and WASTEWATER (see HAZARDOUS WASTE TREATMENT, SEWAGE TREATMENT, WATER TREATMENT), and in the production of chemicals and metal alloys.

Pretreatment   the treatment of WASTEWATER before it enters a SEWAGE TREATMENT plant to remove substances such as grease, fat, lard, large floating objects, CORROSIVE MATERIALS and toxic chemicals that might cause problems were they allowed to enter the plant; also, the treatment of industrial liquid waste to remove substances that are hazardous, toxic, nonbiodegradable or highly corrosive, and the adjustment of the wastewater's PH to within the range of 6.0 to 8.0. Coagulation, SEDIMENTATION, AIR STRIPPING and the flotation and removal of oils via surface skimming are commonly employed pretreatment processes. Air or CHLORINE is often bubbled through the wastewater to encourage OXIDATION.

Pretreatment is also used at sewage treatment plants to remove materials that might damage equipment inside the plant. EFFLUENT generally passes through a grate composed of large metal bars that catch the largest floating material in the wastewater, and is the routed to a SKIMMING TANK where lighter-than-water materials such as grease, fat and lard float to the surface and are skimmed off and a grit tank where small particles of sand, glass and grit settle out and are removed.

Pretreatment Standards   federal standards governing the removal of CONTAMINANTS from industrial EFFLUENT before they can be dumped into a municipal sewer system. The standards apply primarily to chemicals and other materials that could hinder the operation of a municipal SEWAGE TREATMENT plant. Pretreatment standards apply primarily to the emission of toxic chemicals that could kill or disable the BACTERIA used in SECONDARY TREATMENT, and of toxins that could pass through a treatment plant unaltered and thereby pollute SEWAGE SLUDGE or surface water into which treated effluent is dumped. The ENVIRONMENTAL PROTECTION AGENCY establishes guidelines for pretreatment standards in accordance with the CLEAN WATER ACT. Standards may be set either for specific contaminants, or for specific types of industrial processes. Standards for new pollution sources are generally stricter than those governing the addition of pretreatment facilities to an existing source of industrial emissions. The local government agency responsible for operation of the sewage plant sets pretreatment standards, but they must conform with the effluent standards found in the Clean Water Act. See also SEWAGE.

Prevention of Significant Deterioration   a provision in the CLEAN AIR ACT intended to maintain the level of air quality in regions where air is cleaner than required in the NATIONAL AMBIENT AIR QUALITY STANDARDS. Discharges in such areas are governed by *increment standards* that are established in accordance with the area's AIR QUALITY CLASS. Wilderness areas, national parks and similar pristine regions are designated as Class I airsheds, where no deterioration of ambient air

quality is allowed. Air quality in Class III airsheds (primarily industrial and urban areas) is already compromised, and further deterioration is generally allowed as long as national health standards are not violated. The remaining areas are designated as Class II areas, where some deterioration is allowed. An industry planning to build a plant in an area governed by prevention-of-significant-deterioration (PSD) standards must demonstrate that the standard for the area will not be violated as a result of the plant's atmospheric emissions, and that the BEST AVAILABLE CONTROL TECHNOLOGY will be used to remove CONTAMINANTS (see BEST AVAILABLE TECHNOLOGY). State governments that have filed an adequate state implementation plan with the ENVIRONMENTAL PROTECTION AGENCY (EPA) can issue permits for such construction. Otherwise, the permit is issued by the EPA. National parks and forests and Indian reservations are among the areas that have received a Class I designation.

Price-Anderson Act   a federal law passed in 1957 to encourage the commercial development of nuclear energy. The Price-Anderson Act limits the total damage settlements that can be assessed against the owner of a commercial nuclear facility to $560 million, of which the federal government will provide $500 million.

Primary Recycling   see PRE-CONSUMER WASTE

Primary Treatment   the removal of FLOATING SOLIDS or SUSPENDED SOLIDS from WASTEWATER at a SEWAGE TREATMENT plant through the use of physical processes such as screening and settling. Primary treatment removes as much as 60% of the solids and 40% of the organic waste from effluent.

Pretreatment to remove materials that could damage plant equipment is the first stage of primary treatment. In a typical primary treatment process, large floating debris such as twine, cans and bottles and pieces of wood is removed with bars and screens. Wastewater is then routed through a *grit chamber*, where suspended solids such as sand or fragments of metal or glass are allowed to settle to the bottom. In some treatment plants, wastewater is then routed to a *comminutor* where solids are chopped into fragments that measure one-eighth inch or smaller. EFFLUENT passes either from the comminutor or directly from the grit chamber into *primary sedimentation*—also called *clarification*—a *settling basin*, also called a *primary clarifier*, where the remaining organic solids (see ORGANIC MATERIAL) and other materials responsible for *biochemical oxygen demand* are removed. Water flow is slowed enough to allow sediments to settle to the bottom as they pass through the large rectangular or circular sedimentation basin. Sediments settle in accordance with Stokes' Law, which says that when the gravitational force pulling a particle toward the bottom exceeds the horizontal force exerted by the moving water, the particle will settle to the bottom, if given enough time. *Coagulants* such as lime, alum, CHLORINE, clays and polyelectrolytes may be added to the wastewater to increase the percentage of solids that settle out. More solids settle out of wastewater when coagulants are used, so the resulting volume of SLUDGE is greater, which can make the use of coagulants impractical in areas where the disposal of sludge is expensive. Lighter materials such as oils float to the top in quiet water of the sedimentation basin, and are removed with skimmers. Sludge that settles to the bottom is removed mechanically. Primary effluent seeps over the basin's rim and is collected for SECONDARY TREATMENT. Sludge goes through dewatering and further treatment before disposal, often in a LANDFILL. A few decades ago, most effluent was dumped into a lake or stream after primary treatment, but in most modern treatment plants wastewater that has gone through primary treatment must go through secondary and sometimes tertiary treatment before it can be routed to the plant's OUTFALL. See also SEWAGE.

Priority Pollutant   an air or water POLLUTANT (see AIR POLLUTION, GROUNDWATER POLLUTION, WATER POLLUTION) listed by the ENVIRONMENTAL PROTECTION AGENCY (EPA) under the terms of the 1977 amendments to the CLEAN WATER ACT. The list of priority pollutants is updated regularly by the EPA, and the updates are published in the Code of Federal Regulations. Discharge limits for priority pollutants are set by the EPA in accordance with BEST AVAILABLE TECHNOLOGY standards.

Process Chemical   a CHEMICAL that is used as a raw material or a chemical used in an industrial process. See CHEMICAL MANUFACTURING WASTE, HAZARDOUS WASTE, INDUSTRIAL LIQUID WASTE, INDUSTRIAL SOLID WASTE, TOXIC SUBSTANCES CONTROL ACT.

Process Water   WATER that is used in an industrial process. Process water may be used as a raw material, a cleansing agent, an electrolyte, a source of steam or in numerous other ways as part of the manufacturing process. The definition does not generally include water used to cool a power plant or other industrial equipment, although cooling water is also frequently referred to as process water. It takes lots of water to produce modern commodities: 40,000 gallons to produce the STEEL in a typical automobile, 50,000 gallons

to produce the rayon in a living room carpet, 900 gallons of cooling water to produce a kilowatt of electricity at a coal-fired power station and 3,000 gallons to produce one pound of beef. The expansive power of steam, which occupies 1,670 times the space taken up by liquid water, was the dynamic force behind the Industrial Revolution.

Product Disposal Tax   a tax that is imposed on a manufactured product to ensure that it will be properly disposed of. A disposal tax requires that payment for part of the EXTERNALITIES associated with a product's production and use be paid as part of the purchase price. The RESOURCE CONSERVATION AND RECOVERY ACT of 1970 was limited primarily to research and evaluation of potential ways of reducing the flow of reusable resources through the waste stream. A one-cent per pound *resource recovery tax* on items that would normally be thrown away within 10 years of manufacture was proposed but not enacted. The junk vehicle taxes that are a part of licensing fees in many states is a form of product-disposal tax. Several states have imposed disposal taxes on DISPOSABLE DIAPERS. A RECYCLING tax is imposed on all goods manufactured in Germany. See also ENVIRONMENTAL COST, SOCIAL COST.

Product Water   see FINISHED WATER

Public Health Service   the federal agency responsible for the protection and advancement of the public's physical and mental health. The U.S. Public Health Service (PHS)—part of the Department of Health and Human Services—was created by a 1798 act that authorized the creation of marine hospitals for American merchant seamen. The Public Health Service Act of 1944 consolidated federal public health functions in the PHS. State health agencies handle much of the implementation of national health ordinances. The PHS provides resources and expertise to aid the states. The agency also implements international health agreements, policies and programs; conducts and sponsors medical research; and enforces laws concerning the safety of food, drugs, cosmetics, medical equipment. Several PHS agencies work with the release of toxic or hazardous materials.

The *Agency for Health Care Policy and Research* is responsible for research concerning the quality, effectiveness and cost of health care and the publication and dissemination of such information.

The *Centers for Disease Control and Prevention* provides leadership and direction in the prevention and control of disease. The CDC develops and implements programs to deal with threats to environmental health and chemical and radiation emergencies. Research, information and education relating to smoking and health is generated by the CDC as part of the National Health Promotion Program. The National Center for Environmental Health, the National Institute of Occupational Safety and Health (develops occupational safety and health standards) and the National Center for Health Statistics are part of the CDC.

The *Agency for Toxic Substances and Disease Registry* was formed in 1983 to implement the health-related aspects of the COMPREHENSIVE ENVIRONMENTAL RESPONSE, COMPENSATION AND LIABILITY ACT of 1980 and the SOLID WASTE DISPOSAL ACT. The agency generates and disseminates information concerning human exposure to toxic or hazardous substances and assists the ENVIRONMENTAL PROTECTION AGENCY (EPA) in identifying materials to be treated as hazardous waste. It develops procedures for assessing the public health risks resulting from the environmental release of hazardous substances and makes recommendations for the protection of public and worker health and safety.

The *Food and Drug Administration* is the federal agency responsible for protecting the public from impure or unsafe foods, drugs and cosmetics. The Center for Devices and Radiological Health is responsible for controlling human exposure to radiation. It conducts a program intended to limit the radiation emitted by electronic devices and medical equipment, and supports research and testing relating to radiation exposure. The National Center for Toxicological Research conducts research into the biological effects of environmental toxins and implements the National Toxicology Program.

The *National Institutes of Health* is the principal federal agency concerned with biomedical research. The National Cancer Institute supports research on the diagnosis, causes, prevention and treatment of cancer under the auspices of the National Cancer Act. The National Heart, Lung and Blood Institute supports research relating to disease of the respiratory and circulatory systems. The National Institute of Diabetes and Digestive and Kidney Disease supports research into the diagnosis, causes and prevention of disease of the digestive system and kidneys. The National Library of Medicine (NLM) disseminates information generated by other branches of the NIH and is the principal source of medical information within the federal government. The library provides medical information through MEDLINE, TOXLINE and other on-line systems including a toxicology information service for the scientific community. The NLM is also responsible for maintenance and development of the Biomedical Communications Network, which is charged with improving communication within the medical community through the application of advanced technology, and the National Center for Biotechnology Information, which is assigned the task of developing new information technologies to aid in the understanding of molecular processes inherent in disease and health treatment. Address: Public Health Service, 200 Independence Ave., SW, Washington, DC 20201. Phone: 202-619-0257.

Public Utilities Regulatory Policy Act (PURPA)   a law passed in 1978 that required electric utilities to buy power from small, independent power producers. Before the passage of PURPA, utilities had an effective monopoly on electric power production. The law requires utilities to pay their AVOIDED COST for electricity from small installations called *qualifying facilities* (QFs) in the Act, that use renewable sources such as falling water, solar or wind energy to generate power. Avoided cost is defined in the act as the cost of power to the utility if it had to build a new power plant to generate the power. State public utility regulations determine exactly how avoided cost will be computed. In many states, environmental EXTERNALITIES such as the cost of the pollution of air and water (see AIR POLLUTION, GROUNDWATER POLLUTION, WATER POLLUTION), and the depletion of natural resources that are associated with a conventional power generator are included in a utility's avoided cost. Many REFUSE DERIVED FUEL facilities generate electricity for sale under a PURPA-based contract. California has its own version of PURPA and a tax credit for renewables. The legislature also passed a bill that guarantees utilities full cost recovery for investments in QFs. See also COAL, ENVIRONMENTAL COST, NUCLEAR WASTE, SOCIAL COST.

Public Water System   a water system with 15 or more service connections, or that delivers water to 25 or more people, 60 or more days each year. A *community water system* is defined in the SAFE DRINKING WATER ACT as one that serves DRINKING WATER to a resident population. A *noncommunity water system* is defined as one that either serves a transient population, as is the case with hotels, airports and highway rest stops, or a nontransient population, such as that found in a prison, school or factory. Only about 30% of the more than 200,000 U.S. water-supply systems meet these definitions.

Pull-Tab Typology   a classification system for the pull-tabs that were once used on ALUMINUM soda and beer cans. The system was developed in the mid-1970s by students taking part in the University of Arizona's Garbage Project, which was undertaking a study of household recycling practices for the ENVIRONMENTAL PROTECTION AGENCY. To conduct the survey, the school had obtained permission to sort through the garbage cans from a sample group of households in the Tucson area before the trash was hauled to the landfill. During the study, the question of whether it could be assumed that households with no soda or beer cans in their garbage can recycled aluminum. The survey concluded that most people who recycled aluminum cans did not recycle the pull-tabs from the cans, but simply threw them in the trash. By retrieving the pull-tabs from garbage cans, researchers were able to get a good idea of which households recycled aluminum and of how many cans were used. In the process of retrieving and counting the tabs, one of the students noticed that they were not all the same. A pull-tab typology was developed, and by using it researchers were able to discern whether the tabs were from beer cans or soda cans and which brands were used, even for a household with no aluminum cans in the garbage. Pull-tabs are an archaeological HORIZON MARKER in the excavation of landfills because they were produced for only a relatively short time before their use was discontinued out of concern for their effects on birds and animals that tried to eat them. See also FORMATION PROCESS, GARBOLOGY.

Putrefaction   the decomposition of organic material caused by the splitting of proteins by ANAEROBIC BACTERIA and fungi. Putrefaction gives off incompletely oxidized byproducts such as mercaptans and alkaloids that are foul-smelling. See ANAEROBIC DIGESTER. Compare DECAY, OXIDATION.

Pyrite   a MINERAL composed of IRON sulfide (FeS_2) that is usually found in association with COAL deposits and often found with commercial deposits of other minerals. Pyrite, also called "fool's gold" because of its brass-yellow color and metallic luster, is the principal source of the acidity that typifies drainage from mines and the TAILINGS of mines and mills. Pyrite oxidizes readily when exposed to air in a mine or tailings pile, and can release sulfuric acid into water or SULFUR DIOXIDE into the air, depending on the temperature and humidity at which it oxidizes. Bacteria that metabolize iron and SULFUR in the presence of oxygen also break down pyrite, and contribute to acid RUNOFF from mines. Pyrite is used in the commercial production of sulfuric acid. See also ACIDIC WASTE.

Pyrolysis   the chemical decomposition of a compound that results when it is heated to a temperature high enough to break the chemical bonds that hold its molecules together. Pyrolysis is sometimes used to reduce the volume of SOLID WASTE before it is disposed of in a LANDFILL, and can also be used to reclaim metals from solid waste. The process can also be used to break down hazardous compounds that would release toxins into the atmosphere if they were burned in an incinerator (see HAZARDOUS WASTE, HAZARDOUS WASTE DISPOSAL). Scientists at General Motors Corporation are experimenting with using pyrolysis to convert "car fluff"—a mixture of waste products left over from the automobile manufacturing process that consists of PLASTICS, GLASS, rubber, rust and fluids—into gaseous and oil fuels and a black-powder residue than can be used to make car parts, stronger cement, roofing shingles and other products. Combustible materials can be pyrolized if they are heated in the absence of oxygen. Charcoal is produced from the pyrolysis of WOOD, and COKE is obtained from the pyrolysis of COAL. Coke and pyrolized wood burn hotter and cleaner than coal or wood. Pyrolysis occurs in an open fire under the surface of the wood where oxygen can not penetrate. *Synthetic petroleum* can be obtained through the pyrolysis of wood.

R

Radiation Ecology that area of ecology that focuses on radiation, radioactive substances and the environment. Radiation ecology is concerned with the effects of ionizing radiation on humans, plants, animals and the environment. The environmental effects of radioactive fallout from nuclear weapons and power plants, the safe disposal of nuclear waste and the introduction of radioactive effluents into surface water are typical subjects studied by a radiation ecologist. See NUCLEAR REACTOR, RADIOACTIVE WASTE, RADIOACTIVITY.

Radioactive Waste waste that contains radioactive materials (see RADIOACTIVITY). The majority of radioactive waste, also called *nuclear waste* or *radwaste*, is associated with the mining, processing and consumption of URANIUM in nuclear power plants. Military ships powered by nuclear reactors, research and development at national laboratories and universities and the medical use of radiation produce a comparatively modest amount of radwaste by comparison. While the production of nuclear weapons is now responsible for a relatively small quantity of radioactive waste when compared to that from nuclear power plants, the vast quantities of RADIONUCLIDES and other toxic materials that have escaped into the environment from weapons facilities in the past are a big reason for public distrust of nukes (see HANFORD NUCLEAR RESERVATION, OAKRIDGE NATIONAL LABORATORIES, ROCKY FLATS, SAVANNAH RIVER,). Activities associated with nuclear weapons production at about 100 military installations in the United States over the last 50 years has lead to some of the most severe environmental pollution to be seen in the country. The Office of Technology Assessment concluded that there is "evidence that air, groundwater, surface water, SEDIMENTS and soil, as well as vegetation and wildlife have been contaminated at most, if not all, of the Department of Energy nuclear weapons sites." Investigators have blamed the contamination with the haste with which much nuclear-weapons development was undertaken, espe-

cially in the early days of the nuclear era, and on the secrecy involved. Civilian officials—and, in many cases, even high military officials—were simply not aware of the level of environmental destruction that accompanied the production of the bomb. The Department of Energy estimated in 1995 that it will cost from $230 billion to $350 billion to clean up U.S. nuclear weapons plants to a point where they no longer represent a serious environmental threat, and that even after the cleanup, many installations will still be too contaminated to be released for other uses.

The situation is even worse within the boundaries of the former Soviet Union. At the Chelyabinsk-40 weapons plant in the southern Ural Mountains, for instance, HIGH-LEVEL WASTES were simply dumped in the Techa River in the early days of the Soviet nuclear-weapons program. When in 1951 the government traced radioactivity 1,500 kilometers from the plant to the Arctic Ocean, the Soviets started dumping waste into Lake Karachav, which has been called the most polluted spot on the planet. The dumping of nuclear waste in the lake continued into the 1960s. So much radiation is emitted from the lake today that a person standing on its shores for an hour would die within a week of radiation poisoning. The lake dried up during a drought in 1967, and radioactive dust was blown as much as 75 kilometers away, contaminating an estimated 41,000 people with radiation. In 1953, the Soviets started storing reprocessing wastes in underground tanks at Chelyabinsk-40. One of the tanks overheated in September 1957, resulting in an explosion that contaminated hundreds of square miles and resulted in the relocation of 11,000 people.

Classification

A classification system established by the U.S. NUCLEAR REGULATORY COMMISSION (NRC) reflects the relative hazard of nuclear wastes, and specifies how long different types of waste must be isolated (see RADIOACTIVE WASTE DISPOSAL). The NRC classification system is based on the type of radiation emitted, energy level and the concentration and HALF-LIFE of the radionuclides.

Sources

About 73% of the volume and 95% of the radioactivity of the waste from civilian sources came from nuclear power plants. Waste from the mining and milling of uranium has the largest volume and is the least concentrated form of waste in the nuclear fuel cycle. While milling removes most of the uranium from ore, approximately 85% of the radioactivity is left behind. Thorium-230, a decay product of uranium with a half-life of 77,000 years, is the principal source of radioactivity in TAILINGS. It creates the potent CARCINOGENS radium-226 and radon-222, as it decays (see URAVAN URANIUM MILL). Uranium conversion plants process raw uranium yellowcake to *uranium hexafluoride*, which is then processed in a uranium enrichment plant to produce fuel-grade uranium. The Sequoyah Fuels conversion plant in Gore, Oklahoma, was closed in 1991 by the NRC because soil and GROUNDWATER around the plant had become contaminated with uranium and the plant had a long record of mismanaging hazardous materials.

Some MEDICAL WASTE contains radioisotopes that must be stored so that they can decay. Examples cited by Dr. David R. Holley, president of the California Medical Association, include carbon-14 with a half-life of 5,730 years, cesium-137 with a half-life of 30.2 years, and TRITIUM with a half-life of 12.3 years. It takes 57,300 years for carbon-14 to lose 99.9 percent of its radioactivity—123 years for tritium and 302 years for cesium-137. Equipment that contains radioactive waste must occasionally be disposed of by health care and medical research facilities.

Short-Term Storage

Irradiated uranium fuel accounts for 95% of the radioactivity from all U.S. sources of radioactive waste combined: commercial, experimental and military. Fuel rods from commercial power reactors are now stored, usually on-site, in deep water tanks (see NUCLEAR REACTOR) until such a time as one or more repositories for the long term storage of high-level radioactive waste is established. The water-filled storage tanks are not a good solution for long-term storage of the spent fuel rods. They rely on pumps to keep water circulating around the hot fuel rods and filters to keep the water in which the fuel rods are immersed from getting too radioactive and sensors and controls to keep the whole thing running and will eventually break down. One alternative to hauling fuel rods to a repository is to build *dry-cask storage* sites that rely on

passive cooling to keep the spent fuel from overheating at nuclear plants. The NRC estimates that high-level wastes stored in dry-cask storage systems would be safe for 100 years.

Radioactive Waste Disposal the permanent disposal of radioactive materials in such a way that they no longer represent an appreciable environmental threat. An effective method of RADIOACTIVE WASTE disposal must either render the waste materials permanently stable so that no further radiation is emitted, or must provide long-term storage for the millennia necessary for the materials to lose their RADIOACTIVITY naturally. Despite decades of research by all the world's nuclear powers, a safe, practicable means of disposing of radioactive waste has not been devised, although countless tons of nuclear wastes have been disposed of in a temporary manner—often to the detriment of the environment. Once they have entered the environment, the longer-lived radioactive materials will remain a liability until they are cleaned up and either neutralized or stored where they are no longer a threat to living things. The cost of cleaning up nuclear wastes once they have escaped into the environment can be enormous. Even wastes that are currently being temporarily held in storage facilities that are adequate in the short run are just in a holding pattern until a workable solution to the disposal dilemma is devised. The first byproducts of the FISSION reaction—produced by Enrico Fermi in 1942—are now temporarily buried under two feet of soil and a foot of concrete on an Illinois hillside, and a staggering volume of HIGH-LEVEL WASTE has been produced since that time. Even if all the world's nuclear reactors were shut down immediately and no more were built, devising a permanent method of disposing of radioactive waste would still be essential.

Numerous schemes have been proposed for disposing of the international backlog of high-level nuclear wastes, and many seem feasible, at least at first glance. By even the most optimistic estimates, however, wastes must be isolated for 10,000 years before they will constitute no more of an environmental threat than naturally occurring URANIUM deposits. The more conservative estimates of the necessary period of isolation run into the tens of thousands of years, and a lot can happen in that amount of time. Disposal schemes include:

- *Launching wastes into space or into the sun.* The tremendous expense involved and the possibility of a disaster scattering hot wastes over a wide area are drawbacks to the scheme. The idea was pretty much abandoned after the space shuttle *Challenger* disaster.

- *Allowing wastes to melt their way to the bottom of the Antarctic ice cap.* The fact that the ice cap is in constant motion and the difficulty of transporting wastes across the perilous Antarctic Sea shot this proposal down.
- *Drilling a hole through the mantle and pumping wastes into the earth's molten core.* Beyond the fact that the deepest holes drilled haven't penetrated nearly that deep, the expense of drilling and outfitting even one such ultra-deep disposal well and the logistics of transporting massive quantities of radioactive waste to the disposal site(s) make the proposal impractical.
- *Storing wastes in specially constructed guarded buildings.* The near impossibility of monitoring and guarding a waste-storage structure for a period of time that is much longer than human cultures survive mean that few have taken the option seriously.
- *Converting wastes into* ISOTOPES *with a shorter* HALF-LIFE *through transmutation.* This can be accomplished by bombarding the neutrons of the radioactive materials, and the technique is being studied in the United States, Japan, Russia and France. It is uncertain at this point whether the nuclear waste stream would actually be reduced by transmutation, and the method would be extremely expensive if it does.
- *Reprocessing the wastes and recycling the uranium as reactor fuel.* Commercial reprocessing is currently under way in four countries, and 16 countries have either reprocessed high-level wastes or plan to. The necessary chemical processes have already been developed as a way of extracting PLUTONIUM from fission byproducts for use in nuclear weapons. The problem with reprocessing is that the most hazardous elements in the irradiated fuel—radioisotopes such as cesium, iodine, strontium and technetium and elements such as americium and neptunium—remain as waste products after reprocessing is complete. About 97% of the radioactivity found in spent fuel remains as a waste product that must be disposed of as high-level waste, and the volume of waste is much greater than before reprocessing took place. A site 40 miles south of Buffalo, New York, the only U.S. REPROCESSING PLANT, closed in 1972 due to economic and technical problems. It is estimated that it will cost $3.4 billion to contain pollution at the site of the plant.
- *Injecting the wastes into sediments in geologically stable portions of the ocean's floor.* This proposal is being studied by a consortium consisting of 10 countries. The fact that the practice may violate international law banning OCEAN DISPOSAL of radioactive materials, the difficulty and hazards involved in transporting wastes to disposal sites and uncertainty about

what would happen if the wastes leak into the ocean are the principal stumbling blocks.
- *Burying wastes in a specially constructed repository located deep underground.* This would appear to be the most practicable option now available. Underground repositories are being studied by most countries that have nuclear waste to dispose of as a preferred alternative, and a few pilot projects are underway. The possibility that earthquakes or volcanic action may alter the plumbing of groundwater in the area of the repository, or that humans in the distant future will get into the disposal sites in the search for commercial minerals are the principal objections to the idea. The dangers involved in transporting large quantities of high-level waste to such repositories are also formidable. But public opposition to the construction of a waste repository near their city/county/state is the most formidable obstacle (see NIMBY). The Mescalero Apache tribe voted in 1995 to allow the construction of a facility to store processed spent nuclear fuel rods from 30 commercial nuclear power plants on its reservation in southwestern New Mexico.

See also HANFORD NUCLEAR RESERVATION, LOW-LEVEL WASTE, OAKRIDGE NATIONAL LABORATORIES, ROCKY FLATS, SAVANNAH RIVER.

Radioactivity   the emission of radiant energy and charged particles as the result of radioactive decay. Radioactivity comes in the form of ALPHA PARTICLES, BETA PARTICLES and GAMMA RAYS that are released as the result of the spontaneous transformation of one ELEMENT or ISOTOPE into another through the partial disintegration of the atomic nucleus. When a beta particle is released, one neutron in the atomic nucleus in converted to a proton. The release of an alpha particle tears two neutrons and two protons loose from the nucleus. Heat is also released along with alpha and beta particles, which is where most of the heat in a nuclear reactor comes from. Approximately 40 naturally occurring elements and isotopes undergo spontaneous radioactive decay, and other radioactive elements have been created in the laboratory. Emissions of radioactive particles and waves are called IONIZING RADIATION because normal atoms and MOLECULES that are struck by the radiation can be converted into IONS—in other words, radiation physically knocks molecules and atoms that it strikes apart.

The existence of radioactivity was discovered in 1896 by the French physicist Antoine Henri Becquerel, who observed that URANIUM can blacken a photographic plate even though it is separated from the plate by glass or black paper. In 1898 the French chemists Marie and Pierre Curie deduced that radioactivity was associated with atoms, independent of their physical or chemical state. The Curies went on to isolate polonium and RADIUM by chemically treating pitchblende. At the time radiation was discovered, the atom was believed to be the smallest indivisible building block of matter, but the existence of alpha particles and beta particles proved the theory false. The existence of an atomic nucleus was proven in 1911—the first step toward the development of modern atomic theory. The first nuclear FISSION reaction was achieved in December, 1942, in Chicago by physicist Enrico Fermi.

The intensity and type of radiation and the dose received determine their potential for damaging living tissue. Alpha particles are heavier than beta particles and so potentially more destructive, although they cannot penetrate very deeply into living tissue. Beta particles are less ionizing, but are more capable of penetrating tissue. Gamma rays are even less ionizing, but even more penetrating. Alpha and beta particles are generally a health threat only if ingested or inhaled, while gamma rays can penetrate deep enough into tissue to affect health without being inhaled or swallowed. If a molecule of DNA or RNA is damaged, cancer and genetic mutation can result. Damage to other molecules in a living cell may lead to death or disease in the organism. Radiation can damage the cells of a developing fetus, resulting in birth defects or miscarriage. Radiation can also weaken the immune system. Children and fetuses are particularly sensitive to the effects of radiation because their rapidly dividing cells can be more easily damaged. Exposure to 1,000 rems of radiation (see MEASUREMENTS) is lethal to humans, while exposure to 100 rems will cause nausea, vomiting, malaise and limited neural damage. A person getting a chest x-ray is typically exposed to 20- to 30 millirems (thousandths of a rem) of radiation. The average person receives about 300 millirem per year from naturally occurring BACKGROUND RADIATION. The health effects of exposure to very low concentrations of radiation are a subject of controversy, although most scientists agree that any radiation will cause damage to whatever it hits. The ENVIRONMENTAL PROTECTION AGENCY (EPA) has established standards calling for no more than 25 millirems per year of radiation exposure above background levels. The EPA standards are based on a linear model in which it is assumed that exposure to low-level radiation causes damage that is directly proportional to higher doses. RADIONUCLIDES, including uranium and PLUTONIUM, are on the HAZARDOUS AIR POLLUTANTS LIST. In 1989 the EPA issued new regulations governing the emission of radionuclides. The regulations, which were mandated in the CLEAN AIR ACT, cover 7,000 sources of radioactivity including nuclear plants, hospitals, labs, research facilities, uranium mines and processing plants and phosphorous-processing plants. See also HANFORD NUCLEAR RESERVATION, OAKRIDGE NATIONAL LABORATORIES.

Radionuclide   a naturally occurring or manmade element with an inherently unstable atomic structure, that as a result emits radiation (see RADIOACTIVITY). A radionuclide eventually emits enough of its matter or energy that it becomes a stable ELEMENT or ISOTOPE that no longer emits radiation. The amount of time it takes a radionuclide to release half its radioactivity is known as its HALF-LIFE. A *radioisotope* is a radioactive isotope. PUBLIC WATER SYSTEMS are required under the SAFE DRINKING WATER ACT to monitor concentrations of ALPHA PARTICLES, BETA PARTICLES, RADIUM and RADON.

See also NUCLIDE, PLUTONIUM, TRITIUM, URANIUM.

Radium   a radioactive metallic ELEMENT that is a member of the alkaline earth group of metals. Radium is a silver-white METAL that oxidizes instantly on exposure to air. It is almost never handled in its pure form, and is usually handled in the form of radium chloride or radium bromide. Radium is used in the production of luminous paint such as that used on the dial of a clock to make it glow in the dark. A small amount of one of the radium salts is mixed with a compound such as zinc sulfide, which becomes luminous because of alpha-wave bombardment from the radium.

The French chemists Marie and Pierre Curie first isolated radium by chemically treating pitchblende (U_3O_8—one of the two ores from which commercial uranium is obtained) in 1898 (see RADIOACTIVITY). The Curies discovered that one gram of radium gives off 100 calories per hour of radiation. Radium is a decay product of URANIUM. Radium-226—the most common of the four naturally occurring radium isotopes—is formed by the radioactive disintegration of thorium-230, which in turn is the fourth decay product in a series starting with uranium-238. Radium-226 emits ALPHA PARTICLES that form RADON gas. Radium is always present in pitchblende, normally at concentrations of about one part radium to three-million of uranium. Radium is removed from uranium ore by treating it with a barium compound that reacts with the radium to form barium sulfate and radium sulfate, which are then removed from the ore by PRECIPITATION (see PRECIPITATE). The sulfates are then converted into carbonates or sulfides, which are then dissolved with hydrochloric acid. The resulting chloride solutions are then put through repeated crystallization/melting cycles to separate the uranium and the barium.

Radiation from radium can cause cellular damage, and exposure to high levels can lead to skin damage. Lower levels of radiation from radium are used to treat cancer. Cancerous cells are more sensitive to radiation than normal cells, and can be killed through exposure to a controlled level of radiation. Radium chloride or radium bromide is sealed in a tube that is inserted next to the cancerous tissue to be treated.

Radium occurs naturally in GROUNDWATER with the highest concentrations in Iowa, Illinois, Wisconsin and Missouri. PUBLIC WATER SYSTEMS are required under the SAFE DRINKING WATER ACT to monitor concentrations of radium-226 and radium-228—proposed MAXIMUM CONTAMINANT LEVEL (MCL) = 5 picocuries per liter (pCi/L); proposed MCL = 5 pCi/L. Radium (and uranium) can be removed from water by means of lime softening, REVERSE OSMOSIS, ELECTRODIALYSIS and ION EXCHANGE.

Radon   an odorless, colorless, chemically inert gas that is a decay product of RADIUM (which is a decay product of URANIUM). Radon has a HALF-LIFE of just 3.8 days, and transforms fairly quickly into its decay products, called radon daughters, of which two ISOTOPES of polonium are the most health threatening. Radon gas and radon daughters are found in soils that contain particles of uranium, and the gases can pass through cracks in the foundation, joints in concrete slabs and cracks where water pipes and drains pass through the foundation and into sumps to get into indoor air. The gas is heavier than air, and tends to accumulate in the lowest part of a building. GYPSUM wallboard, concrete and even stoneware pottery can emit radon if the material of which they are made contains traces of radon. Radon can also get into indoor air when water contaminated with the gas is aerated (see AERATION) by being sprayed out of a shower, agitated in a washing machine or boiled on the stove. The Washington Energy Extension Service has estimated that about half of an average American's exposure to radiation comes from indoor exposure to radon and radon daughters, with another 25% coming from medical exposure. It is not unusual for concentrations of radon in indoor air to be 100 times higher than found outside. Although the highest radon concentrations encountered by the average person are found in indoor air, workers in some mines are routinely exposed to much higher concentrations.

An ENVIRONMENTAL PROTECTION AGENCY (EPA) survey of 20,000 homes in 17 states found unacceptably high levels of radon in one-fourth. The agency estimates that as many as 8 million homes contain unsafe levels of radon. The highest levels of radon and radon daughters in indoor air occur in areas that have higher-than-average levels of uranium and radium in bedrock

and soils. The Reading Prong, which cuts across New York, New Jersey and Pennsylvania, is one such area, but similar radon zones occur across the country. Most rocks and soils contain at least some uranium, but levels are generally higher in bedrock composed of granite, shale and phosphate and in soils derived from such bedrock. Although areas that are likely to have radon in the soil can be identified, its occurrence in individual structures within an area is highly unpredictable. One house can have high readings while the house across the street has only a trace. An EPA survey conducted during the 1990–91 school year concluded that one school in five exceeds federal standards for radon gas, while 2.7% of the surveyed schools had unacceptably high radon concentrations.

A 1993 poll found that while 78% of respondents were aware that radon in indoor air is a cancer risk, only 9% had tested their homes. There are two principle types of household radon monitors: *charcoal canister* and *alpha track* devices. Charcoal canisters absorb radon gas for up to a week, and are then sealed and sent to a laboratory where they are analyzed for radon content. Such monitors are not highly accurate, but are an inexpensive, quick way of determining if a home has a radon problem. Alpha-track monitors cost a little more than charcoal canisters and must be in place for at least four weeks to obtain results, but the resulting data is more accurate. ALPHA PARTICLES created during the process of radioactive decay strike a specially treated plastic strip inside the monitor where they leave microscopic marks that can later be counted in a laboratory. Since radon gas is heavier than air, monitors must be placed in the lowest room in the house for accurate results. Doors and windows must be closed as much as possible while the monitors are in use to prevent dilution of the gas. High humidity levels and windy weather can invalidate test results.

Radon and radon daughters can cause lung cancer when inhaled. They can enter the lungs either as a gas or attached to airborne particles. Two of the radon daughters, both isotopes of polonium, are especially hazardous if they become lodged in the lungs because they emit alpha particles. The EPA estimates that between 2,000 and 20,000 deaths from lung cancer occur each year in the United States as the result of exposure to radon, with about three-fourths of these deaths resulting from a combination of exposure to radon and tobacco smoke. Individuals who are regularly exposed to 20 picocuries per liter (pCi/L) of radon have about the same risk of contracting lung cancer as smoking two packs of cigarettes a day (see ENVIRONMENTAL TOBACCO SMOKE). The U.S. Surgeon General and the EPA issued a health advisory in 1986 on the dangers of radon in indoor air, and urged that

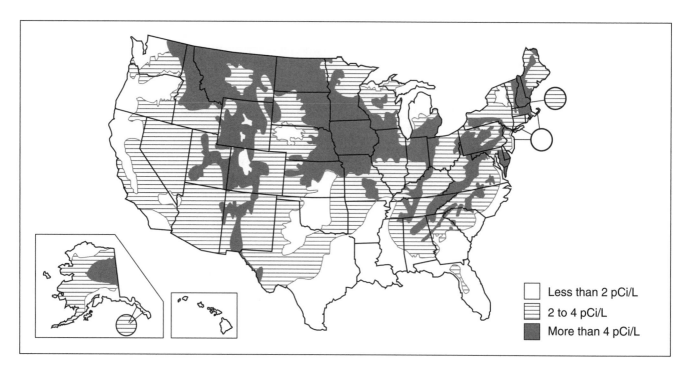

Map of average indoor radon levels (*EPA Journal*)

all homes be tested. A similar advisory was issued soon after for U.S. schools. The U.S. Mines Safety And Health Administration has set the maximum allowable radon exposure for miners at 16 pCi/L. The National Council on Radiation Protection and Measurements advises that a homeowner take corrective action if radon levels in indoor air exceed 8 pCi/L. The EPA's target level of exposure for the general population is 4 pCi/L, and 1.5 pCi/L is the average level found in U.S. homes. The EPA published voluntary radon abatement guidelines during April 1993 that urged builders to seal cracks in the basements of new houses built in about one-third of the country to prevent the INFILTRA-TION of radon. PUBLIC WATER SYSTEMS were required under the SAFE DRINKING WATER ACT to monitor concentrations of radon with a proposed MAXIMUM CONTAMINANT LEVEL of 300 picocuries per liter published by the EPA in 1991. The standard was set aside, however, as the result of controversy about radon's health effects, and a new proposed standard is scheduled for release in FY1995. See also INDOOR RADON ABATEMENT ACT, RADIOACTIVITY. Compare REACTIVE MATERIALS.

Rag Picker individuals who earned a living by removing rags, ferrous metal and other recyclable items from trash cans and municipal dumps and by picking up such items saved for them at residences. Rag picking, though not a glamorous job, provided relatively lucrative work that was performed primarily by those who needed the income the most. Rag pickers were common for more than a century until about the 1940s. The rags that rag-pickers collected were called *thirds and blues*—*thirds* as opposed to firsts, which were new, and seconds, which were used, and *blues* because the rags were blue or light in color, with no blacks or reds allowed. The rags were used by the thriving paper industry in the northeastern United States, and by textile mills in the northeastern U.S. and in England. Rag pickers were paid the equivalent of $350 per ton for rags, comparable to the price now paid for aluminum. The mills produced what was called SHODDY from reprocessed wool rags. The rise of the wood-pulp based paper industry in the 1920s and the WOOL PRODUCTS LABELING ACT in 1939 put an end to the rag picker's trade. See also RECYCLING.

Range Management the management of a piece of land in relation to the rearing of grazing livestock. Increased erosion resulting from overgrazing can lead to SEDIMENTATION and contribute to the EUTROPHICA-TION of lakes and reservoirs. The adoption of grazing techniques that are more sustainable such as limiting the number of animals that graze on a given allotment, rotating grazing areas and keeping cattle in unfenced public grazing land moving to avoid damage to watercourses are techniques that have been used to restore overgrazed rangeland and reduce erosion (see SUSTAINABLE AGRICULTURE).

Rapid Sand Filter a filter that is used to remove small particles and PATHOGENS (down to the size of BACTERIA) from feed water being processed for use as DRINKING WATER. A rapid sand filter consists of a layer of gravel laid over perforated pipe with a layer of sand of uniform size on top the gravel. Feed water is introduced into the top of the filter and is pulled down through the sand and gravel by gravity, sometimes with the assistance of suction from the perforated pipes at the bottom of the filter. Particulates and pathogens are removed from the water being processed by SEDIMENTATION (deposition on sand grains), FLOCCULATION (biological interactions between sand grains that creates masses of that are stuck to one another, and are therefore unable to pass through the available spaces) and *sieving* (particles are left behind when the water being processed passes through a mesh with openings smaller than the particles' diameter). A rapid sand filter is cleaned occasionally to prevent the excessive buildup of particulates and floc inside the filter by backwashing water through the filter at a rate sufficient to agitate the sand particles and wash away the contaminants. Wash water is separated from feed water.

RBMK-1000 Reactor a NUCLEAR REACTOR that uses graphite as a moderator. Most of the reactors operating in the countries making up the former Soviet Union, including those at CHERNOBYL, are of RBMK-1000 design. The reactors are very different from the commercial reactors found in the West. Graphite—in the form of more than 2,000 columnar blocks weighing 1,700 metric tons in the number four Chernobyl reactor—serves as the moderator, a technology that has not been used for commercial reactors since the 1950s in most parts of the world because it is considered too dangerous. Channels through the graphite blocks allow for the circulation of cooling water through the core. In most countries, the only graphite-core reactors still in operation are those that produce FISSION products such as PLUTONIUM and TRITIUM for the production of nuclear weapons. In contrast to the relatively compact bundle of fuel rods and control rods in the core of a modern pressurized water reactor (PWR), the core of a RMBK reactor is spread over a considerable area. The reactor is made up of more than 1,600 zirconium fuel channels. The channels consist of metal tubes 10 meters long and 9 cm in diameter with a 4 millimeter shell made of zirconium and STEEL alloys. The tubes are filled with uranium castings that are enclosed in a zirconium shell and that feature a built-in channel for the passage of water and steam. The uranium in the fuel assemblies is enriched to only about 2% U-235 rather than the normal 3% or 4%. Each channel is essentially a separate nuclear reactor with its own cooling system and separate registers and controls, and heat and steam is generated in each channel. Maintenance is simplified with this arrangement, since individual fuel channels can be removed and replaced without shutting down the entire reactor, and the removal of fission products such as plutonium is thereby made easier. More than 200 neutron-absorbing control rods that run in channels in the graphite blocks regulate the intensity of the nuclear chain-reaction. It takes about 20 seconds for electric motors to move the control rods in the Chernobyl reactors from their fully withdrawn, top position, down to the bottom of the core. (It takes only about one second to accomplish this task in many recently completed reactors.) The temperature of the graphite columns inside an operating RMBK reactor typically reaches 700°C, as opposed to 550°F in a PWR. Another major way in which the Chernobyl reactors differ—one that was to be a major factor in the meltdown—is that steam to drive power turbines is produced inside the core, not in a secondary system as used in a PWR. But perhaps the most striking difference is that there is no containment structure around the reactor core.

Reaction a change in chemical structure that occurs when two or more substances come in contact. Two or more reactive compounds may produce a third, as when HYDROGEN and oxygen combine to form WATER (H_2O), or may interact to produce two new compounds. Heat, pressure or a chemical catalyst may be required to cause a reaction among some compounds. See also CORROSIVE MATERIAL, REACTIVE MATERIAL.

Reactive Material a material that poses an explosion risk. Reactive materials include substances that explode on contact with electric sparks or water or as the result of jarring. Substances that are capable of generating sufficient internal heat to cause spontaneous COMBUSTION are also considered reactive materials. A *reactive liquid* is a liquid that may, under normal temperature and pressure, decompose or spontaneously explode, either by itself or when it comes into contact with water. Such liquids are usually stable, but they are unusually sensitive to small shocks during normal handling. Reactive industrial waste comes primarily from the chemical production and metal finishing industries. See also CORROSIVE MATERIAL, OXIDIZING AGENT. Compare *reaction*, REACTIVITY.

Reactivity the ability of a compound to react chemically with a wide variety of other compounds (see REACTION). Reactivity is one measure of how hazardous a material is (see HAZARDOUS MATERIAL, REAC-

TIVE MATERIAL). CHLORINE is one of the most reactive substances known, and the *transition elements*, which include MERCURY, GOLD, COPPER, CHROMIUM, molybdenium and tungsten, are highly reactive metals. Metallic fumes occur naturally as OXIDES (see OXIDATION, oxidizing agent) because they are chemically reactive and readily combine with free oxygen in the atmosphere. In *destructive distillation*, a complex substance is heated in the absence of oxygen or other reactive compounds to break it down into simpler substances. In the production of PLASTICS, reactive MONOMERS are used as side groups to add rigidity and strength to a POLYMER. See also NITROGEN OXIDES, PARATHION, STYRENE.

Reactor Waste   unusable FISSION products created in a NUCLEAR REACTOR and materials that become contaminated through contact with radioactive decay products or radioactive fuel. A typical 1000-megawatt nuclear plant contains nearly three billion curies of RADIONUCLIDES at the end of its normal fuel cycle. The safe storage and disposal of reactor waste is one of the principal problems facing the nuclear industry (see HIGH-LEVEL WASTE, LOW-LEVEL WASTE, RADIOACTIVE WASTE, RADIOACTIVE WASTE DISPOSAL).

Fuel rods must be replaced after about three years, and the irradiated uranium in spent fuel rods contain most of the radiation found in reactor waste. Used fuel rods from commercial nuclear power plants make up less than 1% of the total volume of all radioactive waste in the United States, but they account for 95% of the RADIOACTIVITY from all U.S. sources—commercial, experimental and military. When a spent fuel rod is removed from the reactor core, it is stored on site in deep tanks filled with water, which acts as coolant and radiation shield. Allowing the fuel rods to cool off in the water tanks for a considerable time before they are transported to a permanent disposal site permits the more short-lived radiation to decay. But until repositories for high-level commercial nuclear waste are built, the temporary on-site storage tanks must serve as long-term storage, and many are nearing capacity at older installations. A typical commercial reactor produces 30 metric tons of irradiated uranium waste per year, with each ton producing 180 million curies of radiation and 1.6 megawatts of heat. Because many of the decay products present in the irradiated fuel have a relatively short half-life, the radiation emitted from the spent fuel declines to about 693,000 curies after one year. After 10,000 years, 470 curies per ton of radiation would remain. Only about 3% of the uranium in fuel rods typically is consumed, and the rest can be reclaimed through reprocessing, and then enriched and used again as a fuel, but this has generally not proven practicable (see REPROCESSING PLANT). U.S. law now forbids the reprocessing of spent fuels, although reprocessing is required in other countries.

A variety of GASEOUS WASTES are generated at a nuclear facility. The water in the fuel rod storage tanks often gets so hot that it boils, so the tanks must include a gas cleansing system. Gases collected from storage tanks include TRITIUM, bromines, iodines, xenon and krypton. Nuclear plants in which air passes through the core emit argon-41, which emits GAMMA RAYS. The ISOTOPE has a HALF-LIFE of only about two hours, so emissions are noticeable only near the plant. The isotope production plants that are found at many large nuclear installations process large quantities of volatile elements such as iodine and tellurium, which necessitates an elaborate ventilation-air cleaning system that generates filters and resins that are generally handled as low-level waste.

LIQUID WASTES are also generated at a nuclear installation. The water in which used fuel rods are stored picks up radioactive decay products and must be continually recycled to keep it from becoming too radioactive. This is usually achieved with a resin-based ION EXCHANGE system. Used resins may be regenerated with acids, alkalis or salts, in which case the regenerants must be disposed of as low-level waste (see LOW-LEVEL WASTE DISPOSAL). If the resins are not regenerated on-site, they are usually disposed of as a solid, powder or slurry. WASTEWATER from plant cleanup and employee showers must be handled as low-level waste. High-level liquid waste is produced when spent nuclear fuel is reprocessed. Most nuclear plants use a lot of water, and are usually located near a large natural body of water that can provide the needed coolant. Although the return water should normally have picked up minimal radioactivity, the fact that it is several degrees warmer than when it entered the plant can lead to problems with THERMAL POLLUTION in susceptible lakes and streams.

Radioactive SOLID WASTES from nuclear plants include glass or ceramic materials that are used to control the activity in high-level liquid wastes; SLUDGES; incinerator ASH; spent filters; and fuel cladding. Worn or failed equipment such as tanks, pipes, valves; used protective clothing and cleanup materials; and—when it comes time to retire a reactor—the containment building, the massive cooling tower and all the associated storage tanks, plumbing, wiring and concrete, all enter the waste stream. The materials carry different levels of radioactive contamination depending on their exposure to radiation and the radionuclides involved. Reactors that use water as a moderator and heat-transfer mechanism in a primary cooling system that is isolated from the secondary system in which steam and power is produced will have fewer heavily contaminated components than one in which the same water that circulates through the reactor core is converted directly to steam to drive a turbine.

The cost of decommissioning a worn-out facility has been a bone of contention between the nuclear industry and environmentalists for decades. As a result, the retirement in 1979 of the world's first commercial power plant, the Department of Energy's 72-megawatt Shippingport Atomic Power Station, which was built on the Ohio River 25 miles northwest of Pittsburgh in 1957, was of special interest. The DOE expected to spend $98.3 million on the project. When completed, according to the DOE, the site would be a grass knoll suitable for any purpose including a school or a playground. By 1991, 58 commercial reactors, most of them relatively small, had been retired after an average life of 16 years. The U.N. International Atomic Energy Agency estimates that an additional 60 reactors, most of them larger power plants will be closed by the year 2000. Dismantling a reactor can produce a larger volume of low-level waste than operating them. A typical commercial thermonuclear power plant will produce about 6,200 cubic meters of low-level waste over a 40-year lifetime, but dismantling the plant can produce 15,480 cubic meters of low-level waste. The decommissioning process also involves the removal of the irradiated fuel stored at most installations and of the tanks in which it is stored.

Reasonably Available Control Technology the level of AIR POLLUTION control required in a NONATTAINMENT AREA, as defined in the CLEAN AIR ACT. The act does not stipulate just what control technology is "reasonably available," although it does say that economic and technological factors must be taken into account. The act calls for efforts to reduce air pollution from *all* air-pollution generators in an area not in compliance with the NATIONAL AMBIENT AIR QUALITY STANDARDS, while no additional cleanup efforts need be undertaken by existing pollution sources. Compare BEST AVAILABLE CONTROL TECHNOLOGY.

Recycling the collection of materials that would otherwise be discarded, and their reprocessing and incorporation in a new product. The ENVIRONMENTAL PROTECTION AGENCY (EPA) estimates that about 20% of the almost 200 million tons of MUNICIPAL SOLID WASTE (MSW) generated each year is recycled, including about 58% of the newspapers, 63% of the ALUMINUM beverage cans, 35% of the GLASS containers, 48% of the STEEL cans and 30% of the PET plastic containers (see POLYETHYLENE TEREPHTHALATE PLASTICS).

Recycling of PRE-CONSUMER WASTE (waste materials that are recycled and used again within the manufac-

turing process) has been well established for decades, but the utilization of POST-CONSUMER WASTE (waste materials that have been used and then recycled to make new products) is a relatively new wrinkle. One fuzzy area in the classification system is magazines that are returned to the printer because they have not been sold. The pulp industry likes them because they are easy to collect and do not have any impurities, but they are officially classified as pre-consumer waste, since they have not been bought and used. But the industry says they should count as post-consumer fiber since they must be put through the same expensive de-inking process as magazines that have been used. Many industrial waste products are recycled as feed stock or process chemicals for another industrial process (see HAZARDOUS WASTE). Although it is more difficult to recycle products after they have been marketed and used, environmental benefits result:

- The volume of garbage dumped is reduced, saving space in LANDFILLS and reducing garbage-removal costs. Recycling is, in other words, a form of SOURCE REDUCTION.
- Natural resources are conserved, and fewer VIRGIN MATERIALS must be harvested through environmentally destructive processes such as mining and logging.
- The energy efficiency of the manufacturing process is, in many cases, increased. It takes only 10% as much energy, for example, to melt down used aluminum cans as it does to produce the same amount of aluminum by processing bauxite ore.

Public support for recycling is strong. A survey conducted by Media General for the Associated Press in 1990 found that 90% of people in communities where curbside recycling is not mandatory favored making it so. Only 600 communities offered curbside pickup programs for recyclables in the late 1980s, but more than 6,000 offered pickup by the mid 1990s. But for many, support of recycling in theory does not extend to the pocketbook. Although almost 30% of respondents in a 1990 poll conducted by the Roper Organization said that they had purchased items because of their recycled-material content, a little more than 30% said they would not pay extra to buy a product that contained recycled materials. The economics of recycling depends on the value of the recycled material. It is easy for a recycling center to make money on aluminum, for example, with as much as $800 a ton paid in some regions. Glass, on the other hand, is inexpensive to produce, and the silica sand it is made from is abundant, so recycling centers have little margin in selling it. NEWSPRINT (see PAPER RECYCLING) is inexpensive to process, but so much is now recycled that a surplus has depressed market prices, forcing some recyclers to give it away or haul it to a landfill. As more newsprint reprocessing facilities are built, and more newspapers switch to recycled PAPER, the situation is expected to improve. Steel is the most recycled material in the United States, thanks primarily to the well-established network of scrap-metal dealers (see METAL RECYCLING). Although community recycling programs often run in the red, if the overall impacts of recycling versus dumping usable materials in a landfill are examined (see RESOURCE DEPLETION COST), the recycling program ultimately comes out as cheaper and more environmentally benign than sending resources to the dump.

A 1993 study by WMX TECHNOLOGIES INC. put the average cost of recycling at $175 a ton and the average cost of the materials recovered at $40 a ton, with homeowners and other recyclers picking up the difference. Curbside recycling, though available to one-third of American houses, is reducing the amount of garbage in U.S. landfills by only 2.5 percent, according to a study by Keep America Beautiful, a Kansas consulting firm that specializes in waste-management issues. The study found that dumping garbage in a landfill is cheapest, costing an average of $87 per ton for a metropolitan area with a population of 500,000 people. COMPOSTING with landfill is $102 per ton; curbside recycling runs the cost up to $104 per ton, and converting to energy and dumping ash in a landfill costs $134 per ton. The report said that curbside recycling typically reduced household garbage by 21% in 1992. It is expected to reach 30% by the end of the decade.

Laws that require the recycling of certain materials and that set optimistic goals for the total percentage of garbage that must be recycled in the future have passed in several states, with California leading the way (see MANDATORY RECYCLING). The results have been impressive. In San Jose, where citywide curbside recycling was introduced in 1989, 49% of all household garbage—primarily paper products, plastics, glass and aluminum—was recycled in 1993, and 28% of all the city's garbage is recycled, San Jose businesses that do not recycle dry materials that don't rot readily must pay a 30% surcharge on the garbage bill. Those that do recycle will pay only for materials ending up in the dump. In 1994 the average San Jose household paid $13.95 a month to have one container of garbage and unlimited recyclables picked up, and recycled an average of 400 pounds of material during that year, or two-thirds more than the year before. Only Jacksonville, Florida, has a higher per capita recycling rate.

Recycling has traditionally required SOURCE SEPARATION—materials to be recycled must be (at least) placed in different containers from garbage destined for the landfill. In many cases, different classes of recyclables such as clear glass, brown glass, different types of plastic, steel and aluminum must each be sorted into a separate container. When recyclables are sent to a MATERIALS RECOVERY FACILITY, however, no source separation is required, because all sorting is undertaken at the facility. A few of the many organizations working with recycling include: Californians Against Waste Foundation, the National Association of Recycling Industries, the NATIONAL RECYCLING COALITION, the National Recycling Congress, the Plastics Recycling Foundation, Recycle America, the Steel Can Recycling Institute, the Institute for Resource Recovery and INSTITUTE OF SCRAP RECYCLING INDUSTRIES. Compare REUSE, WASTE-TO-ENERGY PLANT. See also PLASTICS RECYCLING, RECYCLING MARKETS, REREFINED OIL.

Recycling Markets   buyers of materials that have been collected for RECYCLING—usually manufacturers that use recycled materials to produce new products or fuels. "Money powers recycling as surely as the sun's energy powers the winds; absent the money, and recycling lies becalmed," said William Rathje and Cullen Murphy in *Rubbish*. Businesses and government agencies trying to sell materials that have been collected in municipal recycling programs find out all too frequently that it will cost more to haul the truckloads of bulky materials they have collected to the recycler than they will be paid for the material, or that the buyer is no longer interested in the material at all (see NEWSPRINT). Many such materials end up in a LANDFILL as a result. As Marty Forman, owner of Poly Anna Plastic Products, a small Wisconsin plastics recycling firm, said: "It is not enough to lead people to believe that when they put something in a box at the end of their driveway, they should feel like they've accomplished recycling. Until somebody buys the damn product that was made out of what they put at the end of the driveway, recycling doesn't take place."

Since the revenue received from the sale of recycled materials is often less than anticipated, communities that expect their recycling program to be self-sufficient are often disappointed. The time, effort and investment necessary to actually reprocess materials that have been turned in for recycling and forge them into new materials are much greater than that required to collect them. But if reprocessors of recycled materials and the manufacturers that use them can be assured of a steady supply of raw materials at a reasonable price and a steady market for products that are made from the reprocessed materials, the investments are made. And if the AVOIDED COSTS associated with reducing the flow of solid waste to a landfill or incinerator are taken into account (see SOURCE REDUCTION), recycling starts to look like one of the best bargains available in the waste management field.

The California Integrated Waste Management Act—which requires that at least one quarter of the newsprint used in newspapers published in the state contain at least 40% post-consumer recycled fibers (see POST-CONSUMER WASTE)—is credited with the establishment of more than 400 of the 495 curbside recycling programs that now reach 11 million Californians. A study conducted by Californians Against Waste Foundation (CAWF) found that many profitable business opportunities in recycling, waste reduction and COMPOSTING came about as a result of the passage of the Beverage Container Recycling And Litter Reduction Act of 1986, which required a deposit on GLASS and ALUMINUM beverage containers. The CAWF study identified 1,787 companies operating more than 3,000 facilities across the state that were involved in activities underwritten by the act. Similar regulations calling for the purchase of recycled materials and the recycling of waste materials are being enacted in many states, and with demand for recycled materials becoming more solid as a result, more manufacturers are committing funds to the purchase of the equipment necessary to process recycled materials.

New products designed specifically to use materials collected for recycling are coming on the market to take advantage of the growing supply of recycled raw materials. An agronomy instructor at Penn State University who wanted to provide a consistent market for recycled paper developed a product called "Pennmulch" that is made from 100% post-consumer fiber from newsprint and magazines. George Hamilton, the inventor of Pennmulch, thought that he could develop a solid market for old magazines and similar papers that are seldom recycled because they are difficult and expensive to repulp while offering a better mulch for establishing turf grass than was available at the time. He is now looking at the possibility of using JUNK MAIL to make mulch. Junk mail is not collected for recycling and is not considered marketable waste, because of the difficulty of sorting it. Fifty tons of Pennmulch were produced in 1993, with most sold to turf grass and landscape companies, and the growers were pleased with the results so the manufacturer, Penn Mulch Inc. of Pittsburgh, plans to produce 1,000 tons during 1994. The mulch consists of small green pellets that resemble rabbit food. The pellets have been fortified with FERTILIZER to give the grass a boost after it germinates. The mulch is spread over areas where grass seed has been applied, and the area is watered. The pellets expand when wet, so a thin layer of Pennmulch provides enough ground cover to hold moisture, reduce erosion, stabilize soil temperature and thereby stimulate germination,

according to users. The fact that the mulch is easy to handle and store and does not carry weed seeds are also advantages. Pennmulch can be applied by hand or with a mechanical spreader. See also PAPER, PLASTICS, WOOD.

Red Tide   ocean waters stained red as the result of an algal bloom (see ALGAE). Red tides are caused by a toxic red dinoflagellate that is generally a member of either the *Gymnididium* or *Gonyaulax* genera. A severe red tide can poison fish and shellfish and the birds and humans that eat them. High levels of nutrients and warm water are associated with the algal blooms. The incidence of red tides in Florida has increased along with increases in population and pollution along the coastline. In 1972 a red tide occurred in New England, an area previously thought to be too far north to have red tides. See also EUTROPHICATION, MICROORGANISMS, WATER POLLUTION.

Reduction Plant   a plant in which garbage and animal carcasses are boiled in large vats to extract grease and TANKAGE. A total of 45 reduction plants were started in the United States in the late nineteenth and early twentieth centuries, patterned after plants in Europe (see WARING, GEORGE E., JR.). Reduction plants were backed by the federal government during World War I because they produced substances such as nitric acid and glycerine that were essential to the war effort. Organic wastes were allowed to putrefy as part of the reduction process, and a reduction plant could be strikingly fragrant, especially in hot weather. Lawsuits against the plants were common, and most had shut soon after the end of the war. See also RENDERING PLANT, SWINE FEEDING.

Refillable Bottles   bottles that can be returned to the store and ultimately to the manufacturer to be refilled. Refillable GLASS bottles were common as recently as the 1960s, but are rarely seen today in the United States, having been replaced by the "NO DEPOSIT, NO RETURN" bottle, which in turn has been replaced by the "recyclable" bottle. In most European countries and throughout the Third World, however, refillable bottles are still the norm. Although a can or bottle made from recycled materials (see RECYCLING) is more efficient in terms of energy and resource use (see RESOURCE DEPLETION COST) than one that is made from VIRGIN MATERIALS, a container that is used several times before it is discarded (see REUSE) or recycled is even more efficient. A major revival of the use of refillable bottles in the United States would spell trouble for community recycling centers, however, if it meant that fewer beverages would come in cans, since the money received for ALUMINUM cans is the mainstay of most recycling programs. See also REUSABLE PACKAGING.

Refuse the entire stream of solid waste including dryer materials such as TRASH DEBRIS, WOOD, GLASS, METAL, PLASTICS, PAPER, CONCRETE and other dry solids, as well as wetter material such as ANIMAL WASTES, FECES, FOOD SCRAPS, OFFAL, SEWAGE SLUDGE and YARD WASTE. See GARBAGE.

Refuse Act a statute passed as part of the Rivers and Harbors Act of 1899 that made it illegal to discharge any kind of REFUSE into navigable waters without a permit from the U.S. Army Corps of Engineers. The Refuse Act was widely ignored for decades, but started to be used against industrial polluters in the late 1950s. The Supreme Court ruled in 1960 that the act was valid, but that it did not apply to municipal SEWAGE. A later court decision broadened the definition of "refuse" to include products such as "oils" that were discharged into navigable waterways, and to include other forms of damage to water quality such as THERMAL POLLUTION and LEACHATE from garbage dumps. In 1970, President Nixon asked the Corps to establish a permit process under the authority of the Refuse Act. The resulting Refuse Act Permit Program became the basis of the permit process used in the CLEAN WATER ACT.

Refuse-Derived Fuel a fuel that is derived from solid waste. In a typical refuse-derived fuel (RDF) facility, solid waste enters on a conveyor belt and is fed into a shredder, where it is pulverized by swinging hammers. It then passes under a huge magnet that removes ferrous metals and a blower that removes aluminum. Sand, bits of GLASS and rock and other fine, noncombustible materials are removed by passing the garbage over fine screens. Everything that is left is then passed through another shredder. The papery fluff that remains may be burned on site to create energy or sold to a utility or industrial customer. The processed refuse may also be compressed into pellets of *densified refuse-derived fuel*, which is sold as fuel. A variety of problems have resulted in the failure of many RDF plants. Tough synthetic materials such as twine and pantyhose regularly make it through the sorting process and get tangled up in the shredders, stopping the (dis)assembly line. Explosions caused by pressurized containers and flammable liquids in the waste stream similarly foul the works. A massive explosion in 1984 at an RDF plant in Akron, Ohio, that was processing a mix of sawdust, oil and paint killed three people. It is expensive to run such an installation, and the sale of the recycled materials reclaimed is subject to fickle and competitive RECYCLING MARKETS. Utilities and manufacturers that bought RDF to burn in place of coal frequently complained that the fuel was too wet to burn properly (usually because garbage got soaked by rain before it went to the facility), and many sales contracts were lost as a result. Competition with recycling centers for some of the more combustible portions of the waste stream such as PAPER and PLASTICS, and an inadequate volume of TRASH to operate economically, made the situation for the operators of RDF facilities even more difficult. About a dozen facilities that create RDF for sale and another 20 that burn the fuel on site are now in operation. Operators of RDF facilities have become much more selective about the composition of the refuse that they run through their facilities, and items such as batteries, electronic circuitry and other items likely to contain toxic substances are kept out of the waste stream as much as possible. Many facilities generate electricity and sell it to utilities under the provisions of the PUBLIC UTILITIES REGULATORY POLICY ACT.

Regional Benefits the benefits a project brings to the region in which it is undertaken. The regional benefits of a reforestation project, for example, would include the wages of the tree planters and others directly involved with planting, the money paid to a local nursery for seedlings and the up-and-coming forest and the variety of resources it will provide (see RENEWABLE RESOURCE). The wages of the Forest Service planner who coordinated the project at regional headquarters would not be a regional benefit.

Relic something that is left behind after DECAY, disintegration or disappearance. A relic can be the ruined remains of a vehicle or a building, an ARTIFACT left behind by an ancient culture or the last remaining members of a disappearing race.

Remedial Action an action undertaken to reduce or abate existing POLLUTION. Remedial action for an industrial smokestack emitting illegally high levels of contaminants might include the installation of new pollution-control equipment, and, in some cases, the detoxification of soils in areas immediately downwind from the pollution source.

Rendering Plant an installation where animal bodies are processed or *rendered* to reclaim usable products such as fats that are used in the production of soap, bone meal and dog food. Most rural communities have a rendering plant that will accept the carcasses of farm animals. See REDUCTION PLANT; WARING, GEORGE E., JR.

Renewable Resources resources that will replenish themselves within a relatively short period of time. Natural energy sources such as the sun, the wind, geothermal heat, ocean currents, gravity and falling water are available relatively constantly. A crop grown for food or for *biomass* (corn grown for ETHANOL production, for instance) is renewable once or twice a year, although overproduction can deplete soils. A tree farm that produces fast-growing trees for pulp and PAPER production or for fuel to produce steam or electricity has a crop that is renewable within a decade or less. A forest will provide wood for lumber and heat as long as it is managed on a sustainable basis (see SUSTAINABLE AGRICULTURE, SUSTAINABLE SOCIETY). If overcut, there is at least a decades-long wait until the trees have a chance to regenerate. Even prime-growth trees are renewable, but they may take a century or more to grow, depending on the species and its habitat. In some cases, a forest will not regenerate after cutting, and is therefore no longer a renewable resource. Natural resources such as MINERALS and FOSSIL FUELS, in contrast, are finite and nonrenewable—once consumed they are gone. GROUNDWATER drawn from an AQUIFER that no longer has surface water seeping into it is sometimes called fossil water. Otherwise, water is a renewable resource. GARBAGE, if it is recycled (see RECYCLING, RECYCLING MARKETS), is renewable, although much of its content may be based on nonrenewable minerals and fossil fuels.

A renewable resource may be a resource in many ways. A tree, for example, may be harvested to make lumber, plywood, paper or firewood. Left undisturbed, the tree will provide other kinds of resources including habitat for birds and animals. Its needles or leaves will convert CARBON DIOXIDE and produce oxygen through photosynthesis and produce carbohydrates, and will eventually fall to the ground to help build soil. The shade provided by a tree increases soil moisture and may help a hiker to cool off on a hot day. Its roots hold the soil together and help prevent erosion. Even when the tree dies, it is a resource. While it stands, it will be home to birds, animals and bugs (if it isn't cut for firewood), and when it eventually falls to the ground its trunk and branches will build the forest soil.

The development of renewable resources involves the use of minerals and fossil fuels that are nonrenewable. The construction of a large hydroelectric dam, for instance, will involve huge quantities of CONCRETE and reinforcing STEEL and sophisticated turbines and electronics, the production of which consumes untold nonrenewable energy and materials, not to mention the supertanker load of PETROLEUM energy required to keep all the trucks and other vehicles moving on the job.

California is in a class by itself when it comes to renewables. The state leads the nation in the generation of electricity from every renewable source except hydro. Nonfossil power generation has gone from 5MW in 1979 to almost 10,000 MW today, with a bit more than half that total coming from cogeneration, the rest from renewables. Each of California's three investor owned utilities have at least one approved renewable resource in their plan, and half of that capacity is set aside for renewables. Biomass supplies 3% of California's electricity, the highest rate in the nation. See also RESOURCE DEPLETION COST.

Reprocessing Plant an installation that processes spent fuel rods from nuclear plants producing fuel-grade URANIUM and a variety of radioactive BYPRODUCTS in the process. Only about 3% of the uranium in fuel rods typically is consumed, and the rest can be reclaimed through reprocessing and then enriched and again used as a FUEL. Before reprocessing can proceed, fuel rods are removed from the storage tank in which they are held (see RADIOACTIVE WASTE), their sheath is removed and they are immersed, usually in a strong SOLUTION of nitric acid. Uranium and PLUTONIUM are extracted from the solution into an organic SOLVENT, and the remaining acid solution of fission products must be disposed of as or primary high-level waste (see RADIOACTIVE WASTE DISPOSAL). A considerable quantity of HIGH-LEVEL WASTE is created as part of the reprocessing procedure, and high costs and technical difficulties have put plans for plutonium-fueled fast-breeder reactors (see BREEDER REACTOR) on hold worldwide. As a result, countries with reprocessing plants are having a hard time figuring out how to use the plutonium they produce. Germany, France and Japan are experimenting with mixing plutonium with uranium to form mixed-oxide fuel (or *MOX fuel*) for use in light-water reactors. The International Atomic Energy Agency estimates that only about 20% of France's stockpile of plutonium would be used in MOX fuel, which would leave more than 30 tons of plutonium—enough to build 4,000 atomic bombs—with no use. As such stockpiles become more common, the threat of hijacking or theft becomes more acute.

Rerefined Oil used lubricating oil from engine crankcases that is reprocessed for use as oil or fuel. An estimated 1.4 billion gallons of crankcase oil is changed annually in the United States. Used oil usually is contaminated with HEAVY METALS and PETROLEUM residues that can contaminate water and soil, and many states have prohibited waste oil being disposed of in a LANDFILL. The disposal of oil filters is also a problem, since they are loaded with CONTAMINANTS from the oil they have filtered, and may contain as much as a quart of oil. Recycling programs in many areas will accept used motor oil. In California, a 4-cent tax is levied on each quart of motor oil sold to fund oil recycling programs coordinated by the state's Integrated Waste

Management Board. President Clinton signed an executive order in October 1993 requiring government agencies to start using recycled paper (see PAPER RECYCLING) and to buy recycled motor oil and retread tires for government-owned cars. The order went into effect beginning in 1995.

Quaker State has made an agreement with Interline Resources Inc. to open the first of what may be 30 or more oil recycling plants located throughout North America over the next several years that will make industrial lubricants, GASOLINE and DIESEL fuel from discarded oil. Interline has developed a process that removes contaminants from the oil and reprocesses it for a much lower cost than the original cost of refining crude oil. Many oil recyclers simply filter dirty oil and sell it as a low-grade industrial fuel oil, but contaminants remain, and when the recycled fuel oil is burned, the hazardous materials are emitted into the atmosphere. In addition, the ARSENIC, zinc and LEAD that are often found in used oil are picked up by the recycling plant's filters, which then must be disposed of as a hazardous waste. The Interline process recycles impurities for use in the manufacture of shingles and asphalt.

International Recovery Corporation is a leading recycler of used oil in 16 Southern and Mid-Atlantic states with oil recycling plants located in Delaware, Florida and Louisiana. The company's subsidiary, International Petroleum Corporation, conducts its oil recycling activities. The used oil is converted into light oil, which is used as a fuel in the company's plants and a heavy fuel oil that can be blended to produce several oil grades.

Resistivity   the resistance of a material to the passage of electrical current as measured over a standard distance, usually ohms per centimeter. A material's resistivity is used to determine its vulnerability to *stray-current corrosion* (see CORROSION) when buried in different types of soil. Soils with a high resistivity (such as a well-drained soil consisting of alluvial sand and gravel) have low potential for causing corrosion, while a soil with a low resistivity (such as a poorly drained, clay-rich soil) will have a much higher potential. Metals have a very low resistivity.

Resource Conservation and Recovery Act   a federal law passed in 1976 that addresses the transport, storage, handling, treatment and disposal of solid waste. The Resource Conservation and Recovery Act (RCRA) consists of the SOLID WASTE DISPOSAL ACT of 1965 and a package of amendments. The ENVIRONMENTAL PROTECTION AGENCY (EPA) is charged under the act with developing national standards governing the treatment, handling and disposal of municipal waste and HAZARDOUS WASTE, and is given authority to protect air and water from contamination by hazardous materials. The Office of Solid Waste was established within the EPA by the RCRA to coordinate programs relating to waste disposal. The act calls for tightening the standards governing open dumping that were established in the Solid Waste Disposal Act. The states are responsible for setting and enforcing standards for the materials regulated under the RCRA, and must set standards at least as strict as those established under the act. Congress has called for the EPA to develop standards governing the disposal of used oil, but the EPA has not yet taken action.

The RCRA specifies how hazardous waste is stored, labeled and handled, dictates how facilities that will be handling hazardous materials are designed and requires detailed record keeping of the movement of hazardous materials—from purchase through use and disposal. The EPA must publish and periodically update a list of hazardous materials regulated under the act, and must issue guidelines for their handling and disposal. The agency is required to consider public health and safety, the quality of air and water and aesthetics when establishing standards under the RCRA, and to provide technical assistance and federal funding to the states and municipalities that are setting up hazardous-waste programs and facilities.

Amendments

In the *Hazardous and Solid Waste Amendments of 1984*, Congress required that owners of LEAKING UNDERGROUND STORAGE TANKS take financial responsibility for resulting GROUNDWATER contamination, and that fuels stored underground must be in double-wall, corrosion-resistant tanks with leak-detection systems. Materials defined under the act as hazardous must meet the act's definition of SOLID WASTE—solid particles suspended or in SOLUTION in a liquid and solid PARTICULATES found in containerized gases. A mixture of solid waste and hazardous waste is defined as hazardous waste, but the EPA's rule on mixtures was successfully challenged in court in mid-1992 on procedural grounds, and the agency is considering other ways of regulating such mixtures. Hazardous waste specifically exempt from regulation under the RCRA includes: irrigation return flows (see IRRIGATION WATER, RETURN FLOW), MUNICIPAL SOLID WASTE, coal combustion wastes (see COAL, COMBUSTION), agricultural FERTILIZERS (see AGRICULTURAL WASTES, NUTRIENTS), DRILLING MUDS and BRINES used in oil and gas and geothermal wells, mining wastes (see MINE DRAINAGE, TAILINGS) and DUST from cement kilns (see CONCRETE). WATER POLLUTION regulated under the CLEAN WATER ACT (CWA), and nuclear wastes regulated under the Atomic Energy Act are also exempt. In

1992 the EPA issued the last of the rules mandated in the 1984 amendments to the Resource Conservation and Recovery Act, those governing hazardous wastes disposed of in LANDFILLS and INJECTION WELLS. The new rules called for the treatment of an estimated 30-million tons per year of INDUSTRIAL HAZARDOUS WASTE. Critics of the rules said they were a step back from the RCRA standard of using the BEST AVAILABLE TECHNOLOGY to reduce the material's hazardous qualities. The agency has previously issued rules governing the disposal of SOLVENTS, DIOXIN, CYANIDE, METALS and POLYCHLORINATED BIPHENYLS (PCBs).

Waste Categories

The act is divided into sections called subtitles, and hazardous wastes are regulated under the *subtitle c program*, from generation through use and disposal. Municipal solid waste and INDUSTRIAL SOLID WASTE are managed under the *subtitle d program*, and leaking underground storage tanks are addressed in the *subtitle i program*. A substance must be on one of the three lists of hazardous materials maintained by the EPA to be defined as hazardous waste under the act:

- the *F list* of wastes from nonspecific sources, which includes wastewater treatment sludges from electroplating operations;
- the *K list* of waste from specific sources such as the bottom sediment sludge removed from sedimentation tanks used in the treatment of wastewater from wood preservation processes;
- the *U and P lists* of discarded commercial chemicals, which includes hazardous products that are discarded because they have not been manufactured to specification. The U list includes toxic wastes such as vinyl chloride, while the P list is composed of acutely hazardous materials such as cyanides, and P-list materials are more tightly regulated than other classes of hazardous materials.

The RCRA was up for renewal in 1995 with a good chance of relaxed rather than tightened standards and reduced funding. See also SUPERFUND. More information about the Resource Conservation and Recovery Act can be optained from the RCRA/Superfund hotline—800-424-9346 (382-3000 in Washington DC). See also COMMUNITY RIGHT TO KNOW ACT.

Resource Depletion Cost the overall value of a natural resource, both as a raw material for the manufacture of products and as a part of the natural environment. The resource depletion cost of a tree, for instance, would include both its value as a VIRGIN MATERIAL to be used in the production of lumber or paper and its value as part of the forest ecosystem, which would include its role as habitat, as an erosion-control device and as an aesthetic part of the forest. Resource depletion cost is one of the factors considered when computing AVOIDED COST. For example, part of the cost that would be avoided by recycling newsprint would be the resource depletion cost of cutting a tree to produce paper. Resource depletion allowances, which permit tax deductions by companies that make a product from virgin materials, not only ignore the resource depletion cost involved, but actually give a credit for consuming natural resources. The economics of recycling relatively low-value materials such as plastics and paper are greatly improved when the resource depletion cost avoided by using recycled versus virgin materials is considered. See also ENVIRONMENTAL COST.

Resource Recovery the recovery of some part of the value of a resource that is part of the WASTE STREAM. Resource recovery is most often used in reference to the conversion of liquid or especially SOLID WASTE to energy or to a material that can be used as fuel, although it may also refer to a resource that is recovered through RECYCLING. WASTE-TO-ENERGY PLANTS that burn garbage to produce electricity or steam and hot water, and installations that convert wastes into fuel through processes such as sorting, compaction and PYROLYSIS (see REFUSE-DERIVED FUELS, REREFINED OIL) are common resource-recovery facilities. See also COGENERATION.

Respirable a particle that is small enough to pass into the lungs without being intercepted, but large enough to resist being swept out of the lungs when the breather exhales. The upper respiratory system filters out particles larger than about 10 microns in diameter, and particles smaller than 6 microns are called *respirable suspended particulates* because they are small enough to be inhaled deep into the lungs.

Respirable Suspended Particulate see PARTICULATES

Return Flow WATER that drains off the bottom of irrigated farmland and back into the stream or AQUIFER from which it was drawn. Return flow is often contaminated with SALTS, NUTRIENTS and PESTICIDES, and is a widespread form of DIFFUSE POLLUTION. A certain amount of return flow is necessary to prevent the buildup of salts in irrigated soils. See also IRRIGATION WATER, SALINITY.

Reusable Bottles see REFILLABLE BOTTLES

Reusable Packaging PACKAGING that can be used for other purposes after the product it holds has been removed. Containers can be refilled at many bulk-food stores, but the return of a container to a manufacturer for refilling is an extremely rare (if not extinct) practice in the United States (see REFILLABLE BOTTLES). The majority of reusable packaging was not necessarily designed for REUSE, but is sturdy enough to be reused. Tubs like those used for margarine and yogurt, for instance, are often reused to store anything from food to nuts and bolts. Larger buckets such as those in which paint or drywall compound are packaged can serve as a bucket for years. An increasing number of products are being sold in packaging that was designed for reuse. Plastic zip-lock bags are one example. Public support of RECYCLING, reuse and resource conservation, and manufacturers' belief that "green" product packaging will improve sales, are responsible for the increase in reusable packaging.

Reuse   the use of a previously employed material, either for the function for which it was produced, or for some other purpose. Reuse differs from RECYCLING in that no reprocessing or remanufacture is involved. In terms of energy consumption, POLLUTION and the efficient use of resources (see RESOURCE DEPLETION COST), recycling a material is almost always more efficient than using it once and disposing of it, and reusing it is usually better than recycling it. Reuse also involves maximizing the life of a manufactured product once it is purchased by keeping it in good shape and repairing it if it breaks. By regularly changing a car's oil and thereby extending the life of its engine, both the out-of-pocket expense and the environmental side-effects of the production of yet another car is avoided. Once natural resources and energy have been invested in a product, it makes sense—both financially, and in terms of environmental impact—that it be used until it is consumed.

"Fix it up, wear it out, when it's broken, do without" was a saying popular in the United States during the Second World War, and one that is based on reuse. A manifesto for people wanting to have a light impact on the planet might read: "Don't buy something new that can be purchased second-hand. Keep something until it's worn out, or give or sell it to somebody else so that they can wear it out. Make the purchase of new items with an eye to their durability and repairability. Take good care of things so that they will work better and last longer. Remember that one (hu)man's trash is another's treasure, and give things away rather than throwing them away." Strict adherence to such principals might not be much fun—we do all love our new toys—but if more people considered such factors when making a purchase, natural resources would last longer. Commerce based on PLANNED OBSOLESCENCE

and CONSUMERISM might suffer, but sales of more durable products would thrive, and the ease of achieving an environmentally SUSTAINABLE SOCIETY would be stronger as a result.

Just what value is placed on different natural resources is the critical factor when it comes to computing the environmental cost tied up in a specific product. If PETROLEUM is viewed as a finite resource that is greatly undervalued, products that contain PLASTICS or consume a petrochemical fuel in one way or another will not sound like as good a deal in terms of the environment. If all the pollution caused by oil drilling, transportation, refinement and manufacture into a finished product are viewed as especially serious, petroleum's environmental viability would suffer even more. But the fact is that petroleum is currently cheap and readily available, and that alternatives to the use of GASOLINE and DIESEL fuels or plastics have environmental drawbacks too. Products designed to be used once and discarded fare best when compared to items that must be regularly cleaned or maintained throughout their life—items like an expanded POLY-STYRENE coffee cup or a DISPOSABLE DIAPER. Several studies have reached the conclusion that the "one-way" diaper—although objectionable to some on moral grounds—is at least comparable to the traditional rewashable cotton diaper in terms of energy investment and resource depletion cost (see PLASTICS RECYCLING). The outcome depends, of course, on who's doing the comparing and what resources are seen as most valuable.

Second-hand stores, yard sales, swap meets, auctions and JUNKYARDS all demonstrate the vitality and popularity of the trade in used goods. Items actually recovered from the WASTE STREAM are another avenue of reuse. In Berkeley, California, a business called Urban One salvages BUILDING MATERIALS, furniture, tools and other reusable, resellable items, and markets them at their store. About one-fourth of the merchandise sold in the store comes from the city dump, and the rest is donated by people who are ready to part with the item in question but do not want to pay a dump fee to do so. The store—started by a former college professor—collects and sells old doors and windows, chairs, beds and cabinets, and grossed more than $600,000 in 1989. It is a working example of the popularity and profitability of reuse.

Personally removing material from the waste stream is yet another way in which thrown-away items get reused. From retrieving castaway furniture or appliances from the alley before they are removed by the trash company to bringing back usable things from the dump, some of the more valuable and useful contents of municipal waste are removed before it becomes a permanent part of the landfill. In developing countries, the retrieval of materials from the wastestream is a way of life, as it once was in industrialized nations (see RAG PICKER). Before the days of the strict regulation of landfills, many rural dumps had something similar to the "goody pile" that was a part of the dump in Darby, Montana. Dump patrons disposed of potentially reusable items on the "goody pile," which was up the road a couple of hundred yards from the area where other garbage was disposed of. For many people, a trip to the dump usually involved a quick look at the goody pile to see if anything of interest had shown up. Use of the goody pile was fairly universal, and no stigma was attached to digging through the pile. For the true scavengers in town, the back of the pickup generally had more in it on the trip home than on the trip up.

A junkyard is a bastion of reuse, and most of the parts of a junked automobile that are not sold for reuse are sold to scrap dealers for recycling (see METAL RECYCLING, PYROLYSIS). Tire dealers similarly facilitate reuse by reselling tires that have been traded in for new ones when they are usable and sending worn but functional tire casings in for retreading. Many building materials are not fit for reuse, but there is a ready market for doors, windows, cabinets and other readily removable, reusable building components.

A reusable shopping bag short-circuits the entire process of manufacturing a disposable bag, delivering it to the store and ultimately to the shopper, who either sends it on its way to the landfill or recycles it. A cloth water filter that can be rinsed out and reused saves the expense and environmental costs associated with the production of paper filters in exchange for the inconvenience of having to wash out filters. Reuse works on a more individual basis as well. The durable plastic tubs in which margarine, yogurt and a variety of other products are purchased can last for years as containers for food, nuts and bolts and the like. "Fresh" water is often reused several times. PROCESS CHEMICALS removed from EFFLUENT by pollution control devices are often reused, and waste products from one industrial process are often used as a feed material or process chemical in another (see HAZARDOUS WASTE). See also APPLIANCES, REUSABLE PACKAGING.

Reverse Osmosis a FILTRATION process used to remove CONTAMINANTS from WASTEWATER, DRINKING WATER and PROCESS WATER. It is effective at removing extremely small contaminants such as VOLATILE ORGANIC COMPOUNDS from drinking water and industrial process water. Large-scale reverse-osmosis filters are used to purify industrial wastewater, to process water at a drinking-water treatment plant and to desalinate ocean water for use as drinking water (see DESALINATION). Designated as the BEST AVAILABLE TECHNOLOGY for the removal of BARIUM, CADMIUM, CHROMIUM, MERCURY, NIRTRATES, NITRITES, RADIUM, SELENIUM, synthetic organic chemicals and URANIUM. It is the only filtration technology capable of removing COLLOIDAL SOLIDS from SUSPENSION.

If pure water and a salt solution are on opposite sides of a semipermeable membrane, the pure water will diffuse through the membrane to dilute the salt solution, driven by what is called *osmotic pressure*. When the liquids on the two sides of the membrane have reached equilibrium, the water level will be higher on the salt-water side than on the fresh-water side, a phenomenon called *osmosis*. If the pressure on the saline side of the membrane is raised to a level higher than the osmotic pressure, *reverse osmosis* is initiated, and water moves through the membrane from the salt-water to the fresh-water side of the membrane. Since most of the contaminants in the water are larger than the gaps in the membrane, they are left behind and only fresh water diffuses through the barrier. A variety of membranes are made from POLYMERS for different applications.

Reverse osmosis is energy intensive (although less so than DISTILLATION), and becomes more so as the amount of material suspended and dissolved in the water being processed gets larger and the size of the openings in the membrane becomes smaller. Pressures of 250 to 400 psi are typically required to process brackish water, with 800 to 1200 psi required for seawater. About one third of the seawater introduced into the system passes through the membrane to become fresh water. Several types of reverse-osmosis filters, in addition to those using the original plate configuration in which water is forced through a flat membrane, are now available:

- *Tubular filters* feature a tube-shaped membrane mounted inside a rigid tube or pipe. The membrane is separated from the rigid skin by a permeable membrane. Fresh water drains out of the assembly through holes in the pipe, and is then collected in a permeate channel underneath.
- *Spiral filters* utilize flat membranes like those used in a plate-type filter, but the membrane sheets are rolled up together with permeable membranes into a spiral configuration. The water to be processed is introduced at the center of a module, and flows out through the spiral through one of the permeable layers. Water that has passed through the membrane flows out through the permeable channel for product water and through the holes in the outer rigid skin of the filter for collection.

- *Hollow-fiber filters* introduce saline water on the outside of a fiber membrane and product water flows out of the collector through the hollow of the fiber.

See also HAZARDOUS WASTE TREATMENT, WATER TREATMENT.

Rill Erosion Erosion caused by tiny, ephemeral streams called rills that carry RUNOFF. Thousands of rill channels may form on a hillside during a heavy rainstorm or snow-melt, and their cumulative erosive effects can move a lot of SEDIMENT. Compare SHEET EROSION, SHEETFLOW.

Ringlemann Number see OPACITY

Risk Assessment a process used to assess the public health risk associated with a POLLUTANT. Risk assessment is used by state and federal agencies to prioritize the regulation of individual toxic substances and of individual sources of pollution. By estimating the number of people exposed to a specific toxin or source of pollution, the degree of their exposure, and the TOXICITY of the contaminant in question, a numeric estimation of the degree of health risk involved can be made. This number is usually expressed either as the *odds of death* associated with 70 years of exposure to the pollution source at the maximum point of concentration (as in: "one additional case of cancer for each 100,000 people exposed"), or the *number of deaths* associated with a particular toxin or source of pollution (as in: "an additional three cases of cancer per 1,000,000 people exposed"). It should be noted that risk-assessment numbers reflect *additional* deaths, above and beyond the average rate of occurence. The odds of an American getting cancer in his or her lifetime, for instance are about one in four, or 25,000 in 100,000, so an individual exposed to concentrations of a toxin that cause five deaths for every 100,000 exposures would theoretically increase their overall odds of getting the disease to 25,005 out of 100,000.

A review of available epidemiological and clinical data on the health effects of the toxin in question is the first step of the risk-assessment process. Among the factors considered are the range of individual sensitivities to a given contaminant, the synergistic effect of other substances with which the toxin is likely to be found and the results of laboratory toxicity testing on animals and microorganisms and of epidemiological studies. The health risks resulting from exposure to a toxic substance are highest among the very young, the very old and those whose immune systems are already weakened by disease.

EPA

The ENVIRONMENTAL PROTECTION AGENCY (EPA) is required to compute the costs and benefits of compliance with major environmental regulations, and the assessment of the risk posed by a hazardous material is an essential part of that process. Under the CLEAN AIR ACT (CAA), industrial emissions cannot increase the lifetime risk of cancer by more than one in 10,000. The EPA is required under the act to set standards for air pollutants based on the associated health risks, and to allow an adequate margin of safety in its assessment. An increased health risk of more than one in one million over the lifetime of an individual exposed to emissions of a particular toxin are, as a rule, considered unacceptable by the EPA. The EPA will accept a one in 100,000 risk level in state implementation plans, a number that is in use or under consideration for use as a standard in a variety of other federal programs relating to public health and safety. The risk levels for many toxins is set considerably higher. The standard for POLYCHLORINATED BIPHENYLS (PCBs), for instance is one death per 10,000 exposures.

Risks

The EPA estimates that polluted DRINKING WATER causes as many as 1,000 cancer deaths each year, and that it stunts the development of 240,000 children each year. AIR POLLUTION in American cities accounts for an estimated 60,000 deaths a year, according to an EPA study released 1991. The risk of death resulting from air pollution is greatest among the elderly and people with lung and heart disease. According to a study released by the National Wildlife Federation, an individual who eats one large lake trout from the Great Lakes each week faces an increased cancer risk of one in 10, primarily because of contamination with DDT, dieldrin, CHLORDANE and PCBs. INDOOR AIR POLLUTION is considered to be especially threatening because the average American spends 90% of their time indoors, where the level of contamination is often worse than outdoors. The risk of death from breathing the air in the Los Angeles basin is about one in 10,000, and an estimated 1,600 people die each year as a result of the habit, according to a 1991 study conducted by scientists from several California universities and from the private sector. The study concluded that 12 million people in the area suffered SMOG-related symptoms an average of 17 days each year. The highest risks associated with INDUSTRIAL AIR POLLUTION are from chemical plants in the Gulf states, especially from plants that use butadiene, one of the components of synthetic rubber. A plant owned by Texaco in Port Neches, Texas, increased downwind residents' risks of getting cancer by a factor of one in 10. Emissions from COKE ovens at steel plants expose workers to a cancer risk as high as one in 55, according to the EPA. Risk assessment procedures used by the EPA and other federal agencies are a perennial target of Congressmen looking to weaken environmental regulations (see RISK MANAGEMENT).

Risk Management   the implementation of standards and the establishment of procedures based on the risk associated with public exposure to toxic substances. The availability, efficiency and cost of pollution abatement equipment and supplies, the TOXICITY of the substance in question and the degree of public exposure are all considered in the risk management process. The development of a standard such as a MAXIMUM CONTAMINANT LEVEL (MCL) established in compliance with the SAFE DRINKING WATER ACT is a public process, and review boards, testimony from experts the publication of proposed standards and the evaluation of comments on proposed standards are all part of any such process, so it can take a while.

The ENVIRONMENTAL PROTECTION AGENCY (EPA) developed the risk assessment process during the 1970s as a way of skirting the regulatory quagmire that could result if the agency was forced to debate the merits and dangers associated with each listed substance publicly. By basing management choices on a number that was derived from scientific research, the agency sought to avoid a highly subjective, contentious discussion of the dangers posed by each individual toxin, the cost of controlling it and the resulting public health benefits. Although the numbers that are the end product of the risk assessment process have an aura of solidity and scientific objectivity about them, they are based at least as much on a series of subjective judgements as on solid scientific data. It is a way of keeping public opinion and the political process out of regulatory decisions, one that may be necessary given the volatility of the subject. Risk-management decisions about what constitutes an "acceptable" health risk sound dispassionate because they are based on science, as opposed to the often very passionate opinion of members of the public that even one death or illness among their family and their neighbors is unacceptable.

Unfortunately, the numbers used in the risk-management process as it exists today are really no more than a very well educated guess, given the uncertainties involved in the sampling, testing and analysis of toxic substances. Realistically comparing the risk of exposure and the resulting human health effects of the 30,000 chemicals that are now used in commerce is impossible, due to the lack of basic data on the toxicity and occurence in air and water of the vast majority of the compounds involved. Even less data is available on the cumulative effects of toxins found in the chemical brew that typically exists in the air and water in heavily polluted areas. Two-time EPA director William Ruckelshaus has described risk management as a "kind of pretense" in which everybody imagines that there is sufficient scientific evidence to support regulation decisions that must be made to avoid the "paralysis of protective action" that would result were agencies to wait for more definitive data. Risk assessment ultimately serves as a mechanism to prevent public input on difficult risk management decisions. Risk assessment generally does not take all the EXTERNALITIES relating to the various methods of waste disposal into account. The siting of waste management facilities has generally been presented as the decision of experts rather than a consensus based on a thorough assessment of the risk involved—one of the sources of the NIMBY movement.

To compensate for the lack of basic scientific data on health effects and occurrence of toxins, public health agencies involved with risk assessment try to leave a wide margin of error on the side of public health. If further research indicates a CONTAMINANT is not as toxic as had been assumed, or that it is not found frequently enough to constitute a significant public health threat, an order-of-magnitude error in the assessment of the health threat posed by a toxin is possible. The *upper bound* of a risk assessment is the value arrived at if all the most conservative assumptions about the health effects and occurrence of a substance is assumed to be correct. The billions of dollars spent on the removal of ASBESTOS from schools and public buildings is based on an erroneous reading of the cancer risk posed by commonly encountered airborne fibers of the mineral. A growing number of regulators, toxicologists and pollution-control professionals support a move toward more realistic health effects estimates, at least in part because the current standards leave the entire system of regulation of environmental contaminants open to attack by critics in regulated industries and Congress. The current attack on the cost-benefit formula of the Safe Drinking Water Act, the CLEAN WATER ACT, the RESOURCE CONSERVATION AND RECOVERY ACT and other environmental laws in Congress support such an argument.

Adding to the uncertainty involved in risk management is the fact that toxicology is an uncertain science. Scientists can learn quite a bit about the toxicity of a given substance through tests on animals and microorganisms, but when it comes to the human health effects, they must rely on epidemiological data based on studies of the health effects of the toxin in question on workers who were exposed to high concentrations on a regular basis, and on the residents of communities located downwind from a known source of the contaminant. The range of individual responses to a given toxin is very large, and a dose that would not be noticeable to one individual can be lethal to another. See also AVOIDED COST, CROSS-MEDIA POLLUTION, ENVIRONMENTAL COST.

Rocky Flats   a facility located at the foot of the Rocky Mountain Front, 16 miles northwest of downtown Denver, Colorado, that is the only source of PLUTONIUM triggers for nuclear bombs. Rocky Flats started operation in 1953. In 1988, the Department of Energy (DOE) closed down the main plutonium reprocessing building at Rocky Flats for repeated safety violations, and called Rocky the most environmentally hazardous of the nuclear-weapons-production sites because of contamination of GROUNDWATER and threats to nearby sources of municipal DRINKING WATER. In 1989, 75 agents from the Federal Bureau of Investigation and the ENVIRONMENTAL PROTECTION AGENCY raided the plant after FBI officials learned of illegal handling, treatment and disposal of HAZARDOUS WASTE at Rocky. Agents had evidence that officials of Rockwell International Corp. and the DOE had lied about the situation to cover it up. Rockwell managed the facility for the DOE from 1975 through 1989. When the FBI got wind of the coverup, an investigation with the code name Operation Desert Glow that involved aerial and ground surveillance of the installation was initiated. Rockwell threatened in September, 1989, to close Rocky Flats if the company was not given immunity from prosecution for the illegal dumping. The DOE responded by freezing $5 million in payments to Rockwell and terminating the company's contract to operate the installation. The city of Broomfield was later forced to abandon its water supply in when it was found to contain TRITIUM that had leaked from illegal waste dumps at Rocky Flats.

The day after the raid the Colorado Hazardous Materials and Waste Management Division publicly charged the DOE and Rockwell with improper handling of nuclear waste at the Rocky Flats, and said that hazardous waste had been stored in leaking drums on the site and that groundwater had not been adequately monitored for contamination. Rockwell was fined $18.5 million in federal court on ten counts of illegal disposal of nuclear waste in 1992. The fine was at the time the second largest fine ever assessed for environmental pollution, with the $125 million penalty imposed on Exxon for the 1989 Valdez oil spill a solid first—see OCEAN DISPOSAL.

Rodenticide   a PESTICIDE designed to kill rodents. Most rodenticides are arsenical compounds. Zinc phosphate and other zinc compounds and CHLORINE and THALLIUM compounds are also employed. Rodenticides

tend to be extremely poisonous to other mammals including humans, so extreme caution must be exercised in their use. See also BACTERICIDE, FEDERAL INSECTICIDE, FUNGICIDE AND RODENTICIDE ACT, INSECTICIDE.

Rotating Biological Contactor   a wastewater treatment device used in the SECONDARY TREATMENT of SEWAGE and in the treatment of LEACHATE (see LEACHATE TREATMENT SYSTEM, SEWAGE TREATMENT). A rotating biological contactor is composed of a central shaft on which closely spaced *bio-disks* have been mounted. The disks are covered with zoogloeal, the same kind of gelatinous mass of AEROBIC BACTERIA, ALGAE, fungi, protozoa, rotifera and nematoda (worms) found in a TRICKLING FILTER. The cylindrical assemble is slowly rotated, with about 40% of each disk under water at any given time. When the microbial mass is exposed to the air, it becomes charged with oxygen that will fuel the further oxidation of organics when the disk is again under water. The motion of the disks and the current of the wastewater scrubs away accumulated biological film, and wastewater flow carries the slough off to a secondary settling tank for removal.

Rubber   a white or colorless POLYMER that is used in the production of a variety of products where its elasticity, water-repellant qualities and resistance to the flow of electricity and heat are important. *Pure crude rubber* is a white or colorless HYDROCARBON composed of the MONOMER isoprene (C_5H_8). Rubbers are used in the production of hoses and tubing, TIRES, waterproof clothing, diving gear, seals and liners for tanks holding reactive liquids, electrical equipment, shock and vibration absorbers, balloons, balls, cushions and a variety of hard-rubber products such as bumpers and furniture. Natural rubber is found in colloidal suspension (see COLLOIDAL SOLIDS) in *latex*, a white fluid that is found in many plants. *Synthetic rubbers*—also called *elastomers*—are produced through the polymerization of isoprene and other monomers, and are essentially the same as natural rubber in their chemical and physical characteristics.

Natives of South America and southern North America were using rubber balls in a court ball game when the first European explorers came to the Americas. They also used natural rubber to make waterproof shoes, and to waterproof the fabrics used in coats. The discovery by the British chemist Joseph Priestly in 1770 that pencil marks could be erased by rubbing over them with "rubber" is where the substance got its European name. The Scottish chemist Charles Macintosh established a plant in Glasgow in 1823 to manufacture waterproof cloth and rain gear that still bear his name. The American manufacturer

Chapman Mitchell developed the acid-reclamation process for reclaiming scrap rubber in 1877. Sulfuric acid was used to break down the fabric in the rubber, and the resulting mixture was heated and blended with crude rubber before being incorporated into a new product. Most reclaimed rubber today is obtained by heating recycled rubber and scraps for 12 to 20 hours in the presence of alkali.

About 90% of world natural-rubber production comes from one of two "rubber trees" that are currently tapped to produce natural rubber: *Havea brasiliensis* and related species, which were the source of the rubber used by South American natives, and *H. brasiliensis*, which is cultivated in Indonesia, the Malay Penninsula and Sri Lanka. The vast majority of the world's supply of natural rubber comes from plantations in southeast Asia. Sheets and slabs of crude rubber are shipped to rubber processing plants for further treatment. In the plant natural rubber is cleaned, heated and mixed with varying amounts of scrap and recycled rubber.

Synthetic rubbers are produced by polymerization (see PLASTICS) of isoprene and other monomers. Before World War II, synthetic rubber was expensive, and was used only in applications where special properties were required. Synthetic rubber came into widespread use as the result of a research blitz during the war, and a $700 million wartime effort to build synthetic-rubber plants to produce rubber for TIRES, raincoats and countless other strategic items. The plants were producing 1 million tons of high-quality rubber at a price comparable to that of natural rubber by 1952. The sale of the 29 government plants to private industry was authorized in 1953.

Vulcanized rubber, which is created by heating natural rubber with SULFUR and traces of SELENIUM and tellurium to between 120° and 160°C (248° to 320°F), has more strength and elasticity, has greater tolerance for changes in temperature, becomes impermeable to gases and is more resistant to the effects of heat, abrasion, chemical reaction and electricity. Unvulcanized rubber is used in the production of glues, cements, electrical tape and crepe rubber.

Widely used synthetic rubbers include:

- *Neoprene*, a polymer of the monomer *chloroprene,* is resistant to heat and the action of petrochemicals, and is used inside the hose that delivers gasoline and as an insulator.
- *Buna rubbers* are produced by copolymerization—the polymerization of two monomers, which are called comonomers. The comonomers used to produce buna rubbers are *butadiene* and *natrium*. A buna rubber call *Government Rubber-Styrene* (*Gr-S*), which is a copolymer of butadiene and styrene, was the desig-

nated general-purpose rubber for the U.S. war effort, and is still the principal synthetic rubber produced. *Cold Gr-S* rubber, in which copolymerization occurs at 41°F, is longer-wearing and has other advantages over *regular Gr-S*, which is produced at 122°F.

- *Butyl rubber* is a product of the copolymerization of butadiene or *isoprene* and *isobutylene*. It is not as resilient as natural rubber and is difficult to vulcanize. Butyl rubber is resistant to OXIDATION and the action of corrosive chemicals, and is used in caulking and corrosion-resistant liners. Its low permeability to gases has led to the widespread use of butyl rubber in inner tubes and air bladders.
- *Foam rubber* is produced directly from natural or synthetic latex by adding emulsfying agents and whipping the mixture to a froth and then pouring it into molds where it is vulcanized by heating.

Rubbish   all REFUSE plus construction and demolition debris. See also ANIMAL WASTE FECES, FOOD SCRAPS, GARBAGE, OFFAL, TRASH, YARD WASTE.

Runoff   PRECIPITATION that runs off the surface of the earth into lakes and streams and eventually to the oceans. Channeled runoff is part of stream flow, while overland flow is water running across the land's surface, either in broad sheets called SHEETFLOW, or in tiny, wandering streams called rills. Approximately one-sixth to one-third of the precipitation that falls on the surface of an area is swept away by runoff. The rest either evaporates to become atmospheric water vapor, or soaks into the ground to recharge aquifers. The amount of water that penetrates a soil is a function of the soil's INFILTRATION CAPACITY. Water that infiltrates the soil will eventually become runoff when it enters a surface stream, but it may be delayed as much as several million years if it first enters an AQUIFER. Overland flow occurs only during or directly after a rainstorm and during periods when snow cover is melting. Heavy surface runoff can lead to erosion and the buildup of silt in streams.

Runoff can contaminate surface water with PESTICIDES, NITRATES from FERTILIZER, PETROLEUM products and other toxins (see DIFFUSE POLLUTION). Heavy rain during periods when soil is tilled can lead to excessive runoff and resulting erosion. The percentage of precipitation that becomes overland flow depends on the PERMEABILITY of the surface on which it falls, with almost all the moisture remaining on top of a rocky or frozen surface and very little running off a porous soil (see POROSITY) that is well covered with vegetation and mulch. Rainfall striking the bare ground surface typical of more arid regions compacts the soil and is more likely to become surface runoff. If rainfall is heavy enough, even the infiltration capacity of the most permeable soil will be exceeded, and surface water-flow will result, while lighter rainfall is more likely to soak in. Precipitation that falls directly onto lakes and streams is called *channel precipitation*. *Interflow* is water flowing through the soil, some of which will become part of a surface stream rather than an aquifer. *Baseflow* is water that recharges lakes and streams from springs and seeps from aquifers. See also PERCOLATION.

Rural Clean Water Program   a federal program that provides grants to reduce DIFFUSE POLLUTION from agricultural sources. The program, which was authorized as part of the agricultural appropriations bills of 1980 and 1981, provides technical assistance and up to 75% of the cost of making the improvements. The Agricultural Conservation and Stabilization Service administers the Rural Clean Water Program, and the Soil Conservation Service provides technical assistance. See also CLEAN WATER ACT, RUNOFF.

S

Safe Drinking Water Act federal law that regulates the quality of DRINKING WATER supplied by PUBLIC WATER SYSTEMS. The Safe Drinking Water Act (SDWA) became law in 1974, marking the first enforceable national drinking water standards, other than limited sanctions that applied only to interstate carriers of water. Before that, each state set its own standards. The act applies to public water systems—i.e., water systems with 15 or more service connections, or that deliver water to 25 or more people, 60 or more days each year.

The 1974 act required the ENVIRONMENTAL PROTECTION AGENCY (EPA) to institute National Interim Primary Drinking Water Regulations (NIPDWRs) by March 1975, and Revised Primary Drinking Water Regulations (RPDWRs) by September 1977. NIPDWRs and RPDWRs were renamed NATIONAL PRIMARY DRINKING WATER REGULATIONS (NPDWRs) in the 1986 amendments to the act. The EPA is also required to maintain a drinking-water priority list composed of substances that are a potential public health threat, and to require monitoring for the substances to gather data on their occurrence. Substances on the list may in the future be regulated under the Safe Drinking Water Act. Amendments to the SDWA were passed in 1977 and 1986.

Regulations

Primary regulations specify how clean drinking water must be and specify MAXIMUM CONTAMINANT LEVELS (MCLs) for specific toxins. *Secondary standards*—also known as *secondary maximum contaminant levels* (SMCLs)—are established for nontoxic materials to control factors such as the taste, color and odor of drinking water (see NATIONAL SECONDARY DRINKING WATER REGULATIONS). State governments are given *primacy*—called *primary enforcement responsibility*—in the enforcement of rules relating to drinking water, but state standards must be at least as strict as those specified in the act. A study by the Government

Accounting Office found that the states are generally doing a very poor job of monitoring compliance with the Safe Drinking Water Act. The EPA and state agencies with primacy have taken action in less than 9% of the 20,000 known persistent violations of drinking-water standards by public water systems in the early 1990s, imposing fines in only 41 cases, according to EPA figures.

Amendments

The 1986 amendments to the SDWA included deadlines for the EPA that were intended to speed the act's implementation. The EPA was required to establish standards for a total of 83 contaminants within three years. The amendments required DISINFECTION for all public drinking-water systems, and the FILTRATION of surface water to guard against GIARDIA cysts and other MICROORGANISMS. The amendments also ban the use of LEAD pipe and lead solder in public water systems and household plumbing, and require water utilities to notify the public of MCL violations, failure to comply with a prescribed treatment technique or to perform required monitoring, failure to comply with test procedures defined in the NPDWR, the issuance of a variance or exemption and failure to meet any deadlines required as a condition of the issuance of the variance or exemption. The public notification statute and even the water-testing requirements are often ignored by public water systems and state water-quality agencies.

Variances & Exemptions

State water-quality agencies can give *variances* to water systems that cannot comply with an MCL because of the nature of their water source, provided the utility has already installed BEST AVAILABLE TECHNOLOGY (BAT) and is still in violation of the standard. The variance can be granted only if the state ascertains that it will not represent an unreasonable threat to public health. *Exemptions* can also be granted to water utilities in instances where there are "compelling factors" (usually spelled "d-o-l-l-a-r-s") that prevent the meeting of an MCL, and if the exemption is not found to be a substantial public-health risk. The principal difference between a variance and an exemption is that is not necessary for the water utility to install BAT to qualify for an exemption. The state may impose conditions such as the use of point-of-use water-purification equipment or bottled water when granting an exemption. Systems with fewer than 500 service connections are eligible for renewable exemptions that are granted every two years, if they can show that compelling factors are preventing the action and that public health will not be compromised.

Politics

Volatile politics are at least as much a part of the SDWA as are volatile organic compounds. There have been lawsuits, technical challenges and lobbying every step of the way. The EPA is sued regularly because the pace of implementation is regarded as either too slow or too fast. One of the chief points of controversy is the EPA's RISK ASSESSMENT process, which many water providers say is an inaccurate and inflexible tool that forces water companies and the customers to pay for monitoring and purification equipment that is not necessary. Another difficulty is that there is no provision in the act that allows the EPA to refrain from requiring public water systems to monitor for a regulated contaminant, even if there is virtually no chance that the toxin could be found in water sources in a particular area.

The Safe Drinking Water Act was up for renewal in 1995 and the chances of its being watered down appeared good. Bipartisan support of a strong drinking water law developed in 1996, however, disrailing the attempt to weaken the act. The act is characterized as more of a problem than a help by some municipalities and members of the drinking water industry, who call it an unfunded mandate. Congressional debate over the renewal has included many—often inaccurate—anecdotes about the Safe Drinking Water Act. Rep. Michael Bilirakis, R-Fla., said during House debate on the renewal of the act that "the drinking water act currently limits arsenic levels in drinking water to no more than two or three parts per billion. However, a regular portion of shrimp typically served in a restaurant contains around 30 parts per billion." In fact, the MCL for arsenic is 50 parts per billion (ppb), and the arsenic compounds found in shrimp and other seafood are organic, while much-more-toxic inorganic arsenic compounds are predominant in water. Revisions of the risk assessment and cost-benefit ration aspects of the legislation are likely.

The public generally supports the strong regulation of drinking water, with more than half the respondents to a 1985 survey by the AMERICAN WATER WORKS ASSOCIATION indicating that they would be willing to pay to have even a contaminant that posed a very small health risk removed from their drinking water.

For information about drinking water, call the EPA Safe Drinking-Water Hotline: 800-426-4791 or 202-382-5533.

Salinity   the amount of SALT dissolved in a fluid. Surface water often contains a thousand parts per million of salt or more, and the Great Salt Lake has a 12% salt content (120,000 parts per million). A little more than a quarter pound of salt is found in each gallon of ocean "salt water." DRINKING WATER containing several thousand ppm dissolved minerals is consumed in many parts of the world with no noticeable ill effect. The majority of the minerals found in water are salts. Water softeners replace the calcium and magnesium ions that make water hard with sodium ions (see ION EXCHANGE). Industrial boiler water and PROCESS WATER in the semiconductor industry can contain no more than about 1 ppm salt, and pressurized water reactors, no more than about 10 parts per billion. Water used for cooling thermal power plants can contain up 35,000 ppm.

Increases in the salinity of streams used for irrigation are largely caused by irrigation return water (see IRRIGATION WATER, RETURN FLOW). Naturally occurring salts are common in arid soils, and when they are irrigated, the water picks up some of the salts (see SELENIUM). RUNOFF from mines, industrial emissions and geothermal BRINES can also contribute to a stream's salinity. When salty water is used for irrigation, salt crystals build up around plants' roots, making it increasingly difficult for them to get enough water. Stunted growth or death of the plant can result. To avoid such damage, the salt content of irrigation water should not be more than about 1200 ppm. Ruminating animals can tolerate water containing up to 12,000 ppm salts.

The Colorado River carries six times its natural salt level because of its heavy use for irrigation. Salt-rich soils in some of the river's tributaries give the river's water a high level of natural salinity, and the development of farms in such areas increased the salt load carried by runoff. Forty years of irrigation raised the salinity of much of the land in California's Imperial Valley to the point that it would not support even the desert vegetation that once grew there. The buildup of salts in the Mexicali Valley, Mexico's most productive agricultural region, has seriously reduced the region's productivity.

Groundwater overdraft can lead to *saline intrusion*, which results when a WATER TABLE in an AQUIFER near an ocean or other source of saltwater falls to the point that salt water can move in to replace it. Aquifers that contain highly saline water (with more than 10,000 ppm total dissolved solids) are considered to be ideal for DEEPWELL DISPOSAL, because their high degree of salinity generally indicates that they are isolated from other aquifers.

Salmonella   a genus of pathogenic BACTERIA (see PATHOGEN) that cause typhoid fever and certain kinds of food poisoning. Of the 10 known basic kinds of salmonella, seven are dangerous to humans. Between 30,000 and 40,000 cases are reported in the United States each year, with about half the cases of *salmonellosis* associated with hospitals. Symptoms of salmonellosis, food poisoning caused by the bacterium, *S. enteritidis*, can manifest in as little as eight hours. The bacterium is spread in food and water that have been contaminated, usually associated with FECES and with the flesh of chickens, ducks, turkeys, pigs, rodents, turtles and sometimes cattle. Salmonella bacteria are *motile*—that is, they can move around in a fluid on their own power. Their METABOLISM produces hydrogen sulfide and a variety of antigens. Salmonella can be identified by growing a culture and through microscopic analysis. The numerous COLIFORM BACTERIA that can be found in a normal stool sample make identification difficult, and plates are often treated to inhibit the coliforms and make the salmonella stand out. About 1,400 varieties of salmonella bacteria have been identified.

Salt   a COMPOUND composed of two or more positive and negative IONS rather than complete atoms (although complex salts may include complete atoms and ions). Most salts are formed by the union of a METAL with a nonmetal, as in sodium chloride, or by the neutralization of an ACID with a BASE. (In fact, a base is defined as a substance capable of reacting with an acid to form a salt.) Acids and bases may combine to form *normal salts* (in which neutralization is complete and there are no left over hydrogen (H+) or hydroxide (OH– ions), *acid salts* (in which some hydrogen ions remain) or *basic salts* (in which some hydroxide ions remain). Organic salts, which may be normal, acidic or basic, contain organic ions (see ORGANIC COMPOUND). NITRATES occur as either a salt or an ESTER. Soluble salts dissolve in water or other liquid salts and form electrolytes (see CORROSION, FUEL CELL) on dissolution. Salts are among the most common of the naturally occurring compounds, and they play a role in the metabolism of all living things. Blood and the fluid found inside living cells are electrolytes that are based on common table salt, sodium chloride. Maintaining the correct level of salts in bodily fluids (called the *electrolytic balance*) is essential for health. Heat cramps in muscles can result when salt levels in bodily fluids get too low as the result of excessive sweating. Athletes and workers in very hot climates must sometimes take salt tablets to keep the level of salt in their bloodstream high enough. Those subject to high blood pressure must restrict the level of salts in their diet.

Salt is obtained commercially by evaporating salt water and from mines in underground salt deposits. U.S. companies produced more than 38 million metric tons of salt in 1993. Salts are used to preserve meats, in dyeing and in the production of GLASS and soap. Because they are transparent to infrared radiation, salt crystals are used to make the lenses of instruments that detect infrared energy. An estimated 10 million tons of salts are put on U.S. roadways to melt ice and snow each winter, resulting in CORROSION of metal parts on cars, bridges and building entryways and increased salt levels in RUNOFF. Commercial CHLORINE is processed from salts.

Salts are a major source of WATER POLLUTION (see GROUNDWATER POLLUTION, SALINITY). Salts get into surface water primarily from runoff and RETURN FLOWS from agricultural land. See also HALIDE.

Sanitary Landfill   see LANDFILL

Savannah River   a 315-square-mile federal reservation on the Savannah River near Aiken, South Carolina that is a federal nuclear-weapons production facility. Like the HANFORD NUCLEAR RESERVATION in Washington, the Savannah River site is a major repository of RADIOACTIVE WASTES associated with the production of nuclear weapons. Many of the wastes that have been buried in underground tanks have leaked out and contaminated groundwater (see GROUNDWATER POLLUTION, LEAKING UNDERGROUND STORAGE TANKS). The Savannah River nuclear plant is the only U.S. source of TRITIUM, an ISOTOPE of HYDROGEN that is used to increase the yield of nuclear weapons. Tritium decays at a rate of 5.5% per year, so must occasionally be replenished in nuclear warheads to maintain their potential yield. Westinghouse Savannah River operates the plant. Three reactors at the site were shut down in 1988 because of safety, mechanical and management problems. Construction and operator training costing $1 billion were undertaken before one of the reactors was restarted in 1991. A leak of tritium from the plant later that year forced residents of Beaufort County, downstream from the plant, to switch to an alternate water supply for 10 days. About 150 gallons of water containing an estimated 6,000 curies of radiation were released. The resulting contamination was well below federal standards for RADIONUCLIDES in DRINKING WATER, but public outrage at the release was widespread.

Scat animal FECES. Scat is a key element in the cycling of plant nutrients back to the soil. Scatology is the study of animal droppings, which can furnish information about the eating habits, health and movement of animals. ZOODOO is the composted manure of zoo animals. Earrings made of dried, lacquered moose scat are now available from a company in Maine. Only winter droppings are used because they are composed primarily of wood, reflecting the ruminants' winter diet of twigs and bark. See ANIMAL WASTES.

Scavenger a life form that feeds on dead organic matter. A scavenger may have a vegetarian diet, as does an earthworm, or may be a carnivore, such as a wolf. Scavengers and decay processes play an essential role in breaking down and reducing the volume of dead organic matter, and releasing its nutrients in the process.

Scow Trimmers workers on garbage scows (flat-bottomed beats) who "skimmed" through the garbage and removed recyclable and reusable items before dumping the remaining refuse at sea. In the early days of municipalities such as New York City, garbage scow crews were encouraged to sort through the trash, and were allowed to keep whatever money they earned selling the recovered items. As SOURCE SEPARATION became popular in the era before World War I, the practice died out. See also RAG PICKER, RECYCLING, SHODDY.

Scrubber an AIR POLLUTION abatement system in which contaminants are removed from FLUE GAS through the use of water and other liquids. Scrubbers—also called *stack scrubbers* or *wet scrubbers*—absorb up to 95% of the PARTICULATES and soluble gases from industrial smokestacks (see ABSORPTION), and are one of the most efficient ways of cleaning up air pollution from both STATIONARY SOURCES and POINT SOURCES. Usable materials such as GYPSUM are produced as a byproduct of the scrubbing process. Drawbacks of scrubbers include the high cost of installation, maintenance and operation; the fact that improper disposal of polluted scrubber water can lead to WATER POLLUTION (see GROUNDWATER POLLUTION); and the highly visible vapor PLUME the humid exhaust sometimes causes when it emerges from stack, which makes an installation look like it is producing more smoke even though emissions are being substantially reduced by the scrubber.

There are several types of scrubbers:

- A *packed tower scrubber* is a column filled with rocks or other materials through which water is trickled, while the air or gas to be purified passes through the tower from bottom to top. Impurities are absorbed by the water as the gas moves up through the packing material. A chemical solution, organic solvents, oils and emulsifiers can also be used.

- A *plate tower scrubber* is a cylinder that contains perforated horizontal plates, called bubble caps or sieves. The gas to be cleaned is introduced at the bottom, as it is in a packed tower, and makes contact with the absorbing liquid primarily as it passes through the plates, rather than continuously as it would in a packed tower.

- In a *spray chamber scrubber*, water is sprayed into the top of a cylindrical chamber and then runs down the vessel's walls. Contaminated air is generally introduced at the bottom of the chamber in such a way that it flows in a spiral pattern toward the top of the spray chamber (see CYCLONE), losing impurities to absorption as it goes. Spray chambers rarely become plugged, as can happen with packed towers and plate towers, but they are relatively inefficient.

- In a *cyclone scrubber*, water is sprayed into flue gases, and the resulting mixture is injected at high velocity from angled jets into a circular chamber, creating a vortex inside the chamber. The resulting turbulence thoroughly mixes exhaust gases and water droplets, maximizing contact and absorption of pollutants by the water. Centrifugal force throws both the polluted droplets of scrubber water and particulates to the outer walls of the scrubber chamber, where it is collected.

- In a *venturi scrubber*, a mixture of contaminated air is forced through a venturi (a constriction similar to the narrow middle of an hourglass), at high velocity, and water is sprayed into the passing gases at the throat of the venturi. The turbulence and increased velocity of the gases caused by the constriction at the venturi thoroughly mixes pollutants and the water spray, and contaminants are removed by further processing (see WATER TREATMENT).

- In an *impingement plate scrubber*, flue gases are forced at high velocity through small holes in a plate over which a sheet of water is flowing.

- In an *acid-gas scrubber*, a mixture of lime and water is sprayed into acidic flue gas to form calcium salts including gypsum, and thereby reduce the ACIDITY of exhaust gases.

ELECTROSTATIC PRECIPITATORS and other pollution-control devices that remove pollutants without the use of water are sometimes called "dry scrubbers."

Precipitators are sometimes used to remove particulates bonded to water droplets. Scrubber water sprayed into the stacks of COAL-burning power plants and garbage incinerators is often mixed with powdered limestone. SULFUR DIOXIDE reacts with the limestone in the presence of water to form gypsum, which precipitates out of the air and is collected. See also AIR POLLUTION TREATMENT.

Secchi Disk   a circular METAL or PLASTIC disc that is used to visually estimate the transparency of water (see TURBIDITY). The disc, which is 12 inches in diameter, is divided into four quarter-circles that are painted alternately black and white. It is attached to the end of a weighted, graduated line, and is lowered from a boat or dock down into the water until it is no longer visible. The depth at which the disc disappears is noted by reading the depth marking on the line. The disc is then slowly pulled up until it just reappears, and this depth is also noted. The average of the two depths—the *secchi depth* or *depth of visibility*—is a gauge of the depth to which light will penetrate through a body of water and of its trophic state (see TROPHIC CLASSIFICATION SYSTEM). The secchi depth of an EUTROPHIC body of water can be less than a foot, while it can be more than 150 feet in an *oligotrophic* body of water.

Secondary Recycling   see POST-CONSUMER WASTE

Secondary Treatment   processes used after PRIMARY TREATMENT in a sewage treatment or hazardous waste treatment facility. Secondary treatment relies primarily on MICROORGANISMS, mostly AEROBIC BACTERIA, to convert and metabolize dissolved and colloidal materials responsible for most of the wastewater's remaining BIOCHEMICAL OXYGEN DEMAND (see BIOLOGICAL TREATMENT, METABOLISM). Microbial growth produces particles large enough to be removed through SEDIMENTATION and FILTRATION, which also reduce TURBIDITY. Care is taken to maintain an environment ideal for microbial growth, and the resulting microscopic population explosion creates clumps of organic floc (see SEWAGE), which is removed from wastewater in final settling tanks. Tertiary treatment—also called *advanced treatment*—removes nutrients and traces of organic materials not removed by biological treatment. WASTEWATER is filtered through fine screens with openings less than 0.003 in (0.075 mm) if turbidity may be a problem in the waters receiving the EFFLUENT from the sewage plant. After secondary treatment, wastewater is sometimes used to irrigate municipal parks, golf courses or farmland. An increasing amount of treated sewage effluent is being subjected to further treatment (see WATER TREATMENT), and is purified to the point

that it can be used as DRINKING WATER (see WASTEWATER RECYCLING).

Secure Landfill   a LANDFILL that meets standards for the disposal of HAZARDOUS WASTE (see HAZARDOUS WASTE DISPOSAL). A secure landfill must be sited in an area with impermeable clay or silt soils, and can not be located in the recharge zone of a vulnerable AQUIFER (see AQUIFER SENSITIVITY). Wastes are treated to minimize their hazardous qualities, stabilized to reduce their solubility and cast into tough, monolithic blocks before disposal. A secure landfill must have one or more impermeable liners composed of clay and/or plastic under the entire disposal site, an impermeable cap over each disposal cell and lift, a LEACHATE collection and treatment system (see LEACHATE COLLECTION SYSTEM, LEACHATE TREATMENT SYSTEM) and a system of MONITORING WELLS where leachate would be detected if it makes its way out of the landfill. Some secure landfills exceed these standards with a double impermeable liner covered with an impermeable layer of clay covered, in turn, with yet another plastic liner and a final layer of clay. A leachate collection system is installed under each plastic liner, and groundwater, leachate and air are regularly monitored for contamination.

Sediment   loose material transported by water and eventually deposited on the bed of a lake or stream or on the bottom of an ocean. *Endogenous sediments* are made up of materials that were produced within a body of water, and then settled to the bottom. The shells, excrement and bodies of aquatic organisms and the chemical PRECIPITATES of waterborne IONS are endogenous sediments. The sedimentary rock, limestone, can be formed in either way. *Calcerous limestone* is composed primarily of pure calcium carbonate crystals that have precipitated from solution, while *clastic limestone* is made up primarily of the shells and skeletons of aquatic organisms. *Exogenous sediments* are composed of materials that have been transported by water away from their point of origin before deposition. Sand, silt and gravel washed into streams by erosion resulting from surface RUNOFF makes up the bulk of exogenous sediments, with minor amounts of pollen, humus and plant and animal remains. *Sedimentary rocks* are composed of sediments that have been deeply buried and then have been lithified by heat and pressure. PHOSPHATE rock is a sedimentary mineral composed primarily of *fluorapatite* mixed with sand and clay. *Loess soil*, which is deposited by wind, and *glacial till*, which is deposited by glaciers, are technically termed sediments, although they are seldom referred to as such in popular usage.

See also OAKRIDGE NATIONAL LABORATORIES, OCEAN DISPOSAL, PLACER MINE, SEDIMENTATION, SETTLING BASIN, WASTEWATER, WATER POLLUTION.

Sedimentation the deposition of SEDIMENT on the beds of streams, lakes, reservoirs and other bodies of water. The sedimentation of streams that can result from erosion caused by human activities such as farming, logging and road building. It can devastate fish and other aquatic organisms in streams by sealing off gravel beds that are essential to their survival (see ENVIRONMENTAL COST). A stream's ability to carry sediment decreases as its water slows below the transport speed necessary to keep various materials in suspension. As the speed of water decreases, the size of the particles settling to the bottom decreases. Dams and irrigation diversions can quickly fill with sediments in a stream that is carrying a heavy SUSPENDED LOAD. The deposition of silt sediments (see SILTATION) in Lake Mead, the reservoir created by the Hoover Dam on the Colorado River in 1936, has already filled more than 40 miles at the tail of the 115-mile reservoir, and it is expected to be full of sediment within about 250 years. Riverbed sediments that contain toxins such as RADIOACTIVE WASTE, MERCURY, POLYCHLORINATED BIPHENYLS and ARSENIC can contaminate a river's water when they are stirred up at times of high flow, and can pollute GROUNDWATER in instances where river water is recharging an AQUIFER (see GROUNDWATER POLLUTION). When sedimentation reduces the depth of a lake or other standing body of water, the penetration of light to the lakebed and the average water temperature both increase, leading to EUTROPHICATION.

Sediments settle in accordance with *Stoke's Law*: If the gravitational force that pulls a particle down is greater than the horizontal force exerted by the moving water, the particle will settle to the bottom if given enough time. Coagulants such as lime, alum, chlorine, clays and polyelectrolytes are often added to SEWAGE as it passes into primary sedimentation to increase the settling out of solids. Since more solids are converted to SLUDGE through the use of coagulants, their use may not be practical in areas where the disposal of sludge is expensive. See also SUSPENDED SOLIDS.

Sedimentation Basin see SETTLING BASIN

Selenium an ELEMENT that occurs naturally as a black, gray or red odorless solid commonly found in association with SULFUR. Selenium is a *metalloid*, a nonmetal with some of the characteristics of a METAL. Selenium is an essential trace element for plants, animals and humans, but it is toxic at slightly higher levels than those essential for health. It is used in the manufacture of STEEL, as a vulcanizing agent for RUBBER, in PAINTS and dyes, to clarify GLASS and as a pigment in ruby glass. Electricity flows through selenium in only one direction, and the element is a poor conductor of electricity in darkness. However, when light shines on it, its conductivity increases in direct proportion to the light's intensity. Selenium can also convert light directly into electricity. These characteristics have led to selenium's use in electrical rectifiers and photoelectric cells. Selenium is one of the active ingredients in plain-paper office copiers.

Selenium gets into air and water from COAL and fuel-oil combustion. It gets into water from the TAILINGS of GOLD, SILVER and NICKEL mines and mills, and from natural sources. Selenium salts are common in the drier soils of western North America. When such soils are irrigated, RUNOFF from the fields may be contaminated with selenium, especially if the field is overwatered. Selenium in runoff from irrigated farms in California's San Joaquin Valley killed and deformed thousands of ducks and other protected waterfowl at Kesterson Wildlife Refuge in the early 1980s, resulting in widespread publicity and the closure of the refuge in 1985 for cleanup. Similar selenium contamination has impacted countless wetlands throughout the western United States. Selenium and its compounds have water solubility ranging from low to moderate, but are highly persistent in water, with a HALF-LIFE of more than 200 days. Selenium has high acute TOXICITY to aquatic life and mammals and moderate acute and chronic toxicity to birds (see ACUTE EFFECTS, CHRONIC EFFECTS). Certain plants can accumulate a toxic dose of selenium when they grow in the selenium-rich soils common on western rangeland. "Locoweed" can poison cattle and wild animals because of the selenium content.

Selenium can affect human health when inhaled. DUST or mist containing selenium can irritate the nose, throat and bronchial tubes. Higher levels can cause difficulty breathing, lung irritation (pneumonitis) and headaches. Repeated overexposure can cause garlic breath, metallic taste, irritability, fatigue, pallor, indigestion, increased dental cavities, upset stomach, loss of nails and hair and mood change (depression). There is limited evidence that selenium may damage the developing fetus and decrease fertility in females. Repeated higher exposures may cause liver damage.

PUBLIC WATER SYSTEMS are required under the safe drinking water act to monitor for selenium (MAXIMUM CONTAMINANT LEVEL = 0.05 milligrams per liter). The BEST AVAILABLE TECHNOLOGY for the removal of selenium from DRINKING WATER is activated alumina, lime softening, REVERSE OSMOSIS and coagulation-filtration (selenium IV only). The OCCUPATIONAL SAFETY AND HEALTH ADMINISTRATION has established a permissible exposure limit for airborne selenium of 0.2 milligrams per cubic meter, averaged over an eight-hour workshift. It is also cited in AMERICAN CONFERENCE OF GOVERNMENT INDUSTRIAL HYGIENISTS and Department of Transportation regulations. Selenium is on the Hazardous Substance List, and selenium compounds are on the HAZARDOUS AIR POLLUTANTS LIST. Businesses handling significant quantities of selenium or selenium compounds must disclose use and releases of the compound under the provisions of the EMERGENCY PLANNING AND COMMUNITY RIGHT TO KNOW ACT.

Septic System   a small-scale SEWAGE TREATMENT system used by most rural homes and businesses to purify sewage. An underground septic tank made of concrete or plastic is the large intestine of a septic system. Sewage enters the tank at one end, and suspended solids are allowed to settle out as wastewater slowly moves from the inlet towards the outlet. ANAEROBIC BACTERIA slowly consume the solids in the sewage as it moves through the tank, reducing BIOCHEMICAL OXYGEN DEMAND in the process. WASTEWATER flows out of the septic tank's outlet and is routed into a drainfield or leach field consisting of a perforated pipe laid in a gravel-filled trench that is usually about two feet deep. The drainfield is designed to spread out the EFFLUENT and allow part of it to evaporate and the rest to percolate slowly down through the soil, with remaining pathogens being decomposed by contact with MICROORGANISMS in the soil. The water, in theory, should be purified by the time it seeps into underlying AQUIFERS.

Unfortunately, many septic systems—even new ones—don't work as they are supposed to, for one reason or another, and extensive contamination of soil, GROUNDWATER and surface water is caused by malfunctioning septic systems. Problems that can cause WATER POLLUTION include:

- *improper perc test and/or installation.* The perc test is often performed by the homeowner or the contractor who is installing the system, and test results can easily be altered to avoid the cost of a custom system, or even to allow the homeowner to build at all, on a problem site. Most county sanitation departments do not have the manpower to keep close tabs on all tanks under construction at any given time, and systems are frequently not installed to design specifications. The quality control that can be exercised with a centralized treatment facility is just not possible in scattered rural applications.

- *inadequate maintenance.* Accumulated solids (see SLUDGE) should be pumped out of a septic tank every few years to keep a septic system operating efficiently, but no maintenance is performed on most systems until they start causing problems. Drainfields can become clogged if a tank is not completely digesting solids and lots of suspended material is flowing out of the tank. Enzyme treatments that will restore microbial populations in the septic tank and drainfield may get the system working again, but in many cases, a new drainfield must be installed.

- *too many tanks in an area.* Rural areas with a relatively high population density can put too much wastewater into the soil. If even a small amount of contamination is caused by each system, significant water pollution can result. Most rural residents get their water from wells that tap aquifers under the septic systems, so any contamination can be critical. Polluted wells often force the incorporation or annexation of heavily populated semirural areas so that sewage-collection and drinking water supply systems can be built.

- *too many users on a tank or too large a volume of water periodically going through the tank.* Wastewater should take at least 24 hours to pass through the septic tank to allow time for solids to settle out of the sewage and for anaerobic bacteria to do their work. If too much water is passing through the tank, either on a regular basis or on an occasional basis, on washday for instance, inadequately processed wastewater will flow out into the drainfield.

- *toxic materials are introduced into the septic tank.* Laundry detergents, drain cleaners and other toxins found in household products often go down the drain, and they can easily upset the biological balance in the tank, killing or disrupting the anaerobic bacteria that digest solids leading to system failure.

- *groundwater is too close to the surface or too much runoff is passing through the drainfield.* Both conditions can wash inadequately treated wastewater away into local surface water.

- *soils are impermeable.* Sanitation codes in most areas require a perc test before a permit to install a septic tank is issued. To perform the test, one or more holes about three feet deep are excavated in the area planned for the drainfield, and the holes are filled to a specified depth with water. The time that elapses before all the water in each hole has drained into the soil is noted, and if it takes more than the specified period for the water to drain, the soil is deemed too impervious (usually because of clay content or a shallow layer of hardpan or impervious rock) for a drainfield to work, and no permit is issued.

- *soils are too permeable.* If wastewater percolates down through the soil too quickly, the action of soil microorganisms and the filtering capacity of the soil is bypassed. Surveys of new septic systems that have been installed according to local codes in areas with extremely coarse soils—such as the large-grain sands and gravels often deposited by streams and glaciers—have concluded that almost no water was being distributed through the drainfield because the soil was so permeable that wastewater simply flowed directly down into the aquifer almost immediately after entering the drainfield.

Some sites do not have enough room for a conventional drainfield or adequate biological capacity in soils to process the volume of wastewater produced. An INTERMITTENT SAND FILTER requires considerably less space than a conventional septic tank and drainfield and yields finished wastewater about as pure as that produced by SECONDARY TREATMENT at a sewage treatment plant.

Settling Basin a tank or basin in a SEWAGE TREATMENT plant—also called a sedimentation basin—where suspended materials are allowed to settle out of WASTEWATER. Sewage must remain in the settling basin for at least two hours, during which time it must move no faster than about four feet per minute to keep turbulence low enough to allow settling. In the horizontal-flow settling basins used for PRIMARY TREATMENT, wastewater is introduced into one end of the basin and clarified EFFLUENT is removed from the other. In the vertical-flow settling basins used in SECONDARY TREATMENT, wastewater is piped into the middle of the basin, and FINISHED WATER is removed as overflow over the top. Mechanical scrapers periodically remove accumulated SLUDGE from the bottom of the settling basin, and a skimmer removes OIL, GREASE and FLOATING SOLIDS from the surface.

Sewage domestic WASTEWATER, including human waste and garbage-disposal waste (see BLACK WATER), along with all the rest of the water that goes down household and commercial drains (see GRAY WATER). Sewage consists of materials in SUSPENSION and SOLUTION in a stream of what is typically 99.9% pure water, and generally contains between 350 and 1200 milligrams per liter (mg/L) TOTAL SOLIDS, from 250 to 850 mg/L TOTAL DISSOLVED SOLIDS, and between 100 and 350 milligrams per liter (mg/L) TOTAL SUSPENDED SOLIDS, and a BIOCHEMICAL OXYGEN DEMAND of between 110 and 400 mg/L. Sewage also typically contains between 50 and 150 mg/L of grease, between 4 and 15 mg/L of PHOSPHOROUS and between 20 and 85

mg/L of NITROGEN. Testing the composition of the ever-changing mix of materials found in sewage flowing into a treatment plant is essential if the delicate biological balance of the treatment plant is to be maintained.

Domestic sewage contains relatively nontoxic liquid waste from homes and commercial establishments, while INDUSTRIAL LIQUID WASTES generated by industrial processes and the production and manufacture of goods can be highly contaminated with hazardous and toxic materials. Industrial EFFLUENT often must be treated before it is emptied into a sewer system to avoid damage to the sewage treatment plant (see PRETREATMENT STANDARDS, SEWAGE TREATMENT).

Municipal liquid waste, which often includes wastewater from commercial and industrial sources and LEACHATE from landfills, is called *sewage* although technically it is a mix of liquid wastes. Among the typical components of sewage are:

- *grit*—the sand, cinders, coffee grounds, seeds, metallic shavings and other particles carried by sewage.
- *septage*—a mixture of scum, grit and suspended solids that is pumped periodically from a septic tank.
- *scum*—the oils, fats, soaps, synthetic hygiene products, cigarette filters and other light material that is skimmed off the top of wastewater being treated.

Sewage typically contains between 100 and 300 parts per million of biochemical oxygen demand, carries many BACTERIA, both living and dead, and also contains VIRUSes and other MICROORGANISMS. A variety of toxic materials can be found in sewage, including hazardous materials such as PAINT, SOLVENTS and INSECTICIDES that have been dumped down drains or flushed down toilets and inadequately treated industrial wastes. *Raw sewage* is untreated, in the state that it arrives at the treatment plant. Malfunctions at the treatment plant and storm water overflows can sometimes lead to the discharge of raw sewage into surface water. Most of the odors associated with sewage result from the excretion of HYDROGEN SULFIDE by anaerobic bacteria.

STORM RUNOFF becomes a part of the flow of sewage in a COMBINED SEWER SYSTEM.

Sewage Collection System an underground network of drainage pipes, pump stations and associated controls, manholes, inlets, outfalls, holding basins and other equipment associated with the collection of SEWAGE and its delivery to a SEWAGE TREATMENT plant. A *sanitary sewer system* carries only domestic wastes and commercial wastes that are of similar character. A combined *sewer system* collects both sewage and STORM RUNOFF in the same network of drainage pipes rather than having a separate STORM SEWER. Sewage collection systems—also called *sewers* and *sewerage systems*—are laid out whenever possible so that sewage will flow through them by gravity to the treatment plant, which is located at the lowest point in the system. This is not always possible, however, and lift stations must sometimes be employed when it becomes uneconomical or physically impossible to transport sewage by gravity flow alone. Buildings with deep basements usually must pump sewage up to the level of sewer mains. The volume of WASTEWATER flowing through sewers peaks at breakfast time and again in the evening, and flows also vary according to season, weather conditions and holidays.

The network of pipes that carry wastewater from its point of origin to the treatment plant starts with household drainage pipes four inches or less in diameter and made of PLASTIC or cast IRON, which eventually, lead into sewer mains large enough to drive a truck through. Vitrified clay pipe was once commonly used to carry household waste to sewer mains in the street, and sometimes was also used to construct mains. Vitreous pipes are impervious and highly resistant to CORROSION, but are brittle. Concrete pipe with an inside diameter ranging from four inches to circular or elliptical sewer mains 12 feet in diameter is widely used in new sewer installations. Concrete sewer mains are cast in place for some applications. Fabricated STEEL pipe is built from sheets of corrugated steel that can be shaped to fit the application into circular, elliptical, arch and other shapes. Cast iron, ductile iron and welded steel pipe is used in places such as stream crossings, where strength and tight-fitting joints are critical.

Lots of things can go wrong inside a sewer, which is why manholes are generally placed at the end of every long run and at every point where there is a change in drainage pipe size, direction or slope. Manholes are access holes with heavy cast iron covers that allow workers to get into the sewers for inspection, maintenance and repair. Regular maintenance is a necessity in the "modern" municipal sewer system, which is typically a hodgepodge of connecting pipes from different eras, with century-old mains, sometimes constructed of wood, still in use in some cities. Underwater television cameras are often employed to track down breaks, deteriorating pipe, tree roots and blockages in the collection system. Accumulated debris must regularly be cleaned from sewers and materials in catch basins must be regularly removed in a COMBINED SEWER SYSTEM.

The INFILTRATION of GROUNDWATER into sewers can dramatically boost the volume of wastewater that must be treated. In a relatively new sewerage system with tight joints the flow of wastewater is typically just 60% to 75% of the volume of fresh water entering through the municipal water system. But most U.S. sewer systems aren't that new, and groundwater leaking into a typical sewer increases the total volume of sewage to roughly the equivalent of the volume of fresh water flowing into the area being served. Leaks from cracks in sewer lines often result in groundwater contamination.

Sewage Sludge a semiliquid material containing the solids removed from SEWAGE at a SEWAGE TREATMENT plant, sewage sludge is between 92 and 99.5% water. Sewage sludge is thicker than sewage (which is 99.9% water) because it contains solids that had been held in SUSPENSION (see SUSPENDED LOAD) or in SOLUTION (see DISSOLVED SOLIDS) in the wastewater—solids that were removed during the treatment process. Sewage sludge often contains LEAD and CADMIUM. Most sludge was once dumped either in LANDFILLS or at sea (see OCEAN DISPOSAL), practices that have been largely eliminated because of concerns about water quality. New York City and eight other municipalities dumped sewage sludge at sea until the ENVIRONMENTAL PROTECTION AGENCY (EPA) forced them to switch to land disposal by 1992.

Sludge comes from solids that settle to the bottom in sedimentation tanks, screenings and skimmings. Screenings are solids removed from sewage in the treatment process. The largest screenings include floating debris (see FLOATING SOLIDS) such as wood, bottles, paper, rubber and plastics and heavier solids such as cans, rags and stringy material. Solids are removed from the wastestream with successively smaller screens as the sewage treatment process continues. The volume of screenings is reduced in installations where influent is run through mechanical cutters and shredding devices prior to processing. Sewage sludges are classified as *primary sludge (also called raw sludge)*, which is made up of sewage solids including feces and food scraps that settle out of wastewater in PRIMARY TREATMENT. Primary sludge contains up to 5% solids by weight. The use of household garbage disposal increases the weight of primary sludge an estimated 25% to 40%. Secondary sludge is composed of finer biologic floc that settles out or is removed through flotation (small air bubbles pass up through the sludge where they pick up the solids and float them to the surface, where they can be skimmed off) during SECONDARY TREATMENT.

In the *gravity thickening* process, water is forced out of the sludge because of the weight of overlaying sludge. Gentle stirring releases the water. Additional water can be removed by running sludge through *cen-*

trifugation, the removal of water from the wastestream through centrifugal force in a centrifuge with screens that will hold solids while allowing the passage of liquids. *Supernatant*, the pungent liquid that is removed from sludge in the thickening process, is returned to the sewage plant's inflow for further treatment. *Sludge stabilization* reduces its weight and volume and the concentration of pathogenic and odor-causing MICROORGANISMS through biological, chemical and thermal processes. *Aerobic stabilization* utilizes aerobic microorganisms to oxidize organic matter in sludge into carbon dioxide, water, ammonia and nitrates. Within about three weeks, the sludge has been reduced to a humus-like consistency that is relatively odor free and biologically stable. *Anaerobic stabilization* employs anaerobic microbes to produce methane gas and a humus-like sludge (see ANAEROBIC DIGESTER).

The EPA published rules setting limits for contaminants in sewage sludge in 1993, 15 years after they were mandated in the CLEAN WATER ACT. Before the federal standards went into effect, sludge disposal had been managed on a state-by-state basis. The leather industry is suing the EPA over the limits on CHROMIUM imposed in the new standards.

Sewage Treatment   the control of objectionable, hazardous and pathogenic substances in SEWAGE through the removal of materials in SUSPENSION and in SOLUTION. *Physical processes* such as FILTRATION, SEDIMENTATION, PRECIPITATION and screening; *microbial processes* such as the ability of AEROBIC BACTERIA to metabolize contaminants; and *chemical processes* such as OXIDATION/reduction (called *redox* and ION EXCHANGE (see HAZARDOUS WASTE TREATMENT, WATER TREATMENT) are used to treat sewage. An estimated 30 billion gallons of sewage is processed at municipal treatment plants each day in the United States.

Most U.S. sewage plants dumped effluent into lakes and streams after no more than primary treatment before the passage of the Water Pollution Control Act in 1972 (see CLEAN WATER ACT). The U.S. ENVIRONMENTAL PROTECTION AGENCY estimates it will cost $137.1 billion over the next 20 years to bring sewers and wastewater treatment facilities up to the standards required by the Clean Water Act.

PRETREATMENT involves the removal of material that could foul sewage plant machinery and odors from the waste stream that could make the entire treatment process less savory for workers. The pretreatment of any commercial or industrial liquid waste that might upset this balance is essential (see INDUSTRIAL LIQUID WASTE). Testing wastewater for excess ACIDITY or ALKALINITY and the presence of NONDEGRADABLE POLLUTANTS as it enters the pretreatment process and at

every step thereafter is essential if the delicate balance that allows microbes to thrive in the treatment plant is to be maintained.

Among the materials removed in pretreatment are:

- *grit*—particles of sand, cinders, gravel GLASS, METALS, ceramics and other hard materials that are abrasive and can cause excessive wear in pumps, clog lines, take up space in treatment tanks and damage other equipment used in the treatment process. In a grit chamber, the velocity of wastewater is slowed to a point where suspended solids larger than about 0.008 inch (0.2 mm) settle out. The solids fall onto an inclined belt at the bottom of the chamber that carries them up through the current of incoming wastewater and agitates the particles to shake clinging organic matter loose. The washed grit that remains is largely inert and can generally be disposed of in a landfill.

- *large solids*—floating solids made of wood, plastic and other lighter-than-water materials, and heavier solids including everything from string to diapers, towels and bed sheets can be removed by a *bar rack*, a device consisting of parallel metal bars that catch largest solids on its upstream face and mechanically removes them with a traveling rake. Coarse screens with holes about 0.5 inch (12.5 millimeter) square then remove some of the larger remaining solid debris. This is the first of several screens that wastewater will pass through. Disc-, drum- and band-type screens rotate loaded screen material out of the wastewater and wash away contaminants. *Rotary-drum centrifugal screening systems* are self-cleaning systems that are being installed in many older treatment plants to augment or replace grit chambers and primary settling tanks.

In PRIMARY TREATMENT, suspended solids and floatable materials are removed and wastewater is prepared for biological treatment. Most cities once discharged raw sewage into waterways, and most dumped effluent into waterways after only primary treatment until passage of the CLEAN WATER ACT. Today, the end product of an increasing number of municipal sewage treatment processes is highly treated DRINKING WATER (see WASTEWATER RECYCLING).

SECONDARY TREATMENT relies mostly on MICROORGANISMS, principally aerobic bacteria, to convert dissolved and colloidal materials responsible for most of the wastewater's remaining BIOCHEMICAL OXYGEN DEMAND (see BIOLOGICAL TREATMENT, METABOLISM). *Tertiary treatment* removes nutrients and traces of organic materials not removed by biological treatment.

Wetland Treatment

Many of the SUSPENDED SOLIDS (see TURBIDITY) and the bulk of the biochemical oxygen demand in sewage can be removed in specially constructed wetlands, where water lilies and other aquatic plants and waterborne microorganisms purify wastewater. Fecal COLIFORM BACTERIA, nutrients, solids in colloidal suspension, organic chemicals and HEAVY METALS all can be removed from wastewater through wetland treatment. Calcium hydroxide and hydrogen sulfate are often used to regulate PH. Microbes can also be used to reduce the acidity of wastewater through the biological degradation of organic acids (see BIODEGRADABILITY). See also WASTEWATER RECYCLING.

Sewer see SEWAGE COLLECTION SYSTEM

Sheet Erosion Erosion that removes a relatively uniform layer of soil over a large area. Sheet erosion is the result of SHEETFLOW, which can be caused by heavy rain, and is far less common than RILL EROSION. See also RUNOFF, SEDIMENT.

Sheetflow RUNOFF that is not confined to channels that results when heavy rain exceeds the INFILTRATION CAPACITY of the soil. Sheetflow over bare soils can lead to heavy erosion (see SHEET EROSION). Compare RILL EROSION.

Shoddy wool fabric that was produced from rags. The production of shoddy was largely ended with the passage of the WOOL PRODUCTS LABELING ACT OF 1939. Shoddy was produced by textile mills in the northeastern U.S. and in Yorkshire, England. The rags delivered by RAG PICKERS were cleaned, mixed with virgin wool fibers, respun into new yarn or thread and woven into new fabrics.

Short-Term Turbidity TURBIDITY that will clear up if a water sample is allowed to stand for a specified period, usually seven days. Suspended particles of sand, silt and soil cause most short-term turbidity. Short-term turbidity that results from an isolated event such as erosion from a heavy rainstorm normally causes little environmental damage. If the materials that cause the short-term turbidity are constantly entering a watercourse, as may be the case in an area with extensive farming or logging, streams can become clogged with SEDIMENT and damage to aquatic organisms can result. Compare LONG-TERM TURBIDITY. See also SILTATION.

Sick-Building Syndrome an outbreak of illness of unknown cause associated with bad indoor air quality in a building. Sick-building syndrome typically involves similar symptoms among a significant minority of the workers in an office building. Typical symptoms include irritation of the eyes, nose and throat; congestion of the lungs and sinuses; lethargy, fatigue, irritability, headache, dizziness and poor concentration. In extreme cases, managers have been forced to at least temporarily abandon a new building until healthy conditions can be restored. Inefficient or inadequate ventilation systems, microbial growths and internal POLLUTION sources have been found to be the cause of the syndrome in the instances where a cause has eventually been diagnosed. According to the World Health Organization, 30% of new or remodeled buildings may suffer from sick-building syndrome. A McGill University study printed in the *New England Journal of Medicine* concluded increased ventilation may not reduce symptoms of sick building syndrome. See also INDOOR AIR POLLUTION, VENTILATION.

Sierra Club one of the first environmental organizations, founded in 1892 by John Muir. The interests of the Sierra Club have expanded from the original emphasis on the preservation of wilderness and outings for its members to encompass environmental toxins, the management of public lands, endangered species, population growth, economic issues that relate to the use of natural resources and environmental pollution, recycling, waste disposal and the production of energy and the use of nuclear power. The club's emphasis is on lobbying and testimony, both formal and informal, before government agencies. The organization's principal concerns remain the protection and expansion of wilderness and the preservation of wetlands. As of 1994, the Sierra Club had 550,000 members, 63 chapters and 396 local groups. It conducts 8,000 outings each year for members. Its goal of informing the public on environmental issues is carried out partly through the publication and sale of books and studies. The Sierra Club Legal Defense Fund, an independent, separately funded public-interest law firm, brings litigation on behalf of the club and other local and national organizations. (Address: 2044 Filmore Street, San Francisco, CA 94115. Phone: 415-567-6100.) As one of the largest and certainly the best known of the environmental organizations, the Sierra Club is often attacked by members of more radical environmental groups such as EARTH FIRST! and GREENPEACE as being too conservative, and as having become part of the establishment that it was established to fight. Address: 730 Polk Street, San Francisco, CA 94109. Phone: 415-776-2211. See also ENVIRONMENTALIST.

Silt soil particles larger than clay but smaller than sand. A soil is called silt if more than 80% of its particles are silt. See also SEDIMENT, SILTATION.

Siltation the deposition of silt on the bed of a lake or stream. Rapidly moving water is capable of carrying more—and larger—SEDIMENT particles than slow-moving or standing water, so siltation is most pronounced in reservoirs created on streams, such as the Colorado River, that carry a heavy load of silt.

Silver a white, extremely lustrous metal that conducts heat and electricity better than any other metal. Only GOLD is more malleable and ductile. Silver is harder than gold but softer than COPPER. It is used in the production of jewelry, silverware, mirrors and photographic emulsions. Silver compounds are used as an antiseptic and bactericide and for other medical purposes. Coins in the United States were about 90% silver and 10% copper until the 1960s, when it was eliminated from smaller denominations of coins and reduced to 40% in half-dollar coins. The sterling silver used in tableware is 92.5% silver and 7.5% copper.

Silver is often found in the ores of copper, lead and zinc, and about half of world production is a byproduct of the processing of such ores. It is also frequently found in gold ore. Almost 2,500 metric tons of silver was mined and milled by U.S. industry in 1993, primarily as a byproduct of the processing of other metals. The mining and milling of silver and gold were responsible for massive air and water pollution in frontier America. Silver was removed by the ENVIRONMENTAL PROTECTION AGENCY from the list of substances regulated under the national primary drinking water standards in 1988 because it poses a relatively small threat to health. A Secondary Maximum Contaminant Level (see MAXIMUM CONTAMINANT LEVEL) of 0.09 milligrams per liter has been established for silver (see SAFE DRINKING WATER ACT).

Simple Asphyxiant an inert gas or vapor that is dangerous simply because it can displace so much oxygen that humans and animals can suffer asphyxia (a lack of oxygen and surplus of CARBON DIOXIDE in the body).

Skimming Tank a tank where floating materials are skimmed from the surface of SEWAGE before it enters a treatment plant (see SEWAGE TREATMENT). The flow of WASTEWATER through the plant is slowed adequately in the tank to allow lighter-than-water materials such as wax, small pieces of wood or plastic, grease, fat and lard to float to the surface, and then be removed with a mechanical skimmer. See FLOATING LIQUIDS, FLOATING SOLIDS.

Slag a byproduct of metal smelting that consists primarily of silicates; also called *cinder*. Slags produced by many smelting processes contain toxic materials. See BOILER SLAG.

Slash logging debris. Slash may include uprooted stumps, pieces of bark, limbs, small trees of the type being harvested, larger trees of undesirable species, brush, tree tops and short lengths of commercial sized-timber.

Slash Burning the practice of piling and burning logging waste on site. Slash burning is a major source of particulate pollution (see PARTICULATES, AIR POLLUTION) in timber-producing regions. Attempts are underway to reduce the volume of slash that is burned. The on-site chipping of limbs and tops is one approach. The chips may either be left behind as mulch or hauled away for conversion to wood pulp. Another experimental approach is the combustion of slash in an efficient, high-temperature boiler with provisions for the storage and removal of the energy produced. The utilization of smaller-diameter and shorter-length logs is another means of reducing the particulate pollution resulting from slash burning.

Sludge the congealed mass of solids that are removed from WASTEWATER by SEWAGE TREATMENT processes. A treatment plant produces SEWAGE SLUDGE, which consists of solids (see FECES, GARBAGE DISPOSAL, ORGANIC MATERIAL, SILT) removed during pretreatment, PRIMARY TREATMENT and SECONDARY TREATMENT. ACTIVATED SLUDGE is the microbial mass formed when wastewater is exposed to oxygen and microorganisms for several hours in an aeration tank at a sewage plant. Other types of sludge include the muddy mixture of ground-up rock and water that comes out of a drill hole, the waste from a coal washery, the mucky mix of sediments and water at the bottom of a lake or stream and a variety of waste products produced in the distillation of petroleum. Sludge is a source of CROSS-MEDIA POLLUTION because pollutants that are part of a liquid wastestream (see LIQUID WASTE) are converted to SOLID WASTE. See also HAZARDOUS WASTE TREATMENT, SLUDGE WORMS.

Sludge Worms aquatic worms that live in the bottom SLUDGE of lakes and streams. Sludge worms live inside a tube composed of SILT particles cemented together with mucus, and are therefore also called *tube worms*. Sludge worms are *oligochaetes*, a class of segmented worms that includes earthworms. Oligochaetes are made up of short, tubular segments with a very small or absent head and usually no eyes. They have simple digestive and circulatory systems, and are hermaphroditic—each worm has both male and female reproductive organs. Sludge worms can tolerate the high levels of POLLUTION often found in bottom sediments, and the level of their population in relation to other bottom-dwelling species is often used as an indicator of stream health.

Slurry   a combination of solid particles mixed with enough water so that the mixture will flow. A slurry is a SUSPENSION with enough mass that particles too large to be suspended are carried along (see SUSPENDED LOAD, SUSPENDED SOLIDS). Slurries are transported and stored as a liquid, and can be pumped through a pipeline or into a storage tank. Solids may be moved around as part of an industrial process inside a factory, or may be moved long distances through a pipeline. When the slurry reaches its destination, the particles it carries are removed and dried. A slurry pipeline is an efficient method of moving insoluble minerals such as COAL, and slurry pipelines have also been proposed for shredded garbage. One drawback of slurry pipelines is that they sometimes remove water from one river basin and discharge it into another, which can impair water quality and damage aquatic ecosystems in the stream or lake from which slurry water is drawn. Water can also be polluted at the destination end of the pipeline if it is not adequately treated before discharge. Some slurry pipelines solve these problems by incorporating a closed loop, with slurry going one way and filtered water returning to the source.

Smog   a visible cloud or haze of air POLLUTANTS, usually gray or brown in color. ATMOSPHERIC INVERSIONS tend to multiply problems with smog. Prolonged exposure to smog can damage lung tissue. Individuals with existing respiratory problems may be especially sensitive to the toxins found in smog.

"Old fashioned" smog is a simple mixture of smoke and fog. The potentially lethal nature of such smog has been well known since the dawn of the industrial age, when the COMBUSTION of COAL became widespread. London, England, has had severe problems with such smog for centuries. More than 4,000 persons died there during a period of high pollutant concentration in 1952.

Modern PHOTOCHEMICAL SMOG involves the chemical reaction of the byproducts of the combustion of FOSSIL FUELS with other airborne compounds in the presence of sunlight. Particulate matter, unburned HYDROCARBONS, CARBON MONOXIDE, SULFUR DIOXIDE, various NITROGEN OXIDES, OZONE and LEAD all can play a role in the chemical changes. Smog often has a brownish to grayish tinge. Browner smogs often contain high concentrations of nitrogen dioxide. Blue-gray smogs often contain large amounts of ozone. Smog comes in a kaleidoscope of colors, and blushes of yellow, green and even pink may be seen, depending on the recipe of the chemical soup in question. The Los Angeles basin, with its millions of vehicles and frequent inversions, has had a severe problem with photochemical smog since the 1950s. This is one reason that California now has some of the toughest smog-fighting regulations in the world. A 1992 study published in the journal *Science* estimated that cleaning up southern California's smog would save 1,600 lives and $10 billion each year. See also AIR POLLUTION, AUTOMOTIVE AIR POLLUTION, MOBILE SOURCES, PARTICULATES, VOLATILE ORGANIC COMPOUNDS.

Smoke   the gaseous product of the combustion of carbonaceous materials made visible by the presence of small particles of carbon. Wood smoke from stoves, fireplaces, forest fires and outdoor burning is a leading source of PARTICULATE pollution in many places (see AIR POLLUTION, AIR POLLUTION CONTROL, WOOD), especially areas that regularly experience ATMOSPHERIC INVERSIONS. See also ENVIRONMENTAL TOBACCO SMOKE.

Smoke Tube   a small plastic tube with a rubber squeeze bulb at one end. When the bulb on a smoke tube is squeezed, an inert chemical smoke is emitted. The movement of the smoke traces air currents inside a building, and helps identify areas with inadequate air circulation. Smoke tubes can be used to gauge the efficiency of an exhaust fan or the source of drafts. See AIR POLLUTION MEASUREMENT, INDOOR AIR POLLUTION.

Social Cost   the cost to society of an action or process. The social cost of a product includes environmental degradation caused by resource extraction and processing and manufacturing (see ENVIRONMENTAL COST, RESOURCE DEPLETION COST). The death and injury of workers due to accidents and exposure to toxins, and health damage to members of the public resulting from the contamination of air and water are also part of a product's social cost.

In return for such expenditures, society gets the benefits to be derived from the product or project in question (although these benefits may be hard to make out in certain cases). Social and environmental costs are associated with any process or product, including such essentials as the electric-power grid or the interstate highway system. The AVOIDED COST system is used in many states to balance the benefits derived from electricity with the costs of producing it. The costs and benefits of a highway system can be harder to assess. The benefits are enormous. With inadequate highways, traffic deaths and injuries would sore and the movement of goods, materials and people would slow or clog. Highways are the arteries of the nation's commercial circulatory system. But the costs are also enormous, both in terms of the resources required to build and maintain the system, and in the terms of the NOISE, VISUAL POLLUTION and pollution of air and water associated with a busy highway (see AIR POLLUTION, AUTOMOTIVE AIR POLLUTION, SMOG, WATER POLLUTION).

Soil Water   water that is held in permeable soil or sand between the surface of the ground and the top of the water table. See also DESERTIFICATION, GROUNDWATER, INTERFLOW, RUNOFF, WATER TABLE.

Solid Waste   WASTE that is a solid material, as opposed to a GAS or a liquid (see GASEOUS WASTE, LIQUID WASTE). Solid waste includes: CHEMICALS, WOOD, PLASTIC, ORGANIC MATERIAL, PAPER, METAL, SEDIMENT (see SEDIMENTATION, SILTATION) and rocks (see TAILINGS). Solid waste most commonly refers to MUNICIPAL SOLID WASTE (MSW), which is made up of materials that are primarily dry, although FOOD SCRAPS and other organic waste may contain as much as 90% water. MSW also includes household and commercial GARBAGE, APPLIANCES, FURNITURE, BUILDING MATERIALS, ASPHALT, office supplies, PACKAGING, ANIMAL WASTE, FOOD SCRAPS, YARD WASTE and AUTOMOTIVE PRODUCTS.

Solid materials suspended or dissolved in WATER (see SUSPENDED SOLIDS, WATER POLLUTION), or carried in the atmosphere as a PARTICULATE (see AIR POLLUTION) are a form of solid waste, and removing such contaminants from the waste stream with pollution-control equipment generally produces a solid that must be recycled, reused or disposed of (see AIR POLLUTION TREATMENT, HAZARDOUS WASTE DISPOSAL, SLUDGE, SOLID WASTE DISPOSAL, WATER TREATMENT). Most of the contaminants in SEWAGE (see SEWAGE SLUDGE) and industrial wastewater (see INDUSTRIAL LIQUID WASTE) are solid materials. INDUSTRIAL SOLID WASTE also includes particulates and ASH resulting from combustion processes. Solid waste results from natural-resource extraction processes (see SLAG). Waste that poses a hazard to human health or the environment is regulated as HAZARDOUS WASTE under the RESOURCE CONSERVATION AND RECOVERY ACT (see INDUSTRIAL HAZARDOUS WASTE). SOURCE REDUCTION uses COMPOSTING, REUSE, RECYCLING and other methods to reduce the volume of solid waste.

See also DEBRIS, DISPOSABLE DIAPERS, INTEGRATED WASTE MANAGEMENT, JUNK MAIL, LITTER, REFUSE, RUBBISH, TIRES, TRASH)

Solid Waste Collection   the collection of MUNICIPAL SOLID WASTE. Tributaries of the solid waste stream are wastebaskets and garbage bags that are emptied, along with a variety of other solid waste, into garbage cans and dumpsters, where they are held for pickup by the

moving principal of the wastestream, the GARBAGE TRUCK (see COMPACTION TRUCK). Truckloads of garbage are hauled either to a landfill or incinerator, or to a TRANSFER STATION. The trash is in turn loaded onto larger trucks, barges or railroad cars for transit to a dump or incinerator outside the immediate area.

The handling of solid waste is quite hazardous to sanitation workers, who have a disabling injury rate more than four times the average for other industrial workers. Imploding television tubes, exposure to toxic fumes and hazardous liquids and injuries caused by pieces of spring STEEL flying out of the back of a compaction truck when the hydraulic press is being operated are typical hazards faced by workers. Fires sometimes start inside garbage trucks, usually when ashes that still contain coals are disposed of along with household garbage. Standard procedure in such a case is to dump the garbage into the street, where it will be easier for firemen to put out the fire when they arrive.

The refuse chutes used in some apartment and office buildings and hospitals are a safe, efficient way of collecting solid wastes, but the chutes can aid the spread of fire if not constructed of fire-resistant materials and fitted with automatic fire doors. The chutes can also provide an ideal habitat for microbial growth, which can lead to INDOOR AIR POLLUTION (see SICK BUILDING SYNDROME).

Solid Waste Disposal   the disposal of solid materials that are discarded as worthless or no longer of use (see INDUSTRIAL SOLID WASTE, MUNICIPAL SOLID WASTE, SOLID WASTE). Solid waste materials that are relatively inert can be used as fill material. Municipal solid waste is most frequently hauled to a LANDFILL or an incinerator for disposal (see GARBAGE DISPOSAL, INCINERATION). A WASTE-TO-ENERGY PLANT converts garbage to heat in an incinerator or PYROLYSIS unit, or produces a fuel (see REFUSE DERIVED FUEL). An incinerator breaks down solid waste and produces heat from the combustion of fuels contained in garbage to produce heat, which is in turn used to make steam to drive an industrial process or to drive a steam turbine to generate electricity. The lines between solid, liquid and gaseous waste are far from clear (see CROSS-MEDIA POLLUTION). Solid particles dissolved or suspended in wastewater are liquid waste, at least until they are removed by pollution-control equipment, in which case they become solid waste—although solid waste such as SEWAGE SLUDGE may be more than 90% water (see SOLUBILITY, SUSPENDED SOLIDS, WATER POLLUTION).

Ocean dumping of solid waste was common until it was prohibited by the Marine Protection, Research and Sanctuaries Act of 1972 (see OCEAN DISPOSAL). The SOLID WASTE DISPOSAL ACT of 1965 banned open dumping and provided grants to municipalities to improve landfills. DUST generated in the handling and disposal of solid waste can lead to significant air pollution, which can contain pathogenic MICROORGANISMS, ASBESTOS, LEAD (see CONSTRUCTION WASTES) and other toxic PARTICULATES, and contact with solid waste is also a major source of water pollution.

Siting any kind of waste disposal facility can be very difficult. Getting appropriate permits and meeting statutory standards is time-consuming and expensive, and the not-in-my-backyard syndrome makes matters worse (see NIMBY). But companies such as WMX TECHNOLOGIES and BROWNING-FERRIS INDUSTRIES that can afford the investment have profited by the elimination of smaller landfills and the creation of larger, centralized facilities that meet the standards of the act. As the disposal of solid waste becomes more difficult and expensive, efforts to reduce the volume of solid waste through practices such as COMPOSTING, REUSE, RECYCLING and restructuring of disposal rates have been successful.

The Supreme Court ruled in April 1994 that states could not impose a higher tax on garbage brought in from another state. In May 1994 the court ruled that an ordinance in Clarkston, New York, that prohibited non-recyclable garbage from being hauled out of state was unconstitutional because it interfered with interstate commerce. It further upheld a 1993 appeals court ruling against the city of Chicago finding that ash from its garbage incinerator must be handled as hazardous waste because of the high, leachable levels of lead and CADMIUM that it contains (see ASH).

Solid Waste Disposal Act a federal law passed in 1965 that banned open dumping and established a system of grants to local governments for the creation and maintenance of landfills and recycling centers. The Act ordered the Department of Health, Education and Welfare to study the disposal of HAZARDOUS WASTES and establish standards facilities in which they are disposed. See HAZARDOUS WASTE DISPOSAL, SOLID WASTE, SOLID WASTE DISPOSAL. See also RESOURCE CONSERVATION AND RECOVERY ACT.

Solid Waste Stream the flow of SOLID WASTE from its source through processing to final disposal (see MUNICIPAL SOLID WASTE, SOLID WASTE DISPOSAL).

Solubility the ability of one substance (the *solute*) to be dissolved into another substance (the SOLVENT). The solubility in water of a substance at a given temperature and pressure is usually expressed as the grams of solute that will dissolve in 100 grams of solvent. The solubility of a contaminant in water plays an impor-

tant role in its characteristics as a POLLUTANT (see AIR POLLUTION, WATER POLLUTION). The solubility of a substance in fat—called *lipid solubility*—also plays a key role in the hazardousness of a substance, once it has been released into the environment. Substances that dissolve easily in fat tend to stay there once in SOLUTION and concentrations can slowly build to toxic levels as a result (see BIOACCUMULATION). Increasing temperature generally decreases the solubility of a gas, and increases the solubility of a liquid or solid. Substances with a low solubility in water are often highly persistent once dissolved (see PERSISTENT COMPOUND). Increasing pressure increases the solubility of all substances. See also GROUNDWATER POLLUTION, WATER POLLUTION.

Solution a MIXTURE in which the MOLECULES or IONS of one substance (the *solute*) are distributed evenly throughout the molecules of the second substance (the SOLVENT). Solutions are *homogeneous*—the ratio of solute to solvent is the same throughout the solution. A SUSPENSION can be made up of any combination of solid, liquid or gaseous solute suspended in a solid, liquid or gaseous solvent, although solids suspended in liquid water is the form of suspension most commonly referred to. A solute will not settle out regardless of how long the solution of which it is a part stands, nor can it be removed with a filter or a centrifuge. It is possible to vary the ratio of solute to solvent by adding more of one or the other as long as the solubility of the solute is not exceeded. A solute goes through *solvation* when it is dissolved by a solvent. The individual ions or molecules are pulled from the body of the solute and surrounded by a swarm of molecules from the solvent, which are bonded into place by electrostatic attraction. Most solvents are a liquid, but solutions and solvents may be in gaseous, liquid or solid state (as with metal alloys that are a solution of one solid within another). *Hydration* is the attachment of water molecules to ions of the solute. *Elutriation* is the removal of an absorbed material from a solvent. Compare COMPOUND. See DISSOLVED OXYGEN, NITROGEN, NUTRIENT.

Solvents substances capable of causing another substance to pass into SOLUTION (see SOLUBILITY). Most solvents are HYDROCARBONS and VOLATILE ORGANIC COMPOUNDS. Solvents are among the most common pollutants of water and of indoor and outdoor air. They are used in the production of waxes and wax removers, PAINTS and paint strippers, cleansers, adhesives, PESTICIDES, cosmetics and inks. Solvents are used for industrial degreasing and cleaning, and are a part of virtually every manufacturing process. Commonly used solvents include ALCOHOL, AMMONIA, BENZENE,

CARBON TETRACHLORIDE, CHLOROFORM, ETHYLENE, TOLUENE, METHYL ETHYL KETONE, METHYLENE CHLORIDE, TRICHLOROETHYLENE, and XYLENE. Water is known as the *universal solvent* because of its ability to dissolve a wide variety of substances, creating aqueous solutions in the process. Solvents in gaseous waste from industrial operations can be removed through incineration in an afterburner or ADSORPTION in a filter (see ACTIVATED CARBON). See also OIL.

Sorbent any substance that has the ability to absorb or adsorb another material (see ABSORPTION, ADSORPTION). Sorbents are used to remove pollutants from air and water (see ACTIVATED CARBON, AIR POLLUTION, WATER POLLUTION), and to clean up contaminants after spills of crude oil, chemicals and other toxic substances. Sorbents in streambed sediments can pollute water when they are disturbed. *Imbiber beads* are plastic beads that are capable of absorbing 27 times their volume of organic compounds. They can be woven into blankets, which float on the surface and absorb oil. If imbiber beads are packed between two screens to form a water filter, they will allow normal flow until they are full to capacity when they swell up and restrict water flow.

Source Reduction reducing the flow of solid waste to the landfill. RECYCLING and REUSE both result in source reduction. Increasing the rate charged for trash disposal is one of the most effective means of source reduction. To encourage recycling and reduce the volume of the wastestream, garbage companies in many communities with curbside recycling charge a flat fee for each bag of trash hauled to the dump. When the municipal trash company in Woodstock, Illinois changed to a flat fee of $1.83 for each 32-gallon bag of trash, the volume of the town's MUNICIPAL SOLID WASTE (MSW) quickly fell by 38%. Steps that can be taken in the interest of source reduction include: the purchase of reusable rather than disposable items where practicable, reduction of junk mail through being removed from mailing lists, laws or voluntary guidelines limiting excessive packaging and encouraging the production of packaging that can be reused and/or recycled. An increasing number of communities encourage backyard composting as a way of reducing the flow of organic materials into a landfill by charging a monthly fee for the pickup of YARD WASTE. Other communities collect organic materials and run their own large-scale composting facilities. The LIGHTWEIGHTING of GLASS and PLASTIC products also results in source reduction.

Source reduction also applies to hazardous waste products and pollutants in general. The increasing costs and liabilities associated with the pollution of air, water and land and the high cost of HAZARDOUS WASTE DISPOSAL is forcing waste generators to turn to source reduction. The use of less toxic PROCESS CHEMICALS, improved industrial processes, more efficient equipment, employee training and improved management practices are all ways of reducing the production of hazardous waste before it is created. The Pollution Prevention Act of 1990 establishes source reduction as the preferred way of controlling pollution in the United States, and establishes recycling as the preferred alternative for pollutants that can not be reduced at the source. The ENVIRONMENTAL PROTECTION AGENCY (EPA) developed a Pollution Prevention Strategy in 1991 that is intended to bring all the agency's programs into compliance with the Pollution Prevention Act. For information about the EPA's source-reduction programs, write: Pollution Prevention Division, U.S. EPA (7409), 401 M Street SW, Washington, DC 20460.

Source Separation the separation of different classes of recyclable materials in the home or office. Thorough source separation is especially critical with PLASTICS (see PLASTICS RECYCLING) and with high-quality PAPER (see PAPER RECYCLING), and batches of plastics or paper are frequently rejected by processors because they have not been adequately sorted. GLASS must similarly be sorted by color for recycling. The time and energy involved in sorting out a mixed batch at the recycling plant or processing facility are prohibitive. Resource recovery facilities that pick up unsorted solid waste from schools and businesses and sort it at a centralized location are opening in many areas. About 70% of U.S. cities were running some kind of source separation program by the beginning of World War I, and most recycled paper and rags. See also REDUCTION PLANT, SWINE FEEDING.

Space Junk trash that is left behind in space. From urine bags on the moon to satellites in deep space headed for deeper space, space junk is an integral part of the space age. There are about 7,000 pieces of debris orbiting the earth that are large enough to track by radar, and it is estimated that another 150,000 that are too small to be detected by radar are out there. Even small pieces can be hazardous to spacecraft, however, since they may be moving at 20,000 feet per second. Rocket boosters and other large, expensive pieces of refuse soon are pulled back to earth where they burn up in the atmosphere. One of the larger pieces of space junk that has been discarded is the 400-pound solar panel that was cut loose during repairs of the Hubble Space Telescope.

Special Interest Group a group of people organized to work politically in support or opposition of an issue of common interest. Clubs, associations, neighborhood groups and other citizen's organizations, unions, churches and professional organizations can all be special interest groups. Some groups may be organized to fight a particular battle, such as the construction of a solid-waste facility in the neighborhood, while large national organizations typically include members who are part of a variety of interest groups. Virtually every trade and occupation is represented by an association and by other interest groups that monitor state and national legislation. Environmentalists, steel manufacturers, importers, exporters, truckers, government officials and people from every other conceivable walk of life belong to and support interest groups.

Special interest groups and the lobbyists they employ make it difficult to pass environmental legislation, especially at the national level. Any time a state votes on a bottle bill, interest groups representing the bottling industry spend millions in a media campaign to oppose the measure. See also NIMBY.

Specific Gravity the ratio of the density of a substance compared to the density of a standard control material. If no other conditions are stated, specific gravity—also called *relative density*—is assumed to compare a material's density with that of water at 4°C for comparison to liquids or solids, or air at 20°C for comparison to gases.

Spent Fuel see REACTOR WASTE

Spoil waste material produced by an excavation. *Mine spoil* consists of rocks, dirt and sediments removed to obtain ore (see TAILINGS). They are generally acidic (see ACIDIC WASTE), contain numerous soluble minerals and very few nutrients. Runoff that has seeped through mine spoil is usually polluted with acids and HEAVY METALS. *Dredge spoil* often contains numerous pollutants that have become attached to particles of SEDIMENT (see SORBENT, WATER POLLUTION). Spoil from construction projects and road building is generally not toxic, and can safely be used as fill. See also HEAP LEACHING, PLACER MINING.

State Implementation Plan a plan prepared by state pollution control agencies that outlines enforcement of the CLEAN AIR ACT (CAA) within the state's boundaries. The states are given a large amount of leeway in the preparation of the implementation plans, with the principal requirement being the timetable for compliance with the NATIONAL AMBIENT AIR QUALITY STANDARDS. Point sources in areas that do not meet air-quality standards specified in the act must install REASONABLY AVAILABLE CONTROL TECHNOLOGY under the plan. The ENVIRONMENTAL PROTECTION AGENCY (EPA) can amend the State Implementation Plan if it is felt that it will not bring the air in all parts of the state up to federal standards quickly enough. States are allowed to include stricter standards than those specified in the CAA. If a state refuses to enforce the standards in the implementation plan, the EPA will undertake enforcement within the state.

State Pollutant Discharge Elimination System a system of water-quality regulations developed by state pollution-control agencies intended to ensure compliance with the CLEAN WATER ACT (CWA). The state plan must be at least as strict as the standards in the NATIONAL POLLUTANT DISCHARGE ELIMINATION SYSTEM. A state is granted primacy for the enforcement of the CWA within its boundaries if the ENVIRONMENTAL PROTECTION AGENCY finds that the State Pollutant Discharge Elimination System meets or exceeds the standards established in the act.

State Water Quality Management Plan a plan developed by state pollution-control agencies intended to bring surface waters within the state into compliance with federal AMBIENT QUALITY STANDARDS spelled out in the CLEAN WATER ACT (CWA). Polluted stretches of rivers can be declared *water-quality limited segments*, in which case compliance with CWA water-quality standards is not required. Water quality cannot decline further, however.

Stationary Source a source of AIR POLLUTION that is fixed. A stationary source may be a POINT SOURCE—air pollution with a single point of origin such as an industrial smoke stack or effluent outfall—or an *area source*—pollution issuing from diffuse sources (see DIFFUSE POLLUTION) such as DUST from roads and construction projects or nitrates that get into water from pastures and fertilized fields, and pollution from mobile sources such as vehicle emissions. Emissions from stationary sources can be easier to control than those from a MOBILE SOURCE because contaminants are concentrated in a single location facilitating the installation of pollution control equipment. See also AMBIENT AIR QUALITY STANDARDS, BUBBLE.

Steel an alloy of IRON and carbon. Steel's high strength and durability make it an indispensable part of products from pocket knives to skyscrapers. The ubiquitous "tin" can in which food for humans and their pets is packaged is actually 99% steel. Steel is the stout backbone of industrialized culture. Steel sheets, rods, pipes, beams, brackets, nails and screws are used with one another and in tandem with WOOD, PAPER, CARDBOARD, PLASTIC and GLASS to produce automobiles, trucks, aircraft, ships, trains and railroad rails, buildings, toys, appliances, furniture and other products too numerous to mention. Steel is the most recycled material in the United States, thanks primarily to the well-established network of scrap-metal dealers (see METAL RECYCLING, STEEL RECYCLING). The Steel Recycling Institute estimates that 48% of the steel cans produced in 1993 were recycled. The steel bead wire that gives a radial passenger-car tire its strength makes up about 10% of the tire's weight. The steel in radial TIRES contributes to the chemical process that creates cement, when tires are burned in a cement kiln. The production of steel is a major source of air and water pollution, although today's steel industry is much cleaner than that of a few decades ago.

Carbon steel is fashioned into automobile bodies, ship hulls, structural steel for buildings, bed springs and bobby pins, and accounts for more than 90% of steel production. *Alloy steel* is produced by adding metals such as vanadium, molybdenum, MANGANESE and silicon to molten carbon steel. Alloy steel is used in applications where strength, hardness and long wear are essential, such as gears, axles and engine parts in vehicles and other machines, and knives. Less-expensive alloying agents are used in the production of *high-strength, low-alloy steel*, which is used in applications where carbon steel does not have adequate strength and durability but where alloy steel is too expensive. Most structural steel beams and the thick steel roofing and siding used on railroad cars use high-strength, low-alloy steels. *Case hardened steel* is quenched with a pickling solution containing CYANIDE or other salts, which form a hard surface film of *nitrides* or similar compounds. A large amount of steel sheet is plated with tin in the factory, either electrolytically or by passing through a bath of molten tin, and the resulting *tin-plated steel* is used to make steel containers. *Tin-plated steel foil* laminated to paper or cardboard is used for lightweight containers. Spring steel is produced through the use of alloying elements and tempering.

Steel Production

The production of steel basically involves the burning out of excess carbon and other impurities from pig iron at a temperature higher than the melting point of steel (2,500°F—1,371°C). This is generally accomplished in an open hearth furnace, basic-oxygen furnace or an electric furnace. Alloying elements such as magnesium, CHROMIUM and NICKEL are added to the molten metal as needed to produce *feroalloys* with the desired qualities such as extra strength, hardness and resilience

needed for a particular application. In an *open hearth furnace,* air and fuel gas are preheated in a brick heat-regeneration chamber that is heated by exhaust gases from the blast furnace, thereby making temperatures of 3,000°F (1,649°C) possible inside the furnace. A *basic oxygen furnace* resembles the *Bessemer converter*—a tall, pear-shaped furnace used in the production of pig iron—which can be tilted on its side for pouring melted slag and metal and for recharging. The basic oxygen furnace can also be swiveled for pouring and loading, and has the same basic pear-shaped retort. A high pressure stream of almost pure oxygen rather than air is used to spark OXIDATION, however. After the furnace has been recharged and turned upright, an oxygen lance with a water-cooled tip is lowered to within about six feet of the molten pig iron. Oxygen is blasted out of the nozzle at supersonic speed, initiating a churning action in the pig iron, during which the oxygen combines with carbon and impurities in the pig iron are burned off. It takes less than an hour to produce 300 tons of steel in a basic oxygen furnace. In an *electric furnace,* heat and other conditions inside the tightly sealed furnace can be controlled more closely than is possible than with other furnaces. Electric current arcs from electrodes to the scrap iron load, flows through the iron, and then arcs to another electrode. The heat created by the resistance to the electric current and by the hot electrical arcs quickly melts the load. Other types of electric furnace use heat generated in a resistance coil to melt steel. Electric furnaces are used in the production of stainless steel and other alloys where precise control of temperature is essential.

Molten steel is usually cooled by *quenching* it with water. The quick cooling gives the steel a hard crystalline structure. At least 25,000 gallons of water are used to produce a ton of steel, and most of that is *quench water* used to cool molten steel. Impurities in the iron and pickling salts seriously contaminate quench water, and disposing of the water with extensive treatment can cause serious water pollution. An increasing number of steel mills are processing quench water and reusing it. After quenching, the hot steel is usually rolled and formed into steel ingots that weigh more than two tons each. Working the steel increases its strength by refining its crystalline structure. *Steel sheet* is typically produced by running an ingot of hot steel through a series of rollers. An ingot that is 35 inches wide, 13 feet long and 4 1/2 inches thick can be squeezed out to become a sheet that is 35 inches wide, 1,200 feet long and 0.05 inches thick, in widths up to eight feet wide.

Emissions inside a steel plant, especially those from coke ovens, can pose a significant health risk to workers. In testimony before Congress during debate of the

1990 amendments to the CLEAN AIR ACT, the chairman of Bethlehem Steel said that imposing a standard of one in 10,000 increased risk of death for workers would result in the closure of all but a few of the nation's 40 steel-producing coke ovens.

See also AUTOMOTIVE RECYCLERS ASSOCIATION, JUNK-YARD, INSTITUTE OF SCRAP RECYCLING INDUSTRIES, METAL RECYCLING.

Steel Recycling the reprocessing and reuse of steel that has been discarded as useless. About two-thirds of discarded steel products are recycled. Major sources of steel scrap include:

- *construction and demolition waste* (see CONSTRUCTION WASTES, DEMOLITION WASTES). Of the 63 million tons of steel scrap recycled in 1993, 30 million tons came from construction and demolition sites.
- *steel cans*. About half of the steel cans used are recycled, up from 15% in 1988.
- *steel-drums* The drum reconditioning industry in the United States and Canada cleans 5 to 10 million used drums for recycling, and cleans and processes another 45 million drums each year for reuse.
- *discarded appliances*—called *white goods* in the scrap business. More than 60% of steel appliance parts are recycled, up from just 4% in 1985. More than 1.6-million tons of steel were recycled from discarded home appliances such as refrigerators, freezers, ranges, dishwashers, clothes washers and dryers, air conditioners and water heaters in 1993.

The iron and steel scrap added to a batch of steel helps determine the characteristics of the new steel produced. All steel contains at least 25% to 30% scrap, either generated inside the factory or recycled steel (see PRE-CONSUMER WASTE, POST-CONSUMER WASTE), and may contain 100% recycled metal. For each ton of steel recycled, 2,500 pounds of iron ore, 1,000 pounds of COAL and 40 pounds of limestone are conserved. The use of recycled steel produces 86% less air pollution, 76% less water pollution and 105% less solid waste than the production of new steel from iron ore, according to the EPA. The fact that ferrous metals can be picked up with a magnet makes it easy to remove steel cans from municipal solid waste before they are burned in an incinerator or converted to REFUSE DERIVED FUEL (see MAGNETIC SEPARATOR), and also eases the handling of steel and iron in scrap yards.

Steel Recycling Institute formed in 1988 as the Steel Can Recycling Institute with the support of the American Iron and Steel Institute and eight Canadian and American steel companies. A grassroots effort to promote the recycling of steel is the institute's focus.

Staff members work with public and private recycling centers, ferrous scrap dealers and others involved in the scrap trade to facilitate the recycling of steel cans, appliances, automobiles, construction and demolition materials and other forms of steel. See also INSTITUTE OF SCRAP RECYCLING INDUSTRIES, JUNKYARD, METAL RECYCLING.

Storm Runoff heavy surface RUNOFF—also called *storm flow* or *overland flow*—that results from a storm. Storm runoff can cause severe problems in a COMBINED SEWER SYSTEM if it exceeds the capacity of a SEWAGE TREATMENT plant. The initial wave, or *first flush*, of runoff at the beginning of a storm can contain numerous contaminants picked up from the land's surface, especially if the storm brings the first heavy precipitation in awhile. See STORM SEWER.

Storm Sewer a drainage system designed to carry away RUNOFF. Water enters a storm sewer through curbside storm drains. The INFILTRATION rate of the local soil and the largest amount of STORM RUNOFF that can ordinarily be expected are considered when sizing pipe for a storm sewer. Because the volume of runoff water can be much larger than the volume of effluent in a sewerage system, the drain pipes used must be of a larger diameter. Drainage lines for storm sewers are generally laid parallel to sewer lines, and runoff generally is discharged directly into a lake or stream without treatment. However, in a COMBINED SEWER SYSTEM, the same drainage system carries both sewage and runoff, a practice that can lead to pollution of the body of water into which EFFLUENT is dumped. See also SEWER SYSTEM, SEWAGE TREATMENT.

Stratigraphic Layer a layer of debris deposited in a MIDDEN or other archaeological site during a specific time period. Pollens, ashes, wood and bone fragments and other materials found in the same stratigraphic layer are usually of similar age, allowing archaeologists to date artifacts through *cross dating*. The oldest archaeological remains normally are found in the deepest strata of the excavation, with the next oldest in the next deepest strata, and so forth, a fact that allows the establishment of a rough chronological sequence for events at a given site. *Stratigraphy* is the oldest method of establishing relative dates at an archaeological dig. Stratigraphy is used by the modern garbologist (see GARBOLOGY) to date materials found in the excavation of a landfill. See also ARTIFACT, HORIZON MARKER.

Stratospheric Ozone OZONE found in the stratosphere, which extends from about 10 to 40 kilometers above the earth's surface. Also called the *ozone layer*, this ozone-rich layer forms a protective shell around the earth, shielding the lower atmosphere and surface from the full effects of the sun's ultraviolet radiation. Stratospheric ozone is formed when an oxygen molecule (O_2) is struck and ripped apart by ultraviolet radiation. The freed oxygen atoms so attach themselves to intact oxygen MOLECULES to form ozone (O_3), which is transparent to normal light but nearly opaque to the ultraviolet wavelength. The atmosphere is generally cooler as elevation from the surface increases, but in the upper stratosphere the situation is reversed. Temperature actually increases with elevation as the result of heat energy being released by ozone molecules that have absorbed incoming ultraviolet radiation. The ratio of ozone to oxygen, even in the midst of the ozone layer, is only about one to 100,000.

The condition of the ozone layer has been a subject of much debate, in recent years. At question is the effect of CHLOROFLUOROCARBONS (CFCs) on the concentration of ozone in the stratosphere. CFCs have been used as SOLVENTS, in the production of PLASTICS, and as heat-exchange fluids in air conditioners and refrigerators. CFCs, which have been detected in the stratosphere since the 1970s, can break down when exposed to intense solar radiation and release CHLORINE, which in turn can attack and destroy ozone molecules, thereby allowing more ultraviolet radiation to penetrate to the earth's surface. Methyl chloride (a powerful industrial solvent), METHANE gas (which gets into air from decaying organic matter and from natural-gas leaks) and HALONS also have the potential to destroy stratospheric ozone. To make matters worse, the ozone-depletion theory involves a flywheel effect during which ozone precursors already in the atmosphere concentrate in the stratosphere, so ozone depletion will accelerate for a time regardless of the reduction of CFC emissions in the Troposphere (the lowest level of the atmosphere reaching from the earth's surface out about six miles).

The negotiations that produced the 1987 Montreal Protocol—convened amid growing concern about the thinning of the ozone layer—called for a 50% reduction in the use of ozone-depleting chemicals by the year 2000. At a meeting in London in June 1990, representatives of 54 of the 59 countries that signed the Montreal Protocol agreed to completely phase out CFC use by the year 2000. In 1992, President Bush, citing new evidence of the seriousness of ozone deterioration, ordered that the complete ban of products containing CFCs be moved up to 1995, and U.N. officials proposed making the accelerated ban worldwide, estimating that 4.5 million cases of skin cancer could be avoided as a result.

A marked increase in the rate at which UV radiation penetrates the atmosphere would be critical, since the UV-B ultraviolet wavelength is known to cause human skin cancer, and a large increase could also upset sensitive natural systems. But no widespread increase in surface readings for UV-B that would correspond with the decreasing concentrations of stratospheric ozone have been observed. In fact, a 0.5% decrease was reported to have occurred between 1968 and 1982, according to a study published in *Nature* in September 1989. This surprising statistic is downplayed by some on the grounds that AIR POLLUTION in the lower atmosphere could have intercepted the radiation, throwing off the readings, but average air pollution levels in the U.S. have been declining over the last couple of decades.

Some say that reductions in ozone levels that have been measured over the last decade, and the seasonal holes in the ozone layer above the earth's poles, especially the South Pole, are proof that the predicted damage has already been done, and that the collapse of the ozone layer is imminent. Others say that the fluctuations of ozone levels are just part of normal climatic variation, which includes the constant creation and destruction of ozone, and that CFCs have little or nothing to do with the current ebb in ozone concentration. Only time will tell whether the alarm about the ozone layer and the subsequent worldwide ban on CFCs was justified and prudent, or whether it was an overreaction based on insufficient scientific data. The whole debate proves, at least, the difficulty of predicting the weather—even for tomorrow, but especially for decades to come. See also GLOBAL POLLUTION.

Strip Mine   a mine in which overlying materials are removed to form an open pit out of which ore is blasted and removed. After the ore body has played out, overburden and topsoil are replaced and grass and shrubs are planted to restore the site. Many pit mines are deep enough that they are the below the surface of the local water table, and they quickly become a lake when the mine shuts down and the pumps are turned off. In open-pit coal mines, the coal seam that is being removed is often itself an aquifer, and removing it permanently cuts off the flow of groundwater. Strip mines are a major source of water pollution, and loose materials blowing from strip mines and associated processing facilities are a significant source of air pollution. See also CLARK FORK RIVER, HARD ROCK MINE, HEAP LEACHING, MINE DRAINAGE, MINERAL WASTE, PLACER MINE, SPOIL.

Styrene   an unsaturated HYDROCARBON that occurs as a colorless to yellowish oily liquid with an aromatic odor. When styrene MONOMER is heated, it is converted into the POLYMER, POLYSTYRENE, a clear PLASTIC with excellent insulating properties. Styrene monomer is used in the production of polystyrene and polyester plastics, insulating materials, protective coatings and synthetic rubber. *Styrene oxide* is used as a catalyst and cross-linking agent for epoxy resins, in the production of other chemicals and as a chemical intermediate in cosmetics. Styrene monomer and styrene oxide get into air and water from industrial and municipal waste-treatment-plant discharges and from spills. Styrene monomer is produced primarily from ETHYL BENZENE. Styrene monomer is moderately soluble and nonpersistent in water, with a HALF-LIFE of less than two days. It is flammable (see FLAMMABLE LIQUID) and reactive chemical (see REACTIVE MATERIAL, REACTIVITY) and is a fire and explosion hazard (see EXPLOSIVE MATERIAL). The odor threshold for styrene monomer is 0.08 parts per million (ppm). Styrene monomer has high acute and chronic TOXICITY to aquatic life (see ACUTE EFFECTS, CHRONIC EFFECTS).

Styrene monomer can affect health when inhaled or passed through the skin. Exposure can irritate the eyes, nose, and throat, with higher concentrations causing dizziness, lightheadedness and unconsciousness. Repeated exposure to lower levels can cause problems with concentration and memory, and can affect learning ability, reflexes and the sense of balance. Styrene is a probable CARCINOGEN. There is some evidence that styrene causes leukemia and lymphoma in humans. Styrene is a confirmed cause of lung and stomach cancer in animal tests. Styrene monomer is a MUTAGEN and possible TERATOGEN.

PUBLIC WATER SYSTEMS are required under the SAFE DRINKING WATER ACT monitor for styrene (MAXIMUM CONTAMINANT LEVEL = 0.1 mg/L). The BEST AVAILABLE TECHNOLOGY for the removal of styrene from drinking water is granular ACTIVATED CARBON or stripped-tower AERATION. The OCCUPATIONAL SAFETY AND HEALTH ADMINISTRATION has established a permissible exposure limit for airborne styrene monomer of 50 ppm, averaged over an eight-hour workshift, and 100 ppm, not to be exceeded during any 15-minute work period. It is also cited in AMERICAN CONFERENCE OF GOVERNMENT INDUSTRIAL HYGIENISTS, NATIONAL INSTITUTE OF OCCUPATIONAL SAFETY AND HEALTH and Department of Transportation regulations. Styrene monomer is on the Hazardous Substance List, the Special Health Hazard Substance List and the HAZARDOUS AIR POLLUTANTS LIST. Styrene oxide is on the Hazardous Air Pollutants List. Businesses handling significant quantities of styrene monomer must disclose use and releases of the chemical under the provisions of the EMERGENCY PLANNING AND COMMUNITY RIGHT TO KNOW ACT.

Sublimation a process by which a solid changes directly into a gas without passing through the liquid state. Sublimation occurs when the bonds between MOLECULES in a solid are weak enough to allow some of its surface molecules to escape. The odor of a solid material is caused by airborne molecules that have escaped from the material via sublimation. The gradual disappearance of snow and ice during a period when the temperature is constantly below freezing is also the result of sublimation. Solid CARBON DIOXIDE sublimates so quickly that it does not exist as a liquid under natural atmospheric conditions. Compare EVAPORATION.

Subsidence the fall in the elevation of the land's surface that can result in the depletion of aquifers in certain types of soils. Once the water has been removed from the aquifer, the materials compress. The elevation of Mexico City has fallen by more than 35 feet as the result of subsidence over the last 70 years.

Sulfate a salt or ESTER of sulfuric acid. Sulfuric acid reacts with metals to form sulfates (SO_{4++}). Most sulfates are water soluble, and sulfates get into water from the weathering of rocks, AIR POLLUTION (see AERIAL DEPOSITION, SULFUR DIOXIDE), STEEL mills, pulp mills, textile mills and SEWAGE TREATMENT plants (household detergents contain sulfates). Water that contains more than 750 parts per million (ppm) sulfate can have a laxative effect on humans. Airborne water vapor condenses on sulfur dioxide emitted from industrial sources to form dilute sulfuric acid, which in turn can produce sulfates that end up in watercourses.

Copper sulfate is used in copper plating and the production of dyes. *Iron sulfate* is used in water purification and in the production of ink. *Zinc sulfate* is used in the production of lithopone, which is used as a white pigment in paints. The mineral GYPSUM is hydrated *calcium sulfate*. *Barium sulfate* and *aluminum sulfate* are used in the production of ALUMINUM and BARIUM. A *sulfate process* is used to make pulp from wood chips in a paper mill. PUBLIC WATER SYSTEMS are required under the SAFE DRINKING WATER ACT to monitor concentrations of sulfates (MAXIMUM CONTAMINANT LEVEL GOAL = 500 milligrams per liter). Water utilities could provide public education and alternate water sources such as bottled water or a water purification device installed at the point of use rather than installing central treatment under a new sulfate rule proposed by the ENVIRONMENTAL PROTECTION AGENCY.

Sulfur an odorless, tasteless ELEMENT of very low TOXICITY. Sulfur is the twelveth most common element, making up 5% of the earth's crust. Approximately 10.6

million metric tons of sulfur was produced by U.S. manufacturers in 1993. It is insoluble in water. It is a component of some proteins, and is therefore essential to life. Its principal environmental liability is that it is found as an impurity in COAL and PETROLEUM. When such fuels are burned, the sulfur they contain reacts with oxygen to form SULFUR DIOXIDE and sulfur trioxide. *Sulfides* are inorganic sulfur in its most reduced state (see HYDROGEN SULFIDE). Organic sulfur compounds such as the *mercaptans* are a source of odor in wastewater. See also SULFATE.

Sulfur Dioxide   a heavy, colorless gas with a pungent, irritating odor that is formed when SULFUR reacts with oxygen. Sulfur dioxide (SO_2) and *sulfur trioxide* (SO_3) are the only sulfur oxides that are sufficiently stable to exist outside the laboratory. They are the precursors of ACID RAIN, and are released into the ATMOSPHERE when COAL or PETROLEUM that contain sulfur are burned. Sulfur dioxide is one of the five CRITERIA POLLUTANTS named in the CLEAN AIR ACT. Sulfur dioxide has been employed for thousands of years as a fumigant in wine making and as a preservative and bleach in the processing of grapes, apricots and other fruits and vegetables. Today it is one of several *sulfites* used as a food preservative. Sulfur trioxide is formed when sulfur dioxide and water combine at high temperatures. It has an extreme affinity for water, and dissolves with explosive force to form *sulfuric acid*, one of the most corrosive substances (see CORROSIVE MATERIAL) known. So much heat is released when SO_3 dissolves that the water may boil, releasing toxic sulfuric acid fumes.

When released into the atmosphere, sulfur dioxide reacts with oxidants (see OXIDATION, OXIDIZING AGENT) and PARTICULATES to form *sulfates* and sulfuric acid, which can cause lung damage in humans and animals and are toxic to plants. The American Lung Association cites exposure to SO_2 and its byproducts as the third leading cause of lung disease (after active and passive cigarette smoke). The U.S. Food and Drug Administration requires that sulfites be listed as an ingredient of processed foods after finding that they can cause an allergic reaction and even death when consumed by asthmatics and other sensitive individuals. Short-term exposure to SO_2 fumes can cause constriction of airways for such individuals. Chronic exposure (see CHRONIC EFFECTS) can cause a thickening of the mucus layer of the trachea that is similar to chronic bronchitis. Although sulfates are precursors of acid rain, the majority reach the ground through dry deposition.

The Clean Air Act calls for a reduction in SO_2 emissions to half of 1985 levels by the year 2000. An esti-

mated 19 million tons were released in 1990. The amount of sulfur dioxide emitted by a single power plant is staggering. The Navajo Generating Station near Page, Arizona, for example, produces 12 to 13 tons of sulfur dioxide *per hour*. Sulfur oxide emissions from power plants can be reduced switching to low-sulfur coal or by installing SCRUBBERS in smokestacks. The Clean Air Act also mandated a reduction in the amount of sulfur found in diesel fuel.

See also POLLUTANT STANDARD INDEX.

Sullage   see GRAY WATER

Sunset Scavenger Co.   a San Francisco garbage-collection business founded in 1920. Four men rode on each of the company's GARBAGE TRUCKS, and one or another of the employees was always inside the truck sorting out recyclable materials such as rags, bottles, metals, newspaper and cardboard while the truck made its rounds. The garbage-collection business was highly competitive during the Depression, and the company's management believed it was essential to reclaim as many resellable materials as possible from the waste stream in order to keep rates low enough to remain competitive. Bottles were washed and returned to the distributor for refilling, and other recyclables were sold as scrap. By the early 1950s, the company was making half a million dollars per year selling salvageable materials. By the mid-1950s, however, there was a marked decline in the sale of salvageable materials, due primarily to the advent of the no-deposit, no-return bottle, and the growing popularity of virgin materials in the form of petrochemicals and wood for use in new products. The company soon went out of business, unable to compete with the modern COMPACTION TRUCKS that packed as much garbage as possible into their bed and hauled it directly to the dump.

Superfund   see COMPREHENSIVE ENVIRONMENTAL RESPONSE COMPENSATION AND LIABILITY ACT

Surface Mining Control and Reclamation Act   a federal law governing the strip mining (see STRIP MINE) of COAL. The act, which became law in 1977, sets environmental standards for coal strip mining and for the reclamation of open-pit coal mines, but does not apply to other types of mines (see HARD ROCK MINE, PLACER MINE) or to strip mines for minerals other than coal. Certain categories of especially environmentally sensitive land can be declared unsuitable for mining under the act. The Surface Mining Control and Reclamation Act is enforced by the Office of Surface Mining Reclamation and Enforcement in the Department of the Interior, although state agencies may be given primacy if a state establishes rules at least as strict as those specified in the act. Fees paid for permits and forfeited bonds and fines resulting from noncompliance with the act are placed into an Abandoned Mine Reclamation Fund that is used in the cleanup of strip mines created before the passage of the act.

Surface Tension   a tough outer skin with a tensile strength comparable to that of steel that forms around bodies of water and other liquids. Bodies of water from particles of mist to oceans are enclosed in a tough surface-tension skin. Molecules are packed together more densely on the surface of a liquid because they are pulled toward the center and sides of the mass, but not toward the outside of the liquid. The molecules inside the surface-tension skin are spaced farther apart because they are being exerted to molecular forces coming from all sides. The tightly packed layer of molecules formed by the surface-tension phenomenon forms a tough, resilient surface skin. Some insects have developed feet that do not penetrate the skin on top of a body of water, and can skate around on the surface film. The clinging layer of water that forms on wet skin owes its tenacious grip to surface tension, and the shape of waves, bubbles and water drops are similarly dictated by surface tension.

Surface Water   WATER in lakes and streams. See HYDROLOGIC CYCLE, WATER POLLUTION.

Suspended Load   the total amount of SUSPENDED SOLIDS carried by WATER. See TOTAL SUSPENDED SOLIDS.

Suspended Solids   solids that are held in SUSPENSION in a liquid, usually WATER. Suspended solids include materials that will settle out if the liquid is allowed to stand long enough or is run through a fine enough filter, and COLLOIDAL SOLIDS that can be removed by REVERSE OSMOSIS and DISTILLATION. The *organic fraction* of suspended solids is defined as all materials that evaporate or oxidize at or below 550°C, with the solids that survive termed the *inorganic fraction*. Erosion can lead to an elevated level of suspended solids in a lake or stream and associated SEDIMENTATION.

Suspension   a form of MIXTURE in which very fine suspended particles are dispersed throughout the body of a liquid. Mixtures are usually composed of solids such as SILT suspended in liquids such as WATER, but mixtures of liquid particles suspended in a liquid or liquid or solid particles suspended in a gas also occur. Nonsoluble particles that are held in suspension are said to be dispersed in a liquid *dispersant*, as opposed to a soluble substance, which dissolves in a solvent (see SOLUBILITY, SOLUTION). A colloidal suspension is made up of particles small enough to remain in suspension indefinitely as the result of molecular forces within the liquid in which it is suspended. Much of the waste material found in sewage and INDUSTRIAL LIQUID WASTE comes in the form of SUSPENDED SOLIDS. Compare SOLUTION.

Sustainable Agriculture   an agricultural system that can be carried out indefinitely without depleting the soil, using excessive amounts of energy or petrochemicals or otherwise depleting resources or damaging the environment. Reducing reliance on pesticides through INTEGRATED PEST MANAGEMENT—the control of agricultural pests through a mix of biological, mechanical and chemical means—is one of the cornerstones of sustainable agriculture. The introduction of species that prey on the pest to be controlled is one commonly employed technique. Increasing soil fertility through COMPOSTING and reducing erosion through mulching and the use of cover crops is another important element. For farming to be truly self-sufficient, the amount of FOSSIL FUELS consumed would have to be reduced. Ten times as much energy is typically consumed on U.S. farms as is produced in the form of food. Farmers do have the ability to produce alternative fuels should the supply of petrofuels dwindle or become too expensive (see METHANE, ALCOHOL).

Sustainable techniques are also applicable to forest lands. When properly managed, forests are a RENEWABLE RESOURCE that can provide fuel and building materials ad infinitum while still providing habitat for animals, preventing erosion and building soil. But over-harvesting has led to timber shortages in virtually all parts of the world, and to the almost complete deforestation of many developing countries in just a couple of decades.

See also BIOLOGICAL TREATMENT, DESERTIFICATION, INSECTICIDES, MICROORGANISMS, NATIONAL AGRICULTURAL LIBRARY, SUSTAINABLE SOCIETY.

Sustainable Society   a human culture that uses energy and other natural resources at a rate that can be sustained indefinitely, without irreversible depletion of natural resources or long-term environmental degradation. In an era characterized by a shrinking inventory of natural resources and a rapidly expanding world population, the concept of sustainability becomes especially critical. The way in which resources are now used around the world is inherently inefficient. Laws governing the use of natural resources such as coal or water are archaic, based on a set of social needs and values of an expanding frontier society that existed a century ago. Even in environmentally progressive countries such as the United States, resource use is governed by laws such as the 1872 General Mining Act (see MINING LAW) that was designed to facilitate access to resources on the growing frontier and to reward those who chose to use them. Although many attempts to update the mining law have been made, the political clout of the industries that benefit from the law have so far caused their failure.

The mining act and similar statutes that govern the use of resources such as forests and water are the remnants of a colonial era in which VIRGIN MATERIALS from the hinterlands fueled growth in distant urban centers. The colonies were managed primarily as sources of raw materials to be extracted and shipped back to the mother country. Resources were removed with little regard for the long-term well-being of the people living in the area or for the health of the ecosystem. When the resource base was exhausted, the local people and the local economy were left to fend for themselves. The same kind of process is operative in today's economy. It has led to the deforestation and desertification of many Third-World countries. But while the focus during the past century has been on the control and use of resources such as minerals, lumber, PETROLEUM and COAL, the movement toward a WASTE STREAM based on REUSE and RECYCLING is already well under way.

Moving toward worldwide sustainability in the face of global POPULATION GROWTH that seems certain to lead to a population of more than 10 billion by the year 2100 will require vastly more efficient use of natural resources. Recycling and reuse complete the resource circle in nature, where "waste" such as dead vegetation, FECES and ANIMAL REMAINS are recycled naturally into useful nutrients, and where population is controlled by the natural carrying capacity. To achieve a balance between the demands put on the environment and natural resources by man and the ability of the environment to provide, manmade wastes must similarly be recycled (see COMPOSTING).

Buckminster FULLER referred to the planet as "spaceship Earth"—a vessel with all of humanity on board. He likened the Earth to an egg and to petroleum as the nutrient inside that allows the embryo to develop into a chick with enough strength to break the shell and emerge into the world, where it must fend for itself. In a sustainable society the environmental and financial health of the world community must be weighed when making decisions about resource use. The development of an economy based on the use of secondary materials would mean that raw materials would come less from virgin materials usually acquired in faraway rural areas, but would instead be "mined" from the urban waste stream.

Some of the principal components of sustainability are:

- *control of human population.* Rather than leaving the control of population solely to traditional "remedies" such as war and famine, movement toward a sustainable society could relieve the pressure being put on the natural resource base and resulting human suffering and international squabbles. Long-run sustainability will require an eventual leveling off of human population (see ZERO POPULATION GROWTH).

- *the efficient use of energy.* The development of new technology that makes more efficient conversion of fossil fuels to energy; the application of extensive energy conservation techniques in all sectors; conversion to renewable energy sources such as the sun and the BIOMASS it creates—such measures must be the basis of a sustainable society, not the consumption of finite resources.

- *the development of clean new energy sources.* FUSION reactors and hydrogen fuel cells (see FUEL CELLS, HYDROGEN) and similar technologies that use few finite resources and produce minimal toxic waste will become increasingly important as resources become more scarce.

- *the efficient use of processed materials.* Once the energy investment and ENVIRONMENTAL COST required to convert a raw material into a finished product have been invested, efficiency dictates that the product not be used once and discarded (see CONSUMERISM, PLANNED OBSOLESCENCE). Rather, the product should be durable and reparable enough that it will have a long life, and that the materials in the product be easily recyclable at the end of its useful life.

- *a move to* SUSTAINABLE AGRICULTURE. More efficient ways of growing food and fiber crops and controlling insects, and limiting the rate of the harvest of resources such as forests, fisheries or rangelands to a level that will not deplete them in the long run, must be applied.

- *more efficient use of water.* The development of water-conserving methods of irrigating crops and water conservation and reuse in urban areas will help stretch a shrinking supply that is reducing the productivity of croplands in many parts of the United States and the world.

President Clinton created a 25-member Council on Sustainable Development in 1993, made up of representatives of environmental groups, industry, labor and the general public. The council is charged with developing policies and demonstration programs that will promote sustainable development and informing the public about sustainability. For more information, contact: World Resources Institute, 1709 New York Ave. NW, Washington, DC 20036. Phone: 202-638-6300.

Swill FOOD SCRAPS and YARD WASTE and other organic material found in garbage that can be fed to hogs and pigs. The wet, organic portion of garbage was fed to hogs and other animals as a method of reducing the volume of municipal solid waste in many parts of the United States in the early twentieth century (see SWINE FEEDING).

Swine Feeding the practice of feeding SWILL to hogs and other animals to reduce the volume of MUNICIPAL SOLID WASTE. Swine feeding was not practiced in many of the larger U.S. cities, especially along the Atlantic coast, but many medium-sized towns (such as early Los Angeles) tried it. The practice was encouraged during World War I as a way of increasing food production. More than 40% of the nation's cities engaged in the practice by the mid-1920s. Los Angeles sold the organic portion of its garbage to hog farmers near the city for decades, with more than 9 million pounds of pork produced on the farms each year during the 1940s. The city did not completely give up swine feeding until 1961, and was one of the last in the country to do so. A series of swine epidemics in different parts of the country during the 1950s led to the passage of the Swine Health Protection Act. See also REDUCTION PLANT.

T

Tailings waste rock and debris from mines and ore-processing mills. Tailings often contain PYRITE, which oxidizes readily when exposed to air in a tailings pile, and can release sulfuric acid into water or SULFUR DIOXIDE into the air when it oxidizes. Tailings often pollute surface water with HEAVY METALS such as LEAD, and with RADIONUCLIDES derived from crushed rocks. Thorium-230, a decay product of URANIUM, is the principal source of RADIOACTIVITY in tailings. Homes built on tailings that contain radionuclides can have problems with indoor air quality (see INDOOR AIR POLLUTION). See also ACIDIC WASTE, GROUNDWATER POLLUTION, MINERAL WASTE, RADIOACTIVE WASTE, SEDIMENTATION, WATER POLLUTION.

Tailings Dam an enclosure built to contain a TAILINGS pile and polluted RUNOFF, and to keep LEACHATE from the pile from migrating into surface water.

Tailings Pond a pond constructed to receive TAILINGS from a mining or milling operation. A tailings pond limits the AIR POLLUTION that can be caused by the small particles found in tailings, which go into SUSPENSION in the water and eventually settle to the bottom. Since the water in a tailings pond can be heavily contaminated with metals and other toxins and the water can be highly acidic, the wastewater in the pond must be sealed off from surface water and prevented from seeping into underlying aquifers to prevent WATER POLLUTION. See also HARD ROCK MINE, HEAP LEACHING, MINE DRAINAGE, MINERAL WASTE, PLACER MINE, STRIP MINE.

Tankage one of the products that was produced by a REDUCTION PLANT. Tankage was sold for use as fertilizer and as a raw ingredient for other products.

Tape Sampler a tube through which the air to be tested is drawn by a vacuum pump with CONTAMINANTS intercepted on a movable filter tape mounted at one end. Tape samplers are used to measure the concentration of PARTICULATES in AMBIENT AIR. Air passes through the tube (which is normally one inch in diameter) and the filter tape at a fixed rate—usually about 0.25 cubic feet per minute (cfm). The used portion of the tape is rolled onto a take-up reel at regular intervals (normally two to four hours), and fresh filter tape is positioned over the end of the tube. The tape is later passed over a light source, and the amount of light that passes through the portions that have been used is measured with a photoelectric cell, producing a *coefficient of haze* (COH) for each sampling period. This reading can then be standardized by calculating the COH of a column of air 1000 feet deep (the *COH/1000*). Tape samplers can accurately measure particulate levels while running unattended for weeks or months. Tape samplers are also called American Iron and Steel Institute Smoke and Haze Samplers. See also INDUSTRIAL AIR POLLUTION MEASUREMENT, PARTICULATE MEASUREMENT.

Tell a hill composed of refuse accumulated over centuries. Many tells rose over the sites of ancient cities. For example, archaeologists have estimated that street level in ancient Troy increased by almost five feet per century. See MIDDEN.

Teratogen any substance or organism capable of causing birth defects that alter the development of the embryo or the fetus. CARBON TETRACHLORIDE, CHLOROFORM, DIOXIN and LEAD are a few of the better known teratogens, which means literally "monster causing." Most CARCINOGENS are also teratogenic. Compare MUTAGEN.

Tertiary Treatment see SECONDARY TREATMENT

Tetrachloroethylene see PERCHLOROETHYLENE

Tetraethyl Lead a colorless, oily liquid with a slightly musty odor that is used as an anti-knock additive in GASOLINE. Tetraethyl lead is highly reactive (see REACTIVITY) and explosive (see EXPLOSIVE MATERIAL). It gets into GROUNDWATER from LEAKING UNDERGROUND STORAGE TANKS and fuel spills. Tetraethyl lead can affect health when inhaled or passed through the skin. It is converted to LEAD after it has entered the body, so the symptoms of exposure are the same.

The OCCUPATIONAL SAFETY AND HEALTH ADMINISTRATION has established a permissible exposure limit for airborne tetraethyl lead of 0.075 milligrams per cubic meter, averaged over an eight-hour workshift. It is also cited in AMERICAN CONFERENCE OF GOVERNMENT INDUSTRIAL HYGIENISTS and Department of Transportation regulations. The chemical is on the Hazardous Substance List and the Special Health Hazard Substance List. See also AUTOMOTIVE AIR POLLUTION, CLEAN AIR ACT.

Thallium   a solid, bluish to grayish white metallic ELEMENT that is one of the most toxic of the HEAVY METALS. Thallium commonly occurs in association with potash minerals and as part of minerals such as crookesite, lorandite, hutchinsonite, vrbaite and avicennite. Thallium compounds are used in RODENTICIDES, in the production of alloys, semiconductors, photoelectric equipment, lenses and thermometers. Thallium gets into air and water from industrial discharges and from spills and in the mining, smelting and refinement of thallium compounds, and from lead and zinc smelters and refiners. Thallium and its compounds have water solubility and persistence ranging from low to high, depending on the thallium SALT in question. Elemental thallium is slightly soluble and highly persistent in water. Thallium and its compounds have high acute and chronic TOXICITY to aquatic life (see ACUTE EFFECTS, CHRONIC EFFECTS).

Thallium can affect health when inhaled and passed through the skin. The symptoms of thallium poisoning may not manifest for many hours or even days after exposure. Exposure to high concentrations can cause tremors, delirium, hallucinations, convulsions, coma and death. Survivors of thallium poisoning often suffer permanent brain damage and may have vision loss as well. The effects of repeated exposure to thallium may be delayed for days or weeks after exposure. Fatigue, weakness, poor appetite, insomnia, mood changes, irritability, hair loss, a metallic taste in the mouth and pains in legs and arms are among the symptoms of low-level thallium poisoning. Low-level thallium exposure can also cause nerve damage (see NEUROTOXIN) that can lead to numbness and a "pins and needles" sensation in arms or legs, loss of vision, tremors, abnormal muscle jerking and permanent brain damage.

The OCCUPATIONAL SAFETY AND HEALTH ADMINISTRATION has established a permissible exposure limit for airborne thallium of 0.1 milligrams per cubic meter, averaged over an eight-hour work-shift. It is also cited in AMERICAN CONFERENCE OF GOVERNMENT INDUSTRIAL HYGIENISTS, Department of Transportation and ENVIRONMENTAL PROTECTION AGENCY regulations. Thallium is on the Hazardous Substance List. Businesses handling significant quantities of thallium or thallium compounds must disclose use and releases of the chemical under the provisions of the EMERGENCY PLANNING AND COMMUNITY RIGHT TO KNOW ACT.

Thermal Pollution   the artificial warming of water or air to the point that damage is caused to natural organisms. The principal source of thermal pollution is cooling water from electric power plants. The removal of trees and shrubs along the banks of creeks and small streams through logging and overgrazing can cause higher water temperatures than would exist if more of the water were shaded. Higher water temperatures encourage microbial growth (see MICROORGANISMS) that can lead to the EUTROPHICATION of surface water. Solids are more soluble (see SOLUTION) in warmer water, and a stream can carry more pollutants than it otherwise would. Gases are less soluble in warmer water, so levels of oxygen are lower, and it is easier for the water to become *anoxic* (lacking sufficient oxygen to maintain life). BIOCHEMICAL OXYGEN DEMAND increases as oxygen levels fall because of the growth of ALGAE and other heat-loving aquatic organisms, and because the METABOLISM of all aquatic life speeds up as water temperature increases. An 18°F increase in temperature—not uncommon at a large power plant—is enough to double the metabolism. When water is too warm, many species of fish have trouble reproducing, especially trout and salmon.

Three Mile Island   a nuclear power facility owned by Metropolitan Edison Company located on an island in the Susquehana River, a few miles downstream from Harrisburg, Pennsylvania. Early on the morning of March 28, 1979, a series of equipment malfunctions and operator errors that would make Three Mile Island (TMI) the site of the most serious accident involving a commercial nuclear installation in the United States got under way. The two TMI reactors rely on regular, "light" water as a coolant and as a moderator that controls the rate of fission in their cores. They are termed "pressurized-water reactors" because the water in the primary cooling system that removes heat from the core is pressurized to more than 2,000 pounds per square inch (psi) to keep it from boiling (see NUCLEAR REACTOR). The primary cooling system is routed through a steam generator where it boils the water in the secondary cooling system. The steam created in the steam generator drives turbines, which in turn spin a generator to produce electricity. The steam is then recondensed, passed through a polisher that removes minerals and other impurities through ION EXCHANGE, and is then circulated back to the steam generator.

The sequence of events that led to the accident started when the secondary cooling system on the 880-megawatt number two reactor became blocked at the polisher, causing the pumps that circulated water through the system to stop. The blockage of the secondary cooling system caused the reactor's control rods to automatically be lowered part way into the core to reduce the amount of heat being generated by fission. Two auxiliary pumps were also automatically switched on to feed water into the steam generator so that heat would continue to be transferred from the primary cooling system, but the pumps were unable to move any water because valves in the auxiliary feed water system were inexplicably closed. The steam generator acts as a heat sink for the primary cooling system, so with the water flow in the secondary system stopped, heat and pressure quickly built in the primary system to the point that the reactor was "scrammed"—its control rods started to move into the completely lowered position that would cause a 90% drop in the heat created by fission within about one minute. The reactor's operators then flipped a switch in the control room to close the pressure-relief valve in the primary cooling system that had opened to relieve the pressure buildup caused by the failure of the secondary cooling system. Although the indicator light showed that the valve had closed, it actually stuck part way open. As a result, when the reactor began to cool off, the pressure in the primary coolant system fell rapidly. When it dropped below 700 psi, the water in the primary system started to boil, creating steam pockets. The primary coolant pumps were severely strained trying to move water through the steam pockets, and the operators shut them down to avoid damaging their internal seals. With no flow of water through the primary system, enough cooling water drained out of the reactor core in about a half an hour to expose the top four feet of the fuel assemblies and severely overheat the core. As the core heated, the ZIRCONIUM coatings surrounding the uranium dioxide in the fuel pellets that make up the fuel rods started to oxidize, then to crack. There are 177 fuel assemblies containing about 90 tons of uranium dioxide and more than 100 control rods in the reactor. Although both the primary pumps were switched back on by the end of the day, they were unable to move much water because of the voids that remained in the primary system. Relative stability was restored to the reactor sys-

tem within two days, but nuclear experts from the Nuclear Regulatory Commission and elsewhere feared that hydrogen and oxygen was building up inside the reactor core, increasing the chances of a catastrophic explosion and a complete core meltdown. These fears eased within five days when it was discovered that it was inert xenon gas rather than potentially explosive hydrogen that had been building up inside the core.

Although disaster was averted, public confidence in the nuclear power industry was shaken by the Three Mile Island accident. Reports of releases of radioactive gases and fluids and the spectacle of public and private nuclear authorities issuing contradictory statements about what was happening inside the facility shattered the facade of nuclear power as being safe and reliable. People within a radius of up to 20 miles from the plant were alternately told that there was nothing to worry about and that evacuation was imminent. President Carter and Pennsylvania Governor Thornburgh toured the plant on April 1st to still public fears, even though NRC representatives still believed at the time that an explosion that could breach the containment building might be imminent. As it turned out, only an estimated 15 to 24 curies of radiation were released into the atmosphere by the event (as compared to 50 million curies that were released at CHERNOBYL).

Threshold Limit Value   an industrial indoor-air-quality system based on the concentration of hazardous chemical compounds and physical agents to which most workers can be exposed every day without physical effects. The threshold limit (TLV) value for a substance is its average concentration. The TLV *time-weighted average* is the average airborne concentration of potentially dangerous substances over a 40-hour workweek. Periods when the average is exceeded must be offset by times when concentrations are below the limit in order to reach the average. A TLV *short-term exposure limit* is the maximum concentration of a substance to which workers can be exposed for a short period (up to 15 minutes) without suffering health effects. A TLV *ceiling* is a concentration that should never be exceeded, not even for a short period. Ceiling values are assigned to quick-acting substances with a high degree of acute toxicity. Threshold limit values are set by the AMERICAN CONFERENCE OF GOVERNMENTAL INDUSTRIAL HYGIENISTS (ACGIH). TLVs are specified for toxic materials, mineral dust and noise, and are based on past experience with the substance in question in an industrial setting, and, where available, on testing performed on humans and animals.

Tipping Fee   the fee charged at a solid waste disposal facility to dump a truckload of garbage. The tipping fee charged at most trash-to-energy facilities increased dramatically because of the decrease in wholesale electricity rates (see AVOIDED COST PUBLIC UTILITIES REGULATORY POLICY ACT). Average tipping fees at a garbage incinerator are about $40 per ton (see INCINERATION), while LANDFILL tipping fees average about $25 per ton. See SOLID WASTE DISPOSAL.

Tipping Floor   see GARBAGE INCINERATORS

Tires   the flexible, synthetic-rubber devices that are fitted into the outer rims of wheels. A few modern tires are solid rubber, but most are *pneumatic*—they consist of a tough outer shell that covers an inflatable tube. A pneumatic tire can absorb the shocks to which moving vehicles are constantly exposed and provide the traction needed to turn and stop the vehicle and to transfer the motor's power to the road. About 280 million worn tires are removed from vehicles in the United States each year, and the ENVIRONMENTAL PROTECTION AGENCY (EPA) estimates that approximately 240 million of these are disposed of and that 2 billion worn tires are now stockpiled in the United States. The durability and elasticity that make tires effective while they are in use makes them very difficult to dispose of. When dumped in a landfill, the air trapped inside a tire makes it lighter than surrounding material, which gives them a tendency to float to the top, leaving "worm holes" in the carefully compacted landfill. Tires have been known to erupt out of the surface of a landfill with such explosive force that heavy equipment is upset and workers are injured. For these reasons, tires have been banned from most landfills unless they have first been shredded. The backlog of tires in U.S. tire dumps—estimated at two- to three billion tires—has grown as a result. Dumps holding millions of tires, many of them left behind by fly-by-night operators, have sprung up in every part of the United States, and most states now regulate scrap tires as a result.

A large stack of tires is a threat to public health both because of the significant threat of a catastrophic fire (see WINCHESTER TIRE FIRE), and because the water held in the old tires is prime breeding ground for mosquitoes. The petrochemicals in a typical passenger-car tire have the energy equivalent of about two and a half gallons of oil. A typical radial tire on a passenger car consists by weight of about 50% pure rubber polymer, 10% STEEL bead wire and 5% synthetic fibers, with the rest consisting of carbon black, antioxidants and materials including sulfur that are used to vulcanize, preserve and strengthen the tire rubber. Truck and heavy-equipment tires have a higher percentage of rubber. Burning a pound of the mixture of synthetic rubber yields about 15,000 Btus per pound—more than

either COAL or COKE. The air trapped inside each tire guarantees an ample supply of oxygen to feed the flames, if a fire should get started. Tires have been classified by the EPA as HAZARDOUS WASTE because of danger associated with possible fire and the fact that they can serve as a breeding place for insects and rodents.

Tires are, under normal conditions, practically chemically inert. Only about one-third of the petroleum derivatives in a tire is typically consumed in a fire, however, and much of the rest is freed by COMBUSTION to contaminate water supplies. BENZENE, TOLUENE and SULFUR DIOXIDE are released into the atmosphere when tires burn, as are large amounts of carbon black, which gives the smoke it characteristic dark black color.

About 30 million tires are retreaded each year. In order to be recapped, the tire casing must be in good shape, with no sidewall damage or uneven wear. Some of the best tires traded in for new are resold for further use. The popularity of the radial tire, which is more difficult to retread, has led to a decline in tire retreading. Before the introduction of radials, about 60% of tires sold were retreads, while only about 20% percent are retreads today. President Clinton signed an executive order that went into effect beginning in 1995 that requires federal agencies to buy recycled motor oil and retreaded tires for government-owned cars. A higher percentage of the very expensive tires used on heavy equipment and aircraft are retreaded.

The backlog of junk tires is being reduced by the growing practice of converting industrial boilers to burn tires instead of coal or fuel oil, and through an increasing number of RECYCLING techniques. When burned in a cement kiln, the steel from steel-belted tires contributes to the chemical process that creates cement. Paper mills, municipal WASTE-TO-ENERGY PLANTS and power plants are also joining the move to retired tires as fuel. An estimated 40% of the tires scrapped annually in the United States are used as fuel, and the percentage is growing rapidly. Boiler conversions capable of burning an entire, unshredded tire are being used at some of the more recent installations. The EPA says that tire-fueled boilers with proper emission controls are less polluting than equivalent coal-fired installations, but opponents of the practice say most installations do not have adequate emission controls and that enforcement of air quality standards is generally too lax to catch violations. Businesses are being offered strong financial incentives to burn tires for fuel. Factories typically receive a TIPPING FEE of between $10 and $20 per ton to burn tires, and save in addition the cost of the coal or fuel oil that normally would have been used.

Whole automobile tires have been used to control erosion, to provide habitat for fish and to build houses (with a concrete filling and a stucco finish). Larger tires have been used for everything from watering troughs to playground equipment. But such uses account for only a small fraction of the tires scrapped each year. Tire chips are prepared for recycling by running the tires through shredders, which slice the tires into chips called *crumb rubber*, which is the raw ingredient used as fuel and to produce recycled products including asphalt pavement, concrete with rubber aggregate called *rubcrete*, flooring, gym mats, artificial turf, door mats, trailer bumpers, mud flaps, erosion control nets, artificial reefs and pickup bed liners—even neckties. Crumb rubber is also used to replace some of the rock in gravel roads and septic system drain fields, and as a durable and functional but unsightly mulch. High-quality, long-lived road surfaces can be created by replacing some of the rock aggregate that would otherwise be used with crumb rubber. The equivalent of 16,000 tires worth of crumb rubber can be added to each mile of two-lane road. See also SOLID WASTE DISPOSAL.

Toilet   a plumbing fixture generally consisting of a porcelain bowl with a wood or plastic seat mounted on top the bowl in front of a porcelain water-storage tank. The flush in a *flush toilet* occurs when the stopper in the bottom of the tank is removed, allowing anywhere from one to six gallons of water (depending on the vintage and nationality of the toilet) to rush out of the tank and flush FECES, urine, TOILET PAPER and wastewater down through an opening in the bottom of the bowl, through a S-shaped trap, and into main drain lines that terminate at a SEWAGE COLLECTION SYSTEM or a septic tank (see SEPTIC SYSTEM). The average flush in Scandinavian countries uses only one-third the five gallons typical in the United States. A 30% reduction in the water used for flushing toilets would cut average residential water use in the United States by about 10%. Areas with chronic water shortages such as California often look to the toilet for water savings. Drives to get residents to throw out the old five-gallon flush toilets and replace them with more efficient models can release tens of thousands of old porcelain bowls and tanks into the wastestream. Some are reused as flower planters, chairs and other novelty items, or crushed and used as fill material or aggregate for concrete, but the majority go to a landfill.

Before the days of indoor plumbing and flush toilets, there was an OUTHOUSE in every back yard—anything from a prim privy with a half-moon cut out of the door, to a tilted shack in danger of falling into the pit. Flat pans called *chamber pots*, which were emptied in

the privy after use, were employed for convenience indoors. The *commode* was a piece of furniture featuring a strategically located hole centered over a receptacle that could later be emptied in the outhouse. Outhouses and pit privies are still in wide use in rural areas and in developing nations. Given the proper soils and barring excessive use, even an unlined outhouse can be an environmentally benign way of disposing of human waste, but contamination of both surface water and groundwater by outhouses is common. A composting toilet resembles the traditional privy outwardly, but takes the outhouse technology to a higher level by using two or more water-tight vaults to hold waste. While one side is in use, waste in the other vault is being converted to compost.

Portable toilets are used on construction sites, at concerts or large meetings and other outdoor events. Some are CHEMICAL TOILETS that reduce the volume of feces and solid materials, while others are simply sealed vaults that must be pumped into a tank pump periodically. The U.S. Forest Service and Park Service install permanent sealed-vault outhouses in campgrounds along streams, lakes and other sensitive areas. The Forest Service—which undoubtedly maintains more outhouses than any other agency or business in the United States—has developed a sealed vault outhouse constructed with pre-cast masonry walls with a solar chimney that heats up on sunny days and creates a downdraft through the toilet seat and carries unseemly odors up the chimney and outdoors. See also INCINERATING TOILET.

Toilet Paper   a low grade of paper used for essential bathroom paperwork.

Toluene   a colorless liquid with a sweet pungent odor that is used as a SOLVENT, in aviation gasoline and in the production of perfumes, medicines, drugs, dyes, explosives, detergents and other chemicals. Toluene is an AROMATIC HYDROCARBON that is a FLAMMABLE LIQUID and a fire hazard. Toluene, also called *methylbenzene*, is produced primarily from tar oil. It gets into air and water primarily from industrial discharges. It is slightly soluble and non-persistent in water, with a HALF-LIFE of less than two days. Toluene is employed as a solvent in a variety of household products, and it can get into indoor air when such products are used or spilled.

Toluene can affect health when inhaled or passed through the skin. Light exposure can cause irritation of the nose, throat and eyes, slowed reflexes, trouble concentrating and headaches. Prolonged contact can cause a skin rash. Exposure to higher concentrations can cause dizziness, lightheadedness, unconsciousness and death. Repeated exposure to toluene can damage bone marrow, resulting in a low blood cell count; can cause loss of appetite and nausea; and can damage the liver and kidneys. Toluene is a possible TERATOGEN and MUTAGEN. In May 1993, the California Attorney General threatened to sue manufacturers and distributors of household spray PAINTS unless they warned the public that the toluene they contain can cause birth defects.

PUBLIC WATER SYSTEMS are required under the SAFE DRINKING WATER ACT monitor for toluene (MAXIMUM CONTAMINANT LEVEL = 1 milligram per liter). The BEST AVAILABLE TECHNOLOGY for the removal of toluene from drinking water is granular ACTIVATED CARBON. The OCCUPATIONAL SAFETY AND HEALTH ADMINISTRATION has established a permissible exposure limit for airborne toluene of 200 parts per million (ppm), averaged over an eight-hour workshift, with concentrations not to exceed 300 ppm during any 15-minute work period and a maximum peak allowable concentration of 500 ppm. It is also cited in AMERICAN CONFERENCE OF GOVERNMENT INDUSTRIAL HYGIENISTS, Department of Transportation and NATIONAL INSTITUTE OF SAFETY AND HEALTH regulations. Toluene is on the Workplace Substance List, the Special Health Hazard Substance List and the HAZARDOUS AIR POLLUTANTS LIST. Businesses handling significant quantities of toluene must disclose use and releases of the chemical under the provisions of the EMERGENCY PLANNING AND COMMUNITY RIGHT TO KNOW ACT.

Total Dissolved Solids   all the solids in SOLUTION in a sample of water (see SOLUBILITY, SOLVENT). Water is drawn through a *Gooch crucible*, a clay vessel that allows the passage of materials that are in solution while capturing suspended solids, to measure its load of dissolved solids in a laboratory.

Total Solids   the total load of solid materials in a sample of water. Total solids are the sum of TOTAL DISSOLVED SOLIDS plus TOTAL SUSPENDED SOLIDS. Total solids in a water sample can be measured in the laboratory by boiling away all the water and weighing the resulting residue.

Total Suspended Particulates   the total amount of solid and liquid PARTICULATES suspended in the air at a given time. Total suspended particulates are one of the five CRITERIA POLLUTANTS named in the CLEAN AIR ACT. The NATIONAL AMBIENT AIR QUALITY STANDARDS set a TSP limit of 75 micrograms/cubic meter. The average is about 50. Particulate emissions from U.S. industrial plants were reduced from almost 18 million tons in 1970 to less than 9 million in 1980 as a result of

improved POLLUTION-control technology (see POLLUTANT STANDARD INDEX). See also AERIAL DEPOSITION, ENVIRONMENTAL TOBACCO SMOKE.

Total Suspended Solids   all the solid materials in SUSPENSION in a sample of water.

Toxaphene   an organochlorine INSECTICIDE (see ORGANOCHLORIDE) that has been used to control army worms, grasshoppers and insects that infest cotton crops and to eradicate undesirable fish species. Toxaphene, also called *chlorinated camphene*, is a yellow, waxy solid that is usually dissolved in a liquid. It is not a COMPOUND, but a MIXTURE of about 175 closely related CHLORINATED HYDROCARBONS. Toxaphene gets into air and water from industrial discharges, agricultural RUNOFF, atmospheric deposition and spills. It is slightly soluble and highly persistent in water, with a HALF-LIFE of more than 200 days, and has high acute and chronic TOXICITY to aquatic life (see ACUTE EFFECTS, CHRONIC EFFECTS). Toxaphene tends to accumulate in fats. It can build to considerably higher concentrations in the tissue of fish than is present in the water from which they are taken. The ENVIRONMENTAL PROTECTION AGENCY banned toxaphene in 1982 after it was found in Great Lakes' water, more than 1,000 miles from the nearest area of extensive use.

Toxaphene can affect health when inhaled or passed through the skin. Exposure to concentrated toxaphene can irritate the skin and eyes. Inhaling toxaphene DUST or mist can cause nose, throat and lung irritation. Exposure to higher concentrations can affect the nervous system causing tremors, weakness, dizziness, increased saliva, nausea, vomiting, convulsions and unconsciousness. Exposure to high concentrations or repeated exposure to lower concentrations may cause kidney and liver damage. Toxaphene is a possible human CARCINOGEN that has been shown to cause liver cancer and thyroid tumors in animals. Limited evidence suggests toxaphene is a TERATOGEN.

PUBLIC WATER SYSTEMS are required under the SAFE DRINKING WATER ACT to monitor concentrations of toxaphene (MAXIMUM CONTAMINANT LEVEL = 0.003 milligrams per liter). The BEST AVAILABLE TECHNOLOGY for the removal of toxaphene from drinking water is granular ACTIVATED CARBON. The OCCUPATIONAL SAFETY AND HEALTH ADMINISTRATION has established a permissible exposure limit for airborne toxaphene of 0.5 milligrams per cubic meter, averaged over an eight-hour workshift. It is also cited in AMERICAN CONFERENCE OF GOVERNMENT INDUSTRIAL HYGIENISTS, National Toxicology Program and Department of Transportation regulations. Toxaphene is on the Hazardous Substance List, the Special Health

Hazard Substance List and the HAZARDOUS AIR POLLU-TANTS LIST. Businesses handling significant quantities of toxaphene must disclose use and releases of the chemical under the provisions of the EMERGENCY PLANNING AND COMMUNITY RIGHT TO KNOW ACT.

Toxic Precipitation   precipitation that carries CONTA-MINANTS in SOLUTION and SUSPENSION. Toxic precipita-tion is AIR POLLUTION that has become bonded to rain drops, snow or ice crystals. The aerial application of PESTICIDES, industrial emissions, volatile compounds such as the SOLVENTS in paints that evaporate into the atmosphere, PARTICULATES from emissions, wind ero-sion and volcanoes are some of the principal sources of the contaminants found in toxic precipitation. Contaminants can be carried a long distance, even around the world, before falling to earth as part of pre-cipitation, and land and water in extremely remote locations can become contaminated as a result. DDT and other pesticides and POLYCHLORINATED BIPHENYLS (PCBs) have been found in the polar ice caps where they were deposited by toxic precipitation. Very small, relatively insoluble substances such as DDT and PCBs stay in the atmosphere for a long time. Although the rate at which such PERSISTENT COM-POUNDS wash out of the atmosphere is low, concentra-tions can build up in surface water because of their indestructibility and can build in organisms through BIOACCUMULATION. Studies have estimated that 163 metric tons of POLYAROMATIC HYDROCARBONS get into Lake Superior through polluted precipitation each year—90% of the total. Some of the contaminated pre-cipitation fell directly into the lake, and the rest into its watershed. See also ACID RAIN, AERIAL DEPOSITION, INDUSTRIAL AIR POLLUTION.

Toxic Substances Control Act   federal legislation passed in 1976 designed to protect the public from injury as the result of the use of toxic materials in indus-try. The Toxic Substances Control Act (TSCA) applies to pure chemical substances, impurities and contaminants found in chemicals and incidental reaction products formed when a compound is manufactured or when it is used. Priority is given to CARCINOGENS, MUTAGENS and TERATOGENS. Manufacturers are required under the act to notify the ENVIRONMENTAL PROTECTION AGENCY (EPA) of the introduction of a new chemical, either domestically manufactured or imported, in a *premanu-facturing notification* that must be submitted at least 90 days prior to introduction. The EPA is granted author-ity under the act to restrict or ban any chemical it finds to pose a substantial risk to human health or the envi-ronment or to order a manufacturer to test any such substance if insufficient safety data is available. The

TSCA requires manufacturers to keep records of the purchase, use and disposal of toxic chemical com-pounds, and the act calls for the creation of a national inventory of toxic substances used in industrial processes. The further use of POLYCHLORINATED BIPHENYLS (PCBs) in the United States is specifically prohibited in the act. The post of EPA Assistant Administrator for Toxic Substances was established under the act to oversee TSCA programs. Foods, food additives, drugs, cosmetics, tobacco, pesticides and radioactive materials are exempt from regulation under the TSCA. The INDOOR RADON ABATEMENT ACT is a federal law passed in 1988 as an amendment to the Toxic Substances Control Act that authorized $45 mil-lion in expenditures over a three-year period for research and information dissemination regarding res-idential RADON contamination.

The *TSCA Hotline* is a source of information relating to the TSCA, the ASBESTOS HAZARD EMERGENCY RESPONSE ACT and the Asbestos School Hazard Abatement Act: 202-554-1404.

See also CLEAN AIR ACT, CLEAN WATER ACT, RESOURCE CONSERVATION AND RECOVERY ACT.

Toxic Waste   waste that contains materials harmful to living organisms. Toxic waste is usually thought of as a solid or a liquid, although AIR POLLUTION involves the emission of what amounts to toxic waste into the ATMOSPHERE (see DILUTION). Waste that is especially toxic must be disposed of as a HAZARDOUS WASTE (see HAZARDOUS MATERIALS, HAZARDOUS WASTE DISPOSAL). See also TOXIC SUBSTANCES CONTROL ACT, TOXICITY.

Toxic Waste Disposal   see HAZARDOUS WASTE DIS-POSAL

Toxics Release Inventory   a database containing information about the use and release of toxic sub-stances from industrial sources mandated under the EMERGENCY PLANNING AND COMMUNITY RIGHT TO KNOW ACT. The act covers the release of toxics to air, water or land, and the off-site transfer of wastes for treatment or disposal. In January 1994 the ENVIRON-MENTAL PROTECTION AGENCY (EPA) proposed adding 313 chemicals to the more than 300 toxics originally stipulated in the act. Federal facilities were required to report the release of toxic substances covered by the act by an executive order issued in August 1993. In 1994 the EPA added energy production facilities, waste-management and transportation facilities to the industries already required to comply with the act.

The publication of the first inventory of toxics released from 20,000 U.S. industrial plants and refineries was an embarrassment to industry, with about 10.3 billion pounds of hazardous waste reported. Approximately 3.5 billion pounds of the waste was produced by just 24 plants, with a dozen facilities producing more than 100 million pounds each. INJECTION WELLS and on-site burial (see ON-SITE DISPOSAL) were the most commonly used disposal methods among the largest polluters. The EPCRA has caused a reduction in the amount of reportable waste produced and in the way it is discarded. The 1994 inventory showed an overall 12.6 percent decline from 1993 emissions nationwide. At least part of the reduction, however, simply reflects the fact that more wastes were shipped away from factories for off-site disposal or recycling, and many of these facilities are not subject to the annual reporting requirement. Louisiana ranked first in the 1995 report, with a total of 451 million pounds of reportable chemicals released. Texas was next with 352 million pounds, followed by Tennessee with 188 million pounds, Ohio with 138 million pounds and Mississippi with 118 million pounds.

For information about the TRI database, call: 301-496-6531. To learn more about the EPCRA, call the RESOURCE CONSERVATION AND RECOVERY ACT *Underground Storage Tank-Superfund Hotline* at 800-424-9346 (382-3000 in Washington, DC). To buy a copy of the TRI on CD-ROM, microfiche, diskette, magnetic tape or printed copy, contact the Government Printing Office (710 N Capitol Street NW, Washington, DC 20401—phone: 202-783-3238) or the National Technical Information Service (U.S. Dept of Commerce, 5285 Port Royal Rd., Springfield, VA 22161—phone: 800-553-6847).

Toxicity   the ability of a substance to cause illness or death after it is inhaled or ingested or after it passes through the skin. The toxicity of a substance may vary according to its path of entry into the body. ASBESTOS, for instance, is a known CARCINOGEN when inhaled, but its effects on the digestive system are more subtle or nonexistent. Toxicity is usually expressed as the quantity of a material that will kill 50% of a population exposed to it (see LD$_{50}$).

Toxicity Characteristic Leaching Procedure   a testing procedure in which the TOXICITY of LEACHATE produced by potentially toxic materials such as incinerator ash (see BOTTOM ASH, FLY ASH) or the SLUDGE produced by a DRINKING-WATER treatment facility is measured by passing water through the substance. The procedure is designed to simulate conditions that would be encountered if the material were to be disposed of in a landfill. See HAZARDOUS MATERIAL.

Trace Element   an element that is required in minute quantities for the proper functioning of an organism. Trace elements are, by definition, required in the body at levels of less than 50 parts per million, and most organisms are very sensitive to the level at which they are available. Too much of a trace element commonly causes poisoning, while too little may lead to malnutrition. Many of the actions of trace elements relate either to enzymes or their production. COPPER, SELENIUM, MANGANESE, COBALT, molybdenium, iodine and IRON are essential trace elements.

Transfer Station   a site where GARBAGE TRUCKS dump MUNICIPAL SOLID WASTE for later shipping via semi truck or rail to another site for ultimate disposal. Many rural areas have transfer stations where homeowners dump their garbage in dumpsters that are later dumped into trucks for transport to a LANDFILL or GARBAGE INCINERATOR.

Trash   PAPER, PLASTIC, METAL and other dry objects that are thrown away as useless. See also DEBRIS, LITTER. Compare GARBAGE, REFUSE.

Trash Compactor   a device designed to reduce the volume of domestic and commercial TRASH through compaction. The dense blocks that come out of a compactor are usually disposed of along with other garbage, but blocks made from nonorganic trash are sometimes used for building construction.

Trash-to-Energy Plant   see WASTE-TO-ENERGY PLANT

Trichloroethane   see METHYL CHLOROFORM

Trichloroethylene (TCE)   a clear, colorless nonflammable liquid with a sweet odor that is widely used as a SOLVENT for fats, waxes, adhesives, resins, ores, rubber, paints and varnishes. Trichloroethylene is also used as a degreaser, in the production of other chemicals and in dry cleaning. It gets into surface water from industrial discharges, municipal waste-treatment plants and from spills. TCE often gets into groundwater when it is washed into a floor drain or directly into the earth after being used to degrease tools and equipment. It is also the working ingredient in many septic tank degreasers. TCE is moderately soluble and nonpersistent in water, with a HALF-LIFE of less than two days. Concentrations of TCE in groundwater can build to levels thousands of times higher than those found in surface water because it cannot evaporate.

Trichloroethylene can affect health when inhaled. Symptoms of exposure include lightheadedness, dizziness, visual disturbances, an excited feeling, nausea and vomiting. Liquid TCE may irritate the skin, causing a rash or burning feeling, and may also damage the eyes. Repeated immersion of the hands in trichloroethylene can cause paralysis of the fingers. TCE vapor can irritate the eyes, nose, throat and lungs. Exposure to higher concentrations can cause an irregular heartbeat, brain damage, a buildup of fluid in the lungs, unconsciousness or death. Repeated exposure may cause fatigue, weakness in the arms and legs, memory loss, headache, irritability, intolerance of alcohol, mental confusion and depression. TCE can damage the liver and kidneys. It is a possible human CARCINOGEN that has been linked to liver cancer in animal testing. Limited evidence suggests trichloroethylene is a TERATOGEN in animals, so it is considered a possible human teratogen.

PUBLIC WATER SYSTEMS are required under the SAFE DRINKING WATER ACT to monitor for TCE (MAXIMUM CONTAMINANT LEVEL = 0.005 milligrams per liter). The OCCUPATIONAL SAFETY AND HEALTH ADMINISTRATION has established a permissible exposure limit for airborne TCE of 100 parts per million (ppm), averaged over an eight-hour workshift. The odor threshold for TCE is 28 ppm. It is also cited in AMERICAN CONFERENCE OF GOVERNMENT INDUSTRIAL HYGIENISTS, NATIONAL INSTITUTE OF OCCUPATIONAL SAFETY AND HEALTH and Department of Transportation regulations. Trichloroethylene is on the Hazardous Substance List, the Special Health Hazard Substance List and the HAZARDOUS AIR POLLUTANTS LIST. Businesses handling significant quantities of trichloroethylene must disclose use and releases of the chemical under the provisions of the EMERGENCY PLANNING AND COMMUNITY RIGHT TO KNOW ACT.

Trichloromethane see CHLOROFORM

Trihalomethanes (THMs) a group of compounds that consist of a METHANE molecule (CH_4) with three of its four HYDROGEN atoms replaced by chlorine, fluorine, bromine or iodine. Trihalomethanes are produced when drinking water that contains POLYCYCLIC ORGANIC MATTER is chlorinated (see CHLORINE, DISINFECTION). Bromochloromethane, bromodichloromethane, bromoform, and CHLOROFORM are among the THMs commonly produced. A 1975 ENVIRONMENTAL PROTECTION AGENCY survey of finished drinking water in eighty cities turned up the THM chloroform in all 80 systems and three other THMs present in most systems. A 1980 epidemiological study revealed higher rates of rectal and colon cancer in cities with chlorinated drinking water. Studies by the Harvard School of Public Health and others have identified THMs as the greatest threat to health of any of the toxins commonly found in drinking water. According to the HSPH study, THMs are responsible for

approximately 10,000 additional cases of cancer annually in the United States, including about 20% of rectal cancers and 10% of bladder cancers. PUBLIC WATER SYSTEMS are required under the SAFE DRINKING WATER ACT to monitor for trihalomethanes [interim MAXIMUM CONTAMINANT LEVEL (MCL) = 0.1 milligrams per liter (mg/L); proposed MCL = 0.08 mg/L]. See also WATER TREATMENT.

Trickling Filter a wastewater filter used in SEWAGE TREATMENT plants to remove organic materials. A typical trickling filter consists of a tank filled with layers of rock, SLAG or other inert materials with large surface areas that support microbial growth. Oxygenated wastewater is introduced into the top of the filter, and as it filters down through the gelatinous mass of AEROBIC BACTERIA, ALGAE, fungi, protozoa, rotifera and nematoda (collectively referred to as zoogloeal), growing in the filter medium, the microorganisms metabolize most of the remaining organic material, producing an organic mat full of carbon DIOXIDE, NITRATES and SULFATES, the waste products of microbial action, and suspended particles trapped in the biological filter. *Slough off* is humic mass that breaks away and is washed by wastewater flow through the filter's large voids, and then screened out and removed to a settling tank for capture and disposal of solids. See also PONDING, ROTATING BIOLOGICAL CONTACTOR.

2,4-Dichlorophenoxyacetic Acid (2,4-D) a chlorinated hydrocarbon HERBICIDE (see CHLORINATED HYDROCARBON) used to kill broadleaf plants on farms and in forests, lawns and gardens. 2,4-D, which is a white to yellow crystalline powder, is also used to promote latex production from old rubber trees. The only difference between 2,4-D and 2,4,5-T (see 2,4,5-TRICHLOROPHENOXYACETIC ACID) is that 2,4,5-T has one more chlorine atom per molecule. The compounds have many of the same health effects. It gets into air and water from agricultural and URBAN RUNOFF, spills and industrial waste. 2,4-D is highly soluble and moderately persistent in water, with a HALF-LIFE of 20 to 200 days. The concentration of 2,4-D in fish is normally much higher than that of the water in which the fish lives. 2,4-D is often used as a defoliant.

2,4-D can affect health when inhaled or by passing through the skin. The ACUTE EFFECTS of 2,4-D exposure include flu-like symptoms, sometimes accompanied by muscle tenderness, weakness and twitching and poor coordination. Breathing 2,4-D fumes can irritate the throat and airways, and contact can cause eye and skin irritation. Continued exposure can result in loss of patches of skin pigment, emphysema and kidney and liver damage. A thinning of the lining of the throat called *atrophy* can also result from repeated exposure. 2,4-D is a MUTAGEN and a possible CARCINOGEN. There is limited evidence that it causes breast and other cancers in animals. Epidemiological studies have shown that farmers that use 2,4-D have a higher than average risk of contracting non-Hodgkin's lymphoma. 2,4-D is a possible human and animal TERATOGEN, and it may decrease fertility in males.

PUBLIC WATER SYSTEMS are required under the SAFE DRINKING WATER ACT to monitor concentrations of 2,4-D (MAXIMUM CONTAMINANT LEVEL = 0.07 milligrams per liter). The OCCUPATIONAL SAFETY AND HEALTH ADMINISTRATION has established a permissible exposure limit for airborne 2,4-D of 10 milligrams per cubic meter, averaged over an eight-hour workshift. The BEST AVAILABLE TECHNOLOGY for the removal of 2,4-D is granular ACTIVATED CARBON. It is also cited in AMERICAN CONFERENCE OF GOVERNMENT INDUSTRIAL HYGIENISTS and Department of Transportation regulations. 2,4-D is on the Hazardous Substance List. Businesses handling significant quantities of 2,4-D must disclose use and releases of the chemical under the provisions of the EMERGENCY PLANNING AND COMMUNITY RIGHT TO KNOW ACT. See also PESTICIDE.

2,4,5-Trichlorophenoxyacetic acid (2,4,5-T) a chlorinated-hydrocarbon HERBICIDE that was mixed 50/50 with 2,4-D in Vietnam to form the defoliant known as *Agent Orange*. Between 1965 and 1970 the American military sprayed approximately 11 million gallons of agent orange over 3.6 million acres of what was then South Vietnam. The only difference between 2,4,5-T and 2,4-D (see 2,4-DICHLOROPHENOXYACETIC ACID) is that 2,4,5-T has one more chlorine atom per molecule. The compounds have many of the same health effects. 2,4,5-T is always contaminated with *TCDD*, one of the most toxic forms of DIOXIN. 2,4,5-T degrades fairly quickly in soils, with a HALF-LIFE that ranges between a few weeks to a few months. The herbicide is taken up through the leaves or roots of a plant and then transported to other parts. When it is sprayed on plants from the air, the majority is absorbed through the leaves, leading to quicker action. 2,4,5-T is produced as an ACID, a SALT and an ESTER, and the three forms have slightly different properties. It is more effective than 2,4-D against woody plants.

2,4,5-T is more toxic to aquatic organisms than it is to terrestrial lifeforms. It has been shown to be a a MUTAGEN in tests on mice and hamsters. It was used for clearing brush and right-of-way maintenance until uses were curtailed during the early 1970s. Epidemiological studies of the health effects of exposure to 2,4,5-T by workers at pesticide factories have found that most serious health effects are associated with TCDD.

The ENVIRONMENTAL PROTECTION AGENCY (EPA) issued an emergency suspension order for all uses of 2,4,5-T that was instituted in 1979. It was only the third time in the EPA's history that the agency had taken such an action on a PESTICIDE, and the move was based more on capitulation to public pressure than on science. Spontaneous miscarriages suffered by women in the southern Oregon Coast Range during the 1970s in an area that had been widely sprayed with 2,4,5-T to kill broadleaf vegetation in logged over conifer forests, and concern about 2,4,5-T's role in miscarriages and birth defects among Vietnamese women were largely responsible for the public's interest. Fruits, rice and rangeland were exempt from the ban. The herbicide is no longer manufactured in the United States, and existing stocks are being destroyed.

The OCCUPATIONAL SAFETY AND HEALTH ADMINISTRATION has established a limit for workplace exposure to airborne 2,4,5-T of 10 milligrams per cubic meter. PUBLIC WATER SYSTEMS are required under the SAFE DRINKING WATER ACT to monitor concentrations of 2,4,5-T (MAXIMUM CONTAMINANT LEVEL = 0.05 milligrams per liter. The BEST AVAILABLE TECHNOLOGY for the removal of 2,4,5-T from drinking water is granular ACTIVATED CARBON. The EPA has set a water-quality criterion for 2,4,5-T of 10 micrograms per liter. See also GROUNDWATER POLLUTION, WATER POLLUTION.

Tritium a colorless, odorless gas that is a radioisotope of HYDROGEN (hydrogen-3): It occurs naturally only rarely, but is common as a byproduct of the decay produced by nuclear FISSION (see NUCLEAR REACTOR, RADIOACTIVE WASTE). Tritium—also called *heavy hydrogen*—consists of one proton, two neutrons and one electron. Like hydrogen, tritium reacts readily with oxygen to form water, and the resulting water molecules, called *tritiated water*, are about one-fifth heavier than normal water molecules. Tritiated water is sometimes called HEAVY WATER, but should not be confused with the heavy water created when hydrogen isotope DEUTERIUM combines with oxygen to form water. Most of the tritium released from nuclear facilities and nuclear explosions is tritiated water, in either liquid or vaporous form. Tritium is created in the upper atmosphere when cosmic rays strike air molecules. About twice as much tritium is produced in nuclear reactors each year as is produced by the action of cosmic rays, but only an estimated 4% of this tritium is released to the environment, primarily in nuclear plant cooling water. The atmospheric detonation of nuclear weapons also releases tritium. Tritium is used commercially to produce luminous dials. Tritium gas inside a phosphor-coated glass tube provides emergency illumination for safety signs in buildings. The fusion reaction in which deuterium (also known as *heavy water*) combines with TRITIUM to produce helium and free neutrons is the source of the phenomenal release of energy in a hydrogen bomb. If FUSION power becomes a viable energy option in the future, the rate of release of tritium into the environment is likely to increase. An increase in the use of uranium reprocessing plants (see URANIUM, REPROCESSING PLANT) could also increase releases of tritium.

The SAVANNAH RIVER nuclear plant is the only U.S. source of pure tritium, which is used to increase the yield of nuclear warheads. Tritium decays at a rate of 5.5% per year, so must be replenished occasionally to maintain the warhead's explosive yield. A leak of tritium at the Savannah River plant in 1991 forced downstream drinking-water systems to switch to alternate water supplies for 10 days. Tritium has a HALF-LIFE of 12.3 years, and must be stored for 123 years before it will lose 99.9% of its radioactivity by decaying into helium and BETA PARTICLES. It is collected as a gas from the water tanks in which nuclear fuel rods are stored (see RADIOACTIVE WASTE DISPOSAL).

Since tritium does not concentrate in any particular organ, the symptoms of tritium poisoning are similar to those experienced by the victim of whole-body exposure to any radioactive material. Tritium has the same chemical properties as hydrogen—and it emits a traceable signal as it decays—leading to its use by biologists to trace the movements of hydrogen compounds inside organisms. Hydrogen and therefore tritium compounds are biologically active, and have been distributed throughout the body within an hour of ingestion. The radiation emitted by tritium can only penetrate a few microns of tissue, so the radioactive decay of atmospheric tritium poses little threat.

Trophic Classification System a method of classifying lakes, reservoir bays and other standing bodies of water according to the number of living organisms their water supports. An *oligotrophic* body of water has low biological productivity, a *eutrophic* water body has high productivity and *mesotrophic* is a classification that falls between the two extremes. Trophic classification systems are based on the body of water's *primary*

productivity, the rate at which organic growth occurs. Warmer water tends to support more growth than cooler, and shallower water that receives ample sunlight tends to be more productive than deeper water. The amount of NUTRIENTS in water and in lakebed SEDIMENTS, as well as the concentration of DISSOLVED OXYGEN, are other principal growth-limiting factors. See also THERMAL POLLUTION, SEDIMENT, SILTATION.

Turbidity the presence of suspended materials in WATER (see COLLOIDAL SOLIDS, SUSPENDED LOAD, SUSPENDED SOLIDS). Turbidity reduces the transparency of water and limits the depth to which light can penetrate (see SECCHI DISK). Water with a high degree of turbidity contains organic material that interferes with the DISINFECTION process (see POLYCYCLIC ORGANIC MATTER), producing TRIHALOMETHANES and other contaminants as byproducts. *Inorganic turbidity* is caused primarily by silts and clays, while *organic turbidity* is the result of contamination with MICROORGANISMS and can result from EUTROPHICATION. The fact that available light decreases as microbial populations and resulting turbidity increase limits the degree of turbidity caused by microbial growth.

PUBLIC WATER SYSTEMS are required under the SAFE DRINKING WATER ACT to monitor turbidity, with the turbidity of filtered water never to exceed 0.5 Normal Turbity Units (NTU). Turbidity samples are to be acquired every four hours via grab sampling or continuous monitoring (see POLLUTION MEASUREMENT). No more than 5% of the measurements can exceed 0.5 NTU for conventional treatment or direct filtration, or 1.0 NTU for slow-sand and diatomaceous earth filtration. To avoid the requirement that water systems drawing from surface sources install filtration equipment, a utility must demonstrate that turbidity prior to treatment is less than 5 NTU, based on measurements taken every four hours. Rather than a MAXIMUM CONTAMINANT LEVEL, a treatment process is specified for turbidity. Treatment must reduce the turbidity of the water being processed by at least 80%, or must produce effluent with less than 0.5 NTU of turbidity. See also JACKSON TURBIDITY UNIT, LONG-TERM TURBIDITY, SHORT-TERM TURBIDITY.

U

Uranium a silvery lustrous METAL that emits moderate levels of RADIOACTIVITY as the result of its nucleus being too large to be chemically stable. Uranium is the heaviest naturally occurring element. The HALF-LIFE of uranium-235 is 710,000 years. Decaying uranium is a source of ALPHA PARTICLES. Uranium combines readily with oxygen, and will burn in air at 170°C (338°F) forming several uranium oxides in the process. Because of its affinity for oxygen, elemental uranium does not exist naturally, but is found instead as one of several oxides. Uranium isotopes with atomic weights from 227 to 240 have been produced in the laboratory, but uranium occurs naturally in only three forms: U_{234} with 142 neutrons, U_{235} with 143 neutrons and U_{238} with 148 neutrons. All three isotopes have 92 protons. Uranium's only use is as fuel for NUCLEAR REACTORS and as an explosive for nuclear weapons. U_{235} is a *fissile material*. It is capable of supporting fission, as is U_{233}, a fission product of thorium-232.

Traces of uranium are common in many rocks (see RADON). Commercial ore—*pitchblende* is U_3O_8, and *uranite* is U_2O—usually contains four to five pounds of uranium oxide per ton, of which only 0.02 to 0.03 pounds are U_{235}. At the mill, uranium ore is refined into yellowcake, which is 80% uranium oxide. The uranium oxide is converted into a gas, *uranium heptafluoride*, which is then run through an *enrichment* process which increases the U_{235} content of the fuel from about 0.7% to 3.0 or 4.0%—more for weapons-grade uranium. The uranium gas is then converted back into uranium, which is then shaped into pellets that are packed into the fuel rods for nuclear reactors. Areas with soils and bedrock composed primarily of granite, shale and phosphate rocks and gravels and soils derived from them have the highest concentration of uranium and its decay products. Radon readings are especially important for homes built near the Reading Prong, a uranium rich underground formation that cuts across New York, New Jersey and Pennsylvania.

Approximately 8 million pounds of uranium concentrate was produced by U.S. mines in 1991. About 2 million pounds of that total was exported from the United States, and another 14 million pounds was imported. Spent uranium fuel at commercial nuclear power plants (see RADIOACTIVE WASTE) accounts for less than 1% of the total volume of all radioactive waste in the United States, but accounts for 95% of the radioactivity from all U.S. sources, commercial, experimental and military. Uranium gets into surface water primarily from uranium mines, and from the estimated 200 million tons of TAILINGS at 26 uranium mills located in New Mexico, Utah, Wyoming, Colorado and Texas. Tailings from GYPSUM mines and mills often contain uranium. Uranium gets into groundwater from the inadequate storage of nuclear wastes, such as has occurred at several U.S. military installations involved with the production of nuclear weapons (see HANFORD NUCLEAR RESERVATION, ROCKY FLATS, OAKRIDGE NATIONAL LABORATORIES, SAVANNAH RIVER). PUBLIC WATER SYSTEMS are required under the SAFE DRINKING WATER ACT to monitor concentrations of uranium (MAXIMUM CONTAMINANT LEVEL = 30 picocuries per liter). Uranium can be removed from water by means of alum (see ALUMINUM) and IRON coagulation, lime softening, REVERSE OSMOSIS and ION EXCHANGE.

Uravan Uranium Mill a URANIUM mill located on the San Miguel River, five miles upstream from its confluence with the Dolores River south of Grand Junction, Colorado. The mill started operation in 1915, processing radium, and switched to uranium processing during the Second World War. Uranium from the Uravan Mill was used in the production of the world's first atomic bombs, and, after the war, in nuclear power plants. When the mine was closed in 1984, an estimated 10 million tons of TAILINGS had accumulated at the site. LEACHATE from the tailings piles contaminated both the San Miguel and the Dolores with uranium and other RADIONUCLIDES, and also percolated into the underlying aquifer. Union Carbide agreed in 1987 to clean up the site.

An $80 million cleanup of the Uravan mill site was about half completed in 1996. Uravan is located in one of the least accommodating climates and least populated regions in Colorado. The nearest neighbor of the site lives five miles away. The nearest town, Naturita, population 430, is 13 miles away, and there is no reason to expect much future growth. But when the cleanup is completed, the site will be the kind of place where children can safely play in a backyard or school playground. As a result, Uravan is frequently cited by critics of the Superfund program as an example of the inflexibility and impracticality of cleanup standards.

Urban Runoff   precipitation that falls on city streets, building roofs and other impenetrable surfaces, runs off on the surface and empties into a STORM SEWER system or directly into a lake or stream. Urban runoff is often contaminated with oil and heavy metals, which can pollute water. Polluted surface runoff collected by a COMBINED SEWER SYSTEM can disrupt the metabolism of microbes at the sewage treatment plant, and heavy runoff can overwhelm the system, which can result in the release of raw sewage (see STORM RUNOFF). Urban runoff is a principal source of DIFFUSE POLLUTION.

Urea   see AMMONIA

V

Valley of Gehenna a valley near Jerusalem that was used in Old Testament times as a garbage incineration site. Fires fueled by naturally occurring gas vents in the valley, which is located south of Jerusalem, were used to incinerate the trash. Over a period of time, the name *Gehenna* became a synonym for hell—or a place or state of misery—a meaning it still carries.

Valley of the Drums an illegal, 23-acre hazardous-waste dump near Brooks, Kentucky, about 15 miles south of Louisville at the headwater of Wilson Creek. The A. L. Taylor Drum Cleaning Service started dumping the toxic liquids from the 55-gallon drums that were processed at the site onto the ground in 1967. Although the state cited the company several times during the 1970s for illegal dumping, no action was taken to stop the dumping until 1979, when ENVIRON-MENTAL PROTECTION AGENCY officials became aware of the dumpsite. The ensuing EPA investigation revealed more than 17,000 barrels containing toxic substances stacked on top of the ground, and it was estimated that at least that many more were buried at the site. PHE-NOLS, POLYCHLORINATED BIPHENYLS (PCBs) and a variety of other toxins were found in nearby Wilson Creek at concentrations of up to 100,000 parts per billion. RUNOFF from the site contained BENZENE and HEAVY METALS. Almost 200 toxic substances were identified in the soil at the site. The EPA called for an emergency cleanup at the site to prevent further spread of contamination. The owner and operator of the site, Arthur L. Taylor, died in 1978, the year before the EPA started its investigation of the dump, so no action could be taken against Taylor Drum. The gross contamination found in the Valley of the Drums along with the LOVE CANAL incident helped to alert government officials and the public of the inadequacy of U.S. hazardous-waste disposal ordinances, and helped speed the passage of the Superfund waste-cleanup legislation in 1980 (see COMPREHENSIVE ENVIRONMENTAL RESPONSE, COMPENSATION AND LIABILITY ACT).

Vapor diffused matter that is suspended in the air, altering its transparency. Fog, MIST and SMOKE are commonly encountered vapors. Vapor can also refer to the gaseous state of any compound. See EVAPORATION, VAPOR PRESSURE, VOLATILITY.

Vapor Pressure the pressure at which the number of molecules evaporating from a liquid or solid is equal to the number of molecules being reincorporated so that there is no net loss of volume. Vapor pressure varies widely among different classes of compounds, and is dependent on temperature. See EVAPORATION, VOLATILITY.

Vapor Recovery System an air-handling system designed to trap and store hazardous fumes that evaporate from volatile liquids (see VOLATILITY) stored or used in an industrial process. A typical vapor recovery system consists of a series of hoods over areas where vapors are released connected to hose or ducting that are, in turn connected to a fan or blower that blows the toxic gases back to the tank holding the volatile liquid. In some cases, the air containing the toxic vapors is blown back into the storage vessel to replace the volume of liquid that has been removed. In other instances recovered vapors are routed to a condenser or refrigeration unit where they are reliquified and pumped back into the storage container. See also AIR POLLUTION CONTROL.

Vapor Well see MONITORING WELLS

Ventilation the circulation of fresh air through a building. Ventilation removes stale indoor air and replaces it with fresh outdoor air. Rates of ventilation in homes, businesses and office buildings have been reduced over the last couple of decades as the price of energy for heating and cooling has increased. Higher ventilation rates can translate to higher energy bills since fresh air being introduced into a building must be heated in very cold weather or cooled in very hot weather.

The Arab Oil Embargo in the mid-1970s triggered a widespread reduction in the amount of fresh air going into the nation's buildings. The cost of heating and cooling was escalating quickly because of the rapid inflation of energy prices, and homeowners and building managers were looking for ways to trim utility bills. Since that time, millions of existing houses have been caulked, weather-stripped and insulated and countless storm windows have been snapped into place in an effort to tighten things up. But for every decrease in fresh air, there has been a corresponding increase in the concentration of airborne contaminants—and in the incidence of disease associated with INDOOR AIR POLLUTION.

One of the tenets of outdoor atmospheric pollution (see AIR POLLUTION) is "DILUTION is the solution to pollution." If contaminants are mixed with enough fresh air, in other words, their environmental threat will be minimized. The limits of this maxim are obvious in this era of PHOTOCHEMICAL SMOG, ACID RAIN and global air pollution, but the saying is accurate when it comes to indoor air quality. Ventilation, the dilution of stale indoor air with fresh outdoor air, is the most important solution to problems with indoor air quality. Air purification devices can remove some pollutants from indoor air (see AIR FILTERS, ELECTROSTATIC PRECIPITATOR, HIGH-PERFORMANCE PARTICLE ARRESTING FILTERS, ION EXCHANGE), but only adequate ventilation can ensure good indoor air quality. The introduction of fresh outdoor air into a building reduces the concentration of airborne toxins while it reduces the humidity of indoor air.

Natural ventilation, the unaided movement of air into and out of a building (see INFILTRATION and EXFILTRATION), is driven by differences between indoor and outdoor temperature and pressure and by winds. Air moves naturally into and out of the building through cracks, windows and doors without any mechanical assistance. Natural ventilation can be enhanced by opening windows on the upwind and downwind sides of a building and thereby allowing an easy path for air currents to pass through. Ideally, air will come in through low windows on the upwind side and exit through somewhat larger windows located higher in the wall on the downwind side. While natural ventilation can markedly improve indoor-air quality, it can seldom provide all the necessary ventilation.

Mechanical ventilation, the movement of air into and out of a building by means of fans and blowers, must normally be used to supplement natural ventilation. Air-conditioners, heat pumps, HEAT RECOVERY VENTILATORS and whole-house ventilators all can provide

mechanical ventilation to an entire structure. Smaller blowers in kitchens and bathrooms can remove air that is likely to be full of humidity and contaminants from a single room before the fumes have a chance to spread to the rest of the house. Ceiling fans circulate air within the house, but do not bring fresh air into the building and so are not a source of ventilation.

The capacity of a ventilation system can be measured in one of two ways: in *air-changes per hour* (ac/h), the number of times each hour all the air in the structure is replaced by fresh outdoor air; and in *cubic feet per minute* (cfm), the number of cubic feet of air moved each minute, generally by a mechanical ventilator. Infiltration and exfiltration can be responsible for anywhere from .1 ac/h in an ultra-tight building to 10 ac/h in a drafty structure with ample cracks and holes passing through the walls. A kitchen or bathroom exhaust fan typically has a capacity of 50 cfm, while a large, whole-house ventilator may move 5,000 cfm. The AMERICAN SOCIETY OF HEATING, REFRIGERATION AND AIR-CONDITIONING ENGINEERS (ASHRAE) publish guidelines for ventilation rates. ASHRAE Standard 62-1989 calls for 15 cfm of ventilation per occupant in the living areas of a house, with considerably higher rates recommended in the kitchen, bathroom and garage. See also AIR CURRENT, EXHAUST AIR, HEAT RECOVERY VENTILATOR.

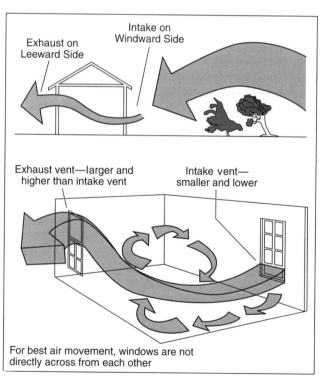

Cross-flow ventilation through windows and doors (*Indoor Pollution*)

Vinyl   any compound containing the vinyl group (C₂H₃Cl–). Vinyls polymerize easily (see POLYMER), and are therefore a constituent of many PLASTICS. Vinyls are CHLORINATED HYDROCARBONS, and as such are an environmental threat. Several plastic materials that have a leather-like surface have come to be called vinyl, although not all such materials actually contain vinyl. See also VINYL CHLORIDE.

Vinyl Chloride   a vinyl MONOMER (see VINYL) used in the production of POLYVINYL CHLORIDE and other polymers and as a refrigerant. It was used as a propellant for aerosol products until banned for that use in 1974. Vinyl chloride is a gas at room temperature, and must be cooled to –14°C (+7°F) before it liquefies. It is highly flammable (see FLAMMABLE GAS) either as a liquid or a gas. It can get into air from industrial discharges and from spills. It can contaminate drinking water contained in PVC water pipes or vessels. Vinyl chloride is moderately soluble and nonpersistent in water. It is extremely volatile: A small amount of the chemical spilled on the skin evaporates so quickly that it can cause frostbite on the affected area.

Vinyl chloride can affect health when inhaled or passed through the skin. It is a human CARCINOGEN that has been linked to cancer of the liver, brain and lungs, and is a possible TERATOGEN in humans and animals. Epidemiological studies have shown that the spouses of workers routinely exposed to vinyl chloride on the job have an increased risk of spontaneous abortion and that more birth defects occur near vinyl chloride processing plants.

The ACUTE EFFECTS of exposure to vinyl chloride include dizziness, lightheadedness and sleepiness, with headaches, nausea, weakness, unconsciousness and death resulting from increasingly higher levels of exposure. Repeated exposure can lead to irreparable damage to the liver, kidneys, nervous system, bones and blood cells, and can cause a skin allergy. A condition called *scleroderma*, in which the skin becomes smooth, tight and shiny, the bones of the fingers erode, and the blood vessels of the hands are damaged, can also result from repeated exposure.

PUBLIC WATER SYSTEMS are required under the SAFE DRINKING WATER ACT to monitor for vinyl chloride (MAXIMUM CONTAMINANT LEVEL = 0.002 milligrams per liter). The OCCUPATIONAL SAFETY AND HEALTH ADMINISTRATION has established a permissible exposure limit for airborne vinyl chloride of 1.0 parts per million (ppm), averaged over an eight-hour workshift with a ceiling of 5.0 ppm that is not to be exceeded during any 15-minute work period. It is also cited in AMERICAN CONFERENCE OF GOVERNMENT INDUSTRIAL HYGIENISTS, Department of Transportation and NATIONAL INSTITUTE OF SAFETY AND HEALTH regulations. Vinyl chloride is on the Hazardous Substance List, the Special Health Hazard Substance List and the HAZARDOUS AIR POLLUTANTS LIST. Businesses handling significant quantities of vinyl chloride must disclose use and releases of the chemical under the provisions of the EMERGENCY PLANNING AND COMMUNITY RIGHT TO KNOW ACT.

Virgin Material   a material that has been processed directly from a natural resource without the addition of any recycled materials. One problem encountered in many RECYCLING operations is that virgin materials cost little more than recycled materials (even less in some instances). Since it is harder for most manufacturers to use recycled materials because they are more likely to contain impurities, it is often difficult to sell materials collected for recycling (see RECYCLING MARKETS). Laws requiring recycling and specifying that new products containing recycled materials be purchased (see MANDATORY RECYCLING) and the end of provisions that subsidize the use of virgin materials (see RESOURCE DEPLETION COST) are beginning to reduce manufacturer's reliance on virgin materials. See also PAPER RECYCLING, PLASTICS RECYCLING, POSTCONSUMER WASTE.

Virus   sub-microscopic agents of infectious disease that are considered to be either the smallest MICROORGANISMS or extremely complex molecules that are capable of reproduction and other attributes of life forms. The latter view is becoming more predominant among biologists. As a result, viruses are generally referred to not as organisms but as *viral particles* or *virions*. Viruses are responsible for most infectious diseases in microorganisms, plants, animals and humans. Public water systems are required under the SAFE DRINKING WATER ACT to monitor for *enteric viruses*. A treatment technology rather than a MAXIMUM CONTAMINANT LEVEL (MCL) is specified for enteric viruses. Treatment, whether through disinfection or filtration, must remove 99.99% of viruses in the water supply. See also CRYPTOSPORIDIUM, INDOOR AIR POLLUTION, LEGIONELLA BACTERIA.

Viscosity   the internal resistance of a fluid to movement. The viscosity of a liquid can be seen primarily in the speed at which it flows. For example, honey, with its high viscosity, pours out of a container much more slowly than water, which is only moderately viscous. The viscosity of a gas is noticeable primarily in its resistance to an object moving through it. The viscosity of air, for instance, creates the drag that is of paramount importance in fuel economy or in the design of a race car. Air's viscosity also creates the lift that makes it possible for an airplane to fly. In a liquid, viscosity is caused primarily by the polar bonds between molecules, while in a gas, it is primarily caused by friction between individual molecules. The *coefficient of viscosity*, or *viscosity index*, is a measure of a fluid's viscosity.

Visibility   the ability to see through the atmosphere. PARTICULATES are responsible for the pollution haze that can reduce visibility in urban areas. A major coal-fired power plant or smelter can cut visibility in an entire region. Before the pollution controls mandated in the clean air act were instituted, visibility in many U.S. cities was frequently limited to as little as half a mile because of SMOG, although conditions have improved somewhat since that time. See also AIR POLLUTION, PHOTOCHEMICAL SMOG.

Visual Pollution   the visible component of air and water pollution (see AIR POLLUTION, GROUNDWATER POLLUTION, WATER POLLUTION) and the general degradation of the landscape as the result of human development and waste disposal. Visual pollution is important because our culture is quick to react to appearances. LITTER along the road, industrial smokestacks belching out SMOKE or a clearcut in the forest—which all can be easily seen—are more likely to arouse public ire than are the invisible emissions coming out of a tailpipe or polluted groundwater. The OPACITY of the smoke coming out of a chimney is a measure of a kind of visual pollution. It gauges the amount of visible material being vented into the atmosphere and the potential for the emission source to cause further visible pollution such as haze or SMOG. Smoke or smog that obscure the view are more distressing to many than the toxic materials that may be involved. The COLOR of the liquid effluent gushing out of a municipal or industrial waste pipe is similarly a measure of the visible contaminants in the effluent and of its ability to cause visible and invisible pollution of the body of water into which it is being introduced.

But visual pollution is not limited to air- or water-borne contaminants. Convenience-food wrappers blowing around in a vacant lot or a junked car parked along a street are both forms of visual pollution, although neither is a threat to air or water quality or to human health. Deciding just what constitutes visual pollution is, in fact, highly subjective. An object that offends the eye of one observer may be seen as beautiful by another. A field full of wind generators, for instance, may be objected to by some as a source of visual pollution, while it may be an inspiration to another. A new factory might similarly inspire someone who wants to get a job there, while his or her neighbor might consider it to be a blight on the landscape.

Volatile Organic Compounds   ORGANIC COMPOUNDS that evaporate easily at a relatively low temperature—abbreviated *VOCs* and also called *volatile organic chemicals*. Solvents, disinfectants, synthetic fabrics and HYDROCARBON fuels are made up primarily of VOCs. FORMALDEHYDE, BENZENE, XYLENE, TOLUENE, STYRENE, and CHLOROFORM are well known VOCs. The concentration of VOCs is typically two to five times higher indoors than outside because of emissions of VOCs from sources such as building materials, household furnishings, carpets, aerosol spray propellants, cleansers, PAINTS and paint strippers, waxes and wax strippers and PESTICIDES (see INDOOR AIR POLLUTION).

Volatile compounds have a vapor pressure of 1,300 Pascals or less. The United States, Canada and nineteen European nations signed a protocol in Geneva in 1992 in which they agreed to curb the VOCs emitted from automobiles (see AUTOMOTIVE AIR POLLUTION, MOBILE SOURCES) in an attempt to reduce OZONE pollution (see AIR POLLUTION) in the lower atmosphere. Another 11 countries that attended the negotiations, which were held under the auspices of the United Nations Economic Commission for Europe, were expected to sign the pact. After that, the governments of at least 16 of the signatories had to ratify the pact before it would go into effect. The protocol covers 85 compounds found in paints, glues and inks, and EVAPORATION during PETROLEUM refining and product distribution.

Clean Air Act

The 1990 amendments to the CLEAN AIR ACT have had a big impact on the formulation of house paint. The ratio of solvents to solids in paints has been tipped in the direction of the solids to reduce the number of volatile compounds that evaporate into outdoor air to form ozone. The regulations governing paint formulations varies widely for different parts of the country. Areas not in compliance with federal limits for ozone must use the new, low-VOC paints, and many states, including California, New Jersey, Illinois, Indiana, Maryland, Massachusetts and Wisconsin, have passed or are considering their own standards that will require the low-VOC paint. The National Paint and Coatings Association favors national adoption of the California paint standards (which limit alkyd coatings to 250 grams per liter of VOCs, as opposed to the 400 grams per liter now emitted by many alkyd paints) to end the hodgepodge of differing regulations regarding paint.

Safe Drinking Water Act

PUBLIC WATER SYSTEMS are required under the SAFE DRINKING WATER ACT to monitor concentrations of volatile organic chemicals. MAXIMUM CONTAMINANT

LEVELS (MCLs) have been established for the following VOCs: 1,1-dichloroethylene (0.0007 milligrams per liter), 1,1,1-trichloroethane (0.2 mg/L), 1,2-dichloroethane (0.005 mg/L), dichloromethane (0.005 mg/L), benzene (0.005 milligrams per liter), TRICHLOROETHYLENE (0.005 mg/L), para-dichlorobenzene (0.075 mg/L), VINYL CHLORIDE (0.002 mg/L) and CARBON TETRACHLORIDE (0.005 mg/L). Water systems in certain areas are also required to monitor for toluene (1 mg/L), xylenes (10 mg/L), styrene (0.1 mg/L), PERCHLOROETHYLENE (0.005 mg/L) and several other VOCs. The BEST AVAILABLE TECHNOLOGY (BAT) for removing all except vinyl chloride from water is packed tower aeration or granular ACTIVATED CARBON, and for vinyl chloride, the BAT is aeration only. Water utilities are required to take one sample per quarter at each point of entry to the distribution system for either surface sources or wells, and sampling must be conducted for one year, regardless of whether VOCs were found. Monitoring for vinyl chloride is required only if its precursors are present in water. VOC samples are analyzed via mass spectrometry and gas chromatography (see POLLUTION MEASUREMENT).

Volatility   the ease with which a liquid or solid evaporates. Volatility is a function of vapor pressure, temperature and air movement at the surface of the liquid or solid in question. For instance, water will evaporate more readily on a hot, windy day than when the weather is cooler and calmer. Volatile substances hold greater potential for damaging living things simply because they convert readily to a vapor, which can move easily through the air and can be inhaled. Once in the respiratory system, a toxin can more readily damage the host organism than when entering through the digestive system. *Volatilization* is the speed at which EVAPORATION (the conversion of a substance from the liquid to the gaseous state) and SUBLIMATION occur. Volatile compounds can accumulate to much higher levels in groundwater than in surface water because they can not evaporate. See also VAPOR PRESSURE, VOLATILE ORGANIC COMPOUNDS.

W

Waring, George E., Jr. a pioneer in the waste-disposal field, who was appointed the Street Cleaning Commissioner of the City of New York in 1895. Waring introduced the curbside RECYCLING of GLASS and NEWSPRINT to a reluctant public shortly after he took the job, but soon abandoned the effort in the face of scant cooperation. Waring started a business, Sanitation Utilization Company, in 1898 to try out a technique commonly employed in Europe at the time in which the volume of the wet fraction of MUNICIPAL SOLID WASTE was reduced. About 15,000 dead horses were removed from New York's streets every year at the time, and the disposal of their carcasses was a big job. At Waring's REDUCTION PLANT, wet garbage and carcasses were stewed in large vats, and the grease that floated to the surface was skimmed off to be sold for the production of soap, candles, glycerin, lubricants and perfume. The remaining stew, called *residuum*, was used to produce fertilizer. Waring's business was so successful that similar ventures soon sprung up across the United States. The pall of smoke from reduction plants became a common (and odorous) source of AIR POLLUTION in U.S. cities, and the molten black RUNOFF they produced was commonly emptied into nearby streams. Because of the decline of animal carcasses following the introduction of the automobile and the malodorous character of reduction plants, most were out of business by the 1920s, although such a plant operated in Philadelphia until 1957.

Waste materials discarded as worthless, damaged, defective, used up or superfluous during or at the end of a process. Waste can be a solid, liquid or gas (see SOLID WASTE, LIQUID WASTE, GASEOUS WASTE), or can come in the form of heat (see THERMAL POLLUTION) or NOISE. Other forms of waste are: the refuse from human and animal habitations and activities (see AGRICULTURAL WASTES, CONSTRUCTION WASTES, DEBRIS, GARBAGE, MEDICAL WASTE, MUNICIPAL SOLID WASTE, REFUSE, SEWAGE, TRASH); materials discarded by a dredging, digging or mining operations (see TAILINGS); wastes from manufacturing operations and the production of power (see CHEMICAL MANUFACTURING WASTE, INDUSTRIAL LIQUID WASTE, INDUSTRIAL SOLID WASTE, RADIOACTIVE WASTE); hazardous and toxic waste (see HAZARDOUS WASTE, TOXIC WASTE); waste heat from a power plant, a building and other process; and rocks and debris loosened by the mechanical and chemical weathering of rocks.

Waste products are excreted by living organisms as the result of the breakdown and loss of living tissue and the digestion of food (see ANIMAL WASTES, FECES, SEWAGE). Waste that degrades the environment into which it is discarded is called a CONTAMINANT or a POLLUTANT. Waste materials are often resources that are out of place (see POLLUTION), and the waste product generated by one process is often used as a NUTRIENT or resource in another process, such as when plants use the CARBON DIOXIDE exhaled by humans and animals in photosynthesis, or when plant nutrients are excreted by AEROBIC BACTERIA in a compost pile. The use of recycled GLASS or PLASTIC waste in the production of new products (see RECYCLING) or the production of refuse derived fuel divert part of the waste stream of human society for useful purposes. See also ACIDIC WASTE, CONCRETE, HIGH-LEVEL WASTE, LOW-LEVEL WASTE, METAL, MINERAL WASTE, OFFAL, PAPER, PATHOGENIC WASTE, PESTICIDE MANUFACTURING WASTES, PLASTIC WASTE, POST-CONSUMER WASTE, PRE-CONSUMER WASTE, REACTOR WASTE, WASTE STREAM, WASTE-TO-ENERGY PLANT, WASTEWATER, WOOD.

Waste Incinerator see GARBAGE INCINERATOR

Waste Management Inc. an international waste-management company, an affiliate of WMX TECHNOLOGIES. Waste Management Inc. (WMI) operates 133 solid-waste landfills in North America, and runs curbside recycling programs in more than 700 communities. WMI's recycling operations reclaimed 10 million tons of steel cans, 16 million tons of aluminum, 87,000 tons of glass and almost 1 million tons of paper in 1992. The company operates 30 landfill-gas-recovery facilities that were used to generate 583 million kwh of electricity in 1992. In addition, WMI owns and operates medical waste disposal facilities including five incinerators and one autoclave, as well as a lab that analyzes groundwater samples from 7,500 monitoring points at treatment and disposal facilities in the United States and Canada. The company's more than 15,000 collection vehicles, including recycling trucks and rear-, side- and front-loading compaction trucks, are used to collect municipal solid waste and recyclables. WMI also offers customers waste-minimization services, and medical-waste disposal. Waste Management International, plc handles solid and hazardous waste in countries outside the United States. WMI also operates 15 waste-tire shredding facilities and leases more than 100,000 portable toilets for use on construction sites and large outdoor gatherings.

Waste Stream   the flow of solid, liquid or gaseous waste products (see SOLID WASTE, LIQUID WASTE, GASEOUS WASTE) from their source to their point of final deposition, or, in the case of air and water pollution, until DILUTION, the action of sunlight, wind, WATER (see GEODEGRADABILITY) and the metabolic processes of living things (see BIODEGRADABILITY) have neutralized the contaminants. (See also AIR POLLUTION, MUNICIPAL SOLID WASTE, WATER POLLUTION).

Waste-to-Energy Plant   a plant where MUNICIPAL SOLID WASTE or other waste products such as wood chips are burned to produce heat, which is in turn used to produce steam that can be used to generate electricity or to drive other industrial processes (see INCINERATION)—or an installation that converts garbage into a solid or liquid fuel (see REFUSE DERIVED FUEL). The first incinerator to produce electricity from the heat produced by burning garbage was in Great Britain in the mid-1890s, and the practice was common in Europe by 1910.

In a typical waste-to-energy operation, GARBAGE TRUCKS dump their loads of municipal solid waste on a sorting-room floor where hazardous and toxic materials and items that might foul processing equipment are removed. A front-end loader scoops up buckets full of waste material and dumps them on a conveyer belt, which carries the waste through a series of hammer mills that reduce it to chunks four inches across and smaller. The pulverized waste is then carried under either a permanent magnet or an electromagnet that removes ferrous metals. At this point, the waste is ready for further processing before transformation either by incineration or PYROLYSIS, or into refuse derived fuel See also GARBAGE INCINERATOR, MEDICAL WASTE INCINERATOR.

Wastebasket   an open-topped container usually made of PLASTIC or METAL and designed for the disposal of trash. Household waste is normally collected in wastebaskets, which are in turn emptied into garbage bags for disposal. There is a considerable risk of fire in a wastebasket, since they usually contain highly flammable trash such as paper and plastic, and a spark or hot ashes from a cigarette or pipe in a wastebasket have caused many fires. Hazardous materials such as solvents and hypodermic needles in wastebaskets can cause injury to the worker or household member emptying it.

Wastewater   water that has been used in some process and discarded, usually along with some of the waste products generated by the process for which it was used. One of the principal tasks long assigned to water has been the removal of wastes—from feces to industrial chemicals. Toxic chemicals, heat, SEDIMENT, microorganisms, salts and nutrients are among the waste products commonly found in wastewater. A large volume of fresh water can become wastewater very quickly when it is not used efficiently, whether in a domestic toilet, on the farm or in an industrial process (see IRRIGATION WATER, PROCESS WATER, WATER CONSERVATION). Some wastewater is nearly as pure as it was before use, while some is so contaminated that it must be handled as HAZARDOUS WASTE. Water with more than 500 parts per million (ppm) of DISSOLVED SOLIDS is considered unfit for human consumption in the United States, while livestock can tolerate water containing up to 2,500 ppm dissolved solids. See also INDUSTRIAL LIQUID WASTE.

Wastewater Recycling   the purification and reuse of WASTEWATER. Both natural and mechanical processes can be used to purify wastewater (see WATER TREATMENT). The purification and reuse of sewage effluent (see SEWAGE, EFFLUENT) for irrigation is nothing new. Golden Gate Park in San Francisco was watered with treated sewage effluent starting in 1932, a practice that was common at the time, and many cities since that time have used treated effluent to water farmland, golf courses and other municipal greeneries. An increasing number of community water systems are turning to highly treated sewage effluent as a reliable "new" water source as a way of avoiding the growing hassle and expense of finding and developing a new water source. Wastewater recycling in industrial processes can significantly reduce the amount of fresh water used while reducing the amount of effluent produced (see INDUSTRIAL LIQUID WASTE).

Recycled Water in Denver

A $30 million pilot project in Denver, Colorado, upgraded an average of 1 million gallons of treated wastewater from the city sewage district to drinking-quality water every day from 1984 to 1991. Denver residents were initially skeptical about using the reclaimed water, but warmed to the idea when they found that the quality of the water produced at the treatment plant was consistently of at least as good a quality as that drawn from traditional water sources. Water recycling is attractive to the Denver Water Department because it costs more than $3,000 to bring an acre-foot of new water from the western slope of the Rockies, and political opposition to water diversions across the continental divide is growing—as is the city's population. Treated wastewater from the city's SEWAGE TREATMENT plant was filtered through beds of sand and coal (see ACTIVATED CARBON), and then aerated with ozone gas, which killed virus and bacteria. The water was then run through a REVERSE-OSMOSIS filter, which removes microcontaminants, and then cascaded through an air stripping tower where dissolved gases evaporated. Finished water was then treated with chlorine dioxide, which removes any remaining harmful byproducts of CHLORINATION.

Miller Time

Southern California, with its swelling population, periodic droughts and scant supply of water, is a leader in wastewater reuse. Drinking water is produced from purified sewage effluent in almost 50 communities in the region. Plans to recycle treated sewage effluent into drinking water in Atwater, California, have run into stiff opposition. The local water district has proposed bringing 8 billion gallons of treated wastewater a year from a sewage plant in Whittier through a nine-mile pipeline and spreading it on the land's surface in an area where it will recharge an underlying aquifer—a practice that has been used in many nearby communities. Miller Brewing Company, which has touted the quality of the San Gabriel Valley well water that it uses to brew beer, is strongly opposed to the project and has filed suit in Los Angeles County Superior Court to stop the project, despite the assurances of health officials that it would not impair water quality in the aquifer. Officials of the Metropolitan Water District of Southern California predict that by 2015, one out of every ten gallons of water used in the region will come from a treatment plant. See also IRRIGATION WATER.

Water an odorless, tasteless compound of HYDROGEN and oxygen with a slight blue tint that is visible only when looking through a relatively thick layer of the gas. Liquid water in oceans and water vapor in clouds is the most prominent feature of the Earth, as seen from space. Water covers about 70% of the Earth's surface to depths of up to seven miles. Of the estimated 344 million cubic miles of water on earth, 315 million are seawater, 7 million are locked in polar ice caps, 9 million are GROUNDWATER; 53,000 fill lakes and streams; 4,000 are atmospheric moisture and 3,400 are part of the bodies of living things. Only about .05 percent of world water supplies are directly usable by man. If all the Earth's water could be removed, it would form a sphere with about half the diameter of the moon, and the dehydrated hulk left behind would be lifeless. One-fifth of the fresh water on the surface of the Earth is in Lake Baikal in southeastern Siberia. More than 300 streams flow into the lake, the world's deepest, covering more than 12,000 square miles.

A clean, abundant supply of water is central to the development of human culture. Some of the world's most remarkable engineering accomplishments are dams, aqueducts, canals and pipelines built to supply water for irrigation and for use in population centers. Agriculture (see IRRIGATION WATER), PUBLIC WATER SYSTEMS (see DRINKING WATER) and industry (see PROCESS WATER) all require large inputs of clean water. Total U.S. water use is about 330 billion gallons per day, a total of about 1,400 gallons per capita, which is a higher rate of water use than in any other industrialized nation. Public water systems and industry each account for roughly one-tenth of total water consumption, with a little less than 40% used to cool electric power plants and a little more than 40% used in agriculture.

An adequate supply of clean water is essential, but water has nevertheless been assigned a low economic value, and can be obtained in water-rich countries like the United States for little more than the cost of development and delivery; users of water from federally subsidized irrigation projects usually pay far less than that. Because the development of water resources has long been heavily subsidized, it is often cheaper and easier to waste water than to make the investment required to use it more efficiently. As new sources of potable water become more scarce, and the cost of delivering fresh water from increasingly distant sources escalates, the value of clean water itself is beginning to be recognized, particularly in arid regions (see DROUGHT). The value of adequate instream flows for the preservation of fisheries, the production of hydroelectric power and the enjoyment of recreationists is also increasing.

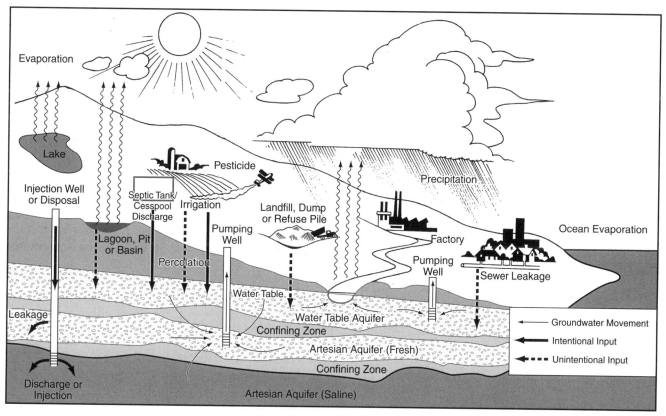

The hydrologic cycle and sources of drinking-water contamination (*EPA Journal*)

But as new sources of potable water become more scarce and the cost of delivering fresh water from increasingly distant sources escalates, the value of clean water itself is beginning to be recognized, particularly in arid regions. The value of adequate instream flows for the preservation of fisheries, the production of hydroelectric power and the enjoyment of recreationists is also increasing. The many uses of water, the inefficiency with which it is typically used and increasing human population (see POPULATION GROWTH) have increased water use in many regions to the point that human use exceeds the rate of freshwater recharge (see CONDENSATION, EVAPORATION, HYDROLOGIC CYCLE), and groundwater is in many cases being overdrafted as a result. Freeing up additional supplies by increasing the efficiency with which water is used (see WATER CONSERVATION) and reuse is becoming a more significant source of additional water supplies than is developing new water sources.

The *consumptive use* of water—the percentage used up by a process—is about 10% for industry, as compared to about 25% for domestic and commercial water users. Only 6% of the water used for cooling in an electric power plant is consumed, while the consumptive use of irrigation water is about 95%.

Water's Unique Properties

Water has several properties that make it unique:

- It is the only substance that occurs as a solid, a liquid and a gas at temperatures normally encountered in the biosphere.
- It is the only liquid that is not at it densest just above its freezing point. Like most liquids, water's density increases as it gets cooler—until about 39°F (4°C), when it starts to become less dense. This means that the coldest layer in a body of liquid water is on its top surface, not on the bottom as it would be with other liquids. If this were not so, ice would start to form on the bottom of lakes, streams and the ocean, not the top, and they would freeze solid as a result.
- Ice is one of the only solids that is lighter than the liquid from which it forms. If this were not the case, ice that formed on the top of a body of water would quickly sink to the bottom.

It is the nearly indestructible bond between the two hydrogen atoms in a molecule of water that is responsible for water's unusual physical attributes. Only an electric current, temperatures of more than 5,400°F, cosmic rays or the action of chlorophyll in living plants is capable of splitting the water molecule. As a result, most of the same water molecules that were around at the origin of the planet are still in existence. The properties of water are used in the development of weights and measures (see MEASUREMENTS).

Water of Life

Life originated in the ocean, and the blood of land animals is similar in composition to seawater. Water plays a central role in the processes that make life possible. Tomatoes are 95% water, and human bodies about 70%—with women averaging between 55% and 65% and men between 65% and 75%. The loss of 15% of the body's moisture is normally fatal. *Hydrolysis*—the metabolic breakdown of proteins, carbohydrates and other essential molecules that occurs continually in living cells—and *photosynthesis* are dependent on water. Water is known as the *universal solvent* because of its ability to dissolve a wide variety of substances. Water also carries materials in SUSPENSION between individual molecules (see SUSPENDED SOLIDS). A suspension that will not settle out is called a *colloid*, with milk being a notable example (see COLLOIDAL SOLIDS). Nutrients are carried to cells and waste materials are carried away by the bloodstream, and blood is a solution carrying dissolved and suspended materials. Water molecules are so small that they can pass through a very fine membrane, the basis of *osmosis*—utilized by cells to screen what compounds enter and exit and to control chemical balance and, in REVERSE OSMOSIS, to purify water. Water combines with salts to form *hydrates*. It reacts with metal oxides to form BASES and with nonmetallic oxides to form ACIDS. It also acts as a CATALYST in many chemical reactions. Water combines so readily with other materials that truly pure H_2O can exist only briefly, under strictly controlled laboratory conditions. Seawater contains 35,000 parts per million of DISSOLVED SOLIDS, the result of eons of erosion and the chemical weathering of rocks. ACID RAIN accelerates the weathering process. Water's eagerness to combine with and carry other materials plays a central role in AIR POLLUTION and WATER POLLUTION. See also ATMOSPHERE, GLOBAL WARMING, HEAVY WATER, VISCOSITY, WATER LAW.

Water Conservation

the installation of equipment and the adoption of practices that increase the efficiency of water use. The low value assigned to clean water and policies that favored the development of new water sources encouraged the wasteful use of water (see WASTEWATER), creating a large potential for stretching existing water supplies and creating "new" water sources through conservation measures. Conservation is certain to become the largest "new" source of water around the world during the twenty-first century. The potential for savings is especially large with water used to irrigate crops and raise livestock—because it represents 90% of the water used for all purposes in the United States and because of the amount of water now wasted (see IRRIGATION WATER). An increasing number of community water systems are paying for the lining of irrigation canals and other efficiency improvements and buying the rights to the water thereby freed up for use as DRINKING WATER. Most water utilities are reducing water wastage by replacing leaky water mains and distribution lines and tightening up the supply system that brings water from its source and stores it for use, and by promoting conservation programs that encourage customers to repair plumbing leaks, install low-flow shower heads and replace water-wasting fixtures such as old, five-gallon flush TOILETS with new, water-conserving models. Installing water meters, raising rates and instituting rate structures that encourage conservation can also significantly reduce per capita water use. Some water systems are also promoting the installation of water conserving plumbing systems in new homes and businesses.

In water-rich countries such as the United States, industrial processes developed around the assumption of an abundant supply of free or cheap water. Water is an essential part of most manufacturing (see PROCESS WATER), as is the pollution of water (see INDUSTRIAL LIQUID WASTE, WATER POLLUTION). As a result, more water is used (and polluted) than is the case in regions where pure water is given a higher value. For instance, 20 times as much water is typically used to make paper in U.S. mills than is used in Scandinavian mills. Industrial processes that use less water and produce less wastewater are being developed in response to shrinking water supplies and increasing penalties for water pollution.

Water Environment Federation

a nonprofit educational and technical federation composed of 64 member associations primarily in the United States and Canada with more than 40,000 members and a staff of 100. Founded in 1928 as the Federation of Sewage Works Associations, the organization was later called the Water Pollution Control Federation. WEF members are engineers, biologists, chemists, students and equipment manufacturers that work with water quality and wastewater treatment. The federation sponsors workshops and seminars and publishes technical works covering all aspects of wastewater treatment. A series of brochures that addresses water-quality issues is available to the general public. Periodicals published by the WEF include *Water Environment and Technology, Water Environment Research, Industrial Wastewater* and *Water Environment Regulation Watch.* The federation also organizes congressional testimony and formal comments of its members concerning water-quality legislation and proposed projects. Address: Water Environment Federation, 601 Wythe Street, Alexandria, VA 22314. Phone 703-684-2400.

Water Farming   the purchase of farmland by urban water utilities, which quit irrigating most of the land, and use the water instead as a source of DRINKING WATER. The water utility gains a permanent water source through the practice, and financially strapped farmers can pick up a significant sum—a typical sale involves a couple hundred acre feet of water that sells for $2,000 an acre-foot or more. But much of the land's value and productivity is lost with its water right, and the local economy normally suffers if the practice is widespread. One of the earliest examples of water farming is California's Owens Valley, where unscrupulous developers working for the city's water utility purchased the rights to most of the valley's water in the early 1900s. Aurora, Colorado, spent $50 million during 1986 to buy water rights from 300 sources in the Arkansas River Valley in southeastern Colorado, thereby increasing the city's water supply by 30%. The amount of irrigated acreage in areas such as central Arizona has similarly decreased as water rights have been sold to thirsty urban areas in the vicinity, and crop yields have declined as a result.

Water Filters   devices that remove contaminants from water through physical processes. Water filters may be composed of fibrous, granular, powdered or modular materials that remove contaminants from water through processes such as ABSORPTION, ADSORPTION, straining and impaction.

The coarsest fiber-type filters remove only the largest suspended particles, while finer fiber filters can remove an intermediate range of particles. Granular ACTIVATED CARBON filters effectively remove very small contaminants such as those responsible for taste and odor from drinking water, and are designated as the BEST AVAILABLE TECHNOLOGY for the control of synthetic organic chemicals in water. REVERSE OSMOSIS is also used to remove very small contaminants such as VOLATILE ORGANIC COMPOUNDS from DRINKING WATER and industrial PROCESS WATER. The process is also used to treat sewage and industrial wastewater, and to render ocean water potable (see DESALINATION). COLLOIDAL SOLIDS cannot be removed from water by any method of FILTRATION. Filtration was required in the 1986 amendments to the SAFE DRINKING WATER ACT for public water systems that use surface water, or that use groundwater that is directly influenced by surface water, for the removal of microbial contaminants.

Only those community water systems that could demonstrate that their water source was too pure to require filtration were exempt, and all others were required to install filtration equipment by the end of 1991. Water utilities wanting to avoid the installation of filtration equipment had to demonstrate that the system meets criteria for total coliform (see FECAL COLIFORM BACTERIA), TURBIDITY and control of pollution sources in the source watershed (see RANGE MANAGEMENT). In addition, DISINFECTION must be practiced by the water system, no outbreaks of waterborne disease can have occurred and the system must be in compliance with the total TRIHALOMETHANE MAXIMUM CONTAMINANT LEVEL. An annual inspection of the water utility's facilities must be conducted by the state or by a third party approved by the state for the utility to maintain the exemption. See also WATER TREATMENT.

Water Law   laws governing the appropriation, quality and use of water. The right to use water is governed by the states, and there is considerable differences in how water rights are handled from state to state. The right to use water is governed in most of the United States by *riparian rights*—which basically means that anyone that owns a piece of property along a stream has a right to use it. Riparian rights are patterned after water use in Europe, and are based on the assumption that there is plenty of water for all. This is not the case, however, in the arid western United States, where water rights are governed by the doctrine of *prior apportionment*, which can be summarized: "first in time, first in right." The doctrine holds that the water user that first diverts water from a stream for a beneficial use has first right to its use, and that all subsequent water users cannot use water if it will mean that the holders of senior rights will not receive their full allotment. This system means that rights to withdraw water often exceed the flow of streams, and that streams are sometimes completely dried up, primarily as the result of irrigation withdrawals. The gross over-appropriation of the Colorado River has caused endless fighting among the southwestern states, Indian tribes and Mexico over rights to its water. Under prior apportionment, if a right to use water is not used for a time, it can be forfeited, a situation that encourages users to draw more water than they need simply to maintain the right to its use. The first water rights in many part of the west belonged to miners, and the "use-it-or-lose-it" stipulation is based on miner's right to develop public land holding minerals, with ownership of the land returning to the public if the claim is no longer being used.

The CLEAN WATER ACT and the SAFE DRINKING WATER ACT are the only national laws that specifically govern

WATER POLLUTION, although numerous other federal regulations have an impact on water quality. The Superfund law (see COMPREHENSIVE ENVIRONMENTAL RESPONSE COMPENSATION AND LIABILITY ACT) governs the quality of groundwater and surface water at hazardous-waste cleanup sites, and the RESOURCE CONSERVATION AND RECOVERY ACT governs water pollution resulting from the disposal of hazardous waste and trash. State governments are generally responsible for the enforcement of the water-quality standards established by federal regulations.

Water Pollution   the presence of harmful CONTAMINANTS in water, from natural sources such as erosion, microbial growth (see MICROORGANISM, RED TIDE) or by manmade compounds and as the result of human activities. The World Health Organization estimates that 10 million people die each year from waterborne disease, primarily residents of the Third World who are infected by biological contaminants. Compared to developing nations, American cities have excellent water quality, and the quality of surface water in the United States has improved over the last couple of decades, due primarily to the implementation of the CLEAN WATER ACT and the SAFE DRINKING WATER ACT. Although most drinking water delivered by U.S. PUBLIC WATER SYSTEMS contains traces of toxic chemicals, lethal pollution by biological contaminants were largely eliminated with the advent of chlorination in most water systems in the early 1900s (see WATER TREATMENT). Outbreaks such as the one in Milwaukee, Wisconsin, in 1993—in which more than 400,000 people got a flu-like illness and more than 100 died as the result of ingesting CRYPTOSPORIDIUM, a water-borne parasite that is found in the intestinal tracts of animals—are extremely rare.

Far more common in U.S. drinking water is contamination with PETROLEUM fuels and petrochemicals, manufactured products such as PESTICIDES and SOLVENTS, byproducts of manufacturing processes such as DIOXINS, airborne contamination such as ACID RAIN and polluted RUNOFF and erosion from roads, parking lots and agricultural lands. Human patterns of consumption have a profound effect on water quality (see CONSUMERISM, ENVIRONMENTAL COST, SUSTAINABLE SOCIETY), as do laws and policies governing resource use (see WATER LAW). The pollution of water is an environmental EXTERNALITY that is part of the production of many products.

A POINT SOURCE of water pollution, such as the OUTFALL from a SEWAGE TREATMENT plant or a factory, can be easier to control than diffuse sources of pollution (see DIFFUSE POLLUTION). A point source is subject to regulation under state and federal water pollution laws, and a relatively large amount of toxin-bearing EFFLUENT is already collected in one place, making treatment comparatively easy. It is often much harder to find out where diffuse water pollution is coming from, and controlling the contamination can be difficult, even after the source is identified. Wastewater from 20 cities' sewage treatment plants and more than 2,000 industrial sources is dumped into the Ohio River. More than 20,000 sites where water contamination has resulted from the disposal of toxic wastes have been identified in the United States.

Other major sources of water pollution include:

- pesticides, nitrates, salts and SEDIMENT in runoff from and the erosion of farmland;
- HEAVY METALS including LEAD, CHROMIUM and ARSENIC from the weathering and erosion of mine and mill TAILINGS. The PYRITE found in most mine tailings oxidizes readily when exposed to air, producing sulfuric acid;
- FUELS released in pipeline ruptures, and vehicle and tanker accidents (see GASOLINE, LEAKING UNDERGROUND STORAGE TANKS);
- polluted air (see ACID RAIN, AERIAL DEPOSITION, DUSTFALL, PARTICULATES, TOXIC PRECIPITATION);
- polluted AQUIFERS (see GROUNDWATER POLLUTION, LEAKING UNDERGROUND STORAGE TANKS). Contaminated water from aquifers can contaminate lakes and streams that are fed by groundwater.

The Mississippi River drains much of the industrial and agricultural heartland of the nation, and it carries a heavy load of contaminants as a result. At least 150 major chemical plants are along the river, and 47 of them discharged an estimated 296 million pounds of toxic chemicals into the river in 1991. More than 600 municipal sewage treatment plants discharge more than 1 billion gallons of effluent into the river and its tributaries annually. DDT, banned in 1972, is still found in catfish caught in the Mississippi. The drinking water for about 18 million people also comes from the river. See also DILUTION, MIXING ZONE, OCEAN DISPOSAL, OPACITY.

Water Pollution Measurement   the measurement of the concentration of contaminants in water sources, in drinking water and in wastewater. The measurement of suspended and dissolved contaminants in water is an essential part of the treatment of drinking water, sewage, industrial waste and hazardous waste. Measurements are commonly taken for PH, TURBIDITY, BIOCHEMICAL OXYGEN DEMAND (BOD), suspended load and for the concentration of HEAVY METALS and synthetic organic chemicals. Grab samples can be taken from a stream, a domestic faucet or a wastewater stream and taken to a lab for analysis (see POLLUTION MEASUREMENT). ACIDITY can be determined directly by observing color change in a piece of litmus paper, and the concentration of many other chemicals can similarly be ascertained by observing a color change based on a chemical reaction. A SECCHI DISK is used to directly observe water's turbidity. U.S. industrial concerns pay $1.5 billion per year to have ten million soil and water samples analyzed to comply with ENVIRONMENTAL PROTECTION AGENCY standards for wastewater discharge and soil cleanup. See also MEASUREMENTS, METRIC SYSTEM.

Water Quality Standard a regulatory standard for water quality. Water quality standards are based on the effect of effluent discharges on aquatic ecosystems and on the water's use as drinking water and irrigation water. The standards are established by the states under the guidelines established in the CLEAN WATER ACT. The ENVIRONMENTAL PROTECTION AGENCY can require a stricter standard in instances where a state's standard is found not to meet the requirements of the act. A standard may be based on: keeping the concentration of specific toxins in the EFFLUENT below a specified level (see EFFLUENT STANDARD); maintaining or improving the water quality of the body of water into which effluent is being dumped so that it is fit for certain uses such as fishing, swimming or use as drinking water); or employing the best technology that it is on the market to remove pollutants from the wastestream (see BEST AVAILABLE TECHNOLOGY, BEST CONVENTIONAL WASTE-TREATMENT TECHNOLOGY). The pollutants that are removed from air or water by a treatment system are typically very toxic, and their disposal can easily lead to further environmental contamination, or CROSS-MEDIA POLLUTION. The next generation of air and water-quality standards may address cross-media pollution by requiring that polluters not only use best management practices to remove contaminants but also switch to more benign process chemicals and otherwise reduce the total amount of pollutants generated. See also NATIONAL POLLUTION DISCHARGE ELIMINATION SYSTEM, ZERO DISCHARGE.

Water Pollution Control Act see CLEAN WATER ACT

Water Table the top surface of the saturated zone in an AQUIFER. Below the surface of the water table, all the voids between particles in the soil are full of water, while above it some of the voids are empty, allowing water to trickle down from void to void. The level of the top of the water table is closely linked to the water level in nearby wells, wetlands, lakes and streams. Water tables that are close to the surface are generally more vulnerable to pollution than those that are more deeply buried (see GROUNDWATER POLLUTION, LEAKING UNDERGROUND STORAGE TANKS). See also GROUNDWATER, PERCHED AQUIFER.

Water Treatment the removal of CONTAMINANTS from WATER. Many of the same basic methods are used to remove contaminants from polluted lakes, streams and GROUNDWATER, DRINKING WATER, industrial PROCESS WATER, SEWAGE, INDUSTRIAL LIQUID WASTE and HAZARDOUS WASTE (see HAZARDOUS WASTE DISPOSAL, SEWAGE TREATMENT). FILTRATION, SEDIMENTATION, AERATION, microbial action (see METABOLISM, MICROORGANISMS), EVAPORATION PRECIPITATION, ION EXCHANGE, REVERSE OSMOSIS, DISTILLATION and DISINFECTION are some of the principal methods of water treatment:

Sedimentation removes suspended particulates by allowing water to slow or stand in a settling tank. Lighter-than-water particles may also float to the surface, where they can be skimmed off.

Aeration breaks down waterborne contaminants by initiating OXIDATION and evaporation. Water is aerated naturally when it splashes over a rock or a waterfall. Packed-tower aeration is used to remove volatile contaminants from drinking water, industrial waste, hazardous waste and sewage. In an EVAPORATION POND, suspended contaminants are allowed to settle and volatile elements are allowed to evaporate. In an oxidation pond, air is bubbled through water or it is introduced through agitation or stirring, promoting the oxidation of contaminants.

Filtration physically removes contaminants from water. Sand, coal, activated carbon, natural and synthetic fibers, soil and streambed sands and gravels can all act as filters. A reverse-osmosis filter removes contaminants by forcing water through minute openings in a membrane.

Ion exchange devices remove contaminants by initiating a chemical reaction in which contaminant ions are exchanged for less-harmful and/or more-controllable ions of the same charge. Water softeners use ion-exchange to replace the calcium and magnesium ions that make water hard with sodium ions.

Reverse osmosis purification systems force water through a membrane with minute pores that filter out contaminants. Large-scale reverse-osmosis filters are used to purify industrial wastewater, and to finish water at some drinking-water treatment plants

Evaporation purifies water because suspended or dissolved impurities are left behind when a water molecule evaporates. Precipitation is caused by the evaporation, primarily of seawater (see HYDROLOGIC CYCLE).

Distillation removes contaminants by boiling water, thereby causing it to evaporate, then recondenses the water from the pure water vapor. Some volatile compounds can evaporate and recondense along with FINISHED WATER, but many distillation systems are capable of venting such contaminants separately. Large-scale water distillation facilities are often combined with a power plant so that waste steam from turbines can be used to heat evaporators. Removing the buildup of mineral deposits that accumulate inside a water distiller is a problem.

Precipitation can be used to remove dissolved heavy metals from water. The formation of PRECIPITATES in wastewater that results when its ability to carry dissolved solids is reduced as a result of the addition of a substance that changes the pH of the wastewater and/or through changing the temperature of the wastewater. Precipitation is used in electronics, steel industries and in the production of inorganic chemicals.

Biological processes such as the growth of microbes and plants can be used to break down contaminants in sewage (see FLOCCULATION) and in contaminated water (see BIOREMEDIATION, GROUNDWATER TREATMENT). Specially designed wetlands have been constructed to remove contaminants from drinking water and sewage. Water lilies and other aquatic plants and microorganisms aid in the treatment.

Disinfection kills most of the biological contaminants such as viruses, bacteria and parasites found in water. CHLORINE is the most widely used disinfectant, but the use of alternatives has increased due to concern about the health effects of TRIHALOMETHANES (THMs), a family of toxic compounds including CHLOROFORM that are formed when POLYCYCLIC ORGANIC MATTER reacts with chlorine. Studies by the Harvard School of Public Health and others have identified THMs as the greatest threat to health of any of the toxins commonly found in drinking water—responsible for approximately 10,000 additional cases of cancer annually in the United States, including about 20% of all rectal cancers and 10% of all bladder cancers. OZONE and *chlorimine*, a combination of AMMONIA and reduced amounts of chlorine that is less toxic to microbes than pure chlorine but less likely to form THMs.

Sewage treatment plants rely primarily on basic processes such as sedimentation, floatation, oxidation, microbial action, precipitation and evaporation to purify water. *Septic tanks* treat sewage generated in relatively small quantities in rural areas not served by a sewer system, with varying degrees of success (see SEPTIC SYSTEM). The ANAEROBIC BACTERIA in a septic tank slowly consume the solids in sewage as it moves through the tank, reducing BIOCHEMICAL OXYGEN DEMAND in the process. The effluent that issues from a septic tank can be at least as pure as that emerging from primary treatment at a sewage treatment plant. Inadequate drain fields, improper design and installation of the septic system and the use of systems in soils that are either not porous enough or too porous can lead to GROUNDWATER POLLUTION with NITRATES, SOLVENTS and ORGANIC COMPOUNDS. See also DESALINATION.

Water-Wall Boiler   see GARBAGE INCINERATORS

Wet Well   see MONITORING WELL

Wheelabrator Environmental Systems   one of the largest international suppliers of equipment and services for processing industrial and municipal water and wastewater. Wheelabrator Environmental Systems owns and operates 14 garbage-fueled incinerators in the United States that generate more than 800 megawatts of electricity, and also has five cogeneration installations. Wheelabrator also operates and sells industrial wastewater, air-pollution treatment and composting facilities, and operates more than 30 municipal sewage plants.

Address: Liberty Lane, Hampton, NH 03842. Phone: 603-929-3000.

Winchester Tire Fire   a persistent blaze started by vandals in a five-acre stack of used tires in Winchester, Virginia, in 1983. Firefighters fought the blaze for eight months before it was extinguished. The tire dump was estimated to contain five to seven million tires piled up to 80 feet deep. The smoke PLUME from the burning tires rose several thousand feet into the atmosphere, and SMOKE from the fire caused air pollution in parts of four states within 50 miles of the site. The intense heat created by the blaze melted rubber so quickly that an estimated 30 to 50 gallons per minute was running off the site, contaminating soil and threatening a nearby creek. The Superfund Emergency Response Program (see COMPREHENSIVE ENVIRONMENTAL RESPONSE, COMPENSATION AND LIABILITY ACT) provided technical assistance in fighting the fire and on air and water pollution resulting from the fire, and coordinated the collection (and subsequent sale) of oil released by the blaze.

WMX Technologies   a corporation with subsidiaries that operate landfills, garbage incinerators, hazardous-waste disposal facilities and toxic-waste site cleanup services. Among partially or wholly owned WMX subsidiaries are the following: WASTE MANAGEMENT, INC. (WMI) operates 133 solid-waste landfills in North America, and runs curbside recycling programs in more than 700 communities. WHEELABRATOR ENVIRON-MENTAL SYSTEMS operates 14 garbage-fueled incinerators in the United States that generate more than 800 megawatts of electricity. CHEMICAL WASTE MANAGEMENT operates 20 facilities for the treatment, recovery and incineration or disposal of hazardous wastes. *RUST International*, owned primarily by Chemical Waste Management and Wheelabrator Environmental Systems, is the largest U.S. pollution remediation contractor, and offers industrial customers waste-management consulting, design and construction services. *CWM Technologies* develops new technologies for the recycling, treatment, neutralization and disposal of hazardous waste.

Address: 3003 Butterfield Rd, Oakbrook, IL 60521. Phone: 708-572-8800.

Windrow Composting   see COMPOSTING

Woburn   a city 12 miles north of Boston that was forced in 1979 to abandon two city wells found to contain TRICHLOROETHYLENE, POLYCHLORINATED BIPHENYLS, ARSENIC, MERCURY, CADMIUM and other toxic chemicals. The contamination was eventually traced to the illegal dumping of solvents and other liquid wastes on adjoining properties in Woburn over a period of decades. The 330 acres with the most serious contamination were designated a Superfund site in 1983.

In 1983, eight families filed suit against three companies they believed to be responsible for the pollution, the first of many similar suits by pollution victims across the nation. The complainants claimed that the companies' negligent actions had caused six leukemia deaths in addition to heart disease, central-nervous-system and immune-system damage among members of their families. In the initial phase of the trial in 1986, W. R. Grace, which manufactured food-production equipment on the site, was found to have contaminated groundwater by allowing employees to dump solvents used in cleaning machine parts on company grounds and down storm sewers. Before the next, more difficult, phase of the trial, in which the plaintiffs would have had to prove that the diseases suffered by their members were caused by the contamination, W. R. Grace agreed to an $8 million settlement, while not admitting it was guilty. The families spent $2 million to press the case. In 1991, six companies agreed to pay $69.4 million in costs associated with the cleanup of the site. ENVIRONMENTAL PROTECTION AGENCY officials estimated that it would take about three years to clean the contaminated soil at the Superfund site, and up to 50 years to purify the groundwater. See also GROUNDWATER POLLUTION.

Wood   a material composed of cells that carry water and dissolved minerals from a tree's roots to its leaves. Wood is used as a fuel, in the production of PAPER and countless items ranging from matches to houses. Lumber, plywood, oriented-strand board, fiberboard and other wood products are used to make furniture, boats, sculptures, toys, crates and pallets for shipping. An estimated 2 to 3 billion shipping pallets are in use at any given time in the United States, and about 500 million are produced to maintain that stock each year. Approximately 40% of U.S. hardwood use is for the production of pallets, more than twice the hardwood use in furniture production. Some pallets are used just once, and others may last for years, depending on the quality of their construction. Discarded pallets are seldom repaired, but are sometimes broken up and burned in wood stoves, or converted to shavings for use as animal bedding and mulch. Plastics have reduced the amount of wood used in many sectors, and shipping pallets made from recycled plastic are beginning to appear.

In *softwoods* (coniferous trees), water moves through long ducts and individual cells tend to be longer and the wood tends to have a more parallel grain. In *hardwoods* (broad-leafed, deciduous trees), water movement occurs from cell to cell, and small vessels often carry resins horizontally through the tree's trunk, so grain direction is more varied. Hardwoods such as maple and oak do, in fact, have very hard wood, but others such as aspen are actually softer than "soft" woods such as Douglas fir. The tough shells on certain nuts is a form of wood. The cutting of trees and the milling of wood are sources of air and water pollution. The roads and skid trails associated with a logging operation are a major source of erosion and the resulting SILTATION of streams and depletion of topsoil. FORMALDEHYDE and other VOLATILE ORGANIC COMPOUNDS emitted by glued wood products are a source of INDOOR AIR POLLUTION. Wood preservatives such as penta and creosote are dangerous environmental toxins. An average American accounts for the use of 18 tons of paper and 23 tons of wood in his or her lifetime, according to estimates by the WORLDWATCH INSTITUTE.

Wood has been used as a FUEL since the dawn of history. Man's mastery of FIRE was one of the first things that set him apart from the animals (see AIR POLLUTION CONTROL, WOOD SMOKE). One indicator of extensive deforestation in ancient times is the absence of wood ash and the presence of dung ash in fire pits. (The disappearance of tree pollen in succeeding strata of soil remnants is another indicator.) Some of the oldest archaeological artifacts are stones that have been repeatedly heated on one side from use in a fire ring. Wood's use in shelters similarly dates to the earliest days of mankind. Analysis of the growth rings in wood beams from ancient buildings is a principal means of dating the construction of prehistoric ruins.

The outermost part of a tree is a dead, corky bark that varies in thickness depending on age and species, and that covers a smoother, living inner bark. Bark was once burned as waste in the tipi burner that was an integral part of sawmills until the 1960s, but it is now used as mulch, as fuel for steam boilers that produce electricity and in certain glued wood products. New wood cells are constantly being formed in a growing tree on the inside of the microscopic cambium layer, located just inside the inner bark, and new bark cells are likewise being grown on the outer side of the cambium. Growth in a tree results entirely from the addition of new layers of cells over older ones. Once a wood cell is formed, no further growth in length or diameter occurs. As a tree's diameter grows, the bark is stretched and it tends to become cracked and ridged. Inside the cambium is anywhere from one inch to six inches of sapwood, depending on climate and species. Sapwood contains only a few living cells, and acts primarily as a conduit for the transport of sap and a storage place for nutrients. Many second-growth trees contain nothing but sapwood. Inside the sapwood in older trees is heartwood—dead cells that have been physically and chemically altered. No sap passes through heartwood as cells are normally plugged with gums and resins. At the center of a tree trunk is a small circle of pith—formed by the elongation of stems and branches and often dark in color. Annual growth rings are formed, in temperate climates, by the differences in the rate of growth and color of new wood formed in different parts of the year. Wood cells, commonly referred to as fibers, are generally elongated and pointed at the ends, and are composed primarily of cellulose. Dried wood is about 50% cellulose by weight. The cells are tightly bound to one another with lignin, which generally comprises about one-fourth the weight of dried wood, and with hemicelluloses.

Once the lignin has been removed, cellulose can be used to produce paper, synthetic textiles, films, lacquers and explosives. XYLENE can be produced through the DISTILLATION of wood tar. Turpentine is distilled from wood waste, and methyl alcohol (also called *wood alcohol*) can be obtained from the *destructive distillation* of wood. *Pyrolization*, a form of destructive distillation that occurs in the first stages of wood combustion, produces charcoal (see PYROLYSIS). Wood stoves and furnaces that pyrolize efficiently are much cleaner burning. Vast new markets are opening for wood chips in the building materials and power production fields. New equipment has been developed to fill the demand and to better utilize the smaller diameter trees available today. A machine called a feller buncher can grasp a tree's trunk with steel claws, snap or saw it off at ground level, and remove the limbs—not even leaving a stump. The log is then fed into a chipper that renders it into chips, which can serve as boiler fuel, the raw material for paper pulp production.

When properly managed, forests are a RENEWABLE RESOURCE that can provide fuel and building materials ad infinitum while still providing habitat for animals, preventing erosion and building soil (see SUSTAINABLE AGRICULTURE, SUSTAINABLE SOCIETY). But over-harvesting has led to timber shortages in virtually all parts of the world, and to the almost complete deforestation of many developing countries in just a couple of decades. Many additional decades must pass before the world's forests have a chance to regenerate, and in some cases, particularly in the tropics, the forests won't revive in the foreseeable future.

More than 2,000 trillion Btus of wood energy are used annually in the United States, with industry consuming about three-fourths of the total and residential space-heating consuming most of the rest. Utilities burned wood containing 12 trillion Btus in the production of electricity in 1990, a figure that is growing rapidly. About 4% of U.S. homes used wood as their principal source of heat in 1990. As much as three percent of wood's weight comes in the form of noncombustible minerals such as calcium, potassium, phosphate and silica, which are left behind after burning in the form of ashes. The minerals are so evenly distributed throughout the wood that its cellular pattern can often still be discerned in the ashes. See also WOOD WASTE.

Wood Smoke   the volatile BYPRODUCTS of the COMBUSTION of WOOD. The toxins CARBON MONOXIDE and BENZO-A-PYRENE are released when wood is burned, as is CARBON DIOXIDE, a greenhouse gas (see GLOBAL WARMING). Smoke from stoves and fireplaces is a major source of winter AIR POLLUTION in colder climates, especially predominant in areas that have ATMOSPHERIC INVERSIONS during winter. Communities with recurring problems with wood smoke pollution have banned the use of wood-burning appliances during pollution episodes, required the use of high-efficiency wood stoves and banned the use of fireplaces. In some cases, they have banned the installation of wood heaters in new homes. Slash burning associated with logging and the clearing of forest land for roads and buildings is a significant source of air pollution in some areas, as are forest fires and brush fires. See also AIR POLLUTION CONTROL.

Wood Waste   wood materials discarded as waste. Logging, milling and the production of new wood products; the trimming or removal of trees and bushes from yards, parks and right-of-ways (see YARD WASTE); and the demolition or remodeling of buildings and other structures (see DEMOLITION WASTES) are major sources of waste wood. Other common sources include old furniture, pallets, wooden shipping crates, old fencing and wood shakes and shingles.

The harvest of trees and the milling and planning of lumber produces a large volume of waste, much of which was burnt in times of more ample wood resources. The tipi burners that once smoldered continually at sawmills have largely been replaced by semi trucks that haul huge loads of wood chips to paper mills, power plant distillers and other large-scale users of wood chips (see PAPER, DISTILLATION, METHANOL, WOOD) or to a railhead or port for export. SLASH—the bushes, small trees, limbs, stumps and bits of bark left behind by a logging operation—is still generally burned, although the conversion of logging waste to energy or to the raw ingredients for the production of building materials is being experimented with. Slash will eventually rot and contribute to the fertility and organic content of the soil.

The construction of buildings, furniture and other wood products produces shavings and sawdust from planing and sanding and assorted scraps of lumber, plywood and particle board. Such wastes were once commonly burned, as was woody yard waste, but since OPEN BURNING can cause significant PARTICULATE pollution under stagnant atmospheric conditions, it has been banned or limited in most areas (see AIR POLLUTION, WOOD SMOKE). Large scraps of plywood and lumber are, in some instances, picked up either at the landfill or on the construction site for resale.

As the demand for wood products increases and the supply of timber decreases, the number of uses for wood chips can be expected to increase dramatically, and the reuse and recycling of wood waste is also expected to increase as a result. Willamette Industries and Lane Forest Products have started a wood recycling venture in Eugene, Oregon, in which waste wood is ground into sawdust and used to produce particleboard at Willamette's mill. See also PAPER RECYCLING.

Wool Products Labeling Act of 1939 a federal law requiring clothing that contained wool or cotton fibers obtained from recycled rags must be so labeled. Before passage of the act, most wool fabric contained fibers from discarded cotton fabric. The necessary supply of waste fabric was supplied by the RAG PICKERS that could be found at most dumps. The resulting material was called SHODDY, a term that had no negative connotations at the time. After passage of the labeling act, however, recycled material was viewed as inferior, and the market in recycled rags virtually ended.

Worldwatch Institute an organization concerned primarily with conducting research and producing publications relating to the supply of natural resource and environmental pollution and with the emergence of a SUSTAINABLE SOCIETY. The Institute publishes a book called the "State of the World" annually. The publication summarizes trends in resource depletion, environmental pollution, population growth, human rights and sustainable technologies.

X

X-Rays electromagnetic radiation produced when a high-velocity electron strikes matter. X-rays have a wavelength shorter than visible light, and they easily pass through materials composed of atoms with a relatively light nucleus and materials of low density. They are partially blocked, however, by matter with higher density and higher atomic weight. LEAD is opaque to x-rays because of its extremely high density and the large nucleus of its atom. When an x-ray hits an atom or a molecule, it can dislodge an electron creating an ION (see IONIZING RADIATION). The shorter the wavelength of an x-ray, the greater is its penetrating power. *Soft x-rays* have a long wavelength and are close to the ultraviolet band of the electromagnetic spectrum, while *hard x-rays* have a shorter wavelength that is close to or shorter than that of GAMMA RAYS. A mixture of x-ray wavelengths is known as *white x-rays*, while *monochromatic x-rays* are made up of a single wavelength.

When an x-ray strikes the light-sensitive silver halides on a photographic negative, it causes a chemical reaction that produces silver. When the exposed film is developed, the resulting *radiograph* displays variations in the rate at which the x-rays passed through the material being analyzed to pass the photographic emulsion on the negative. Bones and teeth or a foreign object such as a bullet absorb more x-rays and thus appear black on a radiograph, while most pass through flesh, which appears light in color. A clear picture of the lungs can be obtained because of the difference in density between air sacs and surrounding tissue. A person getting a chest x-ray is typically exposed to 20 to 30 millirems of radiation (see MEASUREMENTS). Other internal organs can be x-rayed by introducing—either orally or by injection—a medium that is either more or less opaque to x-rays than the tissue to be examined. A CAT (*computerized axial tomography*) scan gives an ultra-clear picture of the inside of the body by shooting a thin beam of x-rays through the body from different angles and recording the percentage of the rays that pass through the cross-section and compiling and displaying the results on a computer. X-rays cause materials such as zinc sulfide to glow, the basis of *x-ray fluorescence*. By using a screen coated with a fluorescent material, direct observation of x-rays passing through the object being inspected is possible.

X-rays have the ability to damage living tissue, although their ability to cause cellular damage is mild compared to that of gamma rays. Excessive exposure to x-rays can cause burn-like lesions that are slow to heal, and can also cause internal cellular damage. Abnormal tissue such as tumors are generally more sensitive to the damaging effects of x-ray radiation than normal cells—a fact that is used in *radiation therapy*. Research in physics, chemistry, metallurgy, mineralogy and biology makes extensive use of x-rays. They are used in industry to inspects castings and welds for cracks or voids, and at airports, schools and other public places to detect concealed weapons. Ultrasoft x-rays are used to confirm the authenticity of works of art and in art restoration. X-rays are usually produced for commercial uses and research by bombarding tungsten with electrons. The x-ray emissions of the sun and other more distant sources are studied in radio astronomy.

Xylene any of a group of three aromatic hydrocarbon ISOMERS (see AROMATIC HYDROCARBON) that are used as solvents and in the production of drugs, dyes, insecticides and aviation gasoline. Xylene, a clear FLAMMABLE LIQUID with a strong odor, is produced through the DISTILLATION of PETROLEUM, wood tar, coal tar and coal gas. Ethylbenzene is frequently present as an impurity along with the three xylene isomers (ortho-, meta-, and para-xylene). Xylenes get into air and water from industrial and municipal waste-treatment plant discharges and from spills. They can pass into groundwater from LEAKING UNDERGROUND STORAGE TANKS. Xylene is moderately soluble and nonpersistent in water, with a HALF-LIFE of less than two days. It has high acute toxicity and chronic TOXICITY to aquatic life (see ACUTE

EFFECTS, CHRONIC EFFECTS). Xylene can get into indoor air from rubber cement and other household products.

Xylenes affects the body when inhaled or passed through the skin. Xylenes are possible TERATOGENS. Symptoms of exposure include irritation of the eyes, nose and throat, headaches, problems with memory and concentration, fatigue and stomach upset. More severe exposure can lead to dizziness, lightheadedness and unconsciousness. Repeated exposure can damage bone marrow, leading to anemia, and may cause permanent eye and stomach damage, and can damage the liver and kidneys. The odor threshold for xylene 1.1 parts per million (ppm).

PUBLIC WATER SYSTEMS are required under the SAFE DRINKING WATER ACT monitor for xylene (MAXIMUM CONTAMINANT LEVEL = 10 milligrams per liter total xylenes). The BEST AVAILABLE TECHNOLOGY for the removal of xylenes from drinking water is granular ACTIVATED CARBON or stripped-tower AERATION. The OCCUPATIONAL SAFETY AND HEALTH ADMINISTRATION has established a permissible exposure limit for airborne xylene of 100 ppm, averaged over an eight-hour workshift, with a ceiling of 150 ppm, not to be exceeded during any 15-minute work period. Xylenes are also cited in AMERICAN CONFERENCE OF GOVERNMENT INDUSTRIAL HYGIENISTS, Department of Transportation and NATIONAL INSTITUTE OF OCCUPATIONAL SAFETY AND HEALTH regulations. Xylenes are on the Hazardous Substance List, the Special Health Hazard Substance List and the HAZARDOUS AIR POLLUTANTS LIST. Businesses handling significant quantities of xylene must disclose use and releases of the chemical under the provisions of the EMERGENCY PLANNING AND COMMUNITY RIGHT TO KNOW ACT.

Y

Yard Waste clippings from grass and bushes, leaves from trees and dead plants that are a byproduct of the maintenance of the suburban (and urban) yard. As a means of reducing the volume of garbage going to landfills (see SOURCE REDUCTION), an increasing number of communities are either charging to haul away yard waste to encourage backyard COMPOSTING, or are hauling yard waste to a centralized, large-scale composting operation. San Jose, California collected an estimated 98,825 tons of leaves and grass clippings during 1993.

Z

Zero Discharge the complete elimination of a given air- or water-pollutant from the EFFLUENT issuing from a particular source. The cost and difficulty of removing a pollutant from the waste stream increases exponentially as the level of the contaminant approaches zero. Achieving zero discharge is impractical for most toxins because of the cost and because of the environmental tradeoffs involved (see CROSS-MEDIA POLLUTION). There comes a point in pollution control where the ENVIRONMENTAL COST of further cleanup exceeds the environmental benefit to be gained. As a result, zero discharge is specified only for carcinogens (see SAFE DRINKING WATER ACT), and for very persistent and toxic compounds such as toxaphene and DDT. Since every method of detecting a contaminant has a lower limit—a concentration that is more than zero—zero discharge is a concept rather than a number. See also CLEAN AIR ACT, CLEAN WATER ACT, TOXIC SUBSTANCES CONTROL ACT.

Zero Population Growth the stabilization of the growth in a given population. The control of the exponential growth of human population (see POPULATION GROWTH) is seen as an essential part of attaining a SUSTAINABLE SOCIETY that doesn't over-tax the world's supply of natural resources. Many industrialized countries have either achieved zero population growth (ZPG) or are approaching it. The rate of U.S. population increase is now 1.04, and Canada's is 1.5. Most European countries have achieved ZPG. Controlling population growth is much more difficult in developing nations, where people tend to want large families because many children die from war, famine and disease. Education and family planning, contraception and sterilization are some of the principal means of achieving ZPG.

Zirconium a metallic element found in igneous rocks that is the nineteenth most abundant in the earth's crust. Zirconium is used as the hard shell around uranium pellets in the fuel rods of a NUCLEAR REACTOR because of its hardness and high melting point (1,852°C/3,366°F). It is also used in the manufacture of STEEL, porcelains, nonferrous alloys and refractories. Zirconium readily combines with gases such as hydrogen, oxygen and nitrogen at high temperatures, and is used to remove gases from vacuum tubes. It is highly corrosion-resistant, and is, as a result, used in pumps, valves, heat exchangers and other equipment that may be exposed to corrosive materials. The principal ores from which zirconium is obtained contain a silicate of the mineral zircon and an oxide of the mineral baddeleyite, which comes primarily from Brazil. Australia is the world's largest producer of refined zirconium. See also CHERNOBYL, THREE MILE ISLAND.

Zoodoo the composted manure of zoo animals. Zoodoo is bagged and sold as a fund-raising effort at many zoos. See ANIMAL WASTES, COMPOSTING, FECES.

SELECTED SOURCES

Access EPA. Information Access Branch, Information Management and Services Division, U.S. Environmental Protection Agency, 1991.

Acid Rain. American Chemical Society, 1991.

Acid Rain and Materials Damage. Electric Power Research Institute, 1990.

Air Properties and Measurement. Carrier Corp., 1991.

Alexander, Judd. *In Defense of Garbage.* Westport, CT: Praeger, 1993.

Alt, David. *Physical Geology.* Belmont, CA: Wadsworth Publishing Company, 1982.

Asbestos in the Home: A Homeowners Guide. U.S. Environmental Protection Agency, April, 1990.

ASHRAE Handbook: 1982 Applications. Atlanta, GA: American Society of Heating, Refrigerating and Air-Conditioning Engineers, 1982.

ASHRAE Handbook: 1984 Systems. Atlanta, GA: American Society of Heating, Refrigerating and Air-Conditioning Engineers, 1984.

ASHRAE Handbook: 1985 Fundamentals. Atlanta, GA: American Society of Heating, Refrigerating and Air-Conditioning Engineers, 1985.

ASHRAE Handbook: 1993 Fundamentals. Atlanta, GA: American Society of Heating, Refrigerating and Air-Conditioning Engineers, 1993.

Ashworth, William. *Encyclopedia of Environmental Studies.* New York, NY: Facts On File, 1991.

Assigning Economic Value to Natural Resources. National Research Council, 1995.

Blumberg, Louis, and Gottlieb, Robert. *War on Waste.* Washington, DC: Island Press, 1989.

Branson, Gary D. *The Complete Guide to Recycling at Home.* Crozet, VA: Betterway Publications, 1991.

Brown, Lester R., ed. *State of the World: 1994: A Worldwatch Institute Report on Progress Toward a Sustainable Society.* New York and London: W. W. Norton & Co, 1994.

Brown, Lester R., ed. *State of the World: 1995: A Worldwatch Institute Report on Progress Toward a Sustainable Society.* New York and London: W. W. Norton & Co, 1995.

Brown, Michael H., *Laying Waste.* New York: Simon and Schuster, 1981.

Brown, Michael H., *The Toxic Cloud: The Poisoning of America's Air.* New York: Harper and Row, 1988.

Canada–United States Air-Quality Agreement. International Joint Commission, 1993.

"Chemical Hazards in Building Materials." *Journal Of American Insurance.* Third quarter, 1990.

Chemical Risk: Personal Decisions. American Chemical Society, 1989.

Cheremisinoff, Nicholas P. *Gasohol for Energy Production.* Ann Arbor, MI: Ann Arbor Science Publishers, 1979.

A Citizen's Guide to Radon. Environmental Protection Agency, 1992.

Coffel, Steve. *But Not A Drop To Drink.* New York: Ballantine Books, 1989.

Coffel, Steve. *Indoor Pollution.* New York: Ballantine Books, 1990.

Cohen, Gary, and O'Connor, John. *Fighting Toxics: A Manual for Protecting Your Family, Community and Workplace.* Washington, DC: Island Press, 1990.

Corman, Rena. Air Pollution Primer. American Lung Association, 1978.

Earthworks Group. *The Recycler's Handbook.* Berkeley, CA: Earthworks Press, 1990.

Economic Implications of Groundwater Contamination to Companies and Cities. Freshwater Foundation, 1989.

Eldred, Bill. "The Specter of Dioxin." *American City and County Magazine.* August, 1986.

Encyclopedia of Environmental Science. New York, NY: McGraw-Hill, 1980.

Energy Efficient New Homes and Indoor Air Pollutants, Bonneville Power Administration, 1989.

Enhancing Our Fish and Wildlife Resources. Bonneville Power Administration, 1984.

Environmental Encyclopedia. New York, NY: Houghton Mifflin Company, 1994.

Epstein, Samuel S., Brown, Lester O., and Pope, Carl. *Hazardous Waste in America.* San Francisco, CA: Sierra Club Books, 1982.

Forman, Dave. *Confessions of an Eco-Warrior.* New York, NY: Harmony Books, 1991.

Freudenberg, Nicholas. *Not in Our Backyards.* New York, NY: Monthly Review Press, 1984.

Gerber, Michele Stenehjem. *On the Home Front: The Cold-War Legacy of the Hanford Nuclear Site.* Lincoln, NE: University of Nebraska Press, 1992.

Global Climate Change and Agriculture. National Agricultural Library, 1992.

Global Warming and the Greenhouse Effect. National Agricultural Library, 1992.

Great Lakes Water Quality Agreement of 1978. International Joint Commission, 1994.

Harte, John, et al. *Toxics A to Z: A Guide to Everyday Pollution Hazards.* Berkeley, CA: University of California Press, 1991.

Hazardous Waste Management. American Chemical Society, 1992.

Healthy House Catalog. Compiled and published by Environmental Health Watch and the Housing Resource Center, Cleveland, OH, 1990.

Home Weatherization and Indoor Air Pollutants. Bonneville Power Administration, 1989.

Godish, Thad. *Indoor Air Quality.* Chelsea, MI: Lewis Publishers, 1985.

Ground Water Recharge Using Waters of Impaired Quality. National Research Council, 1994.

A Guide to the Clean Water Act Amendments. Environmental Protection Agency, 1978.

Hurlburt, Scott. "The Problem with Nitrates." *Water Well Journal*, August, 1988.

Information Digest. Nuclear Regulatory Commission, March, 1994.

The Inside Story: A Guide to Indoor Air Quality. Environmental Protection Agency, 1988.

Introduction to Indoor Air Quality. Environmental Protection Agency, 1991.

Johnson, Arthur H. "Acid Deposition: Trends, Relationships and Effects." *Environment*, May, 1986.

"Knee-deep and Rising: America's Recycling Crisis." *Harvard Business Review*, September, 1991.

Krigger, John, *Residential Energy.* Helena, MT: Quality Books, 1994.

League of Women Voters Education Fund. *The Nuclear Waste Primer.* New York, NY: Nick Lyons Books, 1985.

Martin, Daniel. *Three Mile Island: Prologue or Epilogue.* New York, NY: Balinger Publishing, 1980.

Mazria, Edward. *The Passive Solar Energy Book.* Emmaus, PA: Rodale Press, 1979.

Medvedev, Grigori, *The Truth About Chernobyl.* Basic Books, 1991.

Mitchell, George J. *World on Fire: Saving an Endangered Earth.* New York, NY: Charles Scribner's Sons, 1991.

Mowrey, Mark, and Redmond, Tim. *Not in Our Backyard.* New York, NY: William Morrow and Co., 1993.

Murphy, Pamela. *The Garbage Primer: A Handbook for Citizens.* New York: Lyons and Burford Publishers, 1993.

Nilson, Sten, and Duinker, Peter. "The Extent of Forest Decline in Europe." *Environment*, November, 1987.

Norback, Craig T., and Norback, Judith C., eds. *Hazardous Chemicals on File.* New York, NY: Facts On File, 1991.

Northwest Conservation and Electric Power Plan. Northwest Power Planning Council, 1991 draft.

Paul, J. K. *Methanol Technology and Application in Motor Fuels.* Park Ridge, NJ: Noyes Data Corp, 1978.

Pollution Prevention Directory. Environmental Protection Agency. September, 1994.

Pontius, Fredrick W. "An Update of the Federal Drinking-Water Regs." *Journal of the American Water Works Association.* February, 1995.

Potential Impacts of Global Climate Change on Natural Terrestrial Ecosystems. Electric Power Research Institute, 1993.

Pytte, Alyson. "Clean Air Act Amendments." *Congressional Quarterly*, November 24, 1990.

Radon-Resistant Residential New Construction. Environmental Protection Agency, 1988.

Ranking Hazardous-Waste Sites for Remedial Action. National Research Council, 1994.

Ray, Dixie Lee. *Environmental Overkill: Whatever Happened to Common Sense?* Washington, DC: Regnery Gateway, 1993.

Rathje, William, and Murphy, Cullen. *Rubbish.* New York: HarperCollins Publishers, 1992.

"Recycling." American Chemical Society, 1993.

Roan, Sharon L. *Ozone Crisis.* New York, NY: John Wiley and Sons, 1989.

Sax, N. Irving. *Dangerous Properties of Hazardous Industrial Materials.* Van Nostrand Reinhold, 1979.

Scarce, Rik. *Eco-Warriors: Understanding the Radical Environmental Movement.* Chicago, IL: Noble Press, 1990.

Schneider, Stephen H. *Global Warming.* San Francisco, CA: Sierra Club Books, 1989.

Seventh Biennial Report on Great Lakes Water Quality. International Joint Commission, 1994.

Sitarz, Daniel. *Agenda 21: The Earth Summit Strategy to Save Our Planet.* Boulder, CO: EarthPress, 1993.

Smith, Duane A. *Mining in America.* Lawrence, KS: University of Kansas Press, 1987.

Solving the Hazardous Waste Problem. U.S. Environmental Protection Agency, 1986.

Torrey, S., ed. *Coal Ash Utilization: Fly Ash, Bottom Ash and Slag.* Park Ridge, NJ, Noyes Data Corp, 1978.

Trost, Cathy. *The Elements of Risk.* New York, NY: New York Times Books, 1984.

Udall, James R. "Turning Down the Heat." *Sierra Magazine*, July, 1989.

U.S. Bureau of Census. *Statistical Abstract of the United States, 1994.* U.S. Department of Commerce, 1994.

White, Peter T. "The Fascinating World of Trash." *National Geographic Magazine*, April, 1983.

Whitten, D. G. A., and Brooks, J. R. V. *Dictionary of Geology*. New York: Penguin Books, 1985.

Wood Handbook: Wood as an Engineering Material. U.S. Forest Service Forest Products Laboratory, 1974.

Zakin, Susan. *Coyotes and Town Dogs: Earth First! and the Environmental Movement*. New York, NY: Viking-Penguin, 1993.

Index

This index is designed to be used in conjunction with the cross-references within the A-to-Z entries. The main A-to-Z entries are indicated by **boldface** page references. The general subjects are subdivided by the A-to-Z entries. *Italicized* page references indicate illustrations; "c" following the locator indicates the chronology.